Lecture Notes in Civil Engineering 487

Lecture Notes in Civil Engineering (LNCE) publishes the latest developments in Civil Engineering—quickly, informally and in top quality. Though original research reported in proceedings and post-proceedings represents the core of LNCE, edited volumes of exceptionally high quality and interest may also be considered for publication. Volumes published in LNCE embrace all aspects and subfields of, as well as new challenges in, Civil Engineering. Topics in the series include:

- Construction and Structural Mechanics
- Building Materials
- Concrete, Steel and Timber Structures
- Geotechnical Engineering
- Earthquake Engineering
- Coastal Engineering
- Ocean and Offshore Engineering; Ships and Floating Structures
- Hydraulics, Hydrology and Water Resources Engineering
- Environmental Engineering and Sustainability
- Structural Health and Monitoring
- Surveying and Geographical Information Systems
- Indoor Environments
- Transportation and Traffic
- Risk Analysis
- Safety and Security

To submit a proposal or request further information, please contact the appropriate Springer Editor:

- Pierpaolo Riva at pierpaolo.riva@springer.com (Europe and Americas);
- Swati Meherishi at swati.meherishi@springer.com (Asia—except China, Australia, and New Zealand);
- Wayne Hu at wayne.hu@springer.com (China).

All books in the series now indexed by Scopus and EI Compendex database!

Sheng'an Zheng · Richard M. Taylor ·
Wenhao Wu · Bjorn Nilsen · Gensheng Zhao
Editors

Hydropower and Renewable Energies

Synergistic Integration for Future Energy Systems

Springer

Editors
Sheng'an Zheng
China Society for Hydropower Engineering
Beijing, China

Wenhao Wu
China Society for Hydropower Engineering
Beijing, China

Gensheng Zhao
Department of Hydraulic Engineering
Nanjing Hydraulic Research Institute
Nanjing, Jiangsu, China

Richard M. Taylor
RMT Renewables Consulting Ltd.
London, UK

Bjorn Nilsen
Department of Geoscience
NTNU
Trondheim, Norway

ISSN 2366-2557 ISSN 2366-2565 (electronic)
Lecture Notes in Civil Engineering
ISBN 978-981-97-9186-6 ISBN 978-981-97-9184-2 (eBook)
https://doi.org/10.1007/978-981-97-9184-2

This Springer imprint is published by the registered company Springer Nature Singapore Pte Ltd.
The registered company address is: 152 Beach Road, #21-01/04 Gateway East, Singapore 189721, Singapore

Organization

Editors

Sheng'an Zheng	China Society for Hydropower Engineering
Richard M. Taylor	RMT Renewables Consulting Ltd.
Wenhao Wu	POWERCHINA International Group Limited
Bjorn Nilsen	Norwegian University of Science and Technology
Gensheng Zhao	Nanjing Hydraulic Research Institute

Editorial Board Members

China Society for Hydropower Engineering This work was supported by China Society for Hydropower Engineering and China Renewable Energy Engineering Institute.

Contents

Comprehensive Studies on the Combined Environmental Effects of Integrated Energy Projects

Spotlight on Groundbreaking Sustainable Energy Technologies

Effective Models for the Integration of Green Energy Systems Within Existing Infrastructures

Predictive Insights into the Future Landscape of Energy Technologies

About the Editors

Sheng'an Zheng elected Executive Vice President and Secretary General of the China Society for Hydropower Engineering (CSHE) in late 2022, previously held key positions at POWERCHINA Chengdu Engineering Cooperation Ltd and China Renewable Energy Engineering Institute (CREEI) from 2001 to 2021. As a distinguished hydropower expert, he directed major projects like Xiluodu and Pubugou Hydropower Stations and contributed to the 14th Five-Year Plan on Renewable Energy Development.

Richard M. Taylor renowned in international renewable energy, co-established AMI and later founded the International Hydropower Association (IHA) in 2001. With leadership roles in various organizations, including IRENA and Climate Bonds Initiative, he chairs the HSAC industry chamber, advises the World Bank Group, and consults for the XFLEX HYDRO project. Engaged in UN initiatives, Taylor appraised the Sustainable Water and Energy Initiative.

Wenhao Wu with 38 years of experience serves as Secretary General of IF-CSHE. As a hydropower expert, he studied in Norway, managed international projects, and led river diversion for the Wiquangxi Hydropower project. Awards include the 2nd prize for national engineering for the river diversion.

Bjørn Nilsen Professor Emeritus at the Norwegian University of Science and Technology excels in Geological Engineering, contributing to both academia and practical projects, particularly in rock mechanics and engineering geology. He is an author or co-author of more than 100 scientific papers and member of Editorial Board and regular reviewer of two international, peer-reviewed journals.

Gensheng Zhao Senior Engineer and Researcher coordinates international projects as Managing Director of the Joint Research Center of Water Science and Engineering. As a reviewer for esteemed journals, he has published five books, showcasing his expertise in hydraulic engineering.

Exploration of the Latest Advancements in Hydropower Technology

Research on Dam Crack Identification Method Based on Multi-source Information Fusion

Cun Xin[1(✉)], Dangfeng Yang[1], Xiaodong Liu[1], Yong Huang[1], and Xueming Qian[2]

[1] Power China Northwest Engineering Corporation Limited, Xi'an Key Laboratory of Clean Energy Digital Technology, Xi'an 710065, China
xincun0904@163.com

[2] State Engineering Laboratory of Visual Information Processing and Application, Xi'an Jiaotong University, Xi'an 710049, China

Abstract. Cracks as the main safety concern of dams, high-precision identification of dam cracks is of great application value and scientific significance to ensure the safety of dams. The paper proposes a dam crack identification method based on multi-source information fusion. Specifically, image gray scale and geometric features are extracted based on the image information. And then a single crack identification model based on Support Vector Machine (SVM), Decision Tree (DT), Random Forest (RF), XGBoost, and BP Neural Network are established based on the features, respectively. Finally, a multi-classifier fusion algorithm based on D-S evidence theory is established to identify the presence of cracks by fusing single identification models. Experiments are carried out to compare the proposed method with the existing identification methods based on the evaluation metrics such as accuracy, precision, F1-score, and recall. The results show that the accuracy of crack identification of the proposed method in this paper reaches 98.9%, and the crack identification results are better than the existing methods.

Keywords: Concrete dam · Crack detection · Machine vision · Multi-information fusion · D-S fusion

1 Introduction

As an essential part of water conservancy and hydropower engineering construction, dams have the functions of flood control, power generation, irrigation, navigation, etc. It is of great significance to monitor the safety of dams and find abnormal potential dangers in a timely period [1]. As one of the most common damage of dams, cracks are not easy to ignore even if they are very small, and crack damage can lead to reservoir leakage and dam shutdown for maintenance in the case of light damage, or lead to dam collapse in the case of serious damage, which can lead to the flood disaster in the downstream and threaten the safety of lives and properties of the people in the downstream [2]. Currently, dam crack detection mainly relies on traditional manual visual inspection methods, i.e., hanging baskets on the surface of the dam body for manual visual inspection, or using binoculars for visual inspection [3]. These methods have the problems of high

S. Zheng et al. (Eds.): IHDC 2024, LNCE 487, pp. 3–14, 2025.
https://doi.org/10.1007/978-981-97-9184-2_1

inspection cost, high risk factor, as well as low measurement efficiency, time-consuming and laborious, and at the same time, the measurement effects depend on the personal subjective experience of the engineers, and the measurement effects have a certain degree of randomness [4].

With the development of machine vision technology, image-based crack identification methods have received attention from scholars due to the advantages of speed and convenience. Generally speaking, image-based crack recognition methods are mainly divided into two kinds: texture analysis method and machine learning detection method. In texture analysis methods, it mainly consists of defining various gradient features using gradient filters, e.g., the gray gradient of each pixel in the image is computed by pixel edge detection methods such as Sobel and Canny, and then a binary classifier is used to determine whether an image pixel belongs to a crack region or not [5, 6]. However, texture-based analysis methods usually assume that the cracked region has a large difference with the intact region in terms of image texture features, and the identification accuracy is lower for texture features similar to the cracks, such as rust, scratches, and so on. At the same time, this type of texture-based method needs to judge the crack region with the help of binary method, the recognition effect greatly depends on the selection of binary threshold, the effect is very sensitive to the noise such as light, and the risk of mistaken judgment is high [7, 8].

As a comparison, machine learning-based crack detection methods with high identification accuracy and end-to-end detection advantages have gradually become the mainstream of the current structural crack detection, and this type of method mainly extracts the features in the image through machine learning algorithms such as Support Vector Machine (SVM), Random Forests (RF), XGboost, K Nearest Neighbor Classification (KNN), Convolution Neural Networks (CNN), etc., and then utilizes the extracted features to determine the presence of cracks [9–11]. Mao et al. [12] proposed a dam crack recognition method based on image LBP features and image Gabor features combined with CNN to recognize multiple types of cracks such as horizontal and vertical. Tang et al. [13] proposed a multi-task enhanced dam crack image detection method based on Faster R-CNN, which achieved better identification results in dam crack detection under different lighting environments. Zou et al. [14] proposed a crack detection method for concrete dams based on improved Yolov5s to achieve improved crack detection efficiency and precision. Yi et al. [15] proposed a structural surface crack identification method based on the integration of SVM, which achieved better identification results in experiments with different lighting conditions and different crack morphology. However, for different crack features, different machine learning methods have their own advantages, and a single machine learning method performs differently in different classification situations, e.g., it is presented in the literature [13] that CNN-based crack recognition method outperforms YOLOv5, and it is pointed out in the literature [15] that SVM-based crack recognition method outperforms the BP neural network classifier.

Based on this, this paper proposes a dam crack identification method based on multi-source information. Specifically, image grayscale and geometric features are extracted based on the image information. And a one-crack identification model is established based on the features, such as SVM, DT, RF, and BP neural network, respectively. Finally, a multi-classifier fusion algorithm based on D-S evidence theory is constructed to improve the crack recognition accuracy by fusing multiple single recognition models and reducing the uncertainty of a single model.

2 The Method of Crack Identification

The flowchart of the proposed method is shown in Fig. 1, including feature extraction and D-S multi-information fusion. Specifically, gray scale feature and shape features are firstly extracted based on the image information. And then SVM, DT, RF, XGBoost, and BP Neural Network are established to calculate the classified probability based on the features, respectively. Finally, the multi-classifier fusion algorithm based on D-S evidence theory is constructed to identify the cracks by fusing single identification models. The details of the method are as Fig. 1.

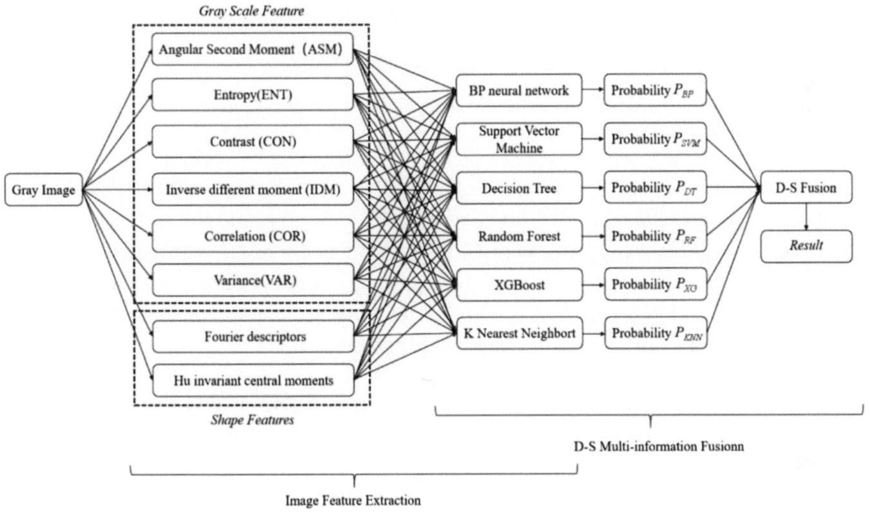

Fig. 1 Flowchart of proposed crack identification

2.1 Feature Extraction

In this paper, the image gray scale features and shape features are measured for identifying the crack. The details of the method are as follow:

2.1.1 Gray Scale Feature Extraction

Gray scale features are computed by the image covariance matrix, which refers to a common method of describing texture by studying the spatial correlation properties of grayscale. As texture is formed by the recurrence of the gray scale distribution at spatial locations, therefore, there will be a certain grayscale relationship between two pixels separated by a certain distance in the image space, i.e., spatial correlation properties of the grayscale in the image. A total of 24 eigenvalues by constructing the grayscale covariance matrices in four directions (0°, 45°, 90°, and 135°) in this paper, as shown in Fig. 2, and calculating six eigenvalues such as angular second moment, entropy, contrast, inverse different moment, correlation, variance, respectively (Fig. 2).

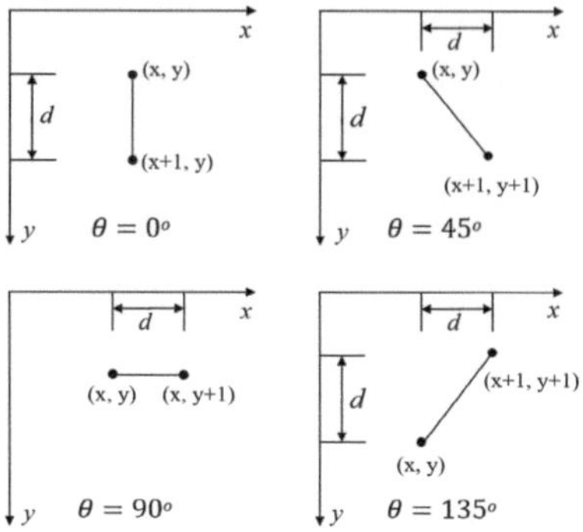

Fig. 2 Method of calculating different directions of the grayscale covariance matrix

Assume an image $f(x, y)$ defined as $L_x \times L_y$, horizontal space and vertical space are described as $Lx = \{1, 2, ..., Nx\}$ and $Ly = \{1, 2, ..., Ny\}$, and the grayscale space is defined as $G = \{1, 2, ..., Ng\}$. That is, each point in $L_x \times L_y$ corresponds to a gray level belonging to G. The probability that pixel (x, y) of gray scale i, with displacement d, orientation θ, and pixel $(x + \Delta x, y + \Delta y)$ of gray scale j appear simultaneously in the image, denoted as $p(i, j, d, \theta)$, can be expressed as:

$$p(i, j, d, \theta) = \{[(x, y), (x + \Delta x, y + \Delta y)]|f(x, y) = i, f(x + \Delta x, y + \Delta y) = j\} \quad (1)$$

Assuming that the gray level of the image is G, the gray matrix p of the image is a matrix of $G \times G$ squares, i.e.:

$$p = \begin{bmatrix} p(0,0) & p(0,1) & \dots & p(0,G-2) & p(0,G-1) \\ p(1,0) & p(1,1) & \dots & p(1,G-2) & p(1,G-1) \\ \dots & \dots & \dots & \dots & \dots \\ p(G-2,0) & p(G-2,1) & \dots & p(G-2,G-2) & p(G-2,G-1) \\ p(G-1,0) & p(G-1,1) & \dots & p(G-1,G-2) & p(G-1,G-1) \end{bmatrix} \quad (2)$$

On the basis of the gray scale covariance matrix of the image, the gray scale features of the image can be calculated as follows:

Feature 1: Angular Second Moment (ASM)

$$ASM = \sum_i \sum_j p(i, j)^2 \quad (3)$$

Feature 2: Entropy (ENT)

$$ENT = -\sum_i \sum_j p(i, j) \log(p(i, j)) \quad (4)$$

Feature 3: Contrast (CON)

$$CON = \sum_i \sum_j (i - j)^2 p(i, j) \tag{5}$$

Feature 4: Inverse different moment (IDM)

$$IDM = \sum_i \sum_j \frac{p(i, j)}{1 + (i - j)^2} \tag{6}$$

Feature 5: Correlation (COR)

$$COR = \sum_i \sum_j \frac{(ij)p(i, j) - u_x u_y}{\sigma_x \sigma_y} \tag{7}$$

Feature 6: Variance (VAR)

$$\sigma = \sum_i \sum_j (i - u_x)(j - u_y)p(i, j) \tag{8}$$

2.2 Shape Features Extraction

Image shape features play a fundamental and important role in image classification, so effective and efficient shape descriptors are the key components of image shape feature extraction and representation. In this paper, two typical shape features, namely Fourier descriptors and Moment descriptors, are utilized. The Fourier descriptor is a series of coefficients generated based on the object contour in the target image after the Discrete Fourier Transform, which can be used to represent the shape features of the target, assuming that there is a closed region of interesting contour containing N data points.

$$s(k) = x(x) + jy(k),\ j = \sqrt{-1},\ k = 0, 1, ..., N - 1 \tag{9}$$

where $s(k)$ is a sequence of complex numbers representing closed contours for which the Fourier transform can be described as:

$$S(w) = \frac{1}{N} \sum_{k=1}^{N-1} s(k) e^{-j2\pi wk/N},\ k, w = 0, 1, ..., N - 1 \tag{10}$$

In addition, region-based shape feature descriptors, such as moments, which are more reliable for shapes with complex boundaries, in particular geometric moments, central moments, and orthogonal invariant moments, have been applied to image shape description and content-based image retrieval. For a two-dimensional digital grayscale image $f(x, y)$, the $(p + q)$ order moments are defined as follows:

$$m_{pq} = \sum_x \sum_y x^p x^q f(x, y)\ p, q = 0, 1, 2, ... \tag{11}$$

The central moment is calculated by the formula:

$$\mu_{pq} = \sum_{p}\sum_{q}(x-\overline{x})^{p}(y-\overline{y})^{q}f(x,y) \tag{12}$$

where $\overline{x} = m_{10}/m_{00}, \overline{y} = m_{01}/m_{00}$, the normalized central moment can be described as:

$$\eta_{pq} = \mu_{pq}/\mu_{00}^{\gamma} \tag{13}$$

where $\gamma = \frac{(p+q)}{2} + 1, p+q = 2, 3, ...$

In this paper, shape features are extracted using second and third order Hu invariant central moments, computed as shown below, to construct a seven-dimensional moment feature vector to represent the target image.

$$\begin{aligned}
&M_1 = \eta_{02} + \eta_{20}\\
&M_2 = (\eta_{02} - \eta_{20})^2 + 4\eta_{11}^2\\
&M_3 = (\eta_{30} - 3\eta_{12})^2 + (3\eta_{21} - \eta_{03})^2\\
&M_4 = (\eta_{30} + \eta_{12})^2 + (\eta_{21} + \eta_{03})^2\\
&M_5 = (\eta_{30} - 3\eta_{12})(\eta_{30} + \eta_{12})\left[(\eta_{30} + \eta_{12})^2 - 3(\eta_{21} + \eta_{03})^2\right] +\\
&\quad (3\eta_{21} - \eta_{03})(\eta_{21} + \eta_{03})\left[3(\eta_{30} + \eta_{12})^2 - (\eta_{21} + \eta_{03})^2\right]\\
&M_6 = (\eta_{20} - \eta_{02})\left[(\eta_{30} + \eta_{12})^2 - (\eta_{21} + \eta_{03})^2\right] + 4\eta_{11}(\eta_{30} + \eta_{12})(\eta_{21} + \eta_{03})\\
&M_7 = (3\eta_{21} - \eta_{03})(\eta_{21} + \eta_{03})\left[(\eta_{30} + \eta_{12})^2 - 3(\eta_{21} + \eta_{03})^2\right] +\\
&\quad (3\eta_{21} - \eta_{03})(\eta_{21} + \eta_{03})\left[3(\eta_{30} + \eta_{12})^2 - (\eta_{21} + \eta_{03})^2\right]
\end{aligned} \tag{14}$$

2.3 D-S Multi-information Fusion

D-S evidence theory is widely used in information synthesis for its superiority in dealing with uncertain information. The Mass (confidence function) function of multiple bodies of evidence is fused by the evidence synthesis rule, and the confidence level of the evidence bodies is determined by the obtained new Mass function thus realizing the fusion of information and reducing the uncertainty of decision-making events. The definition is as follows:

Definition 1: Let θ be a recognizing frame if the function $m : 2^{\theta} \to [0, 1]$ satisfies the following conditions:

1) There is no event with probability 0, i.e.:

$$m(\varphi) = 0 \tag{15}$$

2) The probabilities of all elements of the framework add up to a result of 1, i.e.:

$$\sum_{A \subset \Theta} m(A) = 1 \tag{16}$$

Then m is said to be the basic probability distribution function on the frame Θ, $m(A)$ called the basic trustworthiness number of A, and denotes the trust in A.

Definition 2: Set Θ as a recognizing frame if the function Bel: $2^{\theta} \rightarrow [0, 1]$ conforms:

$$Bel(A) = \sum_{B \subset A} m(B) \tag{17}$$

Definition 3: For, $m_1, m_2, \ldots, m_n$ is a Mass function on the recognition frame Θ, then its evidence synthesis law is:

$$(m_1 \oplus m_2 \oplus \ldots \oplus m_n)(A) = \frac{1}{1-K} \sum_{A_1 \cap A_2 \cap \ldots \cap A_n} m_1(A_1)m_2(A_2)\ldots m_n(A_n) \tag{18}$$

where $K = \sum_{A_1 \cap A_2 \cap \ldots \cap A_n = \phi} m_1(A_1)m_2(A_2)\ldots m_n(A_n)$

3 Experimental Verification

3.1 Data Processing

In this section, the crack identification method proposed in this paper are validated by conducting experiments. The crack images used in this paper are derived from the dam structure, the total number totals 1200, 400 cracked images (shown in Fig. 3) and 800 intact images (shown in Fig. 4).

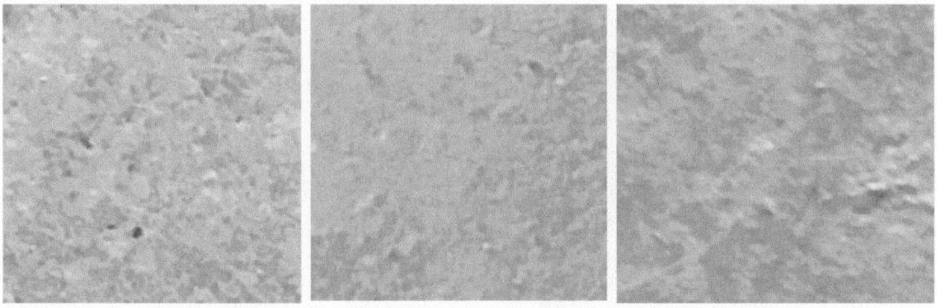

Fig. 3 The intact crack image of dam

In this paper, the gray scale features, Fourier features, and HU invariant moment features of the sample set data are extracted and the extraction results are shown in Table 1.

Based on the calculated features, a single crack identification model based on SVM, DT, RF, XGBoost, and BP Neural Network is established based on the features, respectively. Finally, a multi-classifier fusion algorithm based on D-S evidence theory is established to identify the presence of cracks by fusing single identification models.

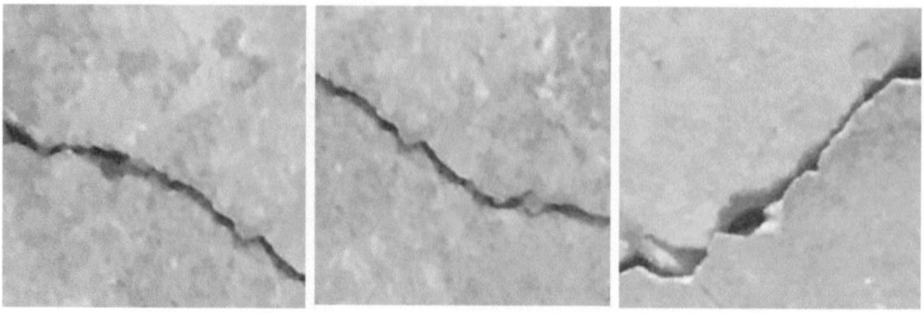

Fig. 4 The crack image of dam

Table 1 The features of samples

Features	Image_1	Image_2	Image_3	Image_4	Image_5
ASM (0°)	0.5394	0.4374	1.4506	−0.0273	−0.5431
ENT (0°)	−0.1693	−0.0264	1.1453	0.3631	−0.1801
CON (0°)	−0.6300	0.4673	1.5976	−0.1395	−0.6321
IDM (0°)	−0.5012	0.2847	1.4479	0.0449	−0.5058
COR (0°)	−0.6317	1.2936	1.0287	−0.1964	−0.6359
VAR (0°)	−0.4783	1.4781	0.6846	−0.1339	−0.4875
M1	−1.5961	−1.6319	−1.6051	−1.5828	−1.5669
M2	−0.7084	−0.6675	−0.7020	−0.7556	−0.7589
M3	−0.3475	−0.3270	−0.3656	−0.2688	−0.3228
M4	−0.3397	−0.3567	−0.3227	−0.3563	−0.3296
M5	1.2871	1.3350	1.3679	1.3195	1.4185
M6	0.3025	0.3162	0.3154	0.2602	0.2663
M7	1.4021	1.3320	1.3122	1.3839	1.2933
S(1)	2.4403	2.4324	2.3707	2.4171	2.4218
S(2)	−0.5094	−0.1361	−0.2986	−0.7485	−0.6862
S(3)	−0.2301	−0.5258	−0.8949	−0.5019	−0.5628
S(4)	−0.5021	−0.4865	−0.3216	−0.2368	−0.3164
S(5)	−0.3737	−0.4009	−0.5787	−0.3014	−0.2555
S(6)	−0.3998	−0.4199	−0.0590	−0.2916	−0.3216
S(7)	−0.4249	−0.4628	−0.2177	−0.3367	−0.2791

3.2 Accuracy Analysis of Crack Identification

The crack images used in this paper are derived from the dam structure and all the data is divided into training set and test set according to 80% and 20%. Gray scale features and shape features of the images were computed using Opencv and the extracted features were normalized and used to train the model. The accuracy of the proposed crack recognition method is validated by utilizing the metrics such as precision, recall, accuracy and F1-score.

$$precision = \frac{n_{tp}}{n_{tp} + n_{fp}} \tag{19}$$

$$\text{recall} = \frac{n_{tp}}{n_{tp} + n_{fn}} \tag{20}$$

$$accuracy = \frac{n_{tp} + n_{tn}}{n_{tp} + n_{tn} + n_{fp} + n_{fn}} \tag{21}$$

$$F_1 = 2\frac{precision \cdot recall}{precision + recall} \tag{22}$$

where n_{tp}, n_{fp}, n_{fn}, n_{tn} denote the number of true positive, false positive, false negative and true negative test images, respectively.

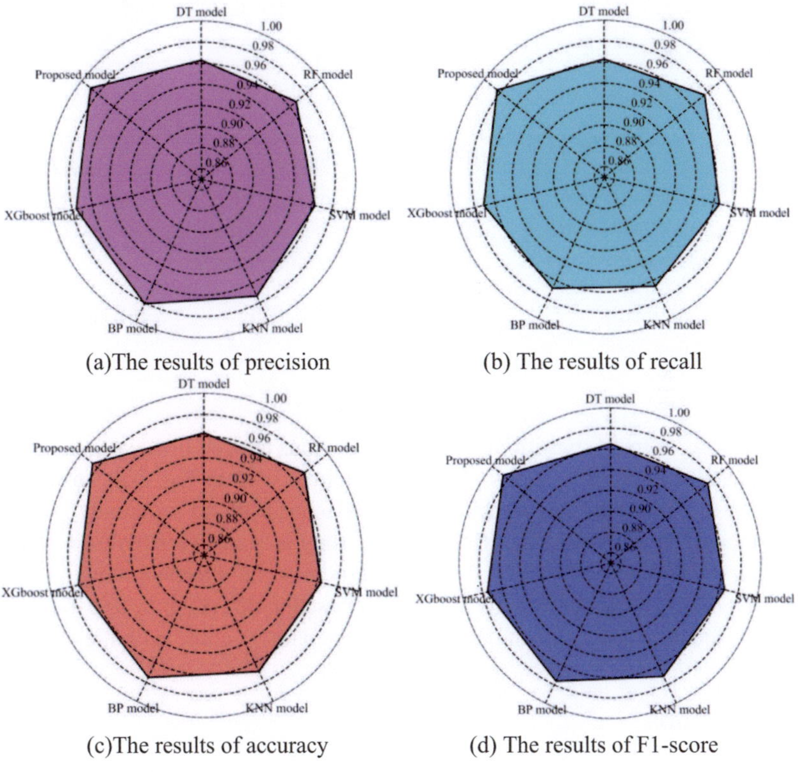

(a)The results of precision (b) The results of recall

(c)The results of accuracy (d) The results of F1-score

Fig. 5 The results of different models

The test results of the four index are shown in Fig. 5, it can be seen that compared with a single classifier, the classification method of multi-information fusion proposed in this paper is performed better than any other crack detection method from the precision, recall, accuracy and F1-score. The crack detection of the proposed method in this paper reaches 98.9%, the results demonstrate that the method proposed in this paper is able to effectively identify the presence or absence of cracks in dam.

3.3 Application of Algorithms to Real Dam Crack Identification

In practical application, special high-resolution cameras or UAVs are used to take images of the surface of the dam, together with computers of certain computing power, to analyze the collected images using the proposed algorithms and to provide guidance to the personnel for on-site surveys. In this paper, the real dam image acquired using camera is shown in Fig. 6(a), further, and the original image is segmented and processed. In which the original image is segmented into 5 × 5 total 25 regions as shown in Fig. 6(b), where the details of the image in region numbered 10 and region 20 are shown in Fig. 6(c).

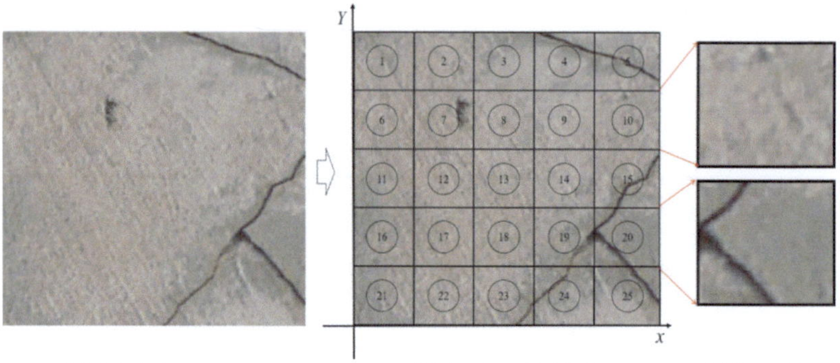

(a)Origin image (b)Region divided (c)Detail Region

Fig. 6 The real image of dam

On the basis of image partitioning, different regions are classified with the proposed algorithm, and the results of crack classification are shown in Fig. 7(a), from which it can be seen that the proposed algorithm in this paper is able to accurately identify the regions of concrete where cracks exist. Further, using the image binarization algorithm, the regions with cracks are processed, and the structure of the binarized region is shown in Fig. 7(b), which indicates that the method proposed in this paper is able to accurately identify the location of cracks on the surface of the dam structure with high accuracy.

4 Conclusions

The paper proposes a method for dam crack identification and morphology measurement based on machine vision and multi-source information fusion, and validates the proposed algorithm in terms of classification accuracy, recall, precision, F1-score. The main advantages are as follows:

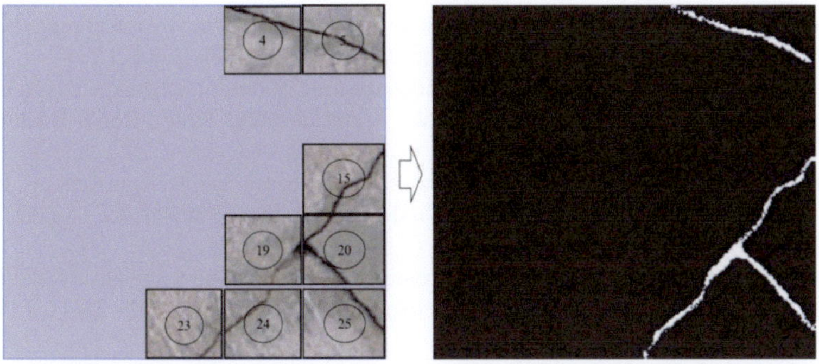

(a)The areas with cracks (b)The crack morphology

Fig. 7 The result of crack identification real image of dam

(1) Compared to the existing manual inspection-based approach, the dam crack identification algorithm proposed in this paper is based on video analysis, which, together with image acquisition equipment such as UAVs and cameras, has the advantages of non-contact and high efficiency.
(2) Compared with a single classifier, the proposed method of the multi-information fusion strategy can reduce the uncertainty of a single model, and the experimental accuracy of crack recognition is 98.6%, which is significantly better than the performance of the traditional single model.

Although the method proposed in this paper can realize the crack identification of dams to a certain extent, the method proposed in this paper has certain disadvantages for specific practical applications: for example, compared with a single model, the proposed multi-model fusion strategy increases the computational cost, and on the other hand, the proposed method can only identify the existence of cracks, and it is difficult to quantify the degree of cracks, etc. In the future, we will make efforts to improve these disadvantages to improve the scope of application of the proposed method.

References

1. Mata, J., et al.: Analysis and interpretation of observed dynamic behaviour of a large concrete dam aided by soft computing and machine learning techniques. Eng Struct **296**, 1–12 (2023)
2. Feng, C.C., et al.: Research on intelligent detection method for crack damage of over flow dam of hydropower station. Automat Instrum **36**(6), 55–60 (2021)
3. Huang, Z.J., et al.: Inspection and treatment of overwater crack of upstream surface of first-stage project of Dan jiang kou dam. Yangtze River **46**(6), 45–48, 74 (2015)
4. Xin, G.F., et al.: A fine extraction algorithm for image-based surface cracks in underwater dams. Meas. Sci. Technol. **34**(3), 1–14 (2023)
5. Fan, W.W., Wang, X.P., Zhu, S.Y.: Crack detection in subway tunnels based on multi-feature analysis. J. Meas. Sci. Instrum. (2024)
6. Chen, M.Y., Yuan, W.Q.: Pavement crack detection algorithm based on gray minimum value. Microprocessors **45**(01), 34–38 (2024)
7. Teshima, Y., Hara, T., Kobori, K.I.: A method of fine crack extraction from concrete structure surface image. J. Inst. Image Electron. Eng. Jpn **36**(3), 243–251 (2007)

8. Xin, C., et al.: Marker-free fatigue crack detection and localization by integrating the optical flow and information entropy. Struct. Health Monit. **22**(2), 1008–1026 (2023)
9. Xu, Y., et al.: Automatic seismic damage identification of reinforced concrete columns from images by a region-based deep convolution neural network. Struct. Contr. Health Monit. **26**(3), e2313 (2009)
10. Zhou, S., Canchila, C., Song, W.: Deep learning-based crack segmentation for civil infrastructure: data types, architectures, and bench marked performance. Autom. Constr. (2023)
11. Shim, S., et al.: Stereo-vision-based 3D concrete crack detection using adversarial learning with balanced ensemble discriminator networks. Struct. Health Monit. **22**(2), 1353–1375 (2023)
12. Mao, Y.C., et al.: Dam defect recognition and classification based on feature combination and CNN. Comput. Sci. (2019)
13. Tang, J., et al.: Multi-task enhanced dam crack image detection based on faster R-CNN[C]//2019 IEEE 4th International Conference on Image, Vision and Computing (ICIVC). IEEE (2019)
14. Zou, Y.Y., et al.: Crack detection of concrete dam based on improved Yolov5s. Autom. Instrum. **2**(11), 1–15 (2023)
15. Li, X., Guo, Y., Li, Y.: Particle swarm optimization-based SVM for classification of cable surface defects of the cable-stayed bridges. IEEE Access. **8**, 44485–92 (2020)
16. Yi, G., Yanxi, Y.: Classification based on multi-classifier of SVM fusion for steel strip surface defects. Chinese Control Conference (2022)

Corona Trials on Rotating Machinery with LuminarHd Ultraviolet Apparatus in Small Hydropower Plants - SHPs Technology and Innovation Company

Afonso Cesar Tavares(✉), Marcelino Santos, and Bruno Dellabeta

CPFL Renováveis, Campinas, Brazil
{afonso.tavares,marcelino.santos,bruno.dallabeta}@cpfl.com.br

Abstract. With the new technologies applied in the Power Electrical System, we enable the shared use of the Ultraviolet Measurement Apparatus, for Corona Measurement (the corona effect is just one type of partial discharge characterized by its visibility when propagating in external mediums due to electric field - potential) in Transmission Lines, into the Assets of Hydropower Plants with rotating machinery from Small Hydropower Plants - SHPs, within Predictive Maintenance. The gain from this Predictive Maintenance brought evaluation benefits in the reception after treatment of generators related to Partial Discharge on the surface, particularly the part related to surface corona effect on rotating machinery. Measurements before and after surface treatment recovery, mainly in coil heads and other parts of the generating unit, show us the results more through frequency spectrum, ranges, and events/minutes measures, associated with a scale already worked for measurement points in Transmission Lines assets, values understood above 5000 events/minutes and mainly the region with the highest concentration of records signaling points to improve or redo specific treatment. It's worth mentioning here that it doesn't substitute tests by offline and online partial discharge devices in peak measures (pC) or nano Coulomb (nC), but it brings the focus spectral sampling region that deserves more attention or reinforcement in treatment to mitigate corona effect, also can be used to evaluate generator cleanliness when intensities are recorded in predictive maintenance. The SBUV camera, Solar Blind filter, has higher sensitivity, compatible wavelength, bi-spectral, and a range from 240 nm to 280 nm (nanometers), which can bring gains in this technological application. The tool, when customized, provides risk assessment for each generator in trials and ensures an operational view of the Asset from the perspective of its health in this regard - Health Index. The new technology is already implemented and has been successfully used in our plant, in its punctual maintenance plan regarding measuring, evaluating own or third-party services, and determining the best time to act in new surface treatments or asset cleaning, mitigating the risk of early burnout due to this phenomenon. Direct gains in application come from the integration of Asset Management with Reliability-Centered Maintenance, optimizing the Maintenance Plan and the concepts of Engineering Applied to the company's Assets.

Keywords: Ultraviolet · Technology · Innovation

S. Zheng et al. (Eds.): IHDC 2024, LNCE 487, pp. 15–24, 2025.
https://doi.org/10.1007/978-981-97-9184-2_2

1 Introduction

The technical work demonstrated here brings the gains applied in predictive maintenance, using Ultraviolet apparatus, corona effect measurement, in rotating machinery of SHPs. This technique, well applied in the electrical assets of substations and high-voltage transmission lines, is customarily used to enhance predictive maintenance associated with the shared use of the OFIL Luminar HD UV apparatus. This initiative has brought benefits on various fronts, such as replacing tests in dark chambers and evaluating the stator asset in its operational conditions during daylight, avoiding scheduling for execution during nighttime. The SBUV camera, Solar Blind filter, has higher sensitivity, compatible wavelength, bi-spectral, and a range from 240 nm to 280 nm (nanometers), which can bring gains in this technological application. The shared use of the tool provides risk assessment for each generator in trials and ensures an operational view of the Asset from the perspective of its health in this regard - Health Index. The new technology is already implemented and has been successfully used in our plant, in its punctual maintenance plan regarding measuring, evaluating own or third-party services, and determining the best time to act in new surface treatments or asset cleaning, mitigating the risk of early burnout due to this phenomenon. Direct gains in application come from the integration of Asset Management with Reliability-Centered Maintenance, optimizing the Maintenance Plan and the concepts of Engineering Applied to the company's Assets.

2 Development

2.1 Data of the Plant/Generators Under UV Measurement - PCH Salto Goes

To illustrate the Ultraviolet trials, Partial Discharge tests were conducted by CEPEL, and all electrical tests of the generating units were conducted by Nishi, detailing the aging of the insulating material. Also, measurements were provided by the Supplier WEG, for commissioning and post-treatment of the generating units, regarding reduction or mitigation of the corona effect, where the results show a significant reduction compared to previous values. Identification of the Project Plant/Company: PCH Salto Góes/CPFL Renováveis SA Municipality: Tangará/SC.Specifications/Generators Quantities 02 (two) units Manufacturer/year: WEG/2012 Type/Model: SH10 1600 Serial Number: G1:1014439195/G2:1014317863 Power/Un/In/rpm 11.11 MVA/13.8 kV/465 A/327.27 rpm (Fig. 1).

2.2 Principle of UV Inspection Corona Effect

Corona Effect is understood as a partial electrical discharge due to the ionization of the air surrounding an electrically charged point where there is an electric field gradient that exceeds a critical value (Ec), therefore, an electric field (by potential) - with the presence of ultraviolet radiation. The point becomes more accentuated by this phenomenon, given parts related to the material medium itself, pollution, humidity, electrical spacing, temperature, at points such as coil heads, and sharp parts where the electric field becomes stronger at sharp and acute edges, making it easier for the material to reach

IDENTIFICAÇÃO DO EMPREENDIMENTO	
Hydro power plant	**PCH Salto Góes / CPFL Renováveis SA**
municipality	**Tangará / SC**
ESPECIFICAÇÃO / GERADORES	
Quantities	**02 units**
Fab. / Year:	**WEG / 2012**
Type/Model:	**SH10 1600**
Number Série:	**G1:1014439195 / G2:1014317863**
Power/ Un/In / rpm	**11,11 MVA/ 13,8 kV/ 465 A / 327,27 rpm**

Fig. 1. Assets identification

the breakdown voltage. (*) IEEE: About 90% of failures in high voltage installations are caused by insulation deterioration (relative to Transmission Lines and Substations). Corona Effect - In Rotating Machinery: • Isolation Problems/Pollution/Air Gap - Free Space • Defected coils/Cracks – Defects (Fig. 2).

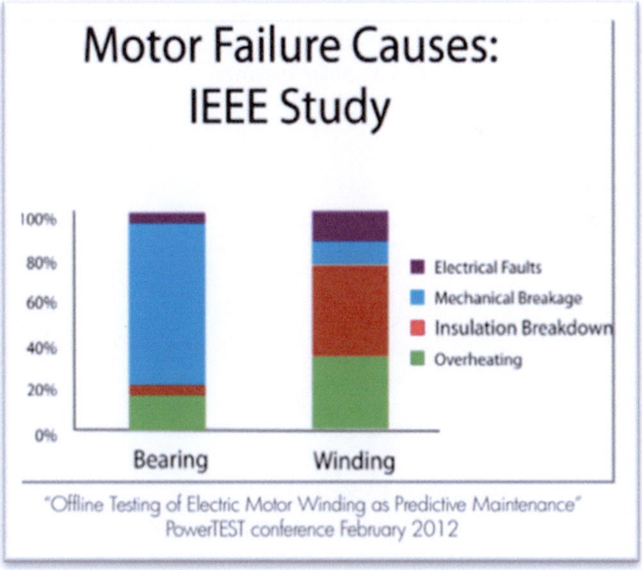

Fig. 2. OFIL Data – Luminar HD

Industry statistics by IEEE indicate that approximately 40% of all machine failures occur due to failure of the stator winding insulation. (See chart). The apparatus used in the measurements: OFIL Technology SBUV – Luminar HD - UV Luminar HD - Light

Weightl Built-in Flashlightl Large LCD OFIL's bi-spectral camera has 2 channels that capture the image of the same object, showing it in 2 different wave spectra: Visible and Ultraviolet. Channel 1: Ultraviolet type C (UVC) wavelength from 240 nm to 280 nm, because in this area 2 conditions converge on the one hand, there is no solar radiation (SBUV - Solar blind zone) and, on the other hand, if there is UVC radiation by Corona. This allows you to work in broad daylight without being bothered by the sun and to see Corona. Channel 2: Visible Captures the same image as Channel 1 but is displayed in the visible spectrum. Finally, the camera mixing both channels allows "the exact location of the source of origin of the Corona."

Thus, we have: Solar Spectrum (Figs. 3 and 4).

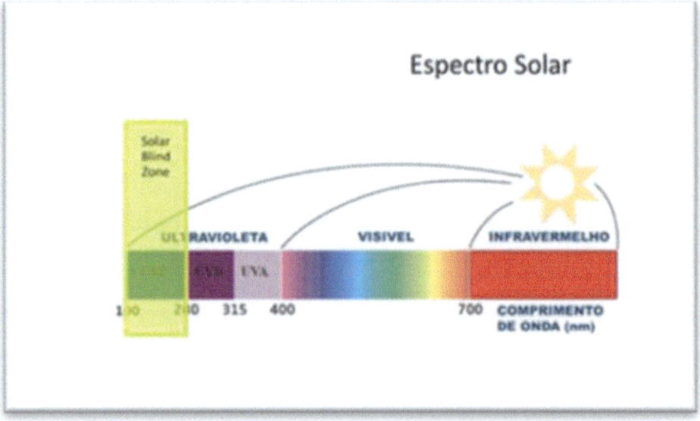

Fig. 3. Solar Spectrum

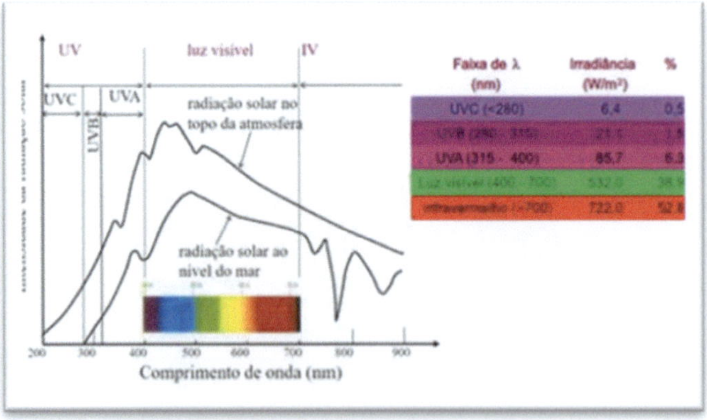

Fig. 4. Wave spectrum in nm, per Ofil

Principle of UV Inspection (Ultraviolet) The Ultraviolet spectrum region is considered in the range of 230 to 405 nm, the equipment used is an Ultraviolet detector model LuminarHD from manufacturer OFIL, mentioned in the figure above, which works in this range. A low level of ultraviolet signal generated by pressure, humidity, temperature, or electrical losses, is captured by the ultraviolet sensor capable of detecting ultraviolet signals at great distances. In the case of measurements in generators, it is graduated up to 4 m. The equipment used is highly directional in identifying losses due to corona effect. It can detect a low-level leakage point up to a distance of 200 m; in the case of rotating machinery, adjustments are made for work up to 4 m. Criteria adopted by the company on criticality (Figs. 5 and 6):

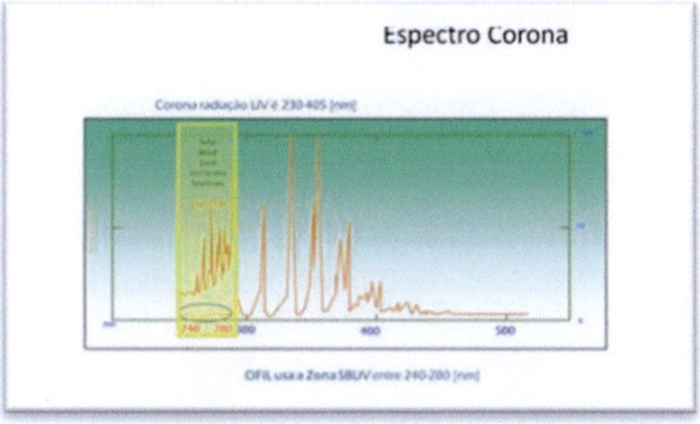

Fig. 5. Wave spectrum in nm, per Ofil

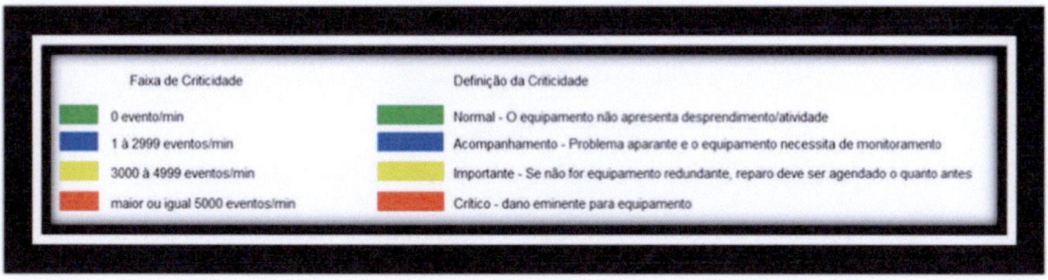

Fig. 6. Criticality Range – events/min (adopted in the company)

2.3 Luminar HD/UV Measurements (Before)

Measurement (BEFORE) of intervention for Surface treatment and reduction/mitigation of corona effect. Anomalies detected are classified following a priority criterion, where they are recorded through evidence, according to events/min (Fig. 7).

The image below shows the location of points with higher concentrations/activities of corona effect: Points with higher concentration of events/min. (Ultraviolet radiation).

For example, in the Salto Goes Plant, in 2018, a UV inspection was carried out on generator UG1, which served as the starting point for this UV apparatus application (Figs. 8 and 9).

The records in the table indicate in the figure below the locations and points of events/min. Captured by the Luminar HD Ultraviolet apparatus, with the region of highest discharges/corona effect highlighted in red (Figs. 10 and 11).

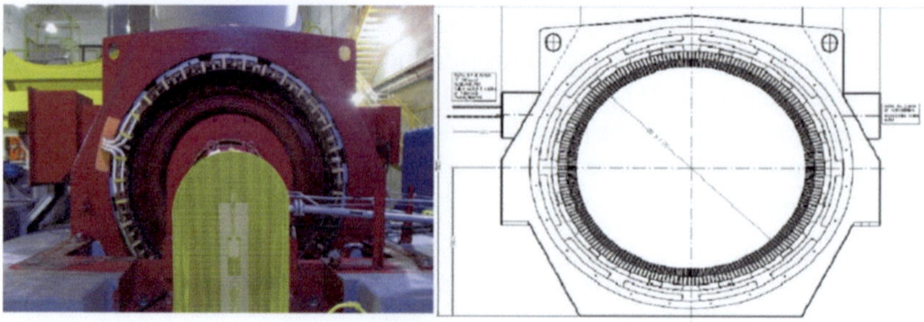

Fig. 7. Frontal view of the generator

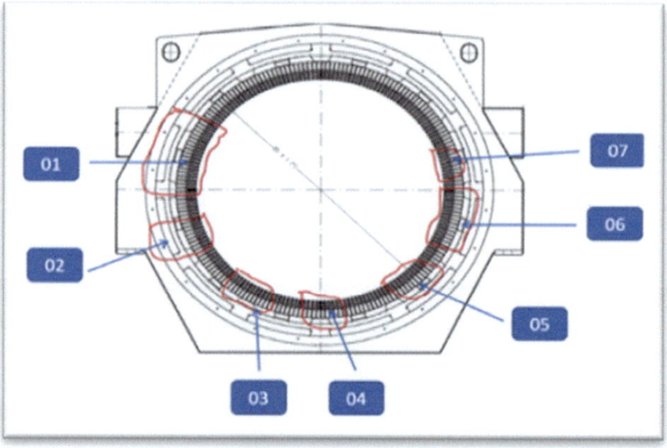

Fig. 8. Frontal view of the generator – measurement/identification points

[illegible]em	no Equipamento	Sensibilidade	por minuto	Criticidade	Ambiente	Relativa do Ar	Gerador	Aproximada	Horário
1	1.1	[illegible]	[illegible]		21 °C				
2	1.2	[illegible]	[illegible]		21 °C				
3	1.3	[illegible]	[illegible]		21 °C				
4	1.4	[illegible]	[illegible]		21 °C				
5	2	[illegible]	[illegible]		21 °C				
6	2.1	[illegible]	[illegible]		21 °C				
7	3	[illegible]	[illegible]		21 °C				
8	3.1	[illegible]	[illegible]		21 °C				
9	3.2	[illegible]	[illegible]		21 °C				
10	3.3	[illegible]	[illegible]		21 °C	[illegible]	[illegible] kV	[illegible]	[illegible]
11	4	[illegible]	[illegible]		21 °C				
12	4.1	[illegible]	[illegible]		21 °C				
13	4.2	[illegible]	[illegible]		21 °C				
14	5	[illegible]	[illegible]		21 °C				
15	5.1	[illegible]	[illegible]		21 °C				
16	6	[illegible]	[illegible]		21 °C				
17	6.1	[illegible]	[illegible]		21 °C				
18	7	[illegible]	[illegible]		21 °C				

Fig. 9. Field Measurement – Criticality: Blue Range

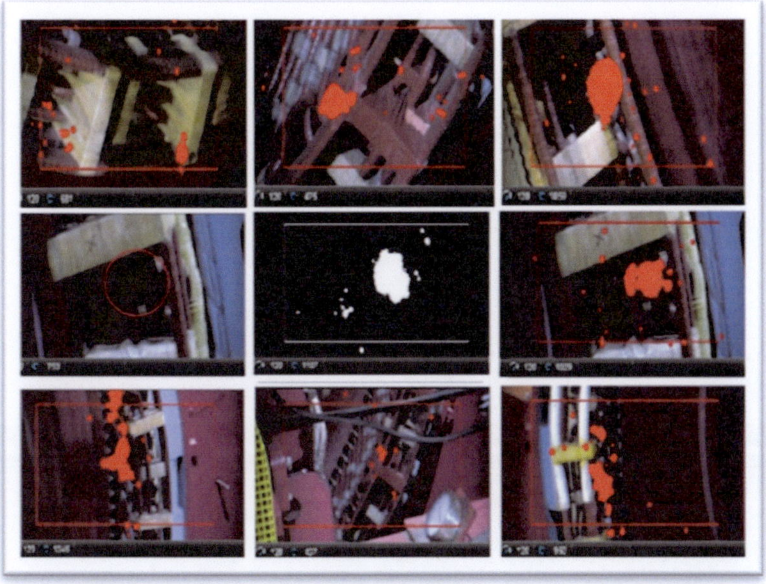

Fig. 10. Analysis of images/spectrum – equipment, spectrum only, combined image (equipment spectrum)

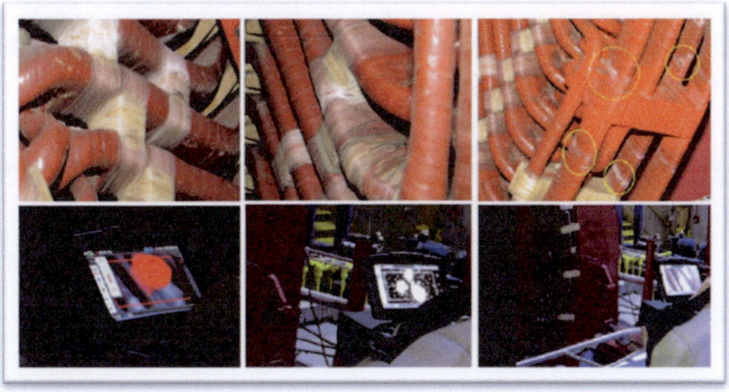

Fig. 11. Analysis of images/spectrum – some evidence (whitened/indicated) detected through UV equipment

2.4 Luminar HD/UV Measurements (After)

Measurement (AFTER) of intervention for Surface treatment and reduction/mitigation of corona effect.

A UV inspection was conducted on generator UG1 (Figs. 12–15).

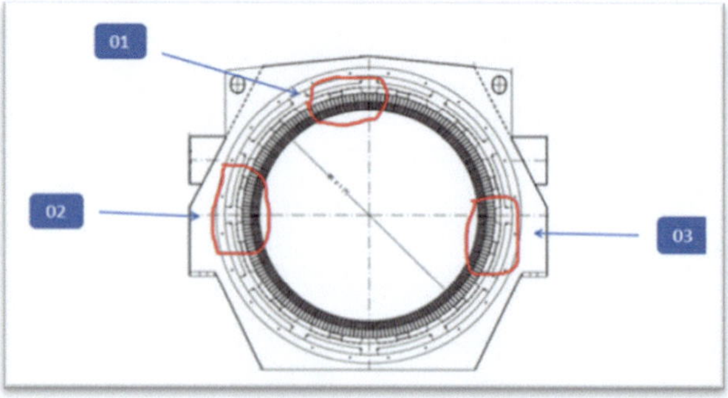

Fig. 12. Frontal view of the generator – measurement/identification points

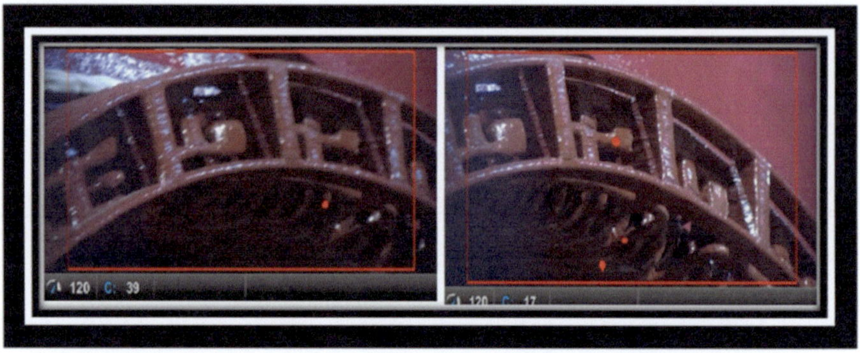

Fig. 13. Point 1 – low measurement of 39 events/min. (previously low)

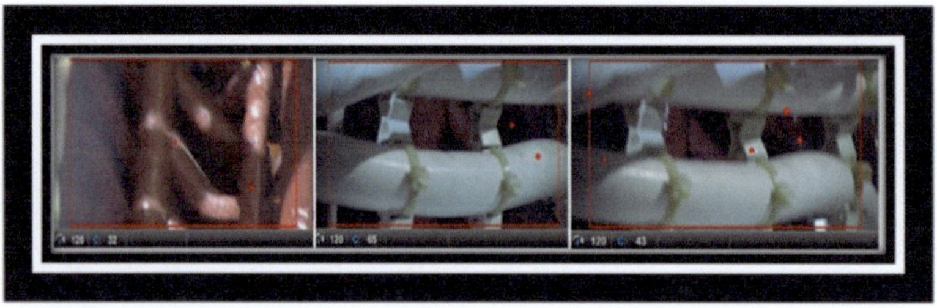

Fig. 14. Point 2 – low measurement of 43 events/min. (previously 2,138)

2.4.1 Important Note

Through the images, it can be verified the minimum events/min. Indices recorded by the OFIL Luminar HD Ultraviolet camera for the reception of contracted surface treatment services to mitigate the intrinsic phenomenon, given that by design, the machine will always be exposed to the constructive phenomenon as it is subject to the electric field of the voltages at its terminals, however, attention should be paid to other improvement points in the installations in order to attenuate the Corona Effect phenomenon, either

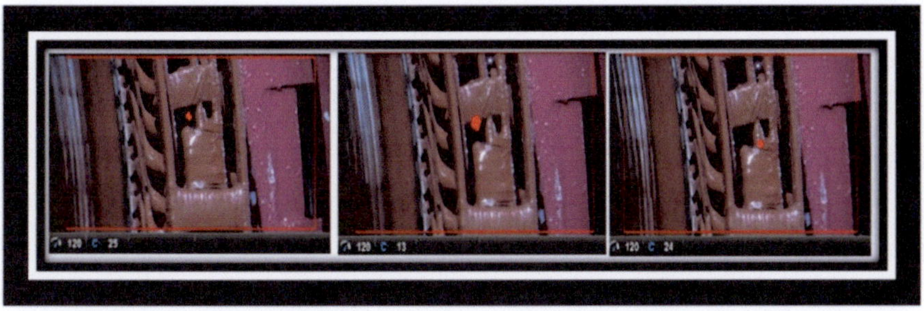

Fig. 15. Point 3 – low measurement of 25 events/min. (previously 712)

periodic cleaning of the winding, installation of filters in the cooling air intakes, closing of moisture entry through the hot air exhaust channel in stopped machinery, but also evaluating the design of the winding construction concept regarding electrical distances within the expected design and subjected to terminal voltage. Thus, not neglecting offline and online tests with partial discharge technique, as here we deal with points related to the Corona Effect in this particularity of technical validation or as a tool in quality control management.

2.5 Conclusion

The tests aimed at the generators, such as Partial Discharges tests for measurement and evaluation of the PCH generators under study, were carried out in the terms of offline and online tests by the company CEPEL, which guided a Technical Specification of Corona Recovery/Mitigation Services in Generators, in the field and/or specialized Workshop, remembering here that the tests for evaluation of the dielectric conditions of the generators, included measurements of capacitances, tangent delta, and the Partial Discharges themselves, which are characterized by a process of ionization in a gaseous environment inside the insulating systems, caused by an intense electric field. Here, we start to record that the corona points must be monitored and/or corrected, but it is important to emphasize that it is difficult to correct, in most cases, under analysis, the central objective would be to delay the degradation process by partial discharges. After treatment and recovery of the generating units, mentioned by the above ET, a significant improvement in the results presented was noticed, and specific recommendation points related to the environment, more precisely local pollution, and other services performed in this context of mitigating and recovering the operational risks of the assets of the PCH under study were mentioned. Now, regarding the use of tests by UV - Ultraviolet apparatus, the central theme of this paper with a tool in use, which does not replace the tests mentioned above, but advocates an evaluation of the phenomenon in question, the Corona Effect, this apparatus, intensively used in Transmission Lines, and which for rotating machinery, we put into practice in the company. In this way, whether replacing the Dark Chamber method or qualifying receipt of post-treatment deliveries and here we demonstrate that the measured results met our quality standards. The technique brings gains, one for the operational risks issue and another for quality, such as measuring the intensities recorded in events/min by UV as a reference and acceptance, where it offers a new level of quality required and perceived when delivering these services and releasing them to Operations of the recovered Assets. The experience disseminated here brings this viable solution, and as shared predictive maintenance between Transmission Lines and Generators, customizing the investments of test kits, as well as periodically monitoring its assets in operation.

References

1. Nakatani, F.T.: Nishi Company, test report on the generating units of the PCH under Study, 10 April 2017
2. Amorim Junior, H.d.P.: CEPEL Company, Offline and online tests report of Partial Discharges in the generators of the Study PCH, 25 September 2017
3. Rataus, Sergio – Ofil Systems – consultation - UV - Ultraviolet apparatus LuminarHd

Numerical Analysis Calculations of Ductile Concrete Gravity Dams Under Seismic Action

Wei Fang, Jingjing He(✉), Yang Yu, Rusheng Hao, and Yan Guo

Power China Northwest Engineering Co., Ltd., Xi'an, Shaanxi, China
Hejing_86@126.com

Abstract. Finite element analysis was conducted on the stress, vertical displacement, and horizontal displacement of ordinary concrete and ductile concrete gravity dams under different working conditions. The results show that: with the decrease of the height of the dam, the stresses on the two concrete dams gradually increase, and the vertical displacements show a decreasing trend during the operation period. Under the effect of earthquakes, the maximum stress of the two kinds of concrete gravity dams appears at the weak point of the dam body. The maximum stress of the normal water level and the falling water level of the ordinary concrete dam reach 9.91 MPa and 9.85 MPa respectively. The maximum stress of two types of concrete dams under seismic conditions is approximately 4 times that of the operation period. Under different working conditions, the maximum stress and displacement of the ordinary concrete are always smaller than that of ductile concrete.

Keywords: earthquake · ductile concrete · ordinary concrete · gravity dam · finite element analysis

1 Introduction

Concrete panel gravity dam has become one of the important structures used in water conservancy projects due to its simple structure, short construction period, strong terrain adaptability and safety. At present, scholars have conducted a lot of research and calculations on the blast resistance, explosion resistance, cracking characteristics and stability of concrete gravity dam [1–3]. At the same time, a lot of experience have been accumulated in structural design.

However, traditional concrete is brittle and easy to crack, resulting in large-scale cracking occurs in the service process. Under the action of earthquakes, dam body is prone to compression damage and large area evacuation [4], which seriously affects the operation of the hydropower station and the service life. Ductile concrete has been applied in engineering practice because its excellent impact resistance, high ductility, permeability and crack resistance [5, 6], but the analysis and calculation of ductile concrete gravity dam is still lacking.

Based on this, this paper adopts ABAQUS software to carry out three-dimensional finite element modelling analysis. By simulating the operation period and seismic conditions of the gravity dam, as well as the stress maps, vertical displacement and horizontal

S. Zheng et al. (Eds.): IHDC 2024, LNCE 487, pp. 25–31, 2025.
https://doi.org/10.1007/978-981-97-9184-2_3

displacement parameters of two kinds of concrete dams, the stress and deformation laws of the concrete gravity dam are analyzed. Finally, the ranges of the properties of the concrete materials to meet the requirements of the different working conditions are obtained. The research results can provide a basis for the design and application of ductile concrete materials, and provide reference for the design of dams in extreme environments.

2 Project Overview and Numerical Calculation Model

Concrete dam is a standard shape, the height of dam is 100 m, the downstream dam face slope is 0.7 m, the dam top width is 7 m, the foundation height is 102 m and the downstream dam surface slope is 0.7. The water depth in front of the dam at normal water level is 95 m, dead water level is 65 m and the falling water level is 30 m.

The combination and selection of working conditions in the numerical calculation process is an important factor affecting the calculation results. This calculation example is based on actual engineering, and the load on the dam body includes self-weight, temperature load, water pressure, and seismic load. The load combinations under different working conditions are shown in Table 1. The dam model is mainly composed of four parts: the dam body, upstream foundation, lower foundation, and downstream foundation.

The material parameters of ordinary concrete and ductile concrete gravity dams are listed in Table 2. Compared to ordinary concrete, ductile concrete has a lower density and better ductility.

Table 1. Calculation conditions and load combinations

Calculation of working conditions	Load/action combinations	Self-weight	Temperature load	Water pressure	Seismic load
Operation period	Normal water level	√	√	√	/
	Falling water level	√	√	√	/
Seismic conditions	Normal water level + seismic	√	√	√	√
	Falling water level + seismic	√	√	√	√

3 Numerical Calculation of the Operation Period

During the operation of the dams, according to the change of water level, it can be divided into two different water pressure modes: normal storage level and falling water level. The stress cloud, horizontal displacement and vertical displacement numerical calculations under different water levels are shown in Figs. 1 and 6, respectively.

Table 2. Calculation conditions and load combinations

Sample	Elastic modulus (MPa)	Poisson's ratio	Density (kg/m^3)	Coefficient of thermal expansion (°C)	Conductivity (W/m·K)
Ductile concrete	3.00×10^4	0.167	2200	1.7×10^5	3.00
Normal concrete	3.45 × 104	0.140	2400	1.0 × 105	1.28

As can be seen from the stress cloud diagram in Figs. 1 and 2, with the elevation of the dam decreases, the stress of the two types of concrete dams gradually increases under different water levels and water pressure. The maximum stress appears near the heel and toe of the dam, and the maximum stress of the ordinary concrete dam is smaller than that of the ductile concrete dam. In addition, the maximum stress of the dam body under ductile water storage level is greater than the falling water level, because the water storage capacity at normal water storage level is large, and the water pressure load applied to the dam body is greater. The maximum stress of ductile concrete dams under normal water level is 2.73 MPa.

From the horizontal displacement diagram of the dam in Figs. 3 and 4, it can be seen that the maximum horizontal displacement of the dams under water pressure occurs at the upstream. The horizontal displacement of ductile concrete dam at the normal water level is 4.67 mm, and the horizontal displacement at the falling water level is 4.66 mm, while both values of ordinary concrete dam are 3.32 mm. The displacement deformation of the ordinary concrete dam is always smaller than that of the ductile concrete dam.

As can be seen from Figs. 5 and 6, with the elevation of the dam decreases, the vertical displacement of two kinds of concrete dams during operation shows a decreasing trend. The maximum vertical displacement of the dam body occurs at the top of the dam, and the dam experiences settlement. The maximum vertical displacement distribution of the ductile concrete dam at normal water level and falling water level is 5.95 mm and 5.90 mm, and that of the ordinary concrete dam is 4.22 mm and 4.19 mm respectively. It can be seen that the effect of water level and water pressure on the vertical displacements during the operation period is small, but the vertical deformations of the ordinary concrete dams are always smaller than that of the ductile concrete dams.

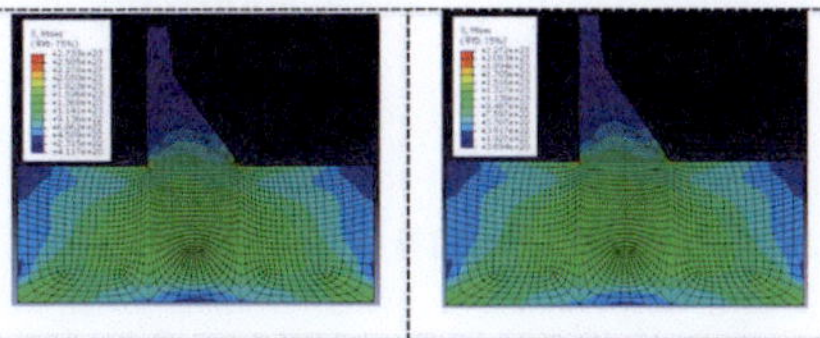

(a) ductile concrete (b) ordinary concrete

Fig. 1. Stress cloud at normal water level

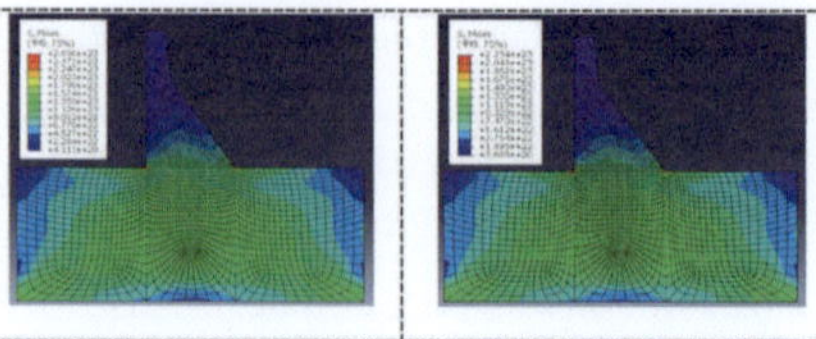

(a) ductile concrete (b) ordinary concrete

Fig. 2. Stress maps for falling water level

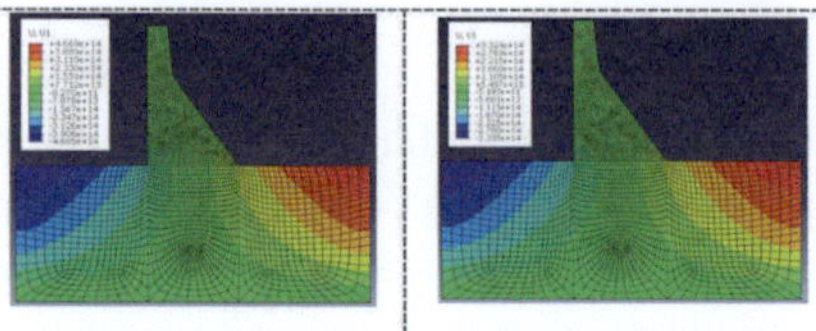

(a) ductile concrete (b) ordinary concrete

Fig. 3. Horizontal displacement at normal storage level

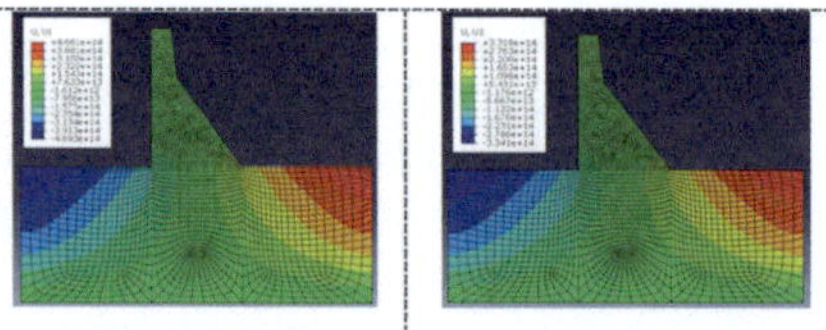

(a) ductile concrete (b) ordinary concrete

Fig. 4. Horizontal displacement for falling water level

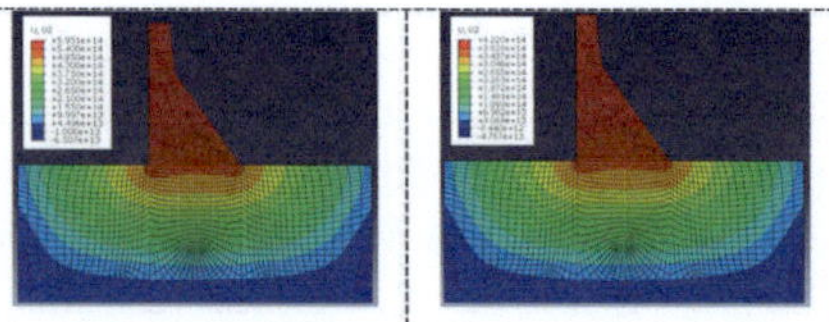

(a) ductile concrete (b) ordinary concrete

Fig. 5. Vertical displacement at normal storage level

4 Numerical Calculation of Seismic Conditions

The seismic dynamic analysis applies the mode decomposition response spectrum method, and the response spectrum is selected from the standard response spectrum curve given by SL203-97 "Code for Seismic Design of Hydraulic Buildings". The maximum spectrum value Bmax is set as 2.0, and the damping ratio is set as 0.05. According

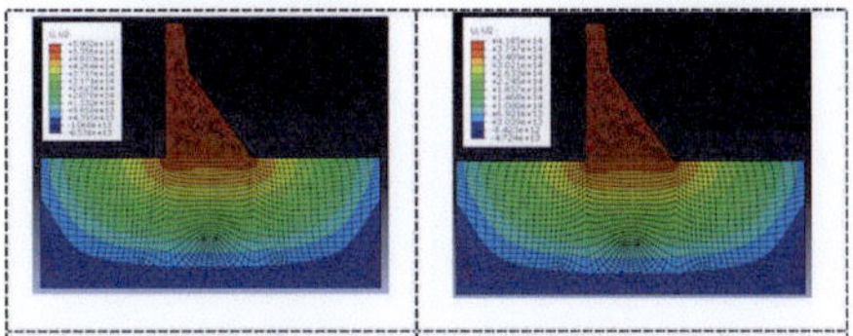

(a) ductile concrete | (b) ordinary concrete

Fig. 6. Vertical displacement for falling water level

to the seismic effects of the calculated vibration modes of each order, the combination of seismic effects is chosen to be the square root of the square sum of squares (SRSS) as stipulated in the specification. The seismic dynamic analysis of dams includes vertical and horizontal seismic action, follows the principle of combination of basic static load and seismic load, and obtains a comprehensive response by combining dynamic response and static response according to the most unfavorable combination principle.

Figures 7 and 8 are the stress maps of the two kinds of concrete dams under seismic conditions at the normal water level and falling water level. Unlike the stress distributions of the dams during the completion period operational period, the maximum stresses of the two dams under seismic conditions occur at the weak points of the dam body and at the connection between the dam body and the rock below. The maximum stresses of ductile concrete dams at normal water level and falling water level is 10.63 MPa and 10.57 MPa, respectively, while those of ordinary concrete are 9.91 MPa and 9.85 MPa. The maximum stress of the two types of concrete dams under seismic conditions is about 4 times higher than those in operation period. Moreover, the dam principal stress at the falling water level is always less than that at the ductile water level.

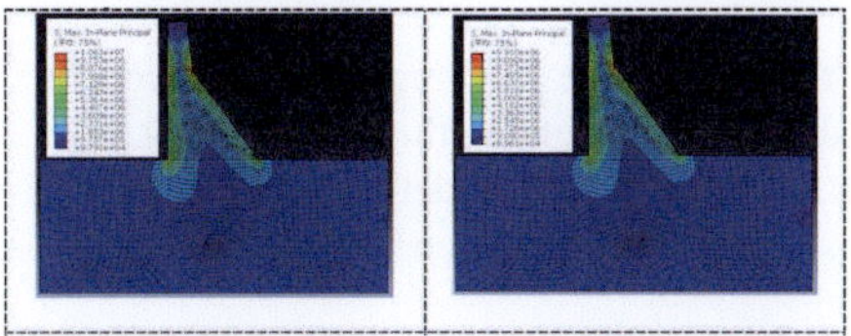

(a) ductile concrete | (b) ordinary concrete

Fig. 7. Stress map of normal water level under seismic conditions

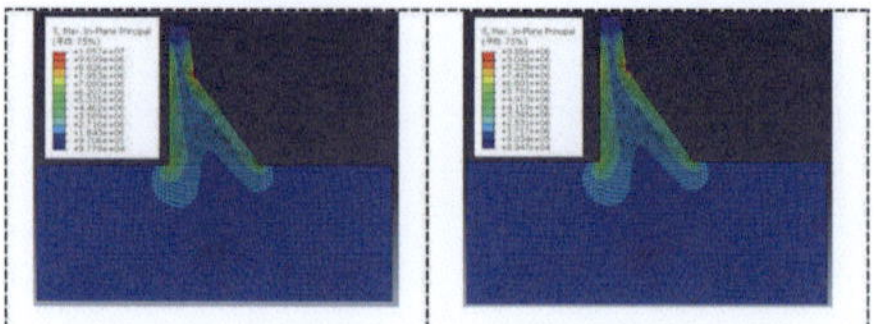

(a) ductile concrete (b) ordinary concrete

Fig. 8. Stress cloud of falling water level under seismic action

5 Conclusion

1. In the and operation period, with the decrease of the dam elevation, the stress of both types of dams gradually increases, but the vertical displacement decreases.
2. The maximum stress of the dam body under normal water level is greater than falling water level. Under normal water level, the maximum stress of ordinary concrete gravity is 2.27 MPa, the vertical displacement is 3.32 mm and the horizontal displacement is 4.22 mm, both of them are smaller than those of ductile concrete gravity dams.
3. Under the action of earthquake, the maximum stress in the two types of concrete gravity dams occurs at the weak point of the dam. The maximum stress of two types of concrete dams under seismic conditions is approximately 4 times that of the operating period. The maximum stress of the ductile concrete dams at normal water level and at falling water level reaches 10.63 MPa and 10.57 MPa, respectively, which both slightly higher than ordinary concrete.

References

1. Zhang, S., Wang, G., Yu, X.: Seismic cracking analysis of concrete gravity dams with initial cracks using the extended finite element method. Eng. Struct. **56**, 528–543 (2013)
2. Patra, B.K., Segura, R.L., Bagchi, A.: Modeling variability in seismic analysis of concrete gravity dams: a parametric analysis of Koyna and Pine Flat Dams. Infrastructures **9**(1), 10 (2024)
3. Xu, B., Wang, S.: Sensitivity analysis of factors affecting gravity dam anti-sliding stability along a foundation surface using Sobol method. Water Sci. Eng. **16**(4) (2023)
4. Sarkar, A., Sharad, G., Bagchi, A.: Performance of 2D-spectral finite element method in dynamic analysis of concrete gravity dams 105770 (2024)
5. Li, V.C.: From micromechanics to structural engineering the design of cementitious composites for civil engineering applications. Doboku Gakkai Ronbunshu **471**, 1–12 (1993)
6. Zhenbo, W., Pen, S., Jianping, Z.: Long-term properties and microstructure change of engineered cementitious composites subjected to high sulfate coal mine water in drying-wetting cycles. Mater. Des. **203**, 109610 (2021)

Experimental Study on the Flexural Performance of Hydraulic High Ductility Concrete

Jingjing He(✉), Wei Fang, Zhi Zhang, Rusheng Hao, and Yan Guo

Power China Northwest Engineering Co., Ltd., Xi'an, Shaanxi, China
Hejing_86@126.com

Abstract. In order to solving the problem of tolerance in the mixing of water workers' consolidation processes, Hydraulic High Ductile Concrete (HHDC) was prepared by using lower cost centimeters and TY-PVA fibers. The effects of sand-binder ratio, fly ash dosage, rubber powder dosage and fibre dosage on the performance of HHDC were investigated. The results show that among the four influencing factors, the rubber powder dosage and sand-binder ratio have a more significant impact on the flexural load, with HHDC-1 having the highest flexural load of 3.49 kN. While the fiber dosage and fly ash dosage have a significant impact on the ultimate deflection, with HHDC-7 having the maximum ultimate deflection of 0.60 mm. Each group of HHDC exhibits ductile failure characteristics during the bending process. The conclusion of the study can provide some theoretical support for the engineering application and promotion of HHDC.

Keywords: Hydraulic High Ductile Concrete · sand-binder ratio · fly ash · fiber · rubber powder

1 Introduction

During the service process, hydraulic concrete inevitably develops cracks that can open up to a millimeter level due to spalling of the concrete protective layer and steel corrosion, which significantly impacts the safety and service life of hydraulic structures. Hydraulic High Ductile Concrete (HHDC) has excellent toughness and durability, and still has the characteristics of strain hardening and small cracks under miscellaneous loads, and its ultimate tensile strain can reach 3%, which is a hundred times of that of concrete [1]. Thus, it could be used to solve the problems of brittle cracking and poor durability of traditional concrete dams.

However, HDC is mainly composed of cement, fly ash, silica fume, slag powder, water and fibers. Its raw materials have a large impact on its performance. In order to achieve the effect of HHDC multi-seam cracking, the design of the materials should not only strictly control the particle size of the sand (particle size less than 150 μm), but also use the excellent performance of Japan Kuraray company produced polyvinyl alcohol (PVA) fibers [2, 3]. Among them, fine silica sand and Japanese PVA fibers are

S. Zheng et al. (Eds.): IHDC 2024, LNCE 487, pp. 32–38, 2025.
https://doi.org/10.1007/978-981-97-9184-2_4

expensive, which makes HDC unable to be widely promoted and applied in hydraulic construction even if it has excellent mechanical properties.

Based on this, this paper uses low-cost TY-PVA fibers and medium sand to prepare HHDC. Orthogonal tests were used to investigate the effects of sand-binder ratio, fly ash dosage, rubber powder dosage and fiber dosage on HHDC compressive performance and bending performance. Finally, a reasonable matrix mix was determined by comprehensive comparison. The results of the study can provide theoretical support for its application and promotion in water conservancy and hydropower engineering.

2 Raw Materials and Test Methods

The cementitious materials include Tianshan cement and Class I fly ash. The particle size of rubber powder ranges from 40 to 80 mesh. The fibers are polyvinyl alcohol fibers (TY-PVA) produced in Changzhou, China, its length is 12 mm, the elastic modulus is 38.3 GPa and the tensile strength is 1680 MPa. In addition, the surface of the fiber is coated with oil to reduce the bonding force between the fiber and the matrix, and play a better bridge effect. Sand is simulated a project fine aggregate grading of the second area of medium sand, fineness modulus of 2.68, stone powder content of 14%. The water is laboratory tap water. The admixtures include superplasticizer and thickening agent.

In this experiment, sand-binder ratio, fiber dosage, fly ash dosage and rubber powder dosage are the main influencing factors, and each factor is designed with four levels. The orthogonal scheme is shown in Table 1.

The molding and curing steps of HHDC specimens are as follows: (1) Mixing matrix: the cementing materials and powder additives were added to the mixer for dry mixing for 30s, after being mixed evenly, the water was slowly added and the mixing is continued for 2 min to obtain the uniform flow of the slurry. (2) the fiber was slowly added to the slurry, and the fiber was fully stirred to ensure the uniform dispersion of the fiber. (3) Pouring specimens: the HHDC slurry was poured into the moulds for 2–3 times, and then the surface was smeared and put into the pre-curing room for curing after vibration. (4) Demoulding and curing: the cast: the cast HHDC was left for 1 day before demoulding, and the specimen was transferred into the standard curing room for curing for 28 days. Before testing, the specimens were taken out from the curing room for related mechanical performance tests.

The flexural specimen is a rectangle of 40 mm $\times$ 40 mm $\times$ 160 mm, and the bending loading rate is 0.15 mm/min. In the experiments, an extensometer is placed below the center of the specimen for real-time recording of specimen deflection values, and the span is 150 mm.

3 Test Results and Analyses

Figure 1 shows the flexural load-displacement curves of each HHDC. As can be seen from the figure, the stress-strain curves can be divided into three stages: (1) Linear elasticity stage: in which the surface of the specimen remains intact and the load increases linearly with the deflection. (2) Strain-hardening stage, in which the load jitter increases with the deflection, accompanied by cracking of the specimen. (3) Strain-softening stage:

Table 1. Orthogonal experimental plan

Sample	Sand-binder ratio	Fiber dosage /%	Fly ash dosage /%	Rubber powder dosage /%
HHDC-1	0.40	1.6	20	0
HHDC-2	0.40	1.8	25	5
HHDC-3	0.40	2.0	30	10
HHDC-4	0.40	2.2	35	15
HHDC-5	0.45	1.6	35	10
HHDC-6	0.45	1.8	20	15
HHDC-7	0.45	2.0	35	0
HHDC-8	0.45	2.2	30	5
HHDC-9	0.50	1.6	30	15
HHDC-10	0.50	1.8	35	10
HHDC-11	0.50	2.0	20	5
HHDC-12	0.50	2.2	25	0
HHDC-13	0.55	1.6	35	5
HHDC-14	0.55	1.8	30	0
HHDC-15	0.55	2.0	25	15
HHDC-16	0.55	2.2	20	10

after reaching the peak load, the main crack continues to open, and the load slowly decreases with the increase of deflection until the specimen fails. In the bending test, due to the toughening and crack resistance effect of internal PVA fibers [4], HHDC exhibits excellent toughness and ductile failure characteristics.

The bending load and ultimate deflection of each HHDC were extracted from the curve and summarized in Table 2. The data in the table are the average values of each group. It can be seen that HHDC-1 has the maximum bending load of 3.49 kN. The maximum ultimate deflection of HHDC-7 is 0.60 mm. The following text will conduct a range analysis of the experimental results.

Table 3 summarizes the results of the extreme variance analysis of flexural loads for each HHDC. It can be seen that the order of the influence of each factor on the bending load is as follows: rubber powder dosage, sand-binder ratio, fly ash dosage and fiber dosage. The rubber powder content has the greatest impact on the flexural load, and as the rubber powder content increases, the flexural load decreases. Analysis suggests that rubber powder not only has low strength, but also the rubber particles are inert organic materials with poor adhesion to the cement matrix. When subjected to loads, cracks are easily generated at the bonding interface, resulting in a decrease in the bending load of HHDC.

In addition, the sand-binder ratio also has a greater influence on flexural load. This is mainly because after the matrix cracks, the fibers between the cracks to play the fiber bridging stress, and then the load continues to increase until it reaches the flexural strength. However, the sand used in this paper is medium sand, the large particle size of the fine sand not only increases the fracture toughness of the matrix, violating the strength criterion in the principle of the design, but also affects the dispersion of fiber uniformity [5], which is not conducive to the fibers to play the role of bridging. As a result, with the dosage of sand increases, the flexural load shows a downward trend. Therefore, if medium sand is used to prepare HHDC, the sand-binder ratio can be properly reduced, and it is recommended that the sand-binder ratio is not greater than 0.45.

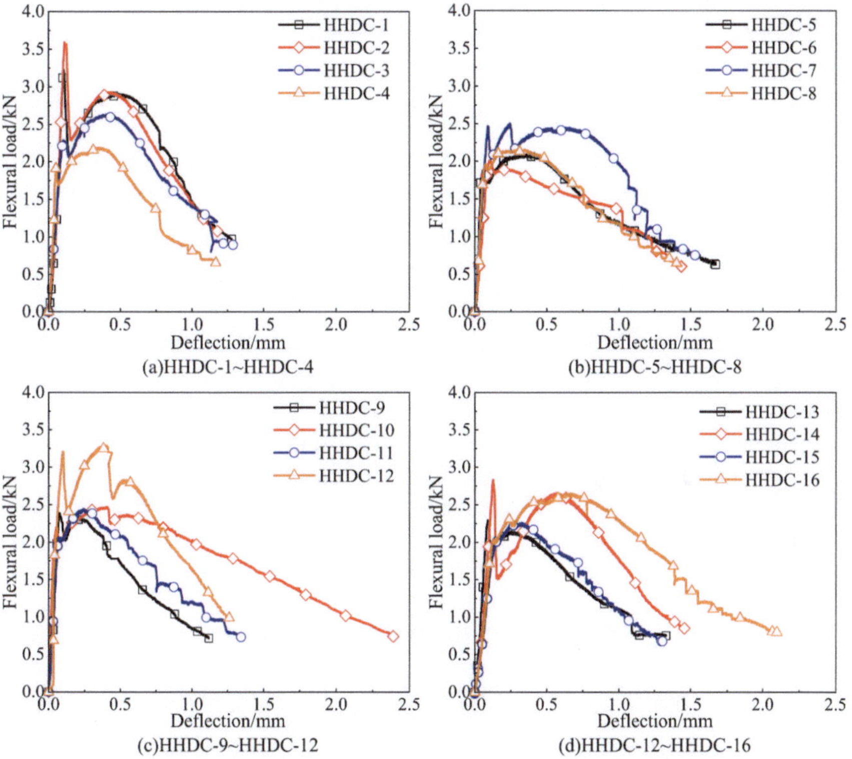

Fig. 1. Bending load displacement curves of each HHDC

Table 4 shows the results of the extreme variance analysis of the ultimate deflection of each HHDC. The ultimate deflection represents the ability to resist bending deformation. The greater the deflection is, the greater the ability of HHDC to resist fracture and deformation is. From Table 4, it can be seen that fiber dosage and fly ash dosage have a greater impact on the ultimate deflection. This is because the fly ash can reduce the bonding of PVA fiber-matrix interface, weaken the damage of the fiber when pulled out, and is more conducive to the fiber to play its bridging role. At the same time, the larger the fiber dosage is, the more fibers are distributed at the cross section of the matrix, and the greater the bridging force of the fiber is. Therefore, with the increase of the content of fiber and fly ash, the HHDC ultimate deflection shows an increasing trend.

Table 2. HHDC orthogonal experimental results

Sample	Bending load/kN	Ultimate deflection/mm
HHDC-1	3.49	0.12
HHDC-2	3.22	0.21
HHDC-3	2.89	0.34
HHDC-4	2.35	0.55
HHDC-5	2.09	0.26
HHDC-6	1.97	0.16
HHDC-7	2.80	0.60
HHDC-8	2.31	0.17
HHDC-9	2.32	0.15
HHDC-10	2.52	0.19
HHDC-11	2.34	0.13
HHDC-12	3.01	0.30
HHDC-13	2.33	0.30
HHDC-14	2.80	0.12
HHDC-15	2.08	0.30
HHDC-16	2.56	0.58

Table 3. Analysis results of bending load range

Influence factor	Sand-binder ratio	Fiber dosage /%	Fly ash dosage /%	Rubber powder dosage /%
K1	2.99	2.56	2.59	3.03
K2	2.29	2.63	2.60	2.55
K3	2.55	2.53	2.58	2.51
K4	2.44	2.56	2.50	2.18
R	0.70	0.10	0.10	0.85

Table 4. Analysis results of ultimate deflection range

Influence factor	Sand-binder ratio	Fiber dosage /%	Fly ash dosage /%	Rubber powder dosage /%
K1	0.30	0.21	0.25	0.28
K2	0.30	0.17	0.26	0.20
K3	0.19	0.34	0.19	0.34
K4	0.32	0.40	0.41	0.29
R	0.13	0.23	0.22	0.14

4 Conclusion

(1) Each group of HHDC exhibits ductile failure characteristics in bending failure, and its load displacement curve can be divided into three stages: linear elasticity, strain hardening, and strain softening.
(2) The rubber powder dosage and sand-binder ratio have a significant impact on the flexural load, with HHDC-1 having the highest flexural load, reaching 3.49 kN.
(3) The fiber dosage and fly ash dosage have a significant impact on the ultimate deflection, with HHDC-7 having the highest ultimate deflection of 0.60 mm.

References

1. Yang, E.H., Li, V.C.: Tailoring engineered cementitious composites for impact resistance. Cem. Concr. Res. **42**(8), 1066–1071 (2012)
2. Li, V.C.: Engineered Cementitious Composites (ECC) Mate-rial, Structural, and Durability Performance. CRC Press, Rotterdam (2008)
3. Li, V.C.: From micromechanics to structural engineering the design of cementitious composites for civil engineering applications. Doboku Gakkai Ronbunshu **471**, 1–12 (1993)
4. Jin, C.Y., Li, Y.Q., Zhu, J.C., Li, Y.: A finite element analysis on compressive properties of ECC with PVA fibers. IOP Conf. Ser.: Mater. Sci. Eng. **544**(1), 012032 (2019)
5. Paul, S.C., van Zijl, G.P.: Mechanically induced cracking behaviour in fine and coarse sand strain hardening cement based composites (SHCC) at different load levels. J. Adv. Concr. Technol. **11**(11), 301–311 (2013)

Economic Analysis of Distributed Photovoltaic Power Generation Projects

Zhang Qian(✉) and Pan Yuwei

Yellow River Engineering Consulting Co., Ltd, Zhengzhou 45003, China
191935830@qq.com

Abstract. With the opportunities brought by China's promotion of achieving the "dual carbon" targets, the technology of China's photovoltaic industry is accelerating improvement, and the scale is steadily expanding. Distributed photovoltaic projects have the advantages of flexible configuration, nearby utilization, low investment, and saving land resources, with huge market space and development potential. Although the photovoltaic industry has enormous growth potential and good market prospects, it also faces many risks and challenges such as consumption problems and unstable income. Therefore, studying the economic viability of distributed photovoltaic projects is of great significance for making project investment decisions and promoting the sustainable development of the industry. This paper conducts the economic analysis of distributed photovoltaic power generation projects, calculates profitability analysis indicators such as financial internal rate of return (IRR) of project investment, financial net present value of project investment, and payback period of project investment. It also conducts preliminary sensitivity analysis on uncertain factors such as construction investment, operating revenue, and operating cost. It conducts in-depth sensitivity analysis on consumption, grid electricity price, and self-use electricity price, and proposes countermeasures to improve the economic efficiency of distributed photovoltaic power generation projects. The research results may provide reference and guidance for similar project investment decisions in more challenging investment environments in the future.

Keywords: Distributed photovoltaics · Economic analysis · Investment decision

According to data released by the National Energy Administration, as of March 2024, China's cumulative installed power generation capacity is about 2.99 billion kilowatts, an increase of 14.5% year-on-year. The total installed capacity of renewable energy is 1.54 billion kilowatts, surpassing the installed capacity of thermal power and becoming the mainstay; among them, the installed capacity of solar power generation is about 660 million kilowatts, a year-on-year increase of 55%, becoming the main incremental entity. Photovoltaic energy, as a clean and renewable energy source, is increasingly preferred, and the continuous growth of the global demand for renewable energy has greatly promoted the development of the photovoltaic industry. With the opportunities brought by China's promotion of achieving the "dual carbon" goals, the technology of China's photovoltaic industry is improving at an accelerated pace and the scale is steadily expanding. Distributed photovoltaic projects have the advantages of flexible

S. Zheng et al. (Eds.): IHDC 2024, LNCE 487, pp. 39–46, 2025.
https://doi.org/10.1007/978-981-97-9184-2_5

configuration, nearby utilization, low investment, and saving land resources, with huge market space and development potential.

Over the past decade, the cost of photovoltaic cells and systems has decreased significantly, making photovoltaic power generation one of the most cost-effective energy solutions in many markets. However, in June 2021, the Development and Reform Price [2021] No. 833 document stipulated that starting from 2021, for newly registered centralized photovoltaic power stations and industrial and commercial distributed photovoltaic projects, the central government will no longer provide subsidies and implement fair grid access; the grid electricity price for new projects in 2021 will be implemented based on the local benchmark price for coal-fired power generation. The "Regulations on the Supervision of Full Guarantee Purchase of Renewable Energy Electricity", which came into effect on April 1, 2024, stipulates that renewable energy generation has changed from full guarantee purchase to guarantee purchase and marketization.

Although the photovoltaic industry has enormous growth potential and good market prospects, it also faces many risks and challenges such as consumption problems and unstable income. Economic analysis is particularly important for investment decisions and sustainable development of photovoltaic projects. This paper takes a rooftop distributed photovoltaic power generation project in Luoyang, Henan Province as an example to conduct economic analysis, propose countermeasures and corresponding measures, and provide reference for investment decisions of similar projects.

1 Economic Analysis Methods

The research methods related to the economic benefits of photovoltaic power generation projects mainly include levelized cost of electricity (LCOE), net present value, investment payback period, internal rate of return, etc. The LCOE model is an internationally recognized method for evaluating the cost of power generation, which is used to estimate the cost of power generation projects in different regions, scales, investment amounts, and power generation technologies. However, the LCOE model only measures costs and cannot determine total benefits. Net present value, internal rate of return, and investment payback period are commonly used economic methods to evaluate the profitability of a project.

1.1 Financial Internal Rate of Return (FIRR)

The financial internal rate of return is the discount rate at which the present value of a project's net financial cash flows accumulates to zero over the calculation period, which satisfies the following expression:

$$\sum_{t=1}^{n} (\text{CI} - \text{CO})_t (1 + \text{FIRR})^{-t} = 0 \tag{1}$$

where FIRR is financial internal rate of return, CI is cash inflow, CO is cash outflow, –$(\text{CI} - \text{CO})_t$ is net cash flow for period t, and n is project calculation period.

When the financial internal rate of return is greater than or equal to the set discrimination benchmark i_c (usually referred to as the benchmark internal rate of return), the project is financially feasible.

1.2 Financial Net Present Value (FNPV)

The financial net present value is the sum of the present values of the net cash flows during the project's calculation period, calculated on the basis of the benchmark rate of return i_c, which can be calculated using the following formula:

$$\mathrm{FNPV} = \sum_{t=1}^{n} (\mathrm{CI} - \mathrm{CO})_t (1 + i_c)^{-t} \tag{2}$$

When FNPV > 0, it indicates that the plan can achieve excess returns in addition to meeting the benchmark return requirements. When FNPV = 0, it indicates that the plan can meet the profitability level required by the benchmark return rate, and the plan is financially feasible. When FNPV < 0, it indicates that the plan cannot meet the profitability requirements of the benchmark return rate, and the technical plan is not feasible.

1.3 Project Investment Payback Period (P_t)

The payback period of a project investment refers to the time required to recoup the project investment based on the net income of the project, usually measured in years. The following formula can be used for the calculation:

$$\sum_{t=1}^{P_t} (\mathrm{CI} - \mathrm{CO})_t = 0 \tag{3}$$

where P_t is the number of years until the cumulative net cash flow for each year is positive or zero for the first time.

A good plan should have a short payback period for investment, a fast payback for surface project investment, and a strong resistance to risk.

2 Profitability Analysis

2.1 Basic Information

The roof usable area of this project is about 100000 square meters, with an installed capacity of 14.5 MWp and an average annual power generation of 14.48 million kWh. It adopts a 10kV connection scheme and adopts a "spontaneous self-use and surplus power grid connection" mode.

The static investment of the project is about 51.3 million yuan, with a unit static investment of 3.54 yuan/Wp. The dynamic investment in the project is about 51.6 million yuan, with a unit dynamic investment of 3.56 yuan/Wp.

According to the relevant national regulations and the actual situation of the industry, the benchmark rate of return is calculated at 7%. The economic evaluation period is 25 years, and the working capital of this project is 30 yuan/kW. The working capital will be randomly put into operation and used and will be recovered in a lump sum at the end of the calculation period.

2.2 Calculation of Operating Costs

Operating costs include material costs, wages and benefits, repair costs, insurance costs, and other expenses.

(1) Material cost: 8 yuan/kW.
(2) Salary and welfare expenses: 150000 yuan
(3) Maintenance cost: within the warranty period, the maintenance rate is calculated as 0.11% of the fixed asset investment of the project, 0.33% for the 6th to 10th year, 0.43% for the 11th to 15th year, 0.53% for the 16th to 20th year, and 0.63% for the 21st to 25th year.
(4) Insurance premium: calculated at 0.25% of the value of fixed assets.
(5) Other expenses: 20 yuan/kW,
(6) Rent and roof repair costs: a one-time payment of 7 million yuan.

2.3 Total Cost Estimation

The total cost includes operating cost, depreciation cost, amortization cost, and interest.

(1) Depreciation expense: According to the straight-line method, the depreciation period is calculated as 25 years, with a net residual value rate of 5%.
(2) Amortization charge: According to the straight-line method, the amortization period is calculated on the basis of 10 years, with no residual value.
(3) Interest: The project funds are obtained from 30% capital and 70% bank loans, with a repayment period of 15 years. The principal and interest repayment method adopts the equal principal method, and the loan interest rate is LPR 3.95%.

2.4 Calculation of Operating Income

(1) Power generation revenue

The access plan for this project is a "spontaneous self-use, surplus power grid" mode, with a self-use ratio of 30%. The self-use electricity price will be discounted by about 90% based on the local industrial and commercial electricity price, with a discount of 0.612 yuan/kWh. The online electricity proportion is 70%, and the online electricity price is the local benchmark coal price of 0.3779 yuan/kWh.

(2) Value added tax: The value-added tax rate is 13%.
(3) Income tax

The corporate income tax rate is 25%. According to relevant national regulations, distributed photovoltaic power generation projects enjoy "three exemptions and three half reductions" of income tax starting from the operation period.

(4) Sales tax surcharges

Sales tax and surcharges include education surcharge, local education surcharge, and urban maintenance and construction tax, which are calculated on the amount of value-added tax, with tax rates of 3%, 2%, and 7%, respectively.

2.5 Profitability Analysis

Based on the above data, this paper prepares the cash flow statement of the project investment and calculates the financial internal rate of return of the project investment to be 7.85%, the financial net present value of the project investment to be 3.9 million, and the payback period of the project investment to be 11.5 years.

The financial internal rate of return of the project investment is 7.85%, which is 7% higher than the benchmark rate of return by 7%, and the financial net present value of the project investment is greater than zero, indicating that the profitability of the project exceeds the profitability level required by the benchmark rate of return, and the financial returns of the project are feasible.

3 Sensitivity Analysis

Since the basic variables used in the economic evaluation of the project are predictions and assumptions about the future, they have uncertainties. Therefore, by analyzing the uncertain factors that have a significant impact on the proposed project, calculating the changes in the basic variables that cause changes in the project's financial indicators, and calculating the sensitivity coefficients, it is possible to identify sensitive factors, estimate the sensitivity of the project's benefits to them, predict the risks that the project can bear, and make investment decisions for the project on a relatively stable basis.

3.1 Preliminary Sensitivity Analysis

Based on the characteristics of the project, uncertain factors, such as construction investment, operating revenue, and operating cost, were selected for sensitivity analysis. The percentage changes of uncertain factors were ± 5% and ± 10%, and the financial internal rate of return was selected as the analysis indicator, as shown in Table 1.

The sensitivity coefficient of operating cost is the lowest, the sensitivity coefficient of construction investment is in the middle, and the sensitivity coefficient of operating revenue is the highest. Therefore, an in-depth analysis of the uncertain factors that have a significant impact on operating revenue can be conducted.

3.2 In-Depth Sensitivity Analysis

(1) According to the policy impact of shifting from full purchase to guaranteed purchase and marketization of power generation, the uncertainty factors include the cancellation of demand and grid electricity prices.
(2) On April 17, 2024, the People's Government of Henan Province once again publicly solicited opinions on the "Notice on Adjusting the Time of Use Electricity Price for Industry and Commerce (Draft for Comments)", which adjusted the peak and valley periods to approximately half of the normal and half of the low periods during distributed photovoltaic power generation. As a result, the user's own electricity price will change, and this is also selected as an uncertain factor.
(3) By changing the above two factors at the same time, a multi-factor sensitivity analysis is conducted.

Table 1. Sensitivity analysis table

Uncertain factors	Change rate (%)	IRR (%)	Sensitivity coefficient
Basic plan	0	7.85	0
Construction investment	−10	9.06	−1.55
Construction investment	−5	8.43	−1.48
Construction investment	5	7.31	−1.37
Construction investment	10	6.81	−1.32
Operating revenue	−10	6.23	2.06
Operating revenue	−5	7.05	2.03
Operating revenue	5	8.63	1.98
Operating revenue	10	9.39	1.96
Operating cost	−10	8.31	−0.58
Operating cost	−5	7.62	0.58
Operating cost	5	7.62	−0.58
Operating cost	10	7.39	−0.58

The financial internal rate of return corresponding to these factors has been lower than the benchmark rate of return, especially when the consumption has decreased by 10% and the grid electricity price has decreased by 10% and the time-of-use electricity price has been adjusted, the financial internal rate of return is only 4.24%, which is low and not financially feasible (Table 2).

Table 2. In-depth sensitivity analysis

Number	Uncertain factors	IRR (%)
1	Reduce consumption by 10% or reduce grid electricity price by 10%	6.91
2	Adjust the time-of-use electricity price	6.19
3	Reduce consumption by 10% and reduce grid electricity price by 10%	6.03
4	Reduce consumption by 10% and reduce grid electricity price by 10% and adjust the time-of-use electricity price	4.24

4 Conclusion and Suggestions

Through economic analysis of distributed photovoltaic power generation projects, profitability indicators such as financial internal rate of return, financial net present value of project investment, and project investment payback period are calculated. Preliminary sensitivity analysis is conducted on uncertain factors such as construction investment, operating revenue, and operating cost. In-depth sensitivity analysis is conducted on consumption, grid electricity price, and self-use electricity price. A 10% change in these factors makes the project financially feasible but not feasible.

Based on the above conclusions, the following countermeasures are proposed to improve the economic efficiency of distributed photovoltaic power generation projects.

(1) Increase energy storage

By increasing the energy storage capacity, surplus power generation can be stored first. On the one hand, it can be used for self-consumption by customers during non-power generation periods, thereby increasing the self-consumption ratio and increasing self-consumption revenue. On the other hand, it can be connected to the grid during peak periods, thereby increasing the grid electricity price and increasing the grid electricity revenue.

(2) Fully market-oriented transactions

Recently, the National Energy Administration proposed a policy that the market-oriented trading of photovoltaic power generation shall not be subject to price limits and shall not be included in the peak and valley time of use electricity prices, which will inject new vitality into the development of the photovoltaic power generation industry. Through market-oriented transactions, photovoltaic power generation enterprises will be able to participate in the market more flexibly, improve market competitiveness, and increase consumption. At the same time, the government also needs to further strengthen the innovation of market mechanisms, establish and improve green electricity trading mechanisms, and gradually expand the scale of green electricity trading.

(3) Technological innovation

Technological innovation is conducive to promoting the development of the photovoltaic industry, optimizing energy consumption and material utilization in the production process, improving battery conversion efficiency, further reducing production costs, and lowering project investment. The future development potential of the photovoltaic industry is enormous, providing valuable technical support for achieving sustainable global energy development.

References

1. National Development and Reform Commission Ministry of Construction: Economic Evaluation Methods and Parameters for Construction Projects, 3rd edn. China Planning Press, Beijing (2006)
2. National Energy Administration. National Energy Administration releases national electricity industry statistics for January to March (2024). https://www.nea.gov.cn/2024-04/22/c_1310772067.htm
3. Wang, F.Y., Quan, C.L., Cong, L.Y., Lu, M.L.: Economic benefits of distributed photovoltaic power generation under different operating modes in situation of subsidy reduction: a case of Beijing-Tianjin-Hebei. J. Arid Land Resour. Environ. **38**(4), 87–94 (2024)

Current Status and Prospects of Dam Safety Monitoring Technology for Hydropower Stations

Bo Jiang(✉), Jinyong Fan, Fuxue Yang, Jian Chen, and Jun Zhou

China Yangtze Power Co., Ltd/Baihetan Power Plant, Yunnan, China
2974437586@qq.com

Abstract. The development of new technologies has greatly promoted the progress of dam safety monitoring technology, mainly reflected in the construction of monitoring norms and standards and technical applications. In terms of monitoring norms and standards construction, this paper mainly introduces the achievements made by China in the construction of monitoring norms and standards for dam safety. In terms of monitoring the application of new technologies, the improvement of monitoring technology and equipment perception is introduced, as well as the construction of monitoring system platforms. Finally, providing an outlook on the areas where current monitoring technologies need to be improved and enhanced.

Keywords: dam safety monitoring · monitoring system platform · standardized construction

In the era of digital economy, various new technologies are flourishing and developing rapidly, such as 5G technology in the communication industry, Internet of Things technology in the information field, intelligent AI technology in the computer industry, and Unreal Engine technology in the gaming industry. While promoting continuous economic development, they are also providing opportunities for the transformation of the traditional water conservancy industry. Safety monitoring of dams is not only one of the key contents of daily management of hydropower stations, but also related to social public safety, which also needs to be upgraded and transformed. On the one hand, the government needs to continuously improve and refine the construction of dam safety monitoring standards to adapt to the integration and development of new technologies, and provide strong institutional guarantees for dam safety monitoring. On the other hand, hydropower enterprises also need to actively upgrade and transform their monitoring systems in accordance with the norms and standards. While further improving the efficiency and ability of safety monitoring work, they should also ensure the safety of dams and social electricity consumption. This paper will elaborate on three aspects: the construction of norms and standards for dam safety monitoring in hydropower stations, the application of new monitoring technologies and technological prospects.

S. Zheng et al. (Eds.): IHDC 2024, LNCE 487, pp. 47–56, 2025.
https://doi.org/10.1007/978-981-97-9184-2_6

1 Construction of Monitoring Norms and Standards

In order to ensure that the work of dam safety monitoring can be carried out in a regulated and systematic manner, China has actively promoted the construction of relevant norms and standards for dam safety monitoring and has achieved certain achievements. According to incomplete statistics, China has issued approximately 84 norms and standards related to dam safety monitoring for hydropower stations from 2000 to 2024, with an average of at least 3 norms or standards issued annually. Table 1 lists 34 norms and standards that are closely related to dam safety monitoring.

From the above listed norms and standards, it can be seen that with the increasing emphasis on dam safety work, the maturity of dam safety monitoring technology in the industry and the continuous development of new technologies, the construction of China's monitoring regulation system has achieved significant results at present, which are mainly reflected in the following three aspects: ① The subdivision of the formulation institution. It can be divided into Chinese national standard GB and industry standard SL/NB/DL from the level of norms, among which GB is a mandatory national standard, and GB/T is a recommended national standard; and industry standard is a recommended standard, which NB, SL, and DL industry standards are respectively formulated by the National Energy Administration, China Water Resources Society, and China Hydroelectric Power Engineering Society, applicable to energy projects, water conservancy projects, and hydropower station projects. ② The refinement of regulatory content. It has been carried out from the stages of monitoring design, construction, and maintenance, and detailed specifications have been formulated for six aspects: instrument standards, design standards, cost standards, construction standards, acceptance standards, and maintenance standards. ③ The proposal of new monitoring concepts, such as the "Technical Standard for Safety Monitoring of Concrete Dams" (GB/T51416-2020), for the first time proposes the concept of whole process management for the entire life cycle of concrete dams, and divides monitoring projects into permanent, long-term, and short-term based on their operating life. Although the relevant norms and standards issued in China are becoming increasingly refined and involved a wider range of aspects, it is still necessary to explore and promote mature practical experience and application theories at home and abroad, and determine them in the form of standards or norms to better guide and promote the normal and orderly implementation of dam safety monitoring in hydropower stations.

2 Application of New Monitoring Technologies

On the one hand, the continuous proposal of new concepts in monitoring standards will promote the in-depth research and application of new monitoring technologies. On the other hand, the research and application of new technologies will also promote the continuous improvement and enhancement of subsequent standards.

2.1 Monitoring Technology and Equipment Perception Improvement

Continuously upgrading and improving monitoring equipment and technology can help expand the monitoring scope of dam safety monitoring in hydropower stations, thereby

Table 1. Partial norms and standards on dam safety monitoring in China

Category	Name	Specification Number of Document
Design Standards	Technical specifications for safety monitoring of earth rock damsn design standards	DL/T5259-2010
	Technical specifications for safety monitoring of concrete dams	DL/T 5178-2016
	Technical specification for strong vibration safety monitoring of hydraulic structures	DL/T 5416-2009
	Technical Specification for Automation of Dam Safety Monitoring	DL/T 5211-2019
	Communication protocol for dam safety monitoring automation system	DL/T 324-2010
	Table structure and identifier standard for dam safety monitoring database	DL/T 1321-2014
	Technical standard for safety monitoring of concrete dams	GB/T51416-2020
Instrument Standards	Dam monitoring instruments - Joint measuring instruments - Part 1: Differential resistance type joint measuring instruments	GB/T 3410.1-2008
	Water conservancy monitoring data transmission protocol Part 1: General principles	SL/T 812.1-2021
	Basic technical requirements for cables for dam safety monitoring instruments	DL/T1735-2017
	Automatic collection device for dam safety monitoring	DL/T 1134-2022
	Technical specification for identification of differential resistance monitoring instruments	DL/T 1254-2013
	Series Type Spectrum of Concrete Dam Monitoring Instruments	DL/T 948-2005

(*continued*)

Table 1. (*continued*)

Category	Name	Specification Number of Document
	Series type spectrum of monitoring instruments for earth rock dams	DL/T 947-2005
Cost Standards	Detailed Rules for Special Investment Preparation of Safety Monitoring System for Hydroelectric Engineering	NB/T35031-2014
Construction Standards	National First and Second Order Leveling Standards	GB/T 12897-2006
	National Triangulation Specification	GB/T 17942-2000
	Technical specifications for construction of safety monitoring system for earth rock dams	DL/T5839-2021
	Specification for measurement of hydropower engineering	NB/T 35029-2014
	Code for construction supervision of dam safety monitoring system	DL/T 5385-2020
	Technical specifications for construction of concrete dam safety monitoring system	DL/T 5784-2019
	Installation standard for dam safety monitoring instruments	SL531-2012
Acceptance Criteria	Acceptance Specification for Dam Safety Monitoring System	GB/T22385-2008
	Practical requirements and acceptance regulations for dam safety monitoring automation system	DL/T 5272-2012
Operation and Maintenance Standards	Evaluation regulation for dam safety monitoring system	DL/T 2155-2020
	Compilation regulations for safety monitoring data of concrete dams	DL/T 5209-2020
	Operation and Maintenance Regulations for Dam Safety Monitoring System	DL/T1558-2016

(*continued*)

Table 1. (*continued*)

Category	Name	Specification Number of Document
	Operation and Management Specification for Safety Monitoring System of Water Conservancy and Hydropower Engineering standards	SL/T782-2019
	Technical specification for identification of dam safety monitoring	SL 766-2018
	Standard for scrapping of dam safety monitoring instruments	SL 621-2013
	Inspection and testing regulations for dam safety monitoring instruments	SL 530-2012
	Compilation regulations for safety monitoring data of earth rock dams	DL/T 5256-2010
	Compilation regulations for safety monitoring data of concrete dams	DL/T 5209-2020
	Code for analysis of dam safety monitoring data	DL/T 2340-2021

obtaining more relevant monitoring information and helping decision-makers better analyze and make decisions. The expansion of monitoring scope not only includes the subdivision of monitoring elements and change of monitoring frequency, but also includes the extension of monitoring space, full coverage of sky, earth and water in the vertical direction, and the expansion of the entire basin from a single water area to the entire river basin in the horizontal direction. The entire monitoring network system is continuously optimized and improved, and the intelligent monitoring and perception ability of coverage is comprehensively improved.

In the field of space, the mainstream technology currently used to improve the accuracy and efficiency of dam deformation monitoring is the use of satellite remote sensing or Beidou positioning technology. For example, the Global Navigation Satellite Technology (GNSS) which is used for dam deformation monitoring and the Spaceborne Synthetic Aperture Radar Interferometry (INSAR) technology which is used for geodesy are combined, and has been deployed in large and medium-sized hydropower stations both domestically and internationally. However, there is still room for improvement in data function modeling, calculation accuracy, and quality analysis of such technologies [1]. In the future, AI data processing technology can be used to quickly and accurately screen and match corresponding function models for collected remote sensing or positioning data, and model algorithms can be used for high-precision calculation and subsequent analysis and evaluation, providing accurate and effective high-quality data for dam safety monitoring systems.

In the sky field, drone technology is widely used. It can replace manual observation of deformation, landslides, and other contents in key high slopes or dangerous areas for daily or encrypted observation. While ensuring the personal safety of monitoring personnel, it can also improve monitoring efficiency and accuracy. It can also carry airborne laser or radar for collaborative measurement with unmanned ships [2], which can to some extent compensate for the low monitoring accuracy of remote sensing monitoring technology in special terrain. In the future, unmanned aerial vehicle (UAV) airborne laser measurement, UAV hyperspectral image remote sensing and other technologies can be combined with AI 3D technologies, such as Mesh-GPT, to accurately and quickly construct high-precision 3D digital reality models that can truly reflect the engineering status of physical entities.

In the process of on-site monitoring, high-precision 3D laser scanning technology in the field of surveying and mapping has been successfully applied to dam safety monitoring. By constructing a point cloud data field of the model, it comprehensively and accurately reflects detailed information about dam deformation, crack development, etc. [3]; The Internet of Things technology in the information field has helped the Xiluodu Hydropower Station to develop an automated system for dam safety monitoring based on the Internet of Things technology framework [4]. In the future, with the continuous application of 5G and Internet of Things technology in dam safety monitoring, intelligent sensor networks based on wireless communication will become a reality, thereby helping monitoring instruments and equipment achieve the transformation from local static perception to large-scale dynamic perception.

In the underwater field, the use of underwater robots can help monitoring personnel complete inspection work and defect localization in the complex water environments. Sonar or radar technology is used for physical inspection and defect localization, then binocular vision or fixed-point camera technology is used to take photos of underwater defects for evidence, achieving timely detection of underwater building defects. In the future, AI image processing technology can be utilized to further enhance the current processing capabilities and effects on radar, sonar images, and underwater photos.

2.2 Construction of Monitoring System Platform

China's monitoring system platform has transformed from the automated monitoring stage, which utilizes advanced sensors, collection devices, and data analysis technologies to the digital stage which is based on modern information technology to convert information into digital formats. In the future, it will achieve intelligence based on artificial intelligence, big data analysis, cloud computing and other technologies [5]. The future intelligent monitoring system platform will have the characteristics of automation and digitization, presenting features such as functional integration, digitization, and unified construction.

2.2.1 Functional Integration

Most of the current dam safety monitoring systems for hydropower stations have implemented online monitoring functions, including online data collection, online inspection and calculation, and online rapid safety assessment. However, a single online monitoring function can no longer meet people's needs for basic monitoring information. At this time, multiple information sources are needed to assist in decision-making and judgment. For example, integrating an engineering video real-time monitoring system to achieve remote video real-time monitoring of important monitoring equipment and facilities in hydropower stations to ensure their operation safety. When a safety accident occurs, the system's video backtracking function can also provide important technical support for subsequent accident cause analysis. The integrated intelligent inspection system compensates for the limitations of blind spots in monitoring instruments by comparing and combining the inspection results uploaded by artificial or intelligent robots with the results of automated monitoring. The two complement and confirm each other, ensuring the authenticity, reliability, and effectiveness of monitoring results.

2.2.2 Information Digitization

The continuous development of the digital economy has given rise to a series of digital technologies such as big data, cloud computing, cloud platforms, and AI, helping traditional and emerging industries achieve digital transformation. In the water conservancy industry, the water conservancy digital twin platform is an innovative product of the integration of digital technology and traditional water conservancy engineering technology. This platform mainly consists of three parts: a data base, a model platform, and a knowledge platform [6]. Among them, the data base actively or passively collects various types of data from monitoring and control networks and river basin information networks, then constructs corresponding basic databases and provides services such as data cleaning, data mining, and data transmission. Its main function is to provide real and effective "data". On the basis of the output results of the data baseboard, the model platform builds a digital virtual model based on the dam or hydropower station entity [7], and conducts systematic training, scientific management, and real-time sharing to dynamically simulate and predict its current and future operating status. Its main role is to provide efficient and accurate "algorithms". The knowledge platform fully utilizes the massive data output from the data base, the simulation and prediction results calculated by the model platform to form a structured water conservancy knowledge base that adapts and learns independently [8], better serving the business and decision-making layers, providing real-time information feedback and optimization suggestions for the current state. Its main function is to provide powerful "examples".

2.2.3 Building Unification

In order to achieve the "Two Hundred Goals" of dam safety monitoring, which is the 100- year data of 100 dams, it is required that the dam safety monitoring systems of each hydropower stations can achieve data sharing. However, due to the different monitoring system architectures designed and constructed by different design institutes for different dam projects, there are problems such as non- universality of basic databases and difficulties in converting various platform interfaces, which has become a major obstacle to the unified development of safety monitoring platforms. The dam safety monitoring and analysis system of China Three Gorges Construction provides a solution for unified construction. The monitoring system follows the three unified construction principles of "unification of data base, unification of technical routes, and unification of platform portals", and includes four large hydropower stations of different types, boundary conditions, and scales in the upper reaches of the Jinsha River basin into the management scope at the same time, achieving the sharing of monitoring information resources related to each power station.

It can be seen that the overall development trend of China's hydropower stations or dam safety monitoring system is shown in Fig. 1: the upgrading and innovation of monitoring equipment and technology have made it possible to integrate monitoring information platforms, and technologies such as digital twins have empowered monitoring systems to transform from automation to digitization, then the integration and digitization of monitoring information also promote the unified development of monitoring information platforms, and are moving towards intelligence.

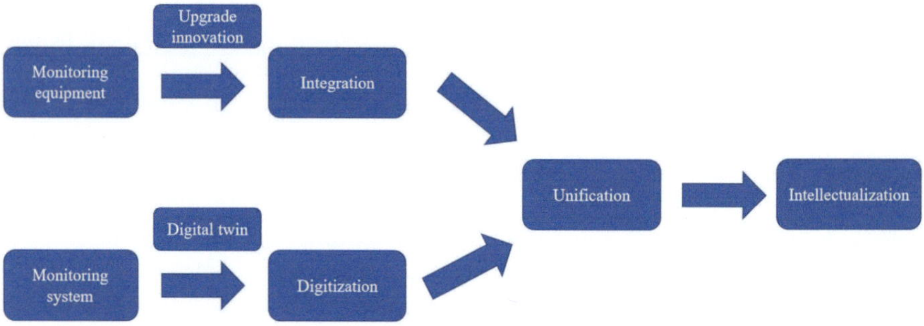

Fig. 1. Overall development of dam safety monitoring system for hydroelectric power stations in China

3 Technical Outlook

The current level of digital transformation of dam safety monitoring systems still falls short of achieving the "Four Predictions" functions required by the Ministry of Water Resources for forecasting, early warning, rehearsal, and contingency plans [9]. The application of artificial intelligence may become an effective way to improve the "Four Predictions" capabilities of dam safety monitoring systems. It is recommended to deeply integrate dam safety monitoring systems with artificial intelligence in the future.

The fusion application of artificial intelligence can be promoted from the following two aspects. ① Promote the collaboration between generative artificial intelligence and model platform applications in twin platforms. Firstly, it is necessary to utilize technologies such as deep learning, transfer learning, and machine learning of generative artificial intelligence to construct a non-linear relationship function for multiple dam monitoring quantities and environmental factors through in-deep analysis of monitoring data, and form an intelligent matching platform that can discover regular features of monitoring data, automatically recognize functions, and match algorithms. Then, the model program is automatically written using the algorithm recognized by the intelligent matching platform, and boundary conditions are automatically processed through data filtering and analysis. Finally, the model is intelligently generated in the model platform application of the twin platform to improve modeling accuracy and practicality and reduce manual workload. Taking the Jinping arch dam as an example, machine learning technology was successfully used to improve models such as index and time through support vector machine algorithm. This not only improved the accuracy and practicality of the model, but also enhanced its prediction accuracy [10]. ② Accelerate the integration of large-scale language models and knowledge platform applications in twin platforms. Firstly, analyze the intrinsic relationship between the effective data of the data base and the output results of the model platform. After being processed by the water conservancy knowledge engine, a structured water conservancy monitoring knowledge graph is formed in the knowledge platform [10]. At the same time, a large language model needs to perform semantic analysis and deconstruction, internal rule cognition, and logical relationship inference on each monitoring element. Then, it is combined with the structured water conservancy monitoring knowledge graph to construct a preliminary intelligent knowledge platform. Finally, continuously train the monitoring knowledge of the intelligent knowledge platform at the basic and expert levels, accelerate its self-learning process, and upgrade it to an efficient intelligent knowledge platform. When answering complex water conservancy problems, it can greatly improve accuracy and relevance, and provide corresponding case support and solutions.

At present, the application of artificial intelligence technology in the improvement and combination of algorithms and models is increasing, such as: Jinping I arch dam and Fengman hydropower station have improved the prediction model and BP neural network algorithm through machine learning and deep learning respectively. Meanwhile, domestic large language models, such as Baidu's Wenxinyanyi, Tsinghua's Zhipuqingyan and Alibaba's Tongyiqianwen are also continuously improving, but the construction of knowledge mapping for water conservancy monitoring is still in further exploration, and the construction of relevant wisdom platform also needs to be improved.

References

1. Chen, Z.: Implementation and application of GNSS based open-pit mine slope monitoring system. Beijing Surveying Mapp. **37**(4), 568–573 (2023)
2. Li, Q.: Application and exploration of three-dimensional measurement technology based on unmanned aerial vehicle airborne laser and unmanned ship multi beam underwater water land integration. Water Conservancy Technol. Supervision **11**, 42–45 (2021)
3. Fan, M., Zhu, Y., Zhang, H., et al.: Research on high-precision dam deformation monitoring technology based on 3D laser scanning. Water Conservancy Tech. Supervision **01**, 32–35 (2023)
4. Yi, H., Han, X., Wang, K., et al.: Application of Internet of Things technology in safety monitoring automation system of large hydropower stations. J. Yangtze River Acad. Sci. **36**(6), 166–170 (2019)

5. Huang, Y., Niu, G., Li, D., et al.: Research and application of intelligent perception and management technology for dam safety monitoring. J. Yangtze River Acad. Sci. **38**(10), 30–32 (2021)
6. Huang, Y., Ren, X., Li, A., et al.: Overall framework and application benefits of digital twin Three Gorges construction. China Water Resour. **2023** (19), 17–22+9
7. Lin, L., Li, Q., Cao, H., et al.: Challenges and measures for water quality safety construction of Digital Twin Danjiangkou. China Water Resour. **11**, 32–36 (2023)
8. Hu, Z., Yuan, J., Li, J.: Practice of digital twin Yongding River construction. China Water Resour. **3**, 13–16 (2024)
9. Peng, W., Liu, X., Huang, W., et al.: Effectiveness and reflection on comprehensive monitoring of operation safety of three gorges project. China Water Resour. **19**, 35–39 (2023)
10. Sun, L., Wang, R., Yuan, R., et al.: Application and prospects of artificial intelligence technology in smart water conservancy. China Water Resour. **3**, 44–51 (2024)

Hydraulic Engineering Safety Platform Under Microservice Architecture - a Case of Shanmei Reservoir Renovation Project

Yanyan Lin(✉) and Wei Ding

Nanjing Research Institute of Hydrology and Water Conservation Automation, Ministry of Water Resources, Nanjing, China
3276931492@qq.com

Abstract. The traditional centralized or monolithic application architecture can no longer meet the needs of modern hydraulic engineering safety supervision, especially when dealing with business expansion and system upgrades, it is easy to encounter bottlenecks. This paper designs a hydraulic engineering safety supervision platform based on microservice architecture. Firstly, the service partitioning was carried out by researching microservice partitioning methods and integrating them with the specific characteristics followed by a detailed introduction of the platform's implementation process. By applying microservice selection in Shanmei renovation project, it has been verified that the platform can significantly improve the efficiency and accuracy of safety monitoring, and provide flexible and scalable solutions for other hydraulic projects.

Keywords: Microservices · Reservoir Safety Monitoring · Hydraulic engineering

1 Introduction

The hydraulic engineering safety supervision platform is responsible for collecting and analyzing operational data of hydraulic projects, ensuring the safe and stable operation of hydraulic facilities, and preventing and reducing disaster risks. With the expansion of hydraulic engineering scale and the improvement of regulatory requirements, traditional centralized or individual application architectures no longer able to meet the needs of modern Hydraulic engineering safety supervision. In this context, microservices architecture has emerged, which can provide more flexible system design, support fast iteration and continuous integration, and help improve system stability and maintainability. This article aims to design a highly reusable hydraulic engineering safety supervision platform based on microservices architecture, apply the platform to the Shanmei reservoir renovation project, and explain how to select and re-develop microservices based on this platform.

The safety monitoring system of Shanmei includes deformation monitoring, seepage monitoring, and environmental quantity monitoring. As a typical case of hydraulic engineering, Shanmei has certain representativeness and research value. Firstly, it has a large

S. Zheng et al. (Eds.): IHDC 2024, LNCE 487, pp. 57–69, 2025.
https://doi.org/10.1007/978-981-97-9184-2_7

scale and involves diverse regulatory content and technical challenges, making it suitable as a research object for microservice architecture applications. Secondly, the climate and hydrological conditions in the area where Shanmei reservoir is located are complex, which puts higher requirements on the safety supervision of hydraulic projects, prompting us to explore more efficient and reliable regulatory methods. Through the case study of Shanmei reservoir, it can provide experience and reference for other similar hydraulic projects, and promote the widespread application of microservice architecture in the field of hydraulic engineering safety supervision.

2 Platform Design and Implementation

2.1 Microservice Division Theory

The microservice partitioning theory refers to the method and principles of dividing microservices into independent services. By following these principles and methods [1, 2], microservice systems can have higher maintainability, scalability, availability, and fault tolerance, thereby better meeting business needs. The following table shows common principles for microservice partitioning (Table 1).

2.2 Platform Microservice Division

According to the requirements of hydraulic engineering safety supervision and combined with the theory of microservice division, the platform can be divided into gateway service, data collection service, basic information service, hydrological monitoring service, safety monitoring service, analysis service, video service, operation and maintenance service, and system permission service. These microservices together constitute the core service system of the hydraulic engineering safety supervision platform. Each microservice has different functions and roles, covering all aspects of the hydraulic engineering safety supervision platform from data collection, processing, analysis, as well as user permission control. Microservices collaborate with each other to jointly complete the supervision and management of hydraulic projects, improving the platform's scalability, maintainability, and reusability.

Gateway service: gateway service is the entrance to the hydraulic engineering safety supervision platform, responsible for receiving and forwarding all external requests, as well as routing and load balancing. It can also provide basic functions such as security authentication, authorization, flow limiting, and logging.

Data collection service: The data collection service is responsible for collecting and summarizing various data from the hydraulic engineering safety supervision platform, including hydrological monitoring data, safety monitoring data, video image data, etc. It can save this data to databases or send it to other microservices for further processing.

Basic information service: this service is responsible for managing the basic information of hydraulic projects, such as basic attributes, location, status, and other information of various facilities such as reservoirs, gates, pumping stations, and hydrological stations. Provide basic information management functions such as querying, adding, deleting, and modifying.

Table 1 Principles of microservice division

Principle Name	Explain
Single responsibility	Each microservice should have a single responsibility and function, and should not assume too much responsibility. This helps improve the maintainability and testability of services, and reduces the coupling between services
High cohesion and low coupling	Each microservice should have a high degree of cohesion internally, meaning that the various functional modules within the service should be highly correlated, while the services should have low coupling, meaning that the dependency relationships between services should be as simple and loose as possible
Domain driven design	Divide microservices by business domain, so that the responsibilities and functions of each microservice match each other and are consistent with the actual needs of the business domain
Scalability	Microservices should have good scalability, which means they can be quickly expanded and reduced according to business needs without affecting the performance and stability of the entire system
High availability and fault tolerance	Microservices should have high availability and fault tolerance, that is, when a single service fails or goes down, it will not affect the operation and stability of the entire system

Hydrological monitoring service: the monitoring service is responsible for collecting, processing, and analyzing hydrological monitoring data, including indicators such as water level, flow rate, and water quality. Implement real-time monitoring, historical data queries, abnormal data alarms, and other functions.

Security monitoring service: security monitoring service is responsible for collecting, processing, and analyzing safety monitoring data of hydraulic engineering, including structural deformation, seepage pressure, vibration and other indicators. Implement real-time monitoring, historical data queries, abnormal data alarms, and other functions.

Analysis service: Analysis service is responsible for analyzing and warning hydrological monitoring and safety monitoring data, timely discovering potential problems in Hydraulic projects, and providing corresponding handling suggestions. They automatically trigger relevant emergency plans based on different warning levels.

Video service: The video service is responsible for collecting, processing, and analyzing video, including real-time images captured by cameras, recordings, and photos. Implement real-time monitoring, historical data queries, abnormal data alarms, and other functions.

Operation and maintenance service: operation and maintenance service can achieve the operation and maintenance of hydraulic projects, including equipment maintenance and inspection management. Provide functions such as equipment management, maintenance records, inspection plans, and maintenance reports.

System permission service: this service isresponsible for managing user permissions and roles, including user management, menu management, role assignment, permission verification, and other functions.

2.3 Service Registration and Service Discovery

The platform utilizes Nacos to achieve service registration and unified configuration management. When service calls are made, each microservice will automatically pull the service list on Nacos. Simply add the @FeignClient annotation on the interface class to implement the definition of the Feign interface, which exposes the public methods of a single service to form a universal interface that can be called by other services. The Feign interface of the platform is mainly divided into two categories:

Business microservice interfaces: some common methods in the interfaces of each business service that can be exposed for other services to call. By defining business interface classes, the interface of a single service is exposed.The platform adopts the form of one microservice corresponding to one interface class,as shown in Fig. 1, IWaterService interface classes for hydrological monitoring microservices, exposing site's time period monitoring data, time period daily data, and other general service interfaces.

```
@FeignClient(name = "waterService" ,url = "http://127.0.0.1:51278",path = "/hydro/water")
public interface IWaterService {

    @GetMapping("/perioddata/getWaterRainStcdData")
    Map<String, Object> getWaterRainStcdData(@RequestParam (value="stcds") String stcds,
                                             @RequestParam (value="st") String st,
                                             @RequestParam (value="et") String et,
                                             @RequestParam (value="type") String type,
                                             @RequestParam (value="interval") Integer interval);

    @GetMapping("/perioddata/getMutiTagPeriodDayData")
    List<WaterData> getMutiTagPeriodDayData(@RequestParam(value="tagIds") String tagIds,
                                            @RequestParam (value="tagType") String tagType,
                                            @RequestParam (value="st") String st,
                                            @RequestParam (value="et") String et,
                                            @RequestParam (value="precision",required = false) Integer precision);

    @GetMapping("/perioddata/getSingleTagPeriodDayData")
    List<WaterData> getSingleTagPeriodDayData(@RequestParam(value="tagId") Integer tagId,
                                            @RequestParam (value="tagType") String tagType,
                                            @RequestParam (value="st") String st,
                                            @RequestParam (value="et") String et,
                                            @RequestParam (value="precision",required = false) Integer precision);

}
```

Fig. 1 Code snippet for hydrological monitoring microservice interface

Third party service interface: as shown in Fig. 2, this type of interface needs to be called after unified authentication. Such as meteorological warning interfaces accessed through third-party meteorological platforms.

```

@FeignClient(url = "https://api.seniverse.com/v3/weather/alarm.json",name="alarmService")
public interface IWeatherAlarmService {

    @GetMapping(value = "")
    public JSONObject  weatherAlarmApi(@RequestParam("key") String key,
                                       @RequestParam("location") String location,
                                       @RequestParam("detail") String detail);

}

```

Fig. 2 Code snippet for meteorological warning service interface

2.4 Service Gateway Implementation

The platform can use the relevant functions provided by the gateway components. This platform mainly uses the following three functions:

(1) API authentication and token renewal: the process of platform authentication and token renewal is shown in Fig. 3, the login interface is in whitelist interface, and the platform directly forwards it to the system permission microservice. The result is returned to the user. At the same time, token and other permission information are stored in Redis, and the token validity period is set. Except for the whitelist, all user requests must carry a token. After passing through the gateway filter, the request retrieves permission information from Redis through the token. At the same time, the token time is extended for interfaces with successful authentication.

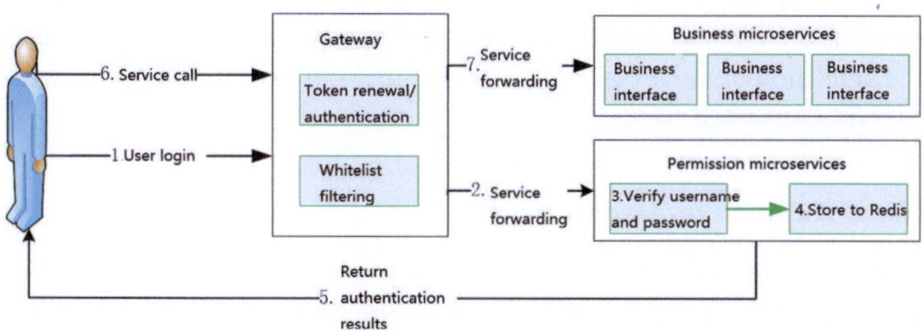

Fig. 3 Gateway authentication and token renewal business

(2) Service forwarding: the platform use spring cloud gateway routes attribute to configure routing rules. Each routing rule consists of attributes such as is, uri, and optional predictions, filters, etc. The is is the unique identifier of the routing rule, and the uri specifies the address of the target service.
(3) Permission parameter injection:the code segment of permission parameter injection is shown in Fig. 4. Implement Gateway filter interface to intercept service interfaces that require parameter injection, automatically parse tokens, and obtain user permission information from Redis. Through this method, permission parameters can be automatically injected into service interfaces.

```java
@Component
public class XAuthRequestFilter implements GlobalFilter, Ordered {

    private Logger log = LoggerFactory.getLogger(XAuthRequestFilter.class);
    private static final String USERINFO = "userInfo:";
    @Autowired
    private RedisService redisService;
    @Autowired
    private AuthRequestConfig authRequestConfig ;
    @Override
    public Mono<Void> filter(ServerWebExchange exchange, GatewayFilterChain chain) {
        ServerHttpRequest request = exchange.getRequest();
        String uri = request.getPath().toString();
        if(!authRequestConfig.getUrilist().contains(uri)){
          return chain.filter(exchange);
        }
        final Object authAddvcd =  jsonObject.get("addvcd");
        final Object authExStcd =  jsonObject.get("exStcd");
        ServerHttpRequest mutateReq = new ServerHttpRequestDecorator(request) {
                @Override
                public URI getURI() {
                    URI uri = request.getURI();
                    if(authAddvcd==null){
                        return uri;
                    }
                    String query = uri.getQuery();
                    if(query == null){
                        return URI.create(uri+ "?authAddvcd="+authAddvcd+"&authExStcd="+authExStcd);
                    }else{
                        return URI.create(uri+ "&authAddvcd="+authAddvcd+"&authExStcd="+authExStcd);
                    }
                }
            };
            return chain.filter(exchange.mutate().request(mutateReq).build());
        }
        return chain.filter(exchange);
    }
}
```

Fig. 4 Gateway parameter injection key code snippet

2.5 Business Microservice Development

A single microservice is developed using Springboot, utilizing the databaseId attribute of the persistence layer framework mybatis to implement a platform's support for multi vendor databases. Based on ShardingSphere, multiple data partitioning strategies are implemented according to project requirements to meet the efficiency of querying large amounts of data. In terms of interface development, fully utilize reflection, database table storage of JSON fields, universal SQL statements, and other methods to implement a universal service interface.

2.6 Platform Automation Deployment

The operation of hydraulic engineering software often accompanies changes in requirements during the implementation of the project. In order to quickly respond to requirements, the platform adopts a continuous integration and delivery approach to achieve automated software construction and deployment. Combining technologies such as version control tools (SVN/GIT), project building tools (such as Jenkins), and deployment tools (such as Docker) to automate the process of building, testing, and deploying software. Automated deployment has improved deployment efficiency and saved labor costs, providing strong support for project implementation.

3 Requirements Analysis for Shanmei Renovation Project

3.1 Project Overview

Shanmei reservoir is located in the middle reaches of Dongxi, a tributary of Jinjiang River in Nan'an County, Fujian Province. It is approximately 50 km south of Quanzhou City and 41 km south of Nan'an County. It is a hydraulic hub project that comprehensively utilizes irrigation, flood control, water supply, power generation, ecological regulation, and aquaculture. The total storage capacity of the reservoir is 655 million cubic meters, with a drainage area of $1023km^2$ above the dam site. The reservoir is mainly composed of the main dam, auxiliary dam, spillway, diversion tunnel, underground power plant, and power station behind the dam. The project scale is large type II reservoir, and the engineering grade is Class II.

The safety monitoring system software for Shanmei consists of data acquisition software,data integration, analysis software. The acquisition unit has been running for nearly 17 years. Due to the technical conditions at that time, this model of product had low excitation voltage, long single point measurement time, few signal output methods and acquisition signal types, and lacked a universal interface, resulting in limited functionality, poor lightning protection and anti-interference capabilities, and can no longer meet the needs of the current development of smart hydraulic. According to regulatory requirements, the maximum service life of the reading instrument or data automatic collection equipment is 9–12 years. Currently, the collection unit has far exceeded its service life and needs to be updated and renovated.

The interface of the operation function of the data integration software is aging, and the software functions cannot meet the needs of reservoir information construction. Due to the large amount of automated monitoring data and the heavy workload of manual export, data input errors are prone to occur, which affects the reliability of the automated monitoring system and has poor compatibility with other application systems; The analysis of monitoring data is relatively simple, and there is an urgent need for upgrading and replacement.

In summary, the main modules of the Shanmei renovation project include:

Data migration: moving old database data and other related safety monitoring data to new data, and push monitoring results to the national large-scale reservoir dam safety monitoring and supervision platform.

Compilation and analysis: there is an automatic generation of charts related to dam safety analysis, such as process lines, correlations, infiltration lines, monthly reports, annual reports, annual data compilation framework, etc.; Functions such as seepage analysis, deformation analysis, on-site inspection information, and comprehensive analysis.

Information management: include data warehousing, monitoring data management, engineering information management, instrument measurement point management, and monitoring information.

3.2 Data Collection and Reporting

The safety monitoring points and data sources for the Shanmei project are shown in Table 2: deformation observation includes 25 surface deformation manual observation points, 13 GNSS measurement points, and 2 opening and closing degree measurement points. Among them, 25 surface deformation observations include vertical displacement observation and horizontal displacement observation. Seven rows are arranged along the dam axis to monitor the vertical and horizontal displacement changes of the dam. The Beidou automatic monitoring system was put into operation in August 2018, with a total of 15 GNSS stations built. The system is based on GNSS navigation satellite positioning technology and can monitor the displacement and settlement of the dam in real time. In addition, to monitor cracks in the wave wall, 11 crack gauges were installed at the crack location of the dam crest wave wall. The crack monitoring instrument for the wave wall is currently damaged and unable to measure normally. It is necessary to add one crack gauge for automatic monitoring on each of the two vertical upward and downward through joints on the right bank. The seepage monitoring adopts a combination of manual monitoring and automated monitoring. Manual monitoring mainly involves pressure tube observation every two weeks, while automated monitoring involves burying pore water pressure gauges in the dam, and the system measures all monitoring points once a day during the set time period. The two measurement methods have the characteristics of mutual verification of monitoring data, simultaneous analysis of monitoring results, and thorough and complete safety evaluation. The environmental quantity includes four measuring points: upstream and downstream water level, air pressure, and rainfall. The monitoring data of upstream water level and rainfall are collected from the old hydrological system, and the downstream water level and air pressure gauges need to be replaced.

Table 2 Data sources for the Shanmei Reservoir

Monitoring items	Observation method	Replace /New	Instrument type	Points number	data sources
deformation	artificial			25	historical database
	automation	new	GNSS	13	
	automation	new	seam gauge	2	
seepage	automation	replace	pressure measuring	26	security monitoring history library+ new collection
	automation	replace	pressure measuring	28	
	automation	replace	pressure measuring	1	
environment	automation		upstream water	1	historical database
	automation		rain gauge	1	
	automation	replace	barometer	1	history library+ new collection
	automation	replace	downstream water	1	

3.3 Monitoring Data Continuity Assurance

In this project, the existing data collection system will be abolished and new monitoring equipment will be introduced to build a brand new data collection system. The new system will seamlessly integrate historical monitoring data accumulated from old instruments, ensuring data integrity and traceability. To achieve this goal, we will take a series of professional measures.

Firstly, it is necessary to compare the measurement range, accuracy, resolution, and other factors of the new and old equipment to determine their corresponding relationships. Then, based on these relationships, convert the monitoring data of the old equipment into the format of the new equipment and perform necessary calibration to ensure data comparability.

Before dismantling old instruments, thoroughly backup their data to ensure the safety and integrity of historical monitoring data. Secondly, the selection of new devices will fully consider compatibility with old devices, in order to achieve smooth data connection during the process of switching between new and old systems. In addition, we will establish a dedicated data migration process to ensure that the new system can accurately and efficiently import monitoring data from the old system.

After the completion of the new system construction, we will strictly calibrate and test the new equipment to ensure the accuracy and reliability of its monitoring data. At the same time, based on the characteristics and functions of the new system, we will organize relevant personnel to receive professional training to improve their operational skills and maintenance level.

4 Shanmei Reservoir Renovation System

4.1 Microservice Selection

In the practice of implementing platform level microservice architecture, selecting and integrating microservice modules is a crucial step. This process requires a deep understanding of the specific requirements of the project, in order to select the most suitable service for the current project from the numerous available microservices. This is not just a simple selection process, but also a comprehensive consideration of various factors such as service functionality, performance, stability, and compatibility with existing systems. Based on the main module requirements of the Shanmei project, we have selected and customized the microservice architecture of the hydraulic engineering safety supervision platform to meet the specific needs of the project and improve the level of safety monitoring.

As shown in the table, the following choices can be made based on the microservice architecture of the hydraulic engineering safety supervision platform:

The data migration push module selects the data collection microservice, which is responsible for data collection and integration. It can be directly used to migrate old database data to customized development needs of new databases. However, due to the need to push data to the national large-scale reservoir dam safety monitoring and supervision platform, the functions of the data collection microservice need to be expanded, and data exchange and synchronization mechanisms with external platforms need to be added.

The integration analysis module selects the analysis and warning microservice, which is responsible for data analysis and warning. It already includes the function of generating various charts and can be directly used to generate analysis charts such as process lines, correlation lines, and infiltration lines. In addition, it may be necessary to customize the generation of specific types of reports (such as monthly reports, annual reports) and reorganization frameworks based on the specific needs of Shanmei.

The information management module selects basic information microservices and information management microservices, which provide basic functions such as data maintenance, monitoring data management, and engineering information management, meeting the requirements of the information management module (Table 3).

Table 3 Microservice selection

Module Name	Microservice selection	Customized development
Data migration push module	Data Collection Microservices	It is necessary to push the data to the national large-scale reservoir dam safety monitoring and supervision platform, and increase the data exchange and synchronization mechanism with external platforms
Reorganization and analysis module	Analysis and early warning microservices	It may be necessary to customize the generation of specific formats and types of reports based on the specific needs of Shanmei Reservoir
Information management module	Basic information microservices, Information Management microservices	

4.2 Visual Display

The system updates the monitoring data overview and sensor equipment status of Shanmei in real time, and uses multiple charts to visually display the seepage, deformation, environmental quantity and other data of Shanmei Reservoir in combination with the monitoring data, analyzing their changing trends. Use contour lines to display real-time collection of key monitoring data such as potential, seepage pressure, temperature, etc., and convert them into raster data format. Subsequently, the contour line algorithm is used to process these data and generate corresponding contour line layers. The application of potential contour lines on the platform can visually display the distribution of gravity potential energy in the reservoir and its surrounding areas, as shown in the following figure. These lines reflect the continuous changes in terrain height and provide an intuitive view for the assessment of reservoir storage capacity and downstream water flow impact. The generation of seepage pressure contour lines reveals the distribution of seepage pressure in the internal and surrounding soil of the dam structure, which helps to identify potential seepage channels and evaluate the impact of seepage on the stability of the dam (Fig. 5).

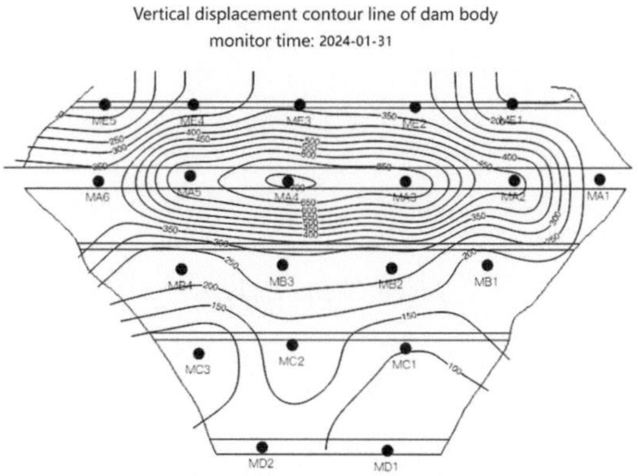

Fig. 5 Isogram function

When monitoring the artificial displacement measurement points of Shanmei, a comprehensive analysis method is adopted to accurately determine the dynamic change trend of the displacement measurement points by drawing a process line graph and integrating deformation analysis and trend analysis techniques. The process line graph is the curve of displacement over time, which can intuitively reflect the historical behavior and current situation of displacement at measurement points. Deformation analysis focuses on interpreting the spatial distribution characteristics of displacement data, while trend analysis focuses on identifying the time series characteristics of displacement data and predicting possible future displacement trends. By combining these two analytical methods, the stability of artificial displacement measurement points in reservoirs can be scientifically evaluated, providing decision-making support for reservoir safety management (Fig. 6).

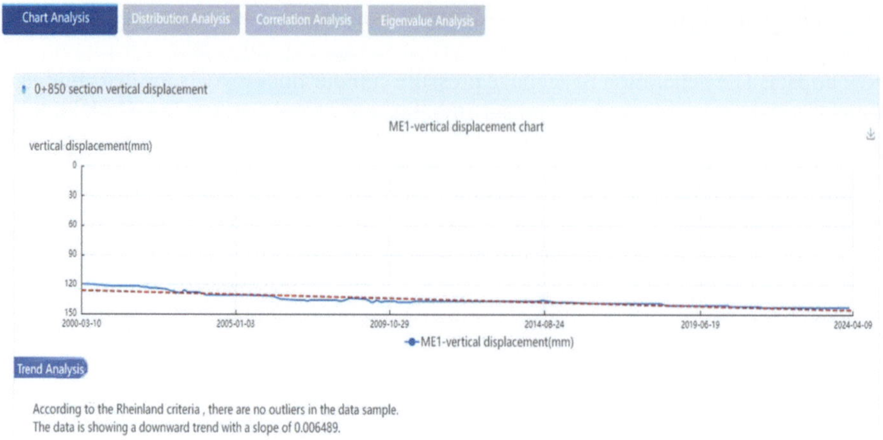

Fig. 6 Trend analysis

5 Conclusion

This article designs a highly reusable microservice architecture for hydraulic engineering safety supervision platform and applies it to the Shanmei renovation project, demonstrating how to select and apply microservices based on specific needs. Through this architecture, not only can the efficiency and accuracy of safety monitoring be improved, but also flexible and scalable solutions can be provided for similar hydraulic projects. Looking ahead to the future, the field of hydraulic engineering safety supervision will develop towards intelligence, automation, and integration. Artificial intelligence, Internet of Things, cloud computing and edge computing, as well as cross sectoral and cross domain data sharing and collaboration will become important trends in future development. The integration and application of these technologies will further improve the efficiency and accuracy of hydraulic engineering safety supervision, and jointly address potential safety challenges.

References

1. Xie, S., Zhu, X., Geng, X.: Research on application of microservice architecture in hydraulic engineering digital twin. Earth Environ. Sci. (2023). https://doi.org/10.1088/1755-1315/475/1/012062
2. Zhang, Y., Wang, L., Dong, W.: Application of microservice architecture in hydraulic digital twin platform. In: 2020 IEEE International Conference on Environmental Engineering (2022).https://doi.org/10.1109/MIC.2020.3006688
3. Xiong, X., Wei, X., Wei, Z.: Application of microservice architecture in digital twin of hydraulic and hydropower projects. In: IEEE International Conference on Big Data (Big Data) (2020). https://doi.org/10.1109/BigData50022.2020.9378284
4. Zheng, H., Lei, X., Chen, Y.: Research on application of microservice architecture in hydraulic digital twin platform. J. Phys. Conf. Series (2021). https://doi.org/10.1088/1755-1315/475/1/012062

5. Liu, Y., Chen, J., Xu, Y.: Application of microservice architecture in digital twin of hydraulic and hydropower projects. In: 2021 IEEE International Conference on Advanced Manufacturing (ICAM 2021) (2021). https://doi.org/10.1109/TIP.2021.3100312

Exploring the Digital Twin System in Slope Engineering

Wu Shu-yu[1](✉), Zhan Zheng-Gang[1], Zhu Huan-Chun[2], Hu Yong-Fu[3], Li Peng-Fei[1], and Deng Yong-Jun[3]

[1] PowerChina Guiyang Engineering Corporation Limited(修改部位), Guiyang 550081, China
wusy_gyy@powerchina.cn

[2] CAN-CN Geo-Digitization Technology Company Limited, Wuhan 430223, China

[3] Lancang River Clean Energy Safe Green Intelligent Construction Technology Innovation Center, Tibet Autonomous Region, Chengdu 854525, China

Abstract. Failure to meet specific standards for slope stability conditions may result in the potential collapse of the slope, either partially or completely. The article focuses on slope engineering and examines the structure and components of a comprehensive digital twin system. It also explores the digital technologies used in this system and assesses its unique features and innovative applications in slope construction. The architecture of the digital twin system for rock slopes must cater to the requirements of multiple professionals, comprehensive process coverage, and data-centricity. It should ensure the separation of software and personnel, with key technologies such as 3D geological modeling and rapid updates, GIM+BIM coupling modeling, and challenging techniques like cutting and sealing. Independent innovation is crucial to address these key technologies. During the slope construction period, the digital twin system of rock slope is utilized to intelligently screen unstable blocks. This innovative approach has been successfully implemented and verified in numerous large-scale rock mass projects, with favorable implementation conditions.

Keywords: slope engineering · Digital twin · GIM+BIM coupling · Digital hydropower

1 Introduction

Over the past two decades, certain sectors within China's infrastructure construction industry have made significant progress in achieving 3D digital delivery of specialized outcomes. The most notable advancements have been made in the development of BIM models for building structures, which serve as prime examples of this technological innovation [1–4]. Additionally, survey professionals have also made strides in creating 3D geological GIM models [5–7] of survey professionals. Its common feature is that the professional objects with special professional characteristics in the project are often completed by the corresponding professionals (such as structural design) using professional software (such as various types of BIM software), which rarely involves the cross-professional collaboration required in the actual workflow, and does not involve process management.

S. Zheng et al. (Eds.): IHDC 2024, LNCE 487, pp. 70–87, 2025.
https://doi.org/10.1007/978-981-97-9184-2_8

The practice of slope engineering, exemplified by the artificial slope, is a specialized structure that is created by transforming natural geological formations. It is a complex entity that combines geological elements with structural components. The digitalization of slope engineering relies on the integration and application of geological GIM and structural BIM technology, which is different from the previous characteristics of the division of labor by profession and single type of technology. Based on GIM and BIM technology, Zhong Heng et al. [8] integrated the geological and supporting structure model of Tuozi landslide in the Three Gorges Reservoir Area Teachers College. They then conducted a thorough safety evaluation using three-dimensional finite element analysis. By the integration of GIS+BIM, Chen Kelen et al. [9] gave full play to their respective strengths and carried out 3D visual design for wind power projects. Li Dechao et al. [10] discussed the future development direction of the integration of BIM and GIS in the field of 3D modeling by analyzing the differences between the data and applications in the two fields. T. Prak et al. [11] developed a system based on BIM+GIS to estimate the construction cost, land acquisition cost and operation and maintenance cost respectively in the feasibility stage of road construction. Amirebrahimi [12] presented a comprehensive framework that utilizes BIM-GIS to construct a detailed building model for damage assessment and conduct visual simulations of the flood process.

The digital twin system of slope engineering is a software system that focuses on the process of slope survey, design, and construction. It establishes a "mirror" relationship and enables data sharing and collaboration among multiple professionals. The construction of this system not only needs to realize the digitalization of all subdivision professional work of slope engineering, but also needs to integrate all these professional data and even methods into one system. Xu Hang [13] developed a the slope information management subsystem that enables intelligent slope management. This system utilizes a BIM engine and 3DGIS engine to create a digital twin model of the slope and facilitate the management of slope information. Fan Qixiang et al. [14] took the cascade hydropower project in the river basin as the main object, recorded and digitally reconstructed the whole elements of the river basin, and built the cascade hydropower project management system based on the digital watershed.

The digital twin system of slope engineering is achieved throughthe integration of "new infrastructure+business digital technology". The new infrastructure primarily includes the Internet, cloud, and other widely used digital technologies. Business digital technology is an essential tool used by professionals in various fields to collect, analyze, and apply data. One of its key components is computer graphics technology, which is relied upon by engineers in tasks such as investigation, design, and calculation. This technology enables engineers to visualize the entire process of slope engineering in three dimensions. Beihang research team from the physical entity, virtual entity, service, twin data and connection and other five levels elaborated the digital twin model structure and application criteria [15, 16]. In a study by Zheng Weihao [17] a modeling scheme was proposed using the digital twin and a sign coding scheme based on the highway traffic facility model. Similarly, Fan Huabing [18] utilized digital twin technology to efficiently plan and design Leishenshan Hospital, ensuring its rapid construction and safe utilization.

This paper focuses on the study of a multi-professional integrated digital twin system for rock slopes. It examines the composition and architecture of the system, along with the key digital technologies involved [19–21]. The characteristics of the slope engineering digital twin system and its innovative use during the slope construction period are analyzed and evaluated, based on previous professional digital technology research and application results.

2 System Composition and Implementation Approach

2.1 Requirements and Objectives

The actual slope can either be a standalone project or a separate component within a larger project. Therefore, the digitalization of slope engineering is the epitome of the digitalization of the industry in the field of infrastructure construction. The construction results not only fully embody the improved professional working methods related to slope engineering, but more importantly, the application process showcases the characteristics of the digital industry and digital economy: For example, there are significant changes being made to the job division, profit model, and other market characteristics. In general, the digital twin system of slope engineering needs to meet the following characteristics and requirements:

1) Considering the engineering application from a multi-specialty perspective to ensure comprehensive process coverage: Multi-specialty refers to the comprehensive process of slope engineering design, which encompasses investigation, design (including calculation), construction and monitoring and early warning. The entire process encompasses the initial stage of the project, the construction and operation period, which is interdependent on various specialties. The initial phase focuses on research and design, while the construction and monitoring experts primarily support the construction and operation phase.
2) From a digital technology standpoint, data is the focal point and is distinct from software and personnel. The engineering application process highlights the importance of mining data value as the core, and plays a crucial role in driving the digital transformation of theindustry.

2.2 System Composition

The digital twin system framework of rock slope, depicted in Fig. 1, has been developed using advanced "new infrastructure+computer graphics" technology. It is a comprehensive solution that addresses the challenges faced by the industry, utilizing multiple software components. The system design takes data as the center, and realizes multi-level project and technical management based on professional data according to the smooth flow of various professional data within and between professional workflows and between various stages of engineering construction.

The data center is composed of professional database and public database. The database organizes the various types of data related to slope engineering, such as surveys, designs, construction, monitoring, and early warning. This data is divided into basic data and achievement data, allowing for the integration and application of multiple professional data. The public database realizes the management of professional documents, general documents, engineering and terminology, personnel and authority and other public data.

The system's front-end utilizes WEB technology to achieve data browsing and visual display. It integrates multi-professional data, enhances application and data mining capabilities, and enables digital management.

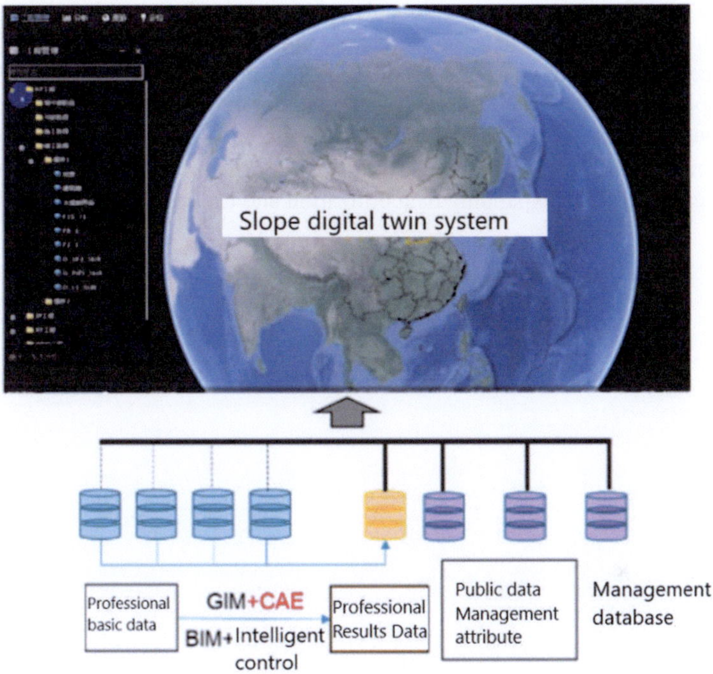

Fig. 1. Composition of slope engineering digital twin system

2.3 Implementation Approach

The digital twin system of slope engineering is built using multiple professional digital subsystems that seamlessly integrate together, following a "plug and play" approach for effective interaction. The plug and play means that each professional subsystem can not only run independently, but also meet the needs of the whole process of the professional work; But also through a simple way (such as API data interface) to achieve integration and interaction between each other, to ensure the data flow between subsystems, to meet the requirements of comprehensive applications.

Figure 2 illustrates the central link of the implementation approach, where all professional subsystems are condensed into four links: data acquisition, data storage, data processing and data application. Data collection is the process of gathering the necessary data that professionals rely on for their work. This includes the outcome data from previous projects, which is one way to foster collaboration among professionals. To this end, the data collected by each major is divided into two categories, the basic data and the outcome data are stored separately. This particular design approach is significant in two ways:

1) The main task of each specialty involved in 3D modeling is to transform raw data into final output data. This allows authorized individuals to use third-party software to meet specific requirements, ultimately achieving the design objective of prioritizing data and reducing reliance on software and human intervention. This approach enhances social collaboration and facilitates the optimization of work distribution.
2) Data application is a broad concept, essentially involving the process of data mining. Efficient research, effective collaboration, meticulous quality control, and performance management are all integral to the data mining applications. These applications play a crucial role in expanding data utilization and driving business growth.

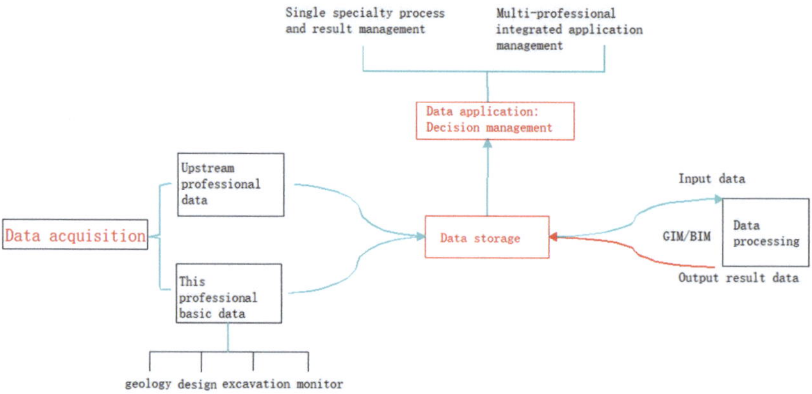

Fig. 2. Realization way of slope engineering digital twin system

3 Key Technologies

3.1 3D Geological Modeling and Rapid Updating Technology

When it comes to 3D geological modeling technology, it's important to take into account the irregularity of geological bodies, the limited amount of available data, and the crucial role of professional analysis and speculation in the geological work process. 3D geological modeling technology needs to realize the conjecture of geologists effectively and construct a highly irregular geological body model [22–25] based on a few known data. Among them, rapid update is the basic requirement, which not only ADAPTS to the realistic needs of repeated speculation but also the basic requirement of updating the preliminary model with construction and excavation information, realizing the coverage of the whole engineering process and creating digital twin geological bodies.

Three-dimensional geological modeling is a practical use of digital technology to analyze natural geological formations. The process involves using mathematical fitting techniques to incorporate geological constraints. The irregularity of geological body determines that continuous mathematical function cannot be used for fitting, and interpolation method based on discrete mathematics is needed. The realization process relies on the interplay of topology, computer graphics, and relevant computer software technology. Typically, the core consists of three main aspects: modeling algorithm, grid technology, and topological data structure.

Figure 3 presents a modeling algorithm that effectively combines constraints and discrete smooth interpolation. Constraints and smooth fitting are two mathematical concepts that are often used in technical writing. The constraints correspond to the known information of geological objects obtained from investigation, such as the position, occurrence and test data, and the modeling should be 100% consistent with the known results as far as possible. Smooth fitting is a way of delineating the connections between known points, which is equivalent to "trends" in geological work.

Figure 3 also demonstrates and describes constraints and smooth fitting using a section line as an example. It is assumed that there are no boreholes arranged in a certain area due to conditions, and the formation morphology of this part is inferred by the positions revealed by the boreholes at both ends. When making predictions, the positions revealed by the boreholes are taken into account as constraints. The goal of the prediction method is to ensure that the trend remains reasonable (Fig. 3 left).

If a hole is arranged in the blank area later, and the revealed position of the stratum is very different from the inferred result, then the position of the stratum revealed by the new borehole is used as the constraint point of the original model (in Fig. 3), and the original model is modified. Although the original borehole's stratum position remains unchanged, the new borehole's position will inevitably alter the shape between the two boreholes. The morphology after smooth fitting, as depicted on the right in Fig. 3, adheres to the principle of "reasonable trend" in geology, rather than relying on any predetermined formula for fitting.

The algorithm for constrained and smooth fitting is specifically designed to work with discrete data structures. Its importance in the geological modeling process is outlined below:

1) To realize the "forward modeling" of any complex geological body, the so-called forward modeling only relies on the known data (survey data), does not rely on the intermediate results (auxiliary section), let alone the copy after the known results (inverted mold);
2) Ensure that the modeling accuracy of exploration points is 100% and that there are reasonable trends among them;
3) Modifying the model based on updated data is crucial for using construction data to adapt the early-stage model and meet the information requirements of the entire life cycle of the professional engineering investigation.

Here is an example of variable grid technology, as shown in Fig. 4. The concept of a "variable" allows for the flexibility to adjust the size and spatial position of the value grid as needed. This is important in order to accommodate the uneven distribution of exploration data and the continuous updating of modeling data throughout the engineering stage. The upper left of Fig. 4 shows the ground layer grid fitted according to the stratigraphic interface exposed by 4 boreholes; The lower left of Fig. 4 shows the newly revealed stratigraphic interface, at which time the in-situ stratigraphic grid needs to be adjusted to adapt to the new exploration data. Given the decrease in exploration spacing due to the encryption of exploration work, it is possible to locally encrypt the accuracy of the grid while making adjustments. This allows for the adjustment of the three-dimensional shape of the grid, resulting in the updated grid as depicted in the lower right of Fig. 4.

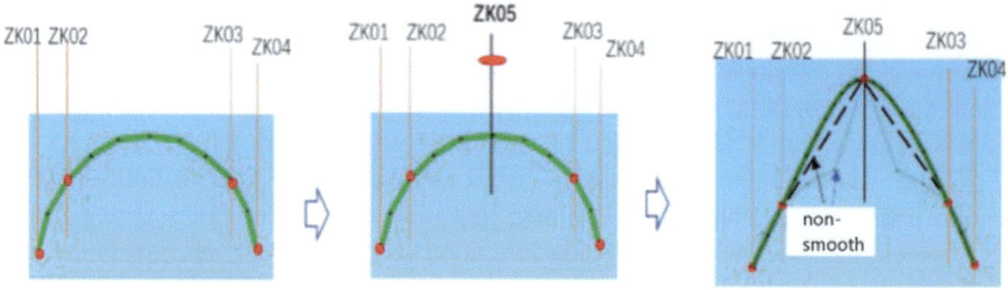

Fig. 3. Conceptual diagram of constraint and discrete smooth interpolation algorithm

Taking lens modeling as an example, Fig. 5 shows the integrated application of DSI algorithm and variable grid technology to realize lens modeling and model updating. It shows the adaptability of core modeling and grid technology to the modeling requirements of any complex geological object.

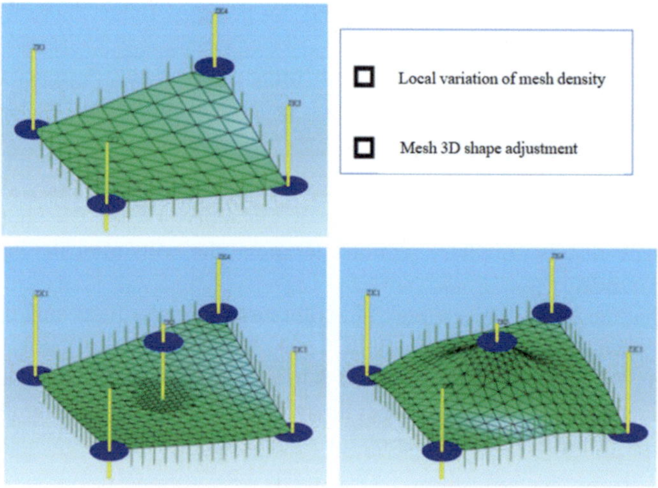

Fig. 4. Example of variable grid technology

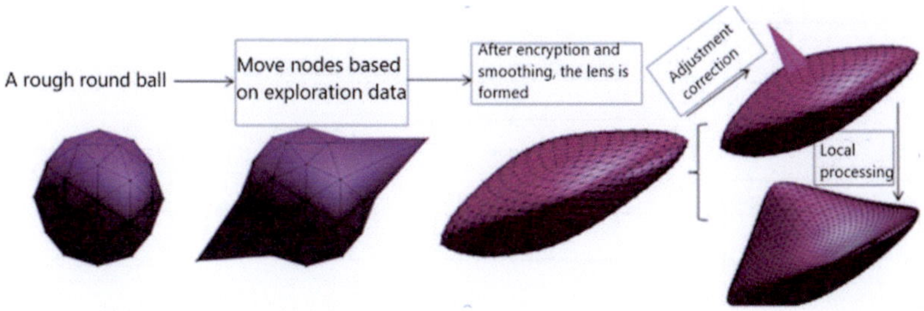

Fig. 5. Model update is realized by using DSI algorithm and variable grid technology

Effective utilization of topological data structures is crucial for programming and implementing core algorithms using grid technology. One of the primary focuses is accurately describing and recording irregular spatial surfaces. By flexibly adjusting the shape of the grid, it becomes possible to achieve high-precision fitting and model updating. The typical characteristics of data structure can be summarized as two points: based on the spatial scatter position and its attribute value (detection or test value), the connection (topology) relationship between the points is recorded.

Figure 6 provides a visual representation of discretization and topology schematically. The geological body contour is divided into spatial points (discretization) and the connection relationship between these points (topology) is recorded at the same time, which determines the object contour. Specifically, these two elements are separate from each other, meaning that one can be modified without impacting the other. Additionally, altering one of these elements has the potential to alter the shape of the object.

Figure 6, left, shows the most basic unit of a discrete, topological data structure, a single point. Each point describes its spatial position with (X, Y, Z) values; Each point can carry a custom index, and is equivalent to the coordinates (X, Y, Z); Multiple points are allowed to form an aggregate.

Have custom index values that function similarly to coordinate values. They serve as the foundation for attribute modeling, professional analysis, and pure discrete point cloud no connection relationship is established between points. Attribute modeling includes tasks like geological interpretation and modeling of geophysical data, while professional analysis involves tasks such as determining resource reserves and mechanical index distribution. When one of the elements is changed, the shape will change. Similarly, a spatial surface is also determined by point coordinates and connection relations, only in this case the connecting unit is a triangular grid instead of a spatial line segment.

3.2 Coupling Technology of GIM and BIM

The adoption of BIM technology in the infrastructure construction industry in our country is heavily influenced by international information technology of manufacturing and construction industry. For over a decade, there has been a proposal to extend the use of BIM software, commonly used in the construction industry, to other professionals involved in infrastructure construction, including surveyors, mappers, and project managers. It is evident that there is a significant technical challenge in implementing this

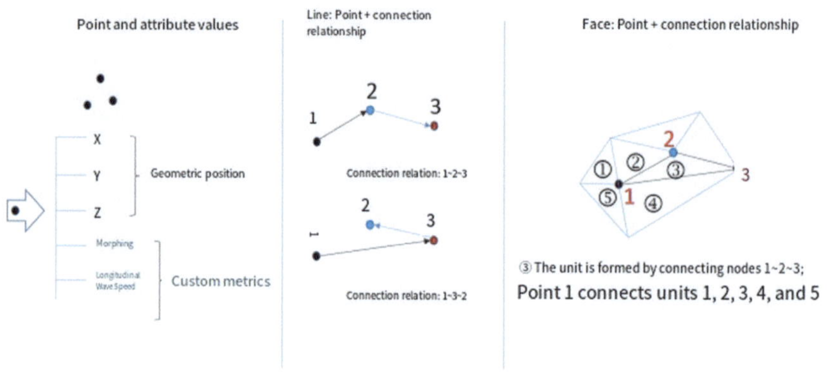

Fig. 6. Conceptual diagram of discrete and topological data structures

approach. The current commercial platforms are unable to cater to the specific requirements of slope, cave, and other design work. This limitation arises from the fact that these platforms are designed for specific types of objects, such as irregular natural geological bodies, large-scale surfaces, and mass information, or artificial construction of regular components. It adopts the corresponding computerized technology (computer graphics algorithm, data structure, functional mode), which can not meet the two different types of objects at the same time.

Rock slope engineering must address the simultaneous presence of two types of objects: the geological body involved in excavation and the regular components for the design of supporting structures. The only feasible way to solve the problem is "integration": Develop a dual-engine software, wherein the dual-engine refers to the coupling modeling [26, 27] of GIM and BIM for irregular geological bodies (GIM) and regular components (BIM) at the same time. Figure 7 shows the modeling technology of excavation slope with the coupling of regular and irregular objects. The starting line of the excavation slope consists of a broken line segment connected by individual points and the circular arc segment described by the mathematical formula. Each slope segment of the excavation slope created is a regular object, but the whole can be an irregular object, so the coupling modeling of the rule and irregular object can be realized.

One way to practically implement this is by integrating the "engine" of another platform into one of the existing platforms. This involves expanding the core technology and functionality to work seamlessly with the standard objects on the geological platform, or modifying the structural platform to be compatible with the irregular geological body. The feasibility of the former is much higher than that of the latter in terms of the technical difficulty and the degree of risk in the implementation process. One of the practical reasons is the "standardization" and "openness" of the structural platform, the "non-standardization" and "privatization" of the geological platform. Combining the underlying technology of the geological and structural platform forms a geotechnical foundation platform, which is a practical method for expanding the underlying geological platform. Through the core technology embedded in the structural platform (based on the mathematical algorithm of continuous function and the technology involved in parametric realization) to achieve compatibility of the two types of objects, to meet the requirements of geotechnical engineering professional information work. The crucial factor lies in the adoption of the underlying technology of the geological platform. Due to its scarcity and exclusivity, only a handful of institutions worldwide have acquired the necessary expertise. On the other hand, the structural platform is well-established and offers accessible resources.

3.3 Cutting and Sealing Technique

When utilizing 3D computer technology to simulate geological phenomena like pinch-out and fault cutting, it requires the cross-cutting of surface to surface, the situation can be categorized into three types, as illustrated in Fig. 8:

1) Completely closed (Air-tight) means that the air-tight co-node is closed, that is, the two intersecting surfaces are completely co-nodal on the intersection line;
2) Generally closed (Water-tight), refers to the water-tight coplanar closure, does not guarantee that the two surfaces of the intersecting node, but ensure that the intersection point falls on the intersection line;
3) Unable to be closed, there is an ongoing issue with an opening.

According to research on 3D technology in the manufacturing and construction industry, the creation of fully enclosed space surfaces has significant practical applications [28]. These closed surfaces define the space unit and enable the following: 1) Filling to create a solid model; 2) The type of geological unit can be identified automatically after arbitrary cutting, and the corresponding cartographic elements like color and pattern can be assigned. 3) It can be Converted into a three-dimensional numerical calculation model grid.

On the contrary, if a completely closed space surface can not be effectively generated, it is difficult to meet the above practical needs in a real sense. The 3D structure platform generally adopts deterministic function to simulate the contour of the object, and can obtain the intersection position of the face-to-surface intersection through accurate calculation, so as to achieve the completely closed intersecting effect. The geological interface is simulated by interpolation method, and the face-to-surface cutting is actually obtained by the intersection operation between discrete grids. Achieving complete closure in the application process of 3D geological models has become a significant challenge, especially for complex or artificially simplified processing. This issue has emerged as a global concern.

When employing 3D computer technology to replicate geological phenomena like pinch-out and fault cutting, it requires the intersection of one surface with another. When creating a 3D geological model, the surface-surface intersection operation can be used to ensure that the geological relationship is correct and meet the requirements of submitting 2D survey results. While implementing the 3D model in practical scenarios, like using numerical simulation to forecast unstable rock slopes, it is necessary for the geological surface to be completely closed after cutting to ensure accurate results.

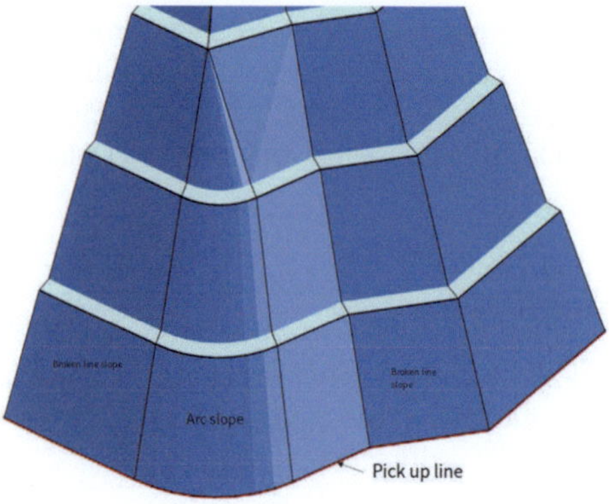

Fig. 7. Modeling technology of excavation slope with coupling of regular and irregular objects

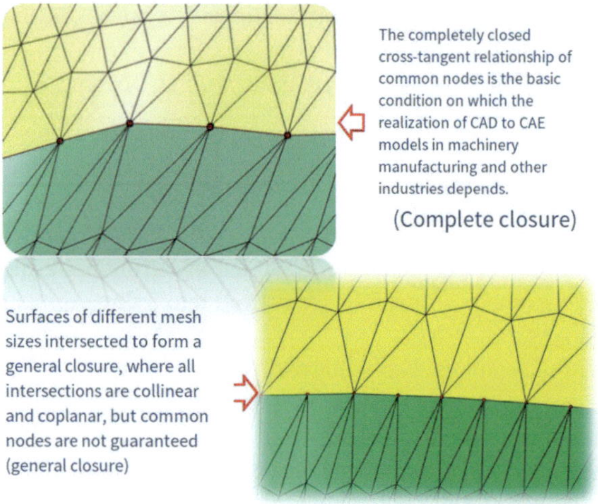

Fig. 8. Closed situation after intersection of space surface (complete closure, general closure)

4 Intelligent Screening of Unstable Blocks

The results of the early stage contained in the slope engineering digital twin system can enable groundbreaking engineering applications by utilizing continuously updated data throughout the construction period. Two notable aspects include the intelligent screening of common rock slope mass failure disaster risk and the implementation of hidden engineering quality control [29–31]. This section solely focuses on introducing the former topic.

4.1 Principle and Process

The concept of achieving "intelligent screening of unstable blocks" in rock slope construction is derived from the impact of digital technology in two key areas: the transformation of randomness into certainty, and the enhancement of technological efficiency:

(1) Random variable certainty means that every structural plane collected from the excavation face is taken as the input data for the analysis of surrounding rock deformation and block stability, to solve the uncertainty problem caused by the "speculation" caused by the sparse exploration results in the early stage. In simple terms, each joint included in the analysis is no longer based on guesswork. Instead, its location and occurrence are precisely determined, transforming it from a random element to a definite one.
(2) Using modern digital technology can greatly improve efficiency in data collection, transmission, processing, and management. This reduces the time it takes to complete the analysis and judgment process. With this technology, prediction results can be obtained within a few hours after recording geological data from excavation and allows for rapid response in field production.

Figure 9 illustrates the principle and process of intelligently screening unstable blocks of rock slope. This process can be divided into three main aspects: 1) Using new infrastructure technology to quickly complete geological data acquisition and transmission of excavation face; 2) Storage and sharing of field data by using cloud data warehouse; 3) Using the deterministic joints revealed by the excavation face to replace the random joints speculated and calculated in the previous exploration, updating the geological model of the structural plane and cutting to generate a deterministic block calculation model, to carry out block stability analysis and quickly identify the potentially unstable blocks in the surrounding rock along the excavation face.

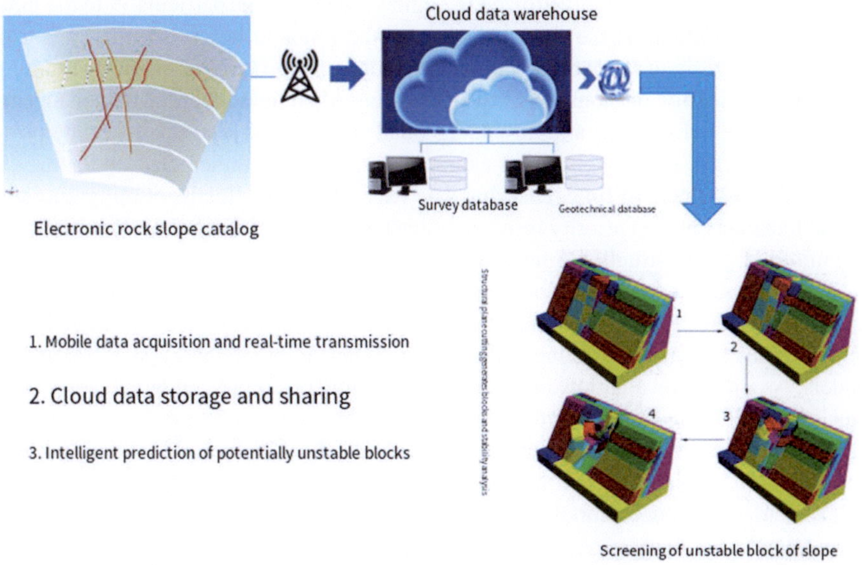

Fig. 9. Principle and flow chart of intelligent screening of unstable blocks in rock slope

The implementation process relies on the 3D digital results of the survey in the initial stage of the project. It continuously enhances and adds to the 3D geological model by utilizing geological catalog data throughout the construction period, ensuring the preservation of the survey data. It is the specific application of digital twin technology in the survey major in the early stage of the project and the construction period, that is, the above process is based on digital twin technology.

Additionally, the potential for deformation of the surrounding rock and instability of the blocks is strongly influenced by the shape of the slope. Therefore, it is necessary to combine the geological and slope structure models in a three-dimensional format. This integration should be further connected with CAE technology, and it often requires the interaction of multiple professionals and the comprehensive analysis of construction and monitoring results. Ultimately, this approach is based on the utilization of 3D digital investigation findings. All these professional objects and information together constitute the "digital twin of rock slope", therefore, the most fundamental principle of geological risk control can be summarized as "the application of digital twin technology in the construction process of rock slope engineering".

4.2 Key Technologies and Application Cases

As shown in Fig. 9, the key technologies involved in the intelligent screening process of rock slope unstable blocks encompasses three main aspects: The theory and method of analyzing block stability, converting geological models into 3D calculation models, and integrating digital results from multiple professionals and visualizing the data.

Among them, the technical reliability of the intelligent screening results depends on the reliability of the block stability analysis results. Up to this point, various technical methods have been extensively utilized in engineering practice, particularly in the initial design demonstration. These methods, such as the efficient key block theory and the discontinuous discrete element method, have reached a high level of maturity. Figure 10 shows the application of block stability analysis based on field recorded data in an engineering test hall (Fig. 10a) and an underground building of a pumped storage power station (Fig. 10b). Both engineering examples successfully anticipated and addressed unstable blocks ahead of time, ensuring safe excavation by effectively mitigating geological hazards. Our top priority is to guarantee the project's safety and efficiency, while also minimizing construction time and costs.

The promotion and application process during the construction period of rock slope relies heavily on the latest advancements in transforming geological 3D models. The following has significantly reduced the time required to create and update calculation models. As a result, evaluation and prediction can now be completed within a few hours after geological recording. Which allows for better application to construction safety management and enables intelligent screening of unstable blocks during the construction period. Figure 11 shows the mature application of geological model rapid transformation of 3D calculation model technology in a large hydropower project, representing the latest technical achievements.

Furthermore, the detection of slope instability blocks is dependent on the 3D digital findings from the initial phase of the project. Throughout the construction period, the

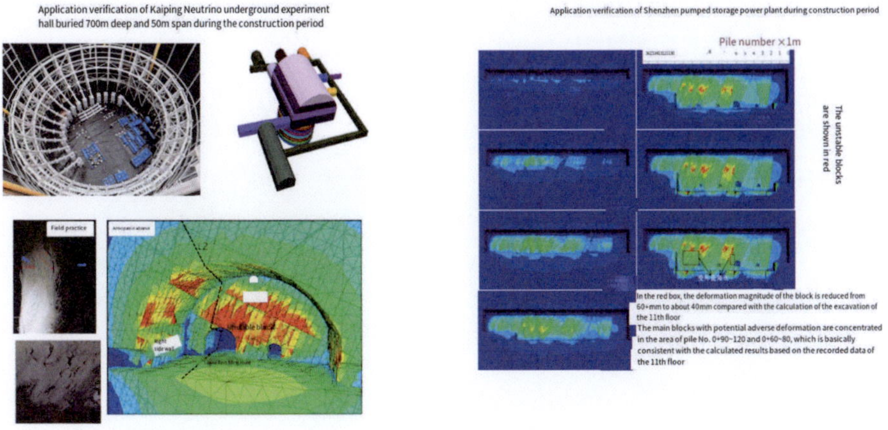

(a) **(b)**

Fig. 10. (a) Example of block stability analysis based on field catalogued data -- a project. (b) Example of block stability analysis based on field catalogued data -- underground powerhouse of a pumped storage power station

geological catalog data is consistently utilized to enhance and expand upon the 3D geological model, ensuring the preservation of survey data. It is the specific application of digital twin technology in the investigation major in the early stage of the project and the construction period, that is, the above process is based on digital twin technology. We must address the crucial aspects of integrating multi-professional digital results and visualizing data. Figure 12 shows the multi-professional integration and data visualization technology realized in the construction of digital twin system and its engineering application verification in a large hydropower station. Currently, the successful integration and visualization of various digital advancements in fields like geology, structure, monitoring, and calculation have established a strong groundwork for the development of a digital twin system for rock slopes.

Overall, the technologies used for intelligent screening of unstable blocks in the digital twin system have been well accumulated, and have been extensively developed and successfully implemented in numerous large-scale rock mass projects. As a result, it is both practical and achievable. Objectively, the intelligent screening of unstable blocks based on digital twin system of rock slope has a good condition for implementation.

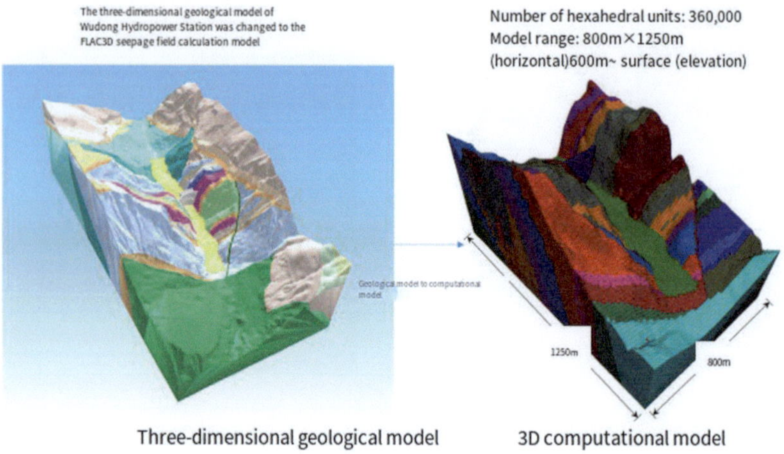

Fig. 11. Key technologies and engineering application verification of geological model to 3D calculation model

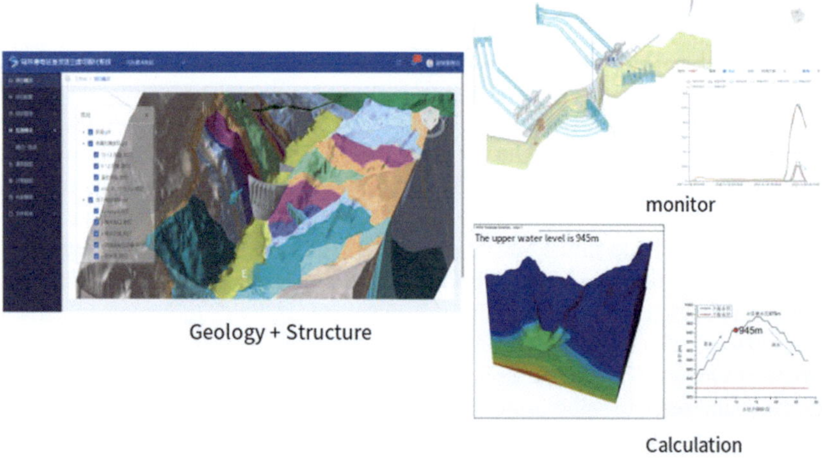

Fig. 12. Verification of key technologies and engineering applications of multi-specialty integration and data visualization

5 Conclusion

Drawing from extensive research and practical application in the field of digital technology, this paper focuses on rock slopes. It examines the structure and components of a comprehensive digital twin system that integrates multiple professions. Additionally, it analyzes and assesses the unique features of the digital twin system for slope engineering and its innovative application during the construction phase. The following conclusions have been obtained:

(1) The architecture of the digital twin system of rock slope needs to meet the requirements of multi-specialty, full process coverage, data center, and separation from software and personnel. The system consists of several professional digital subsystems that follow a "plug and play" approach. These professional subsystems are categorized into four links: Acquiring, storing, processing, and applying data.
(2) The digital twin system of rock slope incorporates several advanced technologies, such as 3D geological modeling and rapid updates, GIM+BIM coupling modeling, and cutting and sealing. These technologies are considered challenging and require independent innovation to be effectively solved.
(3) During the slope construction period, the digital twin system of rock slope is utilized to intelligently screen unstable blocks, showcasing an innovative approach. At present, the key technologies involved have been well accumulated, and have been fully applied and verified in a number of large-scale rock mass projects, and have good implementation conditions.

References

1. Boya, J., Zhenqiang, Q.: Research status of domestic BIM technology. Sci. Technol. Manage. Res. **35**(06), 184–190 (2015)
2. Huahai, Z., Yun, L., Yuanqi, L.: Research and application status of BIM technology. Struct. Eng. **31**(04), 233–241 (2015)
3. Yiming, D., Ren, Y.: Invited Report of the 7th National Geotechnical Engineering Record Exchange Meeting – Discussion on the feasibility of BIM application in engineering investigation. Geotech Eng Technol **30**(01), 6–11 (2016)
4. Xing-Hong, L., Daming, L., Jin, Y., et al.: Application of BIM technology in domestic tunnel engineering. Mod. Tunn. Technol. **57**(06), 25–35 (2020)
5. Mei, L., Jitong, K., Hui, L., et al.: Research on parametric 3D modeling technology of mine roadway based on BIM and GIS. Coal Sci. Technol., 1–12 (2022 August 16)
6. Eryong, T., Yangyang, M., Yunqi, D.: Construction and application of urban geologic body model based on GIM technology. Shandong Transport. Sci. Technol. **03**, 6–9 (2022)
7. Guodong, Z.:. Feasibility study on synergistic aided design of pile Foundation based on GIM system. Fujian Arch. **10**, 96–100 (2020)
8. Heng, Z., Tingwei, L., Guikai, X.: Research on landslide control based on GIM and BIM technology. Northern Commun. **04**, 48–51+56 (2022)
9. Kelen, C., Yaqiang, W.: Research on 3D digital design system of wind farm based on GIS+BIM. Ener. Sci. Technol. **19**(04), 50–53 (2021)
10. Dechao, L., Ruizhi, Z.: Research on application of BIM technology in 3D modeling of digital city. Inf. Technol. Civil Buil. Eng. **4**(01), 47–51 (2012)
11. Park, T., Kang, T., Lee, Y.: Project cost estimation of national road in preliminary feasibility stage using BIM/GIS platform. American Society of Civil Engineers. 2014 International Conference on Computing in Civil and Building Engineering, pp. 423–430 (2014)
12. Amirebrahimi, S., Rajabifard, A., Mendis, P.: A framework for a microscale flood damage assessment and visualization for a building using BIM–GIS integration. Int. J. Digit. Earth **9**(4), 363–386 (2016)
13. Xing, X.: Intelligent management practice of highway High slope. Traff. Eng. **22**(02), 16–20 (2022)

14. Qixiang, F., Heping, J., Wenlin, W., et al.: Design and application of cascade hydropower project management system based on digital watershed. J. Hydroelect. Power Generat. **35**(01), 136–145 (2016)
15. Fei, T., Weiran, L., Meng, Z., et al.: Computer Integrated Manufacturing Systems **25**(1), 1–18 2019 (in Chinese)
16. Fei, T., He, Z., Qinglin, Q., et al.: Theory and Application of Digital Twin Model Construction [J]. Comput. Integr. Manuf. Syst. **27**(1), 1–15 (2021)
17. Weihao, Z., Xingyu, Z., Hongping, W., et al.: Highway traffic digital twin system based on 3D GIS technology. Compu. Integr. Manuf. Syst. **26**(1), 28–39 (2020)
18. Huabing, F., Wentao, L., Xin, W., et al.: Digital twin hospital -- BIM technology application and thinking of Leishenshan Hospital. Huazhong Arch. **38**(4), 68–71 (2020)
19. Sheirong, Z., Peige, J., Zhengqiao, W.: Research progress of lean construction technology in hydropower engineering design and construction integration -- Exploration of digital twin application model. J. Hydroelect. Power Gen. **40**(01), 1–12 (2021)
20. Jian, C., Qian, S., Guoliang, C., et al.: Research progress of digital twin technology in geotechnical engineering. J. Huazhong Univ. Sci. Technol. (Nat. Sci. Ed.) **50**(08), 79–88 (2022)
21. Guogang, W., Wenchao, Z., Yapeng, C., et al.: Analysis on the application scheme of digital twin technology in water conservancy and hydropower engineering geology. Water Conserv. Tech. Supervis. **05**, 309–315 (2020)
22. Guo, X., Changhai, W., Xiaoqin, Z.: 3D geological modeling based on geological features. J. Shenzhen Univ. (Sci. Technol.) **39**(04), 417–423 (2022)
23. Jian, L., Peirong, L., Zhuanxin, L., et al.: Rock and Soil Mechanics **42**(04), 1170–1177 (2021) (in Chinese)
24. Denghua, Z., Xiang, G.: Three-dimensional parametric design method for underground main building of hydropower project. J. Hydroelectr. Power Gen. **28**(01), 171–177 (2009)
25. Lihong, Z., Tianyun, L., Peng, L., et al.: Influence of three-dimensional global model on seismic performance of large tunnel project. J. Hydropow. Gen. **32**(02), 240–245 (2013)
26. Fei, W., Jinfei, L., Xishuang, Y.: Research on digital transfer technology of Sichuan-Tibet Railway tunnel project based on BIM. Sichuan Hydropow. **39**(05), 93–96+102 (2020)
27. Lu Yongdong, D., Sihong, Z.D., et al.: Research progress of BIM and GIS integration in digital and Smart Age: Methods, applications and challenges. Build. Sci. **37**(04), 126–134 (2021)
28. Guang, W.: Research on key technologies in 3D geological modeling [Master's thesis]. Chengdu University of Technology, Chengdu (2015)
29. Fei-Li, W., Shu-Hong, W., Hong-Yan, G., et al.: Instability characterization coefficient of key blocks and evaluation of rock slope stability. J. Traff. Trans. Eng. **18**(04), 44–52 (2018)
30. Zhong Denghua, L., Wenyan, L.J., et al.: Surface block analysis of underground cav based on 3D geological model. Chin. J. Rock Mech. Eng. **30**(S2), 3696–3702 (2011)
31. Minsi, Z., Runqiu, H., Shuhong, W., et al.: Full space block recognition method based on grid division and its engineering applications. Chinese Journal of Geotechnical Engineering **38**(03), 477–485 (2016)

Study on the Impact of Flood Season Operating Water Level on Flood Control of the Three Gorges Reservoir

Yan-wei Zhai(✉), Ding-guo Jiang, Guo-liang Ji, and Zhen-yu Lv

China Three Gorges Corporation, Wuhan 430010, PR China
zhaiyw1992@163.com

Abstract. Climate change has resulted in an increase in extreme weather events, with a sharp rise in droughts and floods. To establish a long-term mechanism for ensuring the safe operation of the Three Gorges Project, it is imperative to utilize hydrodynamic methods to analyze the water level operation mode during the flood season. This is crucial to enhancing the flood control safety and maximizing the overall benefits of this project. Therefore, this study obtained four different frequency design flood processes for each year by utilizing data from typical years. And the influence of different starting water level and discharge flows on the high water level and excess flood volume of flood regulation were revealed under the existing scheduling protocols and a one-dimensional hydrodynamic model. The results indicated that the water level operation modes of floods from different typical years were significantly different under the same scheduling rules, and when faced with the extreme flood conditions, discharge flow is a primary determinant of reservoir safety. Furthermore, the current scheduling scheme for the Three Gorges Reservoir has an extra safety margin. As a result, even without forecasting inflows, elevating the operating water level to 155 m during the flood season effectively mitigates risks from floods with a return period of 100 years or less, while maintaining risk control over floods with 1,000-year or 10,000-year return periods.

Keywords: three Gorges Reservoir · operating water level during flood season · Flood control risk · design flood · reservoir regulation

1 Introduction

The Three Gorges Project serves as a pivotal initiative for managing and protecting the Yangtze River, offering multiple benefits including flood control, power generation, navigation, and water resource utilization. It has achieved a high level in both construction and operational management. The initial design flood for the Three Gorges Reservoir is determined using the natural annual maximum flood series to calculate the characteristic value of design flood, the flood control storage capacity and regulation storage capacity of reservoir with their characteristic water level, and ensure the safety of both the dam and

S. Zheng et al. (Eds.): IHDC 2024, LNCE 487, pp. 88–104, 2025.
https://doi.org/10.1007/978-981-97-9184-2_9

downstream flood defenses (Guo et al. 2018). These design values termed "construction-phase design floods", are chosen from the most severe scenarios, with outsourced values derived from a safety standpoint, and are guided by limiting water level during flood season to manage the operations (Guo et al. 2016).

With the cascade reservoirs in the upper reaches of the Yangtze River, particularly the four massive dams at Wudongde, Baihetan, Xiluodu, and Xiangjiaba, now operated, the hydrological conditions of the Three Gorges Reservoir and its downstream areas have undergone significant changes from those in the preliminary design phase. These changes are attributed to the joint management by these upstream cascade reservoirs, improved inflow and rainfall forecasting capabilities, and other alterations in the hydraulic conditions (Guo and Bao 2016; Dai et al. 2017; Zhang et al. 2019; Wei 2017; Cai 2012; Wang and Guo 2020; Chu et al. 2023; Li et al. 2022). Continuing to use results from the initial design flood to guide the operation of the Three Gorges Reservoir has led to issues such as the lower operating water level during flood season and the need for improved comprehensive utilization efficiency (Guo et al. 2016). In 2022, there was a situation where "discharged flow during dry season, retained water with empty reservoir during flood season, while had no water left for storage after floods" severely limiting the comprehensive utilization of water resources in the Three Gorges Reservoir.

At the same time, current climate change has led to frequent extreme weather events with a sharp increase in droughts and floods. And the changing demand for comprehensive utilization of water resources has raised additional requirements for the operation and scheduling of the Three Gorges Reservoir. Therefore, in order to build a long-term mechanism for the safe operation and maximize the comprehensive benefits of the Three Gorges Project, it is necessary to analyze the water level operation mode during the flood season while ensuring the safety of flood control in the basin.

2 Method

The methods for calculating inflow flood and flood regulation can be categorized into two main approaches: hydraulic method and hydrological method. The hydraulic method is mainly based on the Saint Venant equations, adopting discrete method, and using numerical calculation techniques to calculate the numerical solutions of the equations. Hydrological methods, on the other hand, substitute the continuity and momentum equations of the Saint-Venant equations with water balance equations and reservoir storage equations. By simultaneously solving the reservoir storage equations and water balance equations, these methods compute the outflow process at downstream sections. But for reservoirs with long river reaches, such as the Three Gorges Reservoir (about 757 km from Zhutuo to the Three Gorges Dam), hydrological methods tend to generalize with fewer considerations, making it difficult to accurately analyze complex river segments. Moreover, they fail to clearly depict the spatiotemporal distribution of water levels and flow rates, and the computational accuracy is not superior to that of hydraulic methods.

The one-dimensional hydrodynamic numerical simulation software Telemac-Mascaret, developed by Electricite De France (EDF), is based on the Saint-Venant equations and is applicable for simulating rivers and floodplains. The proposed method established a one-dimensional hydrodynamic model suitable for channel-type reservoirs like the Three Gorges Reservoir based on Mascaret and made the following four improvements to the model. (a) constructed a main and branch river network model to incorporate the topographic influences of the Jialing River and Wu River, the two major tributaries; (b) made the calculated storage capacity consistent with the actual storage capacity of each river section by adjusting the storage capacity by segmenting it along the streamway; (c) proposed a method for determining unsteady flow channel roughness, enabling dynamic adjustment of roughness along the river; and (d) presented a method for calculating flow rates in non-control sections. The one-dimensional hydrodynamic model of the Three Gorges Reservoir was applied to compute several groups of typical flood processes. The improved model calculation accuracy can meet the study needs including flow rate, water level process, flood peak occurrence time and peak value, etc. (as shown in Fig. 1).

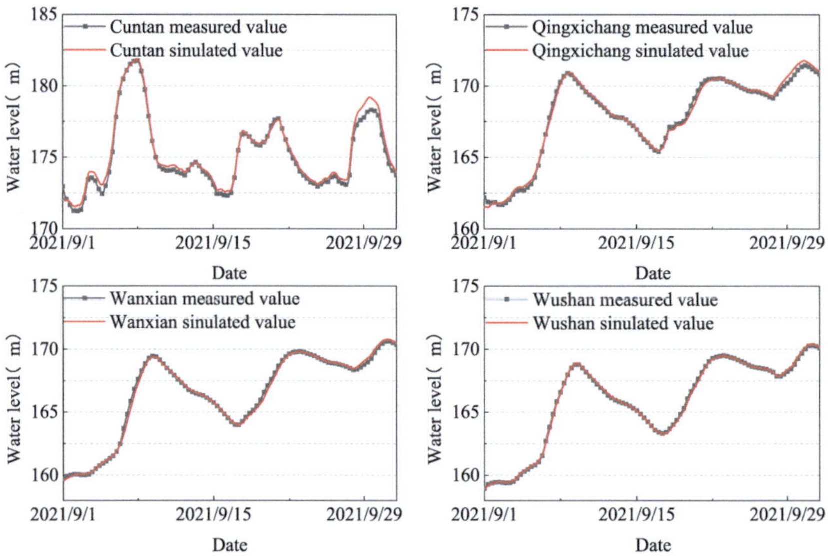

Fig. 1. Comparison between the calculated and measured water levels of typical cross-sections.

3 Typical Flood Process and Scheduling Principles

The core of studying the design floods and flood season operating water levels of the Three Gorges Reservoir lies in determining the flood composition and flood routing of different typical floods above the Three Gorges dam site. The characteristics of major flood years in Yichang are summarized in Table 1.

It can be observed that the major flood years in Yichang are mainly characterized by floods resulting from various factors such as (a) the large inflow of water in the main stream combined with encounters with tributaries; (b) the large inflow of water in tributaries combined with encounters with the main stream; (c) the large inflow of water in specific intervals; (d) the dominant inflow of water in the main stream, and

Table 1. List of typical annual characteristics of Yichang Station.

Type	Year	Flood encounter	The source of flood			Peak type		
		Main stream or tributaries	In main stream	In tributaries	In specific intervals	Sharp thin type	Fat type	
							The main peak is ahead	The main peak is behind
The main stream flood encounters the general flood of Wujiang River	2012	*	*					*
The main stream water encounters the Wujiang River water	1954	*	*	*				*
	1998	*	*	*			*	
The large inflow of water in the main stream	1981		*			*		
	1966		*				*	
The large inflow of water in specific intervals	1982				*			*

occurrences of encounters with the large inflow of water in both the main stream and tributaries, resulting in floods generally classified as "*fat-type*" floods.

Using the same frequency amplification method, the design flood processes for different frequencies such as 5%, 1%, 0.1%, and 0.01% were computed. To investigate the impact of different initial reservoir water levels and discharge flow rates on flood control capacity without considering the adverse effects of pre-release, three scheduling control rules were devised:

1. Rule for moderate floods at 171 m: For design floods with frequencies of 5% and 1%, when the water level ranges from 145 to 155 m, the controlled discharge flow rate is

fixed at $3.0*10^4$ m^3/s. When the water level is between 155 and 171 m, the controlled discharge flow rate is fixed between $5.0*10^4$ m^3/s and $5.5*10^4$ m^3/s, with intervals of $0.1*10^4$ m^3/s. When the water level is between 171 and 175 m, the controlled discharge flow rate equals the inflow rate but does not exceed $7.8*10^4$ m^3/s.

2. Rule for moderate floods at 175 m: For design floods with frequencies of 5% and 1%, when the water level ranges from 145 to 155 m, the controlled discharge flow rate is fixed at $3.0*10^4$ m^3/s. When the water level ranges from 155 to 175 m, the controlled discharge flow rate is fixed between $5.0*10^4$ m^3/s and $5.5*10^4$ m^3/s, with intervals of $0.1*10^4$ m^3/s. When the water level exceeds 175 m, the controlled discharge flow rate equals the inflow rate but does not exceed $7.8*10^4$ m^3/s.
3. Rule for severe floods at 175 m: For design floods with frequencies of 1%, 0.1%, and 0.01%, when the water level ranges from 145 to 155 m, the controlled discharge flow rate is fixed at $3.0*10^4$ m^3/s. When the water level ranges from 155 to 175 m, the controlled discharge flow rate is fixed between $5.0*10^4$ m^3/s and $5.5*10^4$ m^3/s, with intervals of $0.1*10^4$ m^3/s. When the water level exceeds 175 m, the controlled discharge flow rate equals the inflow rate but does not exceed $7.8*10^4$ m^3/s.

The discharge flow rates for each control rule are within the range required by the current regulations. By varying the initial reservoir water level, namely from 145 to 155 m with intervals of 2 m, the high water levels for flood routing and the cumulative excess flood volume exceeding $5.5*10^4$ m^3/s are analyzed using a model.

4 Discussions

4.1 Initial Reservoir Water Level

To investigate the influence of initial reservoir water levels on flood routing high water levels and excess flood volume, the analysis focuses on the results of flood routing calculations with a discharge flow rate of $5.5*10^4$ m^3/s for control rules 1 and 2, and $7.8*10^4$ m^3/s for control rule 3.

4.1.1 The Impact of the Initial Reservoir Water Level on the High Water Level for Flood Routing

Under different control rules, the flood routing high water levels corresponding to initial reservoir water levels of 145 m and 155 m in typical years with a 5% frequency are illustrated in Fig. 2. It is observed that the initial reservoir water levels have varying degrees of influence on flood routing high water levels across different typical years. Notably, in the year 1998, the disparity is the largest, approximately 1.10 m. Conversely, in 1981, the flood routing high water levels remain identical, unaffected by the initial reservoir water levels.

In the results illustrated in Fig. 3, the corresponding maximum flood routing levels in different typical years with 1% frequency are depicted for initial reservoir water levels with 145 m and 155 m. Figure 3a corresponds to control rules 1 and 3, while panel Fig. 3b corresponds to control rule 2. It is observed from the figures that under control rules 1 and 3, the flood routing high water levels do not exceed 171 m, with the maximum

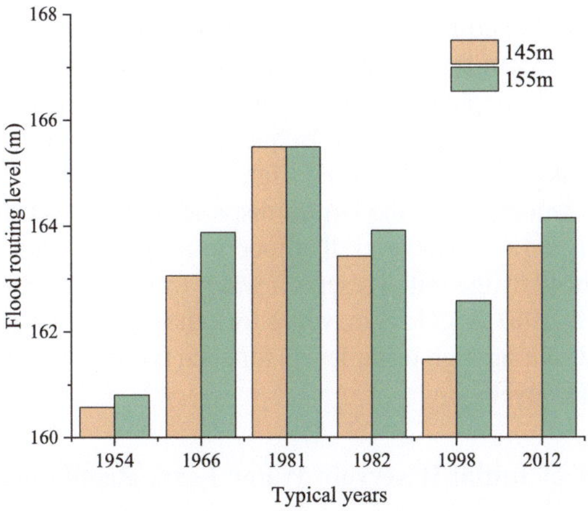

Fig. 2. Corresponding flood routing level for different typical years with a 5% frequency of 145 m and 155 m starting water levels.

disparity occurring in 1982, approximately 1.50 m. However, in 1966, 1981, and 1998, the flood routing high water levels remain identical, indicating no influence from the initial reservoir water level.

Under control rule 2, the flood routing high water levels are all below 175 m, with the largest disparity occurring in 1982, approximately 1.50 m, followed by 1998, approximately 1.43 m. Notably, in 1966 and 1998, flood routing levels are slightly higher than 171 m under control rule 2, while in other typical years, they are below 171 m. Furthermore, under this rule, the maximum discharge is 5.5*10^4 m^3/s, which is less than the prescribed 5.67*10^4 m^3/s. This validates that the current regulations can withstand a 100-year flood with the reservoir water level not exceeding 171 m and suggests that an initial reservoir water level of 155 m is also effective against such an event.

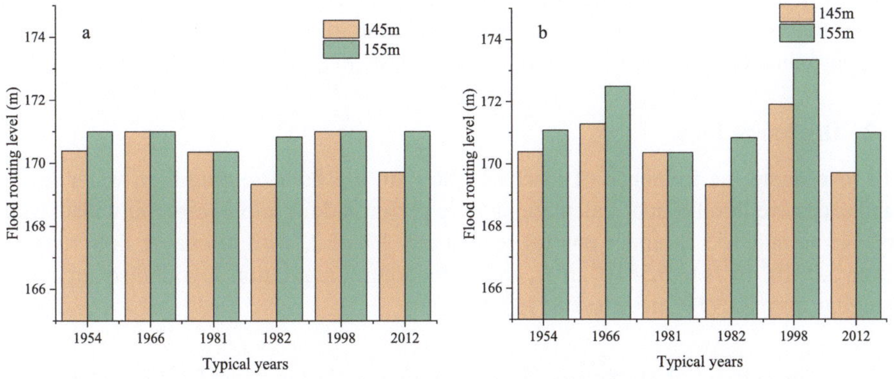

Fig. 3. Corresponding flood routing high water level for different typical years with 1% frequency of 145 m and 155 m initial reservoir water levels: (a) Control rules 1 and 3 and (b) Control rule 2.

The maximum flood routing high water levels corresponding to initial reservoir water levels of 145 m and 155 m for different typical years at 0.1% and 0.01% frequencies under control rule 3 are depicted in Fig. 4. Figure 4a represents the 0.1% frequency, while Fig. 4b corresponds to the 0.01% frequency. It is evident from the figures that for the 0.1% frequency, the flood routing high water levels for each typical year do not exceed 175 m. Among these, the largest disparity in flood routing levels occurs in 1982, approximately 1.13 m. However, the flood routing high water levels in 1981 and 1998 are not influenced by the initial reservoir water level. For the 0.01% frequency, the flood routing level in 1981 is 174.53 m, while for other typical years, it reaches 175 m. Moreover, the flood routing high water levels for each typical year are unaffected by the initial reservoir water level.

4.1.2 The Impact of Initial Reservoir Water Level Regulation on Excess Flood Volume

5% of the frequency does not have any excess floods in any of the rules (discharge flow greater than $5.5*10^4$ m^3/s), and 1% of the control rule 2 does not have any excess floods in typical years. Under various rules, the results of excess flood corresponding to the starting water levels of 145 m and 155 m for different frequencies and typical years are shown in Fig. 5.

Figure 5a representing the 1% frequency, typical years such as 1954, 1981, 1982, and 2012 exhibited minimal occurrences of excess flood volume, largely unaffected by the initial reservoir water level. However, in the typical years of 1966 and 1998, the excess flood volume increased by approximately 800 million m^3 and 900 million m^3, respectively, in response to changes in the initial reservoir water level. Figure 5b representing the 0.1% frequency, except for the typical year 1981, where excess flood volume remained unaffected by the initial reservoir water level, in other typical years, the excess flood volume at an initial elevation of 155 m increased by an average of approximately 11% and 1 billion m^3 compared to an initial elevation of 145 m. Similarly, Fig. 5c representing the 0.01% frequency, except for the typical year 1981, where excess flood volume remained unaffected by the initial reservoir water level, in other typical years, the excess flood volume at an initial elevation of 155 m increased by an average of approximately 10% and 1.7 billion m^3 compared to an initial elevation of 145 m.

4.2 Discharge Flow

To investigate the influence of discharge flow on the flood routing high water levels and excessive flood events, calculations were performed for initial reservoir water levels of 145 m and 155 m under control regulations 1 and 2, with discharge flow ranging from $5.0*10^4$ m^3/s to $5.5*10^4$ m^3/s, and under control regulation 3, with discharge flow ranging from $7.2*10^4$ m^3/s to $7.8*10^4$ m^3/s.

4.2.1 The Impact of Discharge Flow on the High Water Level for Flood Control

Under different control regulations, the flood routing high water levels corresponding to different maximum discharge flow in various typical years for initial reservoir water levels of 145 m and 155 m are presented in Table 2.

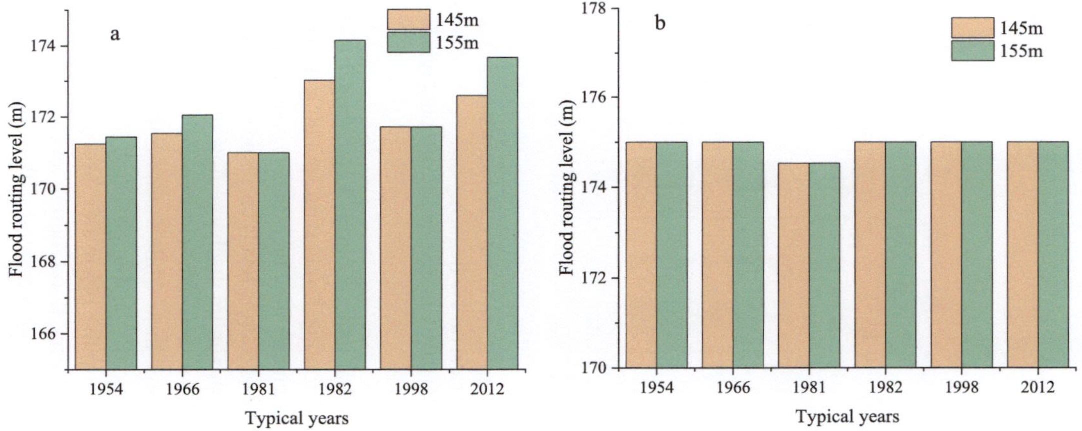

Fig. 4. Different typical years with frequencies of 0.1%, 0.01%, and initial reservoir water levels of 145 m and 155 m corresponding to flood routing high water levels: (a) typical years with frequencies of 0.1% and (b) typical years with frequencies of 0.01%.

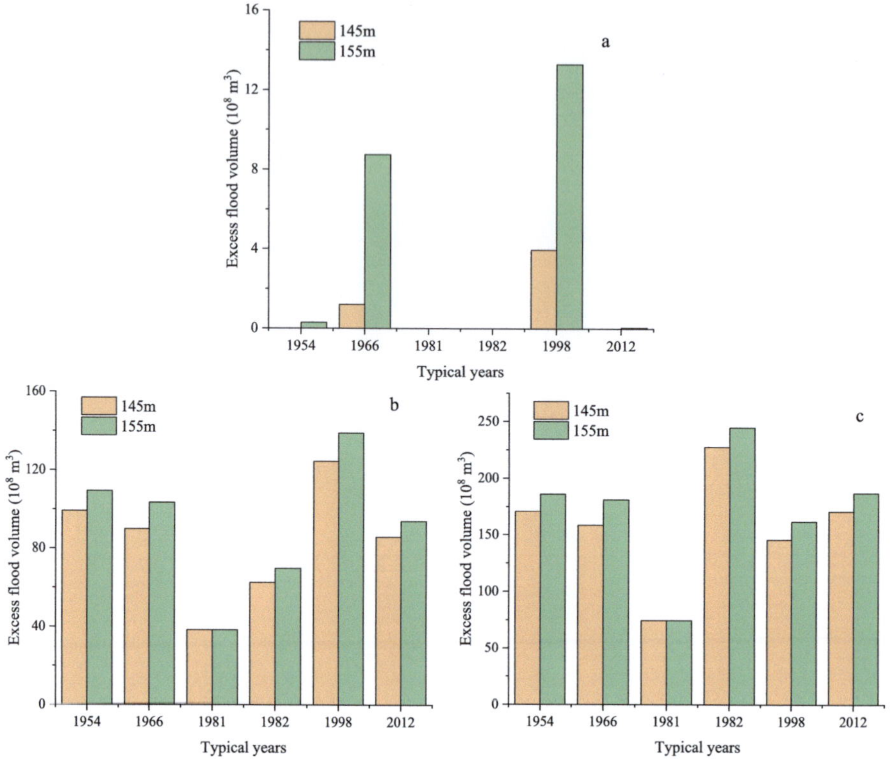

Fig. 5. The excess flood corresponding to the initial reservoir water levels of 145 m and 155 m in different frequencies and typical years: (a) 1% frequency under control rules 1 and 3, (b) 0.1% frequency under control rule 3 and (c) 0.01% frequency under control rule 3.

From Table 2, it is evident that under Control Regulation 1 and a 5% frequency, the process of decreasing flood routing levels corresponding to both initial reservoir water levels as the maximum discharge flow increases is consistent for the typical year 1981. In the typical year 1998, flood routing levels corresponding to both initial reservoir water levels reached 171 m at a maximum discharge flow of $5.0*10^4$ m^3/s, after which flood routing levels for the initial reservoir water level of 145 m decreased uniformly with increasing maximum discharge flow, while those for the initial elevation of 155 m initially decreased slowly, then rapidly, and approached but remained higher than the flood routing levels corresponding to the 145 m elevation at a maximum discharge flow of $5.5*10^4$ m^3/s. For other typical years, flood routing levels corresponding to the initial reservoir water level of 145 m uniformly decreased with increasing maximum discharge flow, while those corresponding to the initial elevation of 155 m were slightly higher at lower maximum discharge flow, then rapidly decreased with increasing maximum discharge flow, gradually approaching the flood routing levels corresponding to the 145 m elevation.

Table 2. Different maximum discharge capacities correspond to the corresponding flood routing high water levels at the starting water levels of 145 m and 155 m in typical years.

Floods Frequency	m m^3/s	Typical year 1954		Typical year 1966		Typical year 1981		Typical year 198		Typical year 1998		Typical year 2012	
		145	155	145	155	145	155	145	155	145	155	145	155
Control Rule 1 with 5% frequency	50,000	167.27	169.04	167.41	168.79	168.00	168.00	166.65	168.48	170.99	171.00	167.94	169.63
	51,000	165.73	167.26	166.53	167.78	167.50	167.50	165.87	167.46	168.95	170.86	166.85	168.34
	52,000	164.30	165.53	165.66	166.77	166.99	166.99	165.22	166.47	166.95	168.70	165.92	167.16
	53,000	162.98	163.86	164.79	165.79	166.49	166.49	164.59	165.53	165.01	166.58	165.10	166.08
	54,000	161.73	162.27	163.92	164.83	165.99	165.99	164.00	164.69	163.14	164.52	164.33	165.08
	55,000	160.58	160.81	163.06	163.88	165.49	165.49	163.43	163.91	161.47	162.58	163.62	164.15
Control Rule 1 with 1% frequency	50,000	171.00	171.00	171.00	171.00	171.00	171.00	171.00	171.00	171.00	171.00	171.00	171.00
	51,000	171.00	171.00	171.00	171.00	171.00	171.00	171.00	171.00	171.00	171.00	171.00	171.00
	52,000	171.00	171.00	171.00	171.00	171.00	171.00	171.00	171.00	171.00	171.00	171.00	171.00
	53,000	171.00	171.00	171.00	171.00	171.00	171.00	171.00	171.00	171.00	171.00	171.00	171.00
	54,000	171.00	171.00	171.00	171.00	170.86	170.86	170.34	171.00	171.00	171.00	171.00	171.00
	55,000	170.39	171.00	171.00	171.00	170.36	170.36	169.34	170.83	171.00	171.00	169.71	171.00
Control Rule 2 with 1% frequency	50,000	175.00	175.00	175.00	175.00	172.91	172.91	175.00	175.00	175.00	175.00	175.00	175.00
	51,000	175.00	175.00	174.96	175.00	172.39	172.39	173.88	175.00	175.00	175.00	175.00	175.00
	52,000	174.78	175.00	174.01	175.00	171.88	171.88	172.66	174.46	175.00	175.00	174.51	175.00
	53,000	173.29	174.54	173.09	174.52	171.36	171.36	171.47	173.19	175.00	175.00	172.77	174.28
	54,000	171.85	172.82	172.18	173.51	170.86	170.86	170.34	171.96	174.03	175.00	171.02	172.42
	55,000	170.39	171.08	171.28	172.49	170.36	170.36	169.34	170.83	171.90	173.34	169.71	171.00
Control Rule 3 with 0.1% frequency	72,000	173.21	173.64	173.58	174.57	171.95	171.95	174.78	175.00	173.94	173.95	174.29	175.00
	73,000	172.77	173.15	173.19	174.12	171.73	171.73	174.48	175.00	173.47	173.47	173.99	175.00
	74,000	172.38	172.72	172.82	173.67	171.52	171.52	174.18	175.00	173.03	173.03	173.70	174.88

(continued)

Table 2. (*continued*)

Floods Frequency	m m^3/s	Typical year 1954		Typical year 1966		Typical year 1981		Typical year 198		Typical year 1998		Typical year 2012	
		145	155	145	155	145	155	145	155	145	155	145	155
	75,000	172.05	172.34	172.47	173.24	171.33	171.33	173.88	175.00	172.63	172.63	173.41	174.58
	76,000	171.75	172.01	172.14	172.82	171.15	171.15	173.59	174.77	172.28	172.28	173.13	174.27
	77,000	171.48	171.71	171.83	172.42	171.01	171.01	173.31	174.46	171.99	171.99	172.86	173.97
	78,000	171.25	171.44	171.54	172.05	171.00	171.00	173.03	174.15	171.72	171.72	172.59	173.67

Under Control Rule 1 and a 1% frequency, only the flood routing levels corresponding to a maximum discharge flow of $5.5*10^4$ m^3/s exhibited slight differences, while flood routing levels for other maximum discharge flow reached or approached 171 m.

Under Control Rule 2 and a 1% frequency, the process of decreasing flood routing levels corresponding to both initial reservoir water levels as the maximum discharge flow increases is consistent for the typical year 1981. For other typical years, flood routing levels corresponding to both initial reservoir water levels were 175 m at a maximum discharge flow of $5.0*10^4$ m^3/s, with flood routing levels decreasing first for the initial elevation of 145 m, followed by gradual decreases for the initial elevation of 155 m as the maximum discharge flow increased.

Under Control Rule 3 and a 0.1% frequency, the process of decreasing flood routing levels corresponding to both initial reservoir water levels as the maximum discharge flow increases is consistent for the typical years 1981 and 1998. In the typical year 1981, flood routing levels stabilized at 171 m when the maximum discharge flow reached $7.7*10^4$ m^3/s or higher, while in the typical year 1998, flood routing levels gradually decreased with increasing maximum discharge flow, showing a trend towards stabilization. For the typical years 1954 and 1966, flood routing levels corresponding to the initial reservoir water level of 145 m decreased with increasing maximum discharge flow, while those corresponding to the initial elevation of 155 m were slightly higher at lower maximum discharge flow, then rapidly decreased with increasing maximum discharge flow, gradually approaching the flood routing levels corresponding to the 145 m elevation. For the typical years 1982 and 2012, flood routing levels corresponding to the initial reservoir water level of 145 m uniformly decreased with increasing maximum discharge flow, while those corresponding to the initial elevation of 155 m remained relatively unchanged at lower maximum discharge flow, then decreased with further increases in maximum discharge flow.

4.2.2 The Impact of Discharge Flow on Excess Flood Volume

Under different rules, the results of excess flood corresponding to the starting water levels of 145 m and 155 m for different maximum discharge flows in typical years are shown in Table 3.

From Table 3, it is evident that under Control Rule 1 and a 1% frequency, the overall trend of excess flood volume corresponding to the two initial reservoir water levels decreases as the maximum discharge flow increases. And it is consistently with the trend in the typical year 1981, and the trend is gradually decelerating in other typical years. Specifically, in the typical year 1954, at a maximum discharge flow of $5.5*10^4$ m^3/s, the excess flood volume corresponding to both initial reservoir water levels are nearly zero. The same circumstance occurred in the typical year 1982, at a maximum discharge flow of $5.3*10^4$ m^3/s with the initial elevation of 145 m, while for the elevation of 155 m, it reaches zero at a maximum discharge flow of $5.5*10^4$ m^3/s. In the typical year 2012, at a maximum discharge flow of $5.4*10^4$ m^3/s, the excess flood volume corresponding to the initial elevation of 145 m is zero, while for the elevation of 155 m, it reaches zero at a maximum discharge flow of $5.5*10^4$ m^3/s.

Table 3. Different maximum discharge capacities correspond to excess flood volume at the starting water levels of 145 m and 155 m in typical years.

Floods Frequency	m m^3/s	Typical year 1954		Typical year 1966		Typical year 1981		Typical year 198		Typical year 1998		Typical year 2012	
		145	155	145	155	145	155	145	155	145	155	145	155
Control Rule 1 with 5% frequency	50,000	32.3	44.6	19.1	30.9	3.7	3.7	17.1	31.2	55.7	60.3	29.0	44.2
	51,000	25.2	36.9	15.8	27.1	2.5	2.5	10.8	23.6	47.1	56.3	20.4	34.1
	52,000	17.8	28.3	12.3	22.9	1.2	1.2	5.1	16.4	36.8	50.4	12.2	24.4
	53,000	10.3	19.1	8.7	18.4	0.1	0.1	0.5	9.7	28.5	38.8	4.8	15.2
	54,000	3.2	9.4	4.9	13.7	0.0	0.0	0.0	3.6	16.4	27.7	0.0	6.8
	55,000	0.0	0.3	1.2	8.7	0.0	0.0	0.0	0.0	4.0	13.3	0.0	0.0
Control Rule 1 with1% frequency	50,000	8.2	20.0	0.0	7.0	0.0	0.0	0.0	6.1	32.2	46.1	3.1	15.4
	51,000	2.0	11.2	0.0	3.1	0.0	0.0	0.0	0.5	27.7	35.1	0.1	6.2
	52,000	0.0	3.1	0.0	0.0	0.0	0.0	0.0	0.0	15.8	28.0	0.0	0.1
	53,000	0.0	0.0	0.0	0.0	0.0	0.0	0.0	0.0	3.7	14.5	0.0	0.0
	54,000	0.0	0.0	0.0	0.0	0.0	0.0	0.0	0.0	0.0	1.7	0.0	0.0
	55,000	0.0	0.0	0.0	0.0	0.0	0.0	0.0	0.0	0.0	0.0	0.0	0.0
Control Rule 2 with1% frequency	50,000	85.5	93.5	77.8	87.1	35.0	35.0	50.6	57.0	111.8	126.1	74.4	80.1
	51,000	88.7	97.0	80.4	90.4	36.0	36.0	52.8	59.0	114.9	129.3	76.7	82.7
	52,000	91.5	100.3	82.8	93.4	36.9	36.9	54.9	61.0	117.7	132.1	78.8	85.2
	53,000	94.1	103.3	84.9	96.3	37.6	37.6	57.0	63.2	120.1	134.6	80.7	87.5
	54,000	96.3	105.8	86.8	98.9	38.1	38.1	58.9	65.4	122.1	136.5	82.6	89.8
	55,000	98.0	107.8	88.4	101.3	38.4	38.4	60.8	67.7	123.4	137.8	84.3	91.9
Control Rule 3 with0.1% frequency	72,000	99.2	109.4	89.8	103.5	38.4	38.4	62.6	69.8	124.6	139.0	85.9	93.9
	73,000	170.6	186.0	158.2	181.1	61.4	61.4	219.8	236.9	145.6	161.6	170.6	187.2

(continued)

Table 3. (*continued*)

Floods Frequency	m m^3/s	Typical year 1954		Typical year 1966		Typical year 1981		Typical year 198		Typical year 1998		Typical year 2012	
		145	155	145	155	145	155	145	155	145	155	145	155
	74,000	170.6	186.0	158.2	181.1	63.6	63.6	220.1	237.0	145.6	161.6	170.6	187.2
	75,000	170.7	186.0	158.2	181.1	65.9	65.9	220.3	237.4	145.6	161.7	170.6	187.2
	76,000	170.7	186.0	158.2	181.1	68.1	68.1	220.6	237.7	145.7	161.7	170.6	187.2
	77,000	170.7	186.0	158.2	181.1	70.2	70.2	221.0	238.1	145.7	161.7	170.6	187.2
	78,000	170.7	186.0	158.2	181.1	72.4	72.4	223.9	241.0	145.7	161.7	170.6	187.2

Under Control Rule 2 and a 1% frequency, the overall trend of excess flood volume decreases with increasing maximum discharge flow, albeit significantly lower in magnitude compared to Control Rule 1. In the typical year 1981, the excess flood volume corresponding to both initial reservoir water levels are zero. The same circumstance occurred in the typical year 1954, at a maximum discharge flow of 5.2*10^4 m^3/s with the initial elevation of 145 m, and at a maximum discharge flow of 5.3*10^4 m^3/s with the initial elevation of 155 m. In the typical years 1966 and 1982, the excess flood volume for the initial elevation of 145 m is zero, while for the elevation of 155 m, it corresponds to zero at a maximum discharge flow of 5.2*10^4 m^3/s. In the typical year 1998, at a maximum discharge flow of 5.4*10^4 m^3/s, the excess flood volume for the initial elevation of 145 m is zero, while for the elevation of 155 m, it corresponds to zero at a maximum discharge flow of 5.5*10^4 m^3/s. In the typical year 2012, at a maximum discharge flow of 5.1*10^4 m^3/s, the excess flood volume for the initial elevation of 145 m is nearly zero, while for the elevation of 155 m, it is almost zero at a maximum discharge flow of 5.2*10^4 m^3/s.

Under Control Regulation 3 and a 0.1% frequency, the overall trend of excess flood volume corresponding to the two initial reservoir water levels increases with increasing maximum discharge flow. In the typical year 1981, the process of excess flood volume corresponding to both initial reservoir water levels increasing with increasing maximum discharge flow is consistent. For the typical years 1954, 1966, and 1998, the trend of excess flood volume corresponding to both initial reservoir water levels increasing with increasing maximum discharge flow gradually slows down. In the typical years 1982 and 2012, the excess flood volume corresponding to both increasing initial reservoir water levels nearly linearly with increasing maximum discharge flow. The net difference in excess flood volume corresponding to the two initial reservoir water levels remains relatively stable across typical years, with minimal influence from variations in maximum discharge flow.

Under Control Rule 3 and a 0.01% frequency, the pattern of excess flood volume change can be categorized into two scenarios: overall increase with increasing maximum discharge flow and no change with variations in maximum discharge flow. In the typical year 1981, the excess flood volume under both initial reservoir water levels increases linearly with increasing maximum discharge flow, while the magnitudes remain essentially identical. In the typical year 1982, the excess flood volume under both initial reservoir water levels initially increases linearly with increasing maximum discharge flow, then accelerates when the maximum discharge flow exceeds 7.6*10^4 m^3/s. In the typical years 1954, 1966, 1998, and 2012, the excess flood volume remains unchanged with variations in maximum discharge flow, with a net difference of approximately 1.6 billion m^3 for the excess flood volume corresponding to the initial elevations of 145 m and 155 m.

5 Conclusions

Based on observed flood processes from six different typical years, this study employed the same frequency amplification method to calculate four types of design flood processes with frequencies of 5%, 1%, 0.1%, and 0.01%, respectively. Building upon the existing scheduling regulations, three scheduling control rules were designed to investigate the impacts of different initial reservoir water levels and discharge flow rates on flood control capacity, without considering the adverse scenarios of pre-release. The findings are as follows:

1. The influence of different initial reservoir water levels on flood routing levels varies with different frequencies of design flood processes. For typical floods occurring once every twenty years, both initial elevations of 145 m and 155 m ensure safe flood discharge, with flood routing levels not exceeding 165.5 m and without increasing downstream flood control pressure. For floods occurring once every hundred years, both initial elevations of 145 m and 155 m ensure safe flood discharge, with flood routing levels not exceeding 171 m; the excess flood volume corresponding to the elevation of 155 m increases by approximately 930 million m^3 at most. If controlled to ensure that the water level in front of the dam does not exceed 175 m, the flood routing level corresponding to the elevation of 155 m does not exceed 173.3 m, with no excess flood volume and no increase in downstream flood control pressure. For floods occurring once every thousand years, both initial elevations of 145 m and 155 m ensure that the flood routing level does not exceed 174.1 m; the excess flood volume corresponding to the elevation of 155 m increases by approximately 0–15%, up to about 1.37 billion m^3 at most. For floods occurring once every ten thousand years, both initial elevations of 145 m and 155 m ensure that the flood routing level does not exceed 175 m; the excess flood volume corresponding to the elevation of 155 m increases by approximately 0–14%, up to about 2.28 billion m^3 at most.
2. The maximum discharge flow has a significant impact on flood routing levels, with higher maximum discharge flow resulting in lower flood routing levels. At the same maximum discharge flow, compared to the initial elevation of 145 m, the elevation of 155 m raises the flood routing level by at most 1.9 m, and the difference in flood routing levels corresponding to the two initial elevations gradually decreases with increasing maximum discharge flow.
3. The influence of maximum discharge flow on excess flood volume varies with different frequencies of design flood processes. For floods occurring once every hundred years, a higher maximum discharge flow leads to a lower excess flood volume. Conversely, for floods occurring once every thousand years, a higher maximum discharge flow leads to a higher excess flood volume. For floods occurring once every ten thousand years, a higher maximum discharge flow results in more excess flood volume for the typical years 1981 and 1982, while the excess flood volume for other typical years remains unaffected by variations in maximum discharge flow.

The current scheduling regulations for the Three Gorges Reservoir have a considerable safety margin. Under the condition of not considering forecasted inflows, elevating the operating water level to 155 m during the flood season can effectively cope with floods occurring once every hundred years or less, while also ensuring that the risks associated with floods occurring once every thousand years or ten thousand years are manageable.

Acknowledgements. This work was supported by the National Key Research and Development Program of China (2022YFC3203900).

References

1. Cai, Q.H.: Scientific scheduling of the Three Gorges Reservoir should accurately grasp three important prerequisites. China Water Resour. **14**, 4–6 (2012). (in Chinese)
2. Chu, M.H., Li, R.B., Yan, Y.L.: Optimal operation practices and reflections on the Three Gorges Reservoir. China Water Resour. **22**, 22–26 (2023). (in Chinese)
3. Dai, M.L., Wang, J., Zhang, M., Chen, X.: Impact of the Three Gorges Project operation on the water exchange between Dongting Lake and the Yangtze River. Int. J. Sediment Res. **32**(4), 506–514 (2017). https://doi.org/10.1016/j.ijsrc.2017.02.006
4. Guo, L., Bao, Z.F.: On the inflow change of the Three Gorges Reservoir since its completion. Hydropower New Energy **5**, 34–36 (2016). https://doi.org/10.13622/j.cnki.cn42-1800/tv.1671-3354.2016.05.009. (in Chinese)
5. Guo, S.L., Liu, Z.J., Xiong, L.H.: Advances and assessment on design flood estimation methods. J. Hydraul. Eng. **47**(3), 302–314 (2016). https://doi.org/10.13243/j.cnki.slxb.20150913. (in Chinese)
6. Guo, S.L., Xiong, L.H., Xiong, F., Yin, J.B., Chen, K.: Theory and method of design flood in reservoir operation period. J. Water Resour. Res. **7**(4), 327–338 (2018). (in Chinese)
7. Li, X.N., Fu, Q.P., Zhang, S., He, X.C., Zou, M., Ding, Y.: Study on operating water level during flood season of Three Gorges Reservoir. Yangtze River **53**(02), 21–26+40 (2022). https://doi.org/10.16232/j.cnki.1001-4179.2022.02.004. (in Chinese)
8. Wang, J., Guo, S.L.: On Three Gorges Reservoir control water level and operating conditions in flood season. Adv. Water Sci. **31**(4), 473–480 (2020). https://doi.org/10.14042/j.cnki.32.1309.2020.04.001. (in Chinese)
9. Wei, S.Z.: The strategy of flood control and disaster mitigation of the Yangtze River in the new period. Yangtze River **48**(4), 1–7 (2017). https://doi.org/10.16232/j.cnki.1001-4179.2017.04.001. (in Chinese)
10. Zhang, D.D., Dai, M.L., Xu, G.H., Deng, X.P.: Research on change of the outflow of Dongting Lake during the refill period of the Three Gorges Reservoir. Adv. Water Sci. **30**(5), 613–622 (2019). https://doi.org/10.14042/j.cnki.32.1309.2019.05.001. (in Chinese)
11. Zhang, S.H., Jing, Z., An, W.J., Zhang, R.Q., Yi, Y.J.: Flood-control ability of the Three Gorges Reservoir and upstream cascade reservoirs during catastrophic flooding. Scientia Sinica (Technologica) **52**(05), 795–806 (2022). (in Chinese)

Deep Learning-Based Multi-Model Coupled Flood Season Daily Runoff Prediction Model

Xiaoyu Ye, Dong Wang(✉), Chenlu Yu, Zhuo Yang, and Along Zhang

School of Earth Science and Engineering, Nanjing University, 163 Xianlin Avenue, Qixia District, Nanjing 210023, China
502023290072@smail.nju.edu.cn, wangdong@nju.edu.cn

Abstract. Accurate runoff forecasting is of great significance for flood control, drought prevention, reservoir scheduling, and ecological protection. To explore the applicability of deep learning networks combined with signal processing techniques in runoff forecasting, an ICEEMDAN-VMD-CNN-LSTM daily runoff forecasting model for the flood season was developed. First, the original runoff series was decomposed using the Improved Complete Ensemble Empirical Mode Decomposition with Adaptive Noise (ICEEMDAN). Then, the complex series was further decomposed using Variational Mode Decomposition (VMD) to reduce data complexity. Next, each mode component was input into a Convolutional Neural Network (CNN) - Long Short-Term Memory (LSTM) combined model to extract local features of the data and capture long-term dependencies of the time series. Finally, the predicted values were reconstructed to obtain the final prediction results. Using the measured daily runoff data from the Hekou station in the Diaojiang basin as an example, the results showed that the ICEEMDAN-VMD-CNN-LSTM achieved testing MAE and NSE of 5.232 m^3/s and 0.977, respectively, demonstrating excellent forecasting accuracy.

Keywords: Runoff Forecasting · ICEEMDAN · VMD · CNN · LSTM · Diaojiang Basin

1 Introduction

Accurate runoff forecasting is essential for water resource planning and management, reservoir optimization, and environmental protection (Lei et al., 2018). Surface runoff is typically influenced by various factors such as precipitation, evaporation, solar radiation, vegetation cover, and human activities (Zheng et al., 2023). These factors contribute to significant nonlinearity and high uncertainty in runoff patterns, posing a major challenge for accurate forecasting (Deb et al., 2019).

Over the past few decades, numerous forecasting models have been proposed. These models can be broadly classified into two categories: process-driven models and data-driven models. Process-driven models, based on hydrological concepts and physical processes, simulate the temporal and spatial variations of runoff, enabling accurate forecasting (Aqil et al., 2007). Data-driven models are a black-box approach, primarily

S. Zheng et al. (Eds.): IHDC 2024, LNCE 487, pp. 105–114, 2025.
https://doi.org/10.1007/978-981-97-9184-2_10

focusing on considering state variables as the model's input and output, to establish the optimal mathematical relationships describing their correlation, with relatively less consideration for physical processes (Feng et al., 2020). Compared to process-driven models, data-driven models do not require extensive and hard-to-obtain hydrometeorological data as inputs. They also do not need to account for the differences in physical processes under various conditions, demonstrating strong adaptability and generalization capabilities (Yaseen et al., 2016). Therefore, in time series forecasting, data-driven models possess unique advantages. With the advancement of artificial intelligence technology, the enhancement of computational power, and the reduction in difficulty of acquiring hydrological data, deep learning is gradually gaining widespread attention and application in the field of hydrology and water resources (Shen, 2018).

Due to environmental influences, the surface runoff process exhibits strong randomness and uncertainty, making it difficult for a single forecasting model to achieve satisfactory prediction results. Signal processing techniques can effectively remove noise from complex signals, decomposing the original series into a set of more stable components. Combining signal processing techniques with deep learning models can significantly reduce prediction difficulty and improve forecasting accuracy (Li et al., 2021). Currently, common signal processing techniques include Empirical Mode Decomposition (EMD), Ensemble Empirical Mode Decomposition (EEMD), Complementary Ensemble Empirical Mode Decomposition (CEEMD), and Complete Ensemble Empirical Mode Decomposition with Adaptive Noise (CEEMDAN) (Yeh et al., 2010; Wang et al., 2014; Lu and Ma, 2020). Scholars in various fields have attempted this combination and achieved promising results. For instance, Ni et al. combined Wavelet Decomposition (WD) with Convolutional Neural Networks (CNN) and LSTM, constructing two prediction models: WD-LSTM and CNN-LSTM. Their research demonstrated the applicability of combining WD, CNN, and LSTM in the context of forecasting (Ni et al., 2020). Previous studies have indicated that the Improved Complete Ensemble Empirical Mode Decomposition with Adaptive Noise (ICEEMDAN), compared to other signal processing techniques, effectively reduces residual noise and spurious components, further enhancing model performance (Colominas et al., 2014).

To explore the applicability of deep learning combined with signal processing techniques in runoff forecasting, we establish a flood season daily runoff prediction model named ICEEMDAN-VMD-CNN-LSTM, based on ICEEMDAN and VMD signal processing techniques as well as CNN and LSTM deep learning technologies. Firstly, ICEEMDAN is applied to decompose the original runoff time series, reducing data complexity. Subsequently, VMD is used to further decompose the highest-frequency and most complex component, lowering complexity and extracting hidden information. Finally, CNN-LSTM is employed to model and predict the components, effectively extracting local features of the data and capturing long-term dependencies in the components. Validation and analysis were conducted using measured daily runoff data from the Hekou station in the Diaojiang basin, confirming the model's excellent performance in flow prediction, especially in predicting peak flow values. This validates the applicability of deep learning combined with signal processing techniques in runoff forecasting.

2 Methodologies

2.1 Improved Complete Ensemble Empirical Mode Decomposition for Adaptive Noise (ICEEMDAN)

ICEEMDAN is an improvement upon the CEEMDAN algorithm, where white noise is added to each component based on the components obtained after EMD decomposition. In each iteration, the new IMF is obtained by calculating the difference between the residual from the previous step and the white noise signal calculated in the current step, as well as the difference from the average of the residuals from the previous step (Colominas et al., 2014). The combination of these two improvement methods effectively addresses the issues of spurious components and residual noise in CEEMDAN decomposition. For detailed calculation steps, please refer to the literature (Colominas et al., 2014).

2.2 Variational Mode Decomposition (VMD)

Unlike EMD, VMD can determine the number of decompositions of a time series based on actual conditions. It divides the original signal into k components (referred to as VMF), each with a limited bandwidth and centered frequency. Additionally, it minimizes the sum of the estimated bandwidths of each component. The calculation process of VMD is quite complex, and the specific calculation methods can be found in (Dragomiretskiy and Zosso, 2014).

2.3 Long Short-Term Memory (LSTM) Neural Networks

LSTM is a special type of recurrent neural network designed to tackle the problems of gradient explosion and vanishing gradients that often arise during the training of long time series (Hu et al., 2020).

The structure of an LSTM unit mainly consists of the input gate i_t, the forgetting gate f_t, the output gate O_t, the cell state C_t and the hidden state h_t. The formula can be updated as follows:

$$f_t = \sigma(W_f \cdot h_{t-1} + W_f \cdot x_t + b_f) \tag{1}$$

$$i_t = \sigma(W_i \cdot h_{t-1} + W_i \cdot x_t + b_i) \tag{2}$$

$$O_t = \sigma(W_o \cdot x_t + W_o \cdot h_{t-1} + b_o) \tag{3}$$

where W is the weight matrix of different cell gates, b is the corresponding bias vector, h_{t-1} is the hidden state at moment t−1, x_t is the current input, $\sigma(.)$ represents the logistic sigmoid function.

2.4 Convolutional Neural Network (CNN)

CNN is a type of neural network that can extract data features, mainly composed of convolutional layers, pooling layers, and fully connected layers. The convolutional layer is the core component of CNN, which performs convolution operations with input data using convolutional kernels to extract data features. The pooling layer comes after the convolutional layer and is used to reduce the output of the convolutional layer while preserving important features, thereby performing secondary feature extraction. The fully connected layer is located after the pooling layer, where the pooled data is flattened into one-dimensional form for subsequent classification or regression tasks. These layers of CNN can be stacked in the network to more effectively learn and extract features from input data.

2.5 ICEEMDAN-VMD-CNN-LSTM Model Construction

This paper presents the ICEEMDAN-VMD-CNN-LSTM (I-V-C-L) flood season daily runoff prediction model that integrates ICEEMDAN and VMD signal processing techniques with CNN and LSTM neural networks. The detailed steps are as follows:

1. Apply ICEEMDAN to decompose the processed runoff data into several IMFs.
2. Apply VMD to decompose the most complex IMF, resulting in several VMFs and a residual sequence (Res).
3. Perform data normalization on each component.
4. Split the dataset into training and testing sets. Train the CNN-LSTM model based on the training set and make predictions on the testing set.
5. Aggregate the predictions of all components obtained in step (4) and obtain the predicted results of the original runoff time series.

2.6 Accuracy Verification

Evaluate the model performance using the Nash-Sutcliffe efficiency coefficient (NSE) and mean absolute error (MAE). The calculation formula for each indicator is as follows.:

$$NSE = 1 - \frac{\sum_{i=1}^{n} (y_{obs} - y_{pred})^2}{\sum_{i=1}^{n} (y_{obs} - \overline{y_{obs}})^2} \tag{4}$$

$$MAE = \frac{1}{n} \sum_{i=1}^{n} |y_{pred} - y_{obs}| \tag{5}$$

where y_{obs} represents the observed value, y_{pred} represents the predicted value, $\overline{y_{obs}}$ denotes the average of observed value, n is the number of observations.

3 Results and Discussion

The daily runoff time series from the Hekou station for the flood seasons from 2006 to 2017 was selected for the experiment. Located in Jiuxu Town, Hechi City, Guangxi, the Hekou Station has a catchment area of 1044 km^2. It is a representative station in the

Diaojiang basin and a national principal hydrometric station, playing an important role in flood forecasting and warning. Due to the relatively small and less volatile daily runoff during the non-flood seasons (January-May and October-December), and the larger and more volatile daily runoff during the flood season (June-September), which is a critical period for flood management, the predictions focus solely on the runoff from June to September.

3.1 Runoff Time Series Decomposition

Apply ICEEMDAN to decompose the original runoff time series, with the decomposition results shown in Fig. 1. As observed in Fig. 1, the complexity and volatility of the time series from IMF1 to IMF9 gradually decrease, while the Res is very small and can be ignored. It is evident that IMF1 is the component with the highest complexity. Therefore, VMD is used to further decompose IMF1 in order to reduce its complexity.

3.2 IMF1 Decomposition

Among various VMD parameters, the number of decomposition level k is crucial. Extracting too few VMFs can lead to poor signal component extraction from the original time series, while extracting too many VMFs can cause duplicate information to be included among signal components. After multiple experiments, it was found that when $k = 8$ with other parameters set to default values, the decomposition effect is better. VMD is applied to decompose IMF1 into 8 VMFs and 1 Res, as shown in Fig. 2.

3.3 Model Testing Results and Comparison

Through experiments, it was determined that the CNN-LSTM model consists of 1 input layer, 2 convolutional-pooling layers, 2 LSTM layers, and 2 fully connected layers. The time step is set to 12, and the number of training iterations depends on the specific characteristics of the components. The dataset is divided into a training set comprising the first 80% of the data and a testing set comprising the remaining 20%.

In the same experimental setup, this study also includes five comparative models: LSTM, CNN-LSTM (C-L), VMD-CNN-LSTM (V-C-L), ICEEMDAN-CNN-LSTM (I-C-L) and VMD-ICEEMDAN-CNN-LSTM (V-I-C-L). LSTM and C-L do not decompose the original time series; V-C-L decomposes the original time series using VMD; I-C-L decomposes the original time series using ICEEMDAN; V-I-C-L decomposes the original time series using VMD and further decomposes the residual of VMD using ICEEMDAN. The testing set predictions for each model are shown in Fig. 3 (Obs represents the original runoff time series), and the performance evaluation indicators can be found in Table 1.

Table 1 demonstrates that compared to other models, I-V-C-L exhibits superior metrics on the testing set, indicating the strongest overall performance among the models. Compared to LSTM and C-L, I-V-C-L shows an increase in NSE of 59.24% to 66.14% and a decrease in MAE of 60.18% to 64.67%. Compared to I-C-L and V-C-L, I-V-C-L demonstrates an increase in NSE of 8.28% to 13.55% and a decrease in MAE of 37.41%

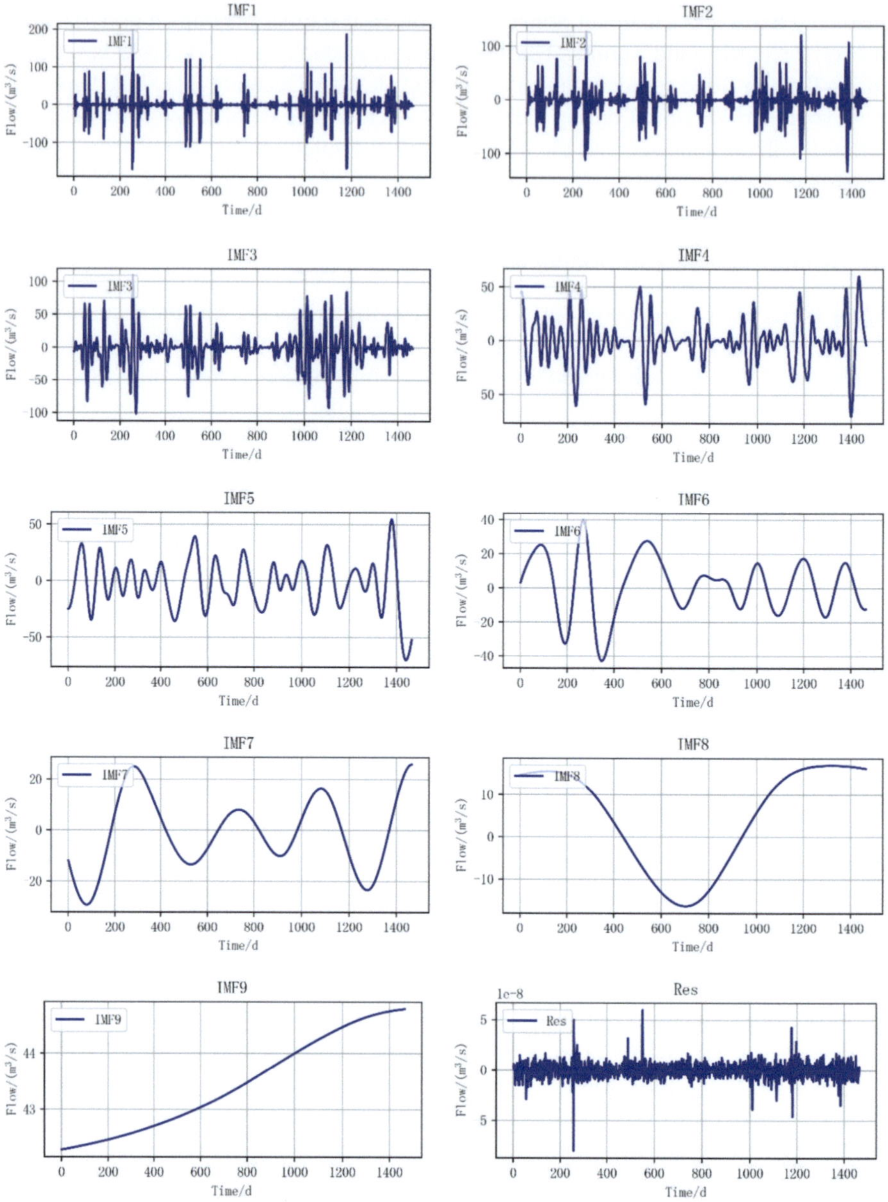

Fig. 1. Results of ICEEMDAN decomposition

to 66.27%. In comparison to the second-order decomposition model V-I-C-L, I-V-C-L shows an increase in NSE of 1.13% and a decrease in MAE of 29.25%.

A comprehensive comparison of the prediction results among the models reveals the following insights:

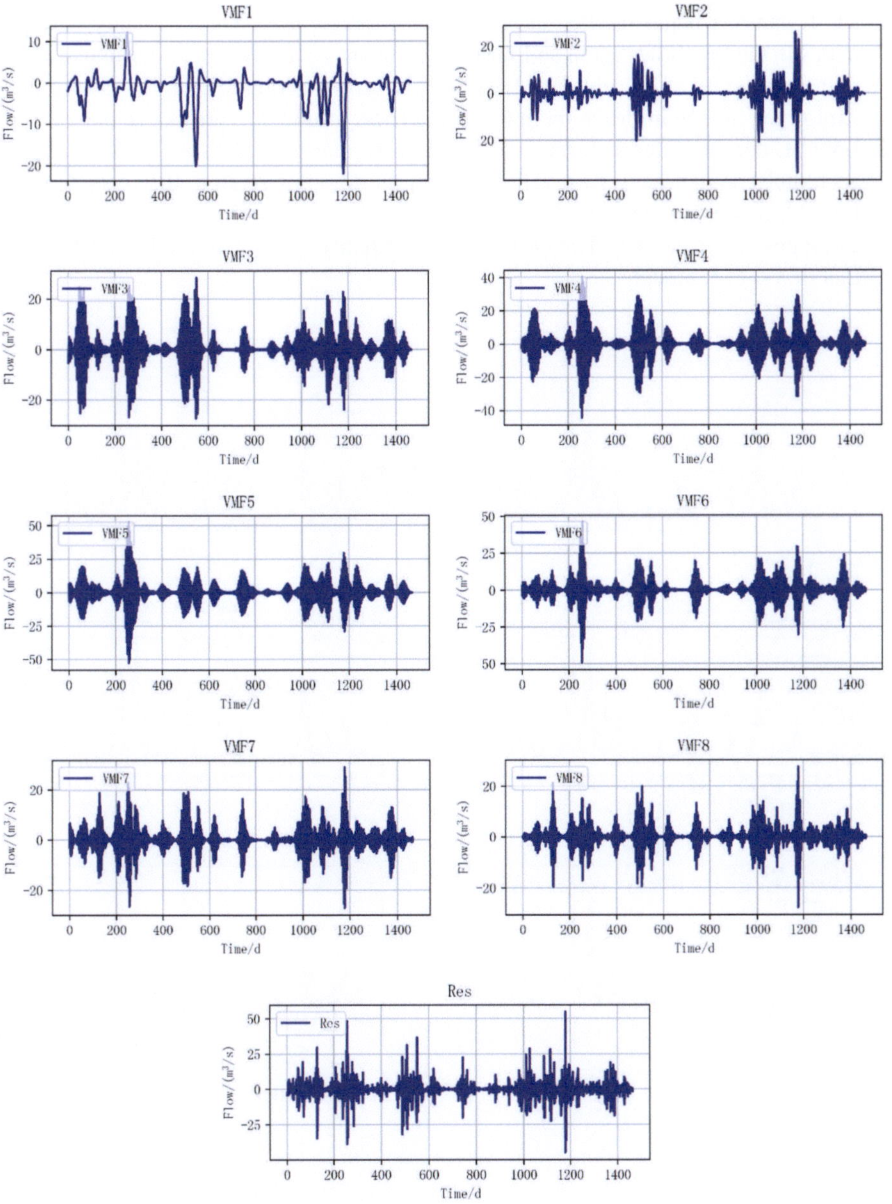

Fig. 2. Results of VMD decomposition

1. LSTM outperforms C-L in prediction performance. This may be due to the fact that when the time series is not sufficiently decomposed, C-L is influenced by the complexity of the time series, unable to fully learn the fluctuation interval information of the time series, leading to lower prediction accuracy compared to LSTM.

2. I-C-L exhibits better prediction performance than V-C-L. V-C-L notably overestimates in the low-flow region and shows significant volatility. This is attributed to the high complexity of Res obtained from VMD decomposition, where C-L fails to fully exploit the hidden information in the time series, resulting in prediction biases.
3. I-V-C-L demonstrates superior prediction performance over V-I-C-L. Both models accurately predict peak flow values, but in the low-flow region, the predictions of I-V-C-L model are closer to the observed values, while V-I-C-L model slightly overestimates in this interval. Therefore, the performance of I-V-C-L model is better than that of V-I-C-L model.

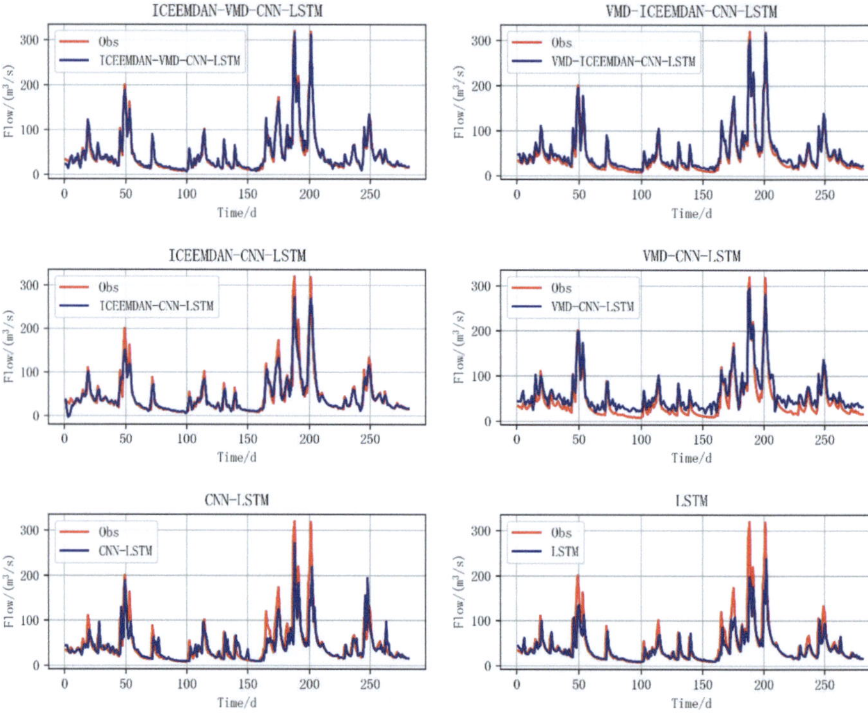

Fig. 3. Prediction results of each model

4 Conclusions

This study investigates the applicability of deep learning networks integrated with signal processing techniques in runoff forecasting. To this end, we propose the ICEEMDAN-VMD-CNN-LSTM hybrid model for daily runoff prediction during the flood season. The model is validated using daily runoff data from the estuary station during the flood season, with MAE and NSE selected as the evaluation metrics. Additionally, five comparative models, such as LSTM, are constructed for performance evaluation. The results indicate

Table 1. Performance evaluation indicators for each model

Models	Training set		Testing set	
	MAE(m^3/s)	NSE	MAE(m^3/s)	NSE
LSTM	10.058	0.551	13.136	0.613
CNN-LSTM	7.447	0.749	14.808	0.588
VMD-CNN-LSTM	9.314	0.921	15.512	0.860
ICEEMDAN-CNN-LSTM	8.110	0.888	8.358	0.902
VMD-ICEEMDAN-CNN-LSTM	9.008	0.926	7.395	0.966
ICEEMDAN-VMD-CNN-LSTM	3.118	0.991	5.232	0.977

that the ICEEMDAN-VMD-CNN-LSTM model performs well in runoff forecasting, with a testing MAE of 5.232 m^3/s and an NSE of 0.977. The main conclusions are as follows:

1. The introduction of two signal processing techniques effectively extracts hidden information from the runoff time series, reduces time series complexity, and significantly improves prediction accuracy. Compared to other models, the ICEEMDAN-VMD-CNN-LSTM shows markedly enhanced prediction accuracy for both low and peak flow values, and the prediction curve is more stable.
2. The prediction accuracy of the ICEEMDAN-VMD-CNN-LSTM is superior to that of the VMD-ICEEMDAN-CNN-LSTM. This is because the complexity of the Res from VMD decomposition is too high for ICEEMDAN to effectively reduce. This indicates that the order of decomposition methods applied to the time series is also crucial, significantly impacting the model's performance.

The ICEEMDAN-VMD-CNN-LSTM model demonstrates the applicability of deep learning integrated with signal processing techniques in runoff forecasting, showcasing excellent prediction accuracy. However, this model pays less attention to the parameters of VMD decomposition, and as a result, the decomposition effectiveness may not reach its optimal state. Therefore, the next step in our work plan is to introduce optimization algorithms to optimize the parameters of VMD decomposition and the hyperparameters of the CNN-LSTM model, aiming to further enhance the model's predictive capabilities.

References

Aqil, M., Kita, I., Yano, A., et al.: A comparative study of artificial neural networks and neuro-fuzzy in continuous modeling of the daily and hourly behaviour of runoff. J. Hydrol. **337**(1–2), 22–34 (2007)

Colominas, M.A., Schlotthauer, G., Torres, M.E.: Improved complete ensemble EMD: A suitable tool for biomedical signal processing. Biomed. Signal Process. Control **14**, 19–29 (2014)

Deb, P., Kiem, A.S., Willgoose, G.: Mechanisms influencing non-stationarity in rainfall-runoff relationships in southeast Australia. J. Hydrol. **571**, 749–764 (2019)

Dragomiretskiy, K., Zosso, D.: Variational Mode Decomposition. IEEE Trans. Signal Process. **62**(3), 531–544 (2014)

Feng, Z., Niu, W., Tang, Z., et al.: Monthly runoff time series prediction by variational mode decomposition and support vector machine based on quantum-behaved particle swarm optimization. J. Hydrol. **583**, 124627 (2020)
Hu, Q., Cao, S., Yang, H., et al.: Daily runoff predication using LSTM at the Ankang Station, Hanjing River. Prog. Geogr. **39**(4), 636–642 (2020)
Lei, X.H., Wang, H., Liao, W.H., et al.: Advances in hydro-meteorological forecast under changing environment. J. Hydraul. Eng. **49**(1), 9–18 (2018)
Li, F., Ma, G., Chen, S., et al.: An Ensemble modeling approach to forecast daily reservoir inflow using bidirectional long- and short-term memory (Bi-LSTM), variational mode decomposition (VMD), and energy entropy method. Water Resour. Manage **35**(9), 2941–2963 (2021)
Lu, H., Ma, X.: Hybrid decision tree-based machine learning models for short-term water quality prediction. Chemosphere **249**, 126169 (2020)
Ni, L., Wang, D., Singh, V.P., et al.: Streamflow and rainfall forecasting by two long short-term memory-based models. J. Hydrol. **583**, 124296 (2020)
Shen, C.: A transdisciplinary review of deep learning research and its relevance for water resources scientists. Water Resour. Res. **54**(11), 8558–8593 (2018)
Wang, Y., Yeh, C., Young, H.W.V., et al.: On the computational complexity of the empirical mode decomposition algorithm. Physica A: Statistical Mechanics and its Applications **400**, 159–167 (2014)
Yaseen, Z.M., Jaafar, O., Deo, R.C., et al.: Stream-flow forecasting using extreme learning machines: A case study in a semi-arid region in Iraq. J. Hydrol. **542**, 603–614 (2016)
Yeh, J.R., Shieh, J.S., Huang, N.E.: Complementary ensemble empirical mode decomposition: a novel noise enhanced data analysis method. Adv. Adapt. Data Anal. **2**(2), 135–156 (2010)
Zheng, X., Chen, J., Hu, Z., et al.: Runoff Prediction in the Middle and Upper Reaches of Yuanjiang River Based on EEMD-PSO-GRNN Model. Journal of China Hydrology **43**(4), 96–103 (2023)

Research and Application of Balanced Rise of Concrete High Arch Dam

Zhang Junhong(✉)

Sinohydro Engineering Bureau 4 Co., Ltd, Kunming City, Yunnan Province, China
1094538144@qq.com

Abstract. In the construction of domestic concrete high arch dams, there have been varying degrees of situations where the height difference between adjacent dam sections and the entire dam exceeds the design indicators. Due to the complex structure, large scale, high temperature control requirements for the dam body concrete, strict control requirements for the height difference between adjacent dam sections and the entire dam body, and complex boundary conditions, the rising speed of the bank slope dam section, non-orifice dam section, and orifice dam section is not consistent during the construction process, resulting in uneven rise of the dam concrete construction. In response to this situation, the resources need to be fully considered according to the simulated schedule of dam concrete construction. In this research, the finite element, PKPM and other software were used to study technical measures for rapid rise of the bank slope dam section and orifice dam section, and achieved the goal of balanced rise of the Baihetan concrete arch dam.

Keywords: Concrete high arch dam · Synchronous rise · Research application

1 Introduction

Due to the complex structure, complex construction boundary conditions, high temperature control requirements for dam concrete, strict control requirements for adjacent dam sections and the entire dam height difference, various factors combined make it difficult to organize on-site construction production and increase construction costs. It required the owner to study forward-looking technical measures before pouring the dam concrete, and required the constructor to use mature technical means to study practical and feasible technical measures before and during construction.

2 Overview

2.1 Project Overview

The total concrete volume of the concrete double curvature arch dam of Baihetan Hydropower Station is about 8.1 million cubic meters, with a dam crest elevation of 834.00m and a maximum dam height of 289.0m. The dam body is equipped with 6 spillway surface holes, 7 spillway deep holes, and 6 spillway bottom holes. The arc length

S. Zheng et al. (Eds.): IHDC 2024, LNCE 487, pp. 115–136, 2025.
https://doi.org/10.1007/978-981-97-9184-2_11

of the dam crest axis is 709m, with a total of 30 transverse joints divided into 31 dam sections. The spacing between the transverse joints along the upstream dam surface arc length is 20.0–24.2m, with a maximum of 24.2m.

2.2 Domestic Research Level

The bottom hole is mainly used for high-quality pouring of the channel bottom plate to ensure the flatness requirements of the bottom plate; The side walls are constructed quickly using large formwork, while the top plate of the flow channel is constructed using cross disc fasteners, scaffolding pipes, embedded positioning cones, and truss structures. The brackets are constructed usingmethods such as internal pulling and external bracing + combined steel formwork. Quick installation of deep hole steel lining, etc. The surface hole gate pier adopts standardized steel formwork for rapid preparation and pouring, the overflow surface concrete adopts sliding mode for rapid and high-quality construction, and the surface hole support beam adopts pre-embedded positioning cone + truss, support column + truss, pre-embedded positioning cone/steel section + bailey frame support structure for rapid formation. The bank slope dam section adopts the method of low rise layer(3m) + internal tension and internal support, which results in a large workload and difficulty in clearing the foundation surface, small segment area, and slow preparation speed.

3 Practical Basis for Research

Using a schedule simulation plan to analyze the construction progress of the dam concrete, analyze the construction time and difficulties faced by each part of the segment, mainly the slow rise of the bank slope dam section and the orifice dam section, and the fast rise of the other non-orifice dam sections, resulting in uneven rise of the dam concrete.

Using finite element, PKPM and other software to design high lift formwork support, deep hole support beams, inverted "T" beams for surface holes, truss prefabricated formwork, hydraulic self-climbing formwork, steel substrate support system, gate slot high lifting truck, etc. For the slope dam section, a comprehensive and detailed analysis of each construction link of bottom holes, deep holes, and surface holes is carried out, refining each process, resource, and implementation situation, providing theoretical, technical, and management support for the balanced rise of concrete high arch dams.

4 Research Content

4.1 Research on Rapid Construction Technology of Diversion Bottom Hole

1) Formwork

The upstream bracket adopts a truss style prefabricated formwork, which is easy to install and does not require dismantling.

The downstream brackets are cast in situ, making it difficult to install and remove the formwork. The basic steps for installing the formwork are as follows:

1. Pre-embedded on the segment surface #16 channel steel columns and anchor bars, welded channel steel columns and anchor bars with a diameter of 20mm pull rod;
2. Loosen the bolts at the positioning cone of the formwork, and use a segment mounted truck crane to lift the formwork to the installation elevation with a special "C" - shaped beam;
3. Adjust the position of the template bottom with a 5t chain block, connect the formwork back frame to the lower formwork back frame, and use the rotating kit on the back frame to adjust the angle of the formwork, so that the slope of the formwork surface is approximately the same as that of the inclined surface of the suspended body;
4. Install the formwork steel pipe purlin, weld the tension rod between the formwork and the channel steel reinforcement column, and adjust the tension rod to meet the installation accuracy requirements of the formwork before embedding the panel positioning cone.

2) Concrete pouring

The allowable interval time for the concrete covering of the upper and lower layers should be controlled within4 hours. When unloading from the cable crane material tank, the free fall height of concrete shall not exceed 1.5m, and when there is a horizontal steel mesh, it shall not exceed 1.0m. A safe distance of 3m shall be maintained between the cable crane material tank and the formwork, and the distance between the edge of the material pile and the formwork shall not be less than50cm. The unloading of materials on both sides of corridors, openings, and other areas should be synchronized and evenly raised. Special parts such as formworks, waterproofing (slurry) sheets, thermometers, corners, and densely reinforced areas shall be manually fed. It is strictly prohibited to directly feed at the waterproofing, cooling upper pipes, thermometers, or the boundary between vibrated and non-vibrated areas. Please refer to Fig. 1 for details.

Due to the fact that the anti-abrasion concrete around the bottom hole channel is only 60cm thick, when pouring each layer, the ordinary concrete on both sides is first cut and leveled, and then the anti-abrasion concrete is cut to a position 3m away from the formwork, and the leveling machine is pushed to the anti-abrasion concrete area.

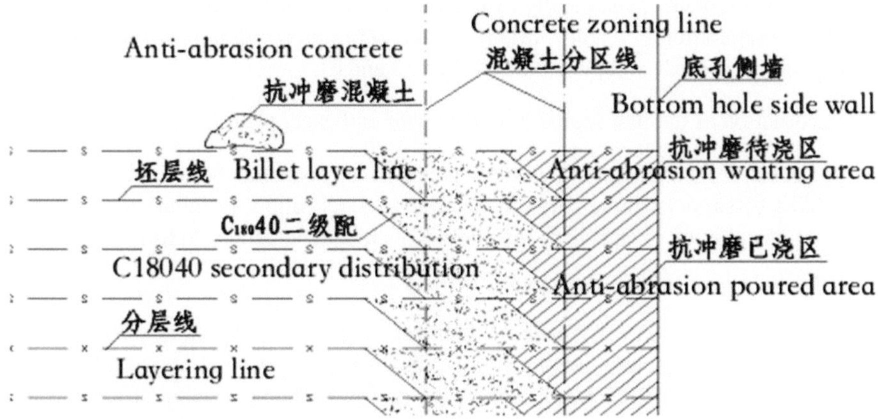

Fig. 1. Pouring Method of Anti-abrasion Concrete on Both Sides of Bottom Hole Flow Channel

4.2 Research on Rapid Construction Technology for Deep Discharge Holes of High Arch Dams

#1–7 flood discharge deep holes are—the upward curved (or downward curved) type with pressure discharge holes. The inlet is trumpet shaped, with elliptical curves on the top and sides, circular curves on the bottom, and symmetrical bottom sills. The inlet is equipped with emergency gate slots and maintenance gate slots, and both inlet and outlet are inverted cow leg structures. The deep hole body has a rectangular cross-section with dimensions of 4.8 × 12.0m (upstream section) and 5.5 × 12.0m (downstream section). The entire cross-section of the hole body is protected by steel lining, and the size of the deep hole opening is 5.5 × 8.0m. The elevation of the outlet bottom sill is724.00m, and the outlet is equipped with a working arc gate slot and arc gate support hinge beam. At present, the construction effect of similar projects in China is relatively poor, and the controllability of construction progress and quality is poor. By relying on the Baihetan Dam to conduct rapid construction technology research on high arch dam flood discharge deep holes, the node goal of "passing through deep holes in 100 days" has been achieved.

1) Segment pouring preparation design

The main content of segment pouring design includes: (1) Segment features, segment elevation, materials and corresponding quantities of each part, and configuration of segment surface diagram; (2) Features of quantity. Fill in the Quantity Form based on the specific segment with respective quantity involved; (3) Segment preparation process. Based on the process, combined with work efficiency, work constraints, and intermittent requirements, prepare a segment preparation work plan, including: logical relationship of operations, start and end time, work team and number of people, equipment, etc. (4) Material entry and exit segment plan, including material types, lifting and stacking locations, entry and exit quantities, methods, time, etc. (5) The main concerns forsegment pouring preparation includes technology, quality, production, acceptance, and protection. (6) The key points of quality control for segment preparation include templates, steel bars, and embedded parts. (7) Safety, environmental protection, and occupational health. (8) Evaluation on segment preparation. Comparing the design and implementation of the segment, dynamically evaluating the segment preparation progress, analyzing the preparation efficiency, evaluating the entire preparation process, and giving corresponding rewards and punishments (Figs. 2 and 3).

2) Construction of the first segment of deep hole with steel plate lining bottom

The construction content of the first segment of the deep hole bottom plate includes: embedding of support frame parts, installation of steel lining, installation of grouting pipelines, installation of steel bars, installation of embedded parts, and formwork lifting.

(1) Deep hole steel lining construction

The steel lining plate is made of stainless steel composite steel plate, and the cladding material (concrete flowing surface) is austenitic ferritic duplex stainless steel 022Cr23Ni5Mo3N (S22053) with a thickness of 4mm. The base material is hot-rolled low-alloy high-strength structural steel Q345C with a thickness of 20mm; The outer rib plate of the steel lining and ventilation hole is Q345C, with a plate thickness of

Fig. 2. Pre-shift Segment Design Disclosure

Fig. 3. Post-shift Progress Briefing Meeting

20mm; The external anchor bars of the steel lining are C36mm steel bars with a length of 1500mm.

1. Steel lining stack section planning

According to the steel lining design blueprint and the lifting requirements of two cable cranes, two stacked sections will be completed in the factory. If the lifting and transportation requirements are not met, single section lifting will be adopted. The steel lining stacked sections are shown in Fig. 4.

Fig. 4. Steel lining stack section

2. Installation of steel lining
A. Overall planning of installation sequence for steel lining
a. According to the simulation plan for the installation of deep hole steel lining in the dam, the installation of 3# deep hole steel lining will begin first, and finally the installation of 2# deep hole steel lining will be completed.
b. According to the structural form and characteristics of the steel lining for deep holes 1# to 7# of Baihetan Hydropower Station, the installation of deep holes 1 #, 2#, and 3# will be completed in one time;
c. The steel lining of the upstream overhanging section of the 4 # deep hole is installed in two stages due to its maximum overhanging height of 6m. The horizontal section is installed during the interval between the first layer of concrete pouring, and the overhanging section is installed after the first layer of concrete pouring.
d. The downstream section of the 5# steel lining hole has a relatively high upward protrusion, resulting in a significant height difference during one-time installation. It will be installed in two stages; The 6# and 7# steel lining hole sections are basically horizontally constructed and can be installed in one time.
B. Installation process of steel lining

When the main concrete of the dam section where the steel lining is located is poured to a distance of about 2 m from the steel lining, embedded steel lining support rails and reinforcement required embedded parts such as steel bars and iron stools are embedded; After the concrete pouring stopped, install the steel lining support track and set up relevant control point lines in a timely manner, and use a double machine lifting method to lift the steel lining into place; Finally, the reinforcement, joint pressing, welding, and anti-corrosion repair of the steel lining are completed. After the steel lining is accepted as qualified, the main concrete pouring is carried out in a timely manner. The steel lining supports the track as shown in Fig. 5.

a. Support track for steel lining installation

When installing the steel lining, use a single cable crane or two cable cranes to lift and transport it to the installation hole position on the I-beam track. Based on the support steel frame and track, the steel lining is accurately positioned using jacks and chain hoists. Follow the sequence of adjusting the gap between track sections (mileage adjustment) → adjusting in the left and right directions (relative to the direction of water flow) → adjusting elevation → comprehensive adjustment.

Fig. 5. Installation of supporting track

b. Measurement and layout of benchmark control points and lines for installation

Set control sample points at the projection point position of the lower center point of each pipe end to control the installation position of the steel lining, and establish elevation control points to control the elevation of the steel lining.

c. Installation of initial section

The initial installation section of the hole body is located in the middle, and it extends from the initial section to the upstream and downstream ends for section by section, as shown in Fig. 6.

During installation, adjust the center first, use a jack to adjust the steel lining, and monitor it using a plumb bob method to align the projection point of the lower center point of the steel lining with the embedded control point, and adjust the steel lining to the required elevation. After passing the inspection, wedge iron shall be inserted between the steel lining support legs and the track gap, and the center, elevation, and mileage shall be rechecked and adjusted. After reinforcement, conduct another inspection of the center, elevation, and mileage.

Fig. 6. Adjustment of the initial installation section in place

d. Installation of steel lining on the side of the door slot

During the installation of the embedded parts of accident maintenance door slot of the deep hole, the steel lining on the side of the door slot is installed. The steel lining on both sides is lifted separately, and the steel lining on one side is lifted in two layers. The internal support is provided by a high lifting truck, and the external support is constructed according to the reinforcement form of the inlet section.

1) Construction of gate piers and hinged beams at the exit section

The structure around the flood discharge deep hole is complex, with dense steel reinforcement, multiple embedded parts, multiple construction processes, narrow segment surface, high quality requirements, and a maximum overhanging cantilever of 48m for the gate pier. It is equipped with a circular pre-stressed anchor cable, and the installation of metal structures is large and intersecting with concrete pouring. At the same time, during the stage of passing through the orifice, the number of lifting hooks by the cable crane increases, and the storage period is extended, which affects the overall lifting speed of the dam.

The traditional support system construction process is complex and occupies the straight-line construction period of the segment, resulting in slow construction progress. In order to quickly construct the dam section passing through the orifice, the construction technology of synchronously pouring and lifting the hinged main beam and the dam body is adopted, relying on the combination of embedded steel trusses and steel lining at the bottom of the main beam as the bottom formwork for the construction of the main beam, making the concrete construction progress controllable.

2) Operation points
1. Simultaneous pouring and layering

The layering of the hinged main beam should be kept as consistent as possible with the overall layering elevation of the dam body, while taking into account the strength deformation and overall stability of the steel truss embedded in the main beam, the bottom steel lining, and the support frame of the two sides of the gate pier to meet the requirements of the specifications.

2. Support frame and truss system

The steel lined bottom plate of the hinged beam and the upstream side plate are assembled as a whole outside the site. The steel lining embedded truss, steel lining, anchor cable sleeve, and steel reinforcement are welded as a whole outside the segment, and lifted into place by a cable crane double machine, as shown in Fig. 7.

Fig. 7. Overall assembly of steel lining and embedded truss

3) Formwork construction
1. The bottom and upstream side of the first phase concrete of the hinged beam are lined with steel and serve as formwork. The second phase concrete section is made of standardized wooden trusses and formwork, as shown in Fig. 8.
2. The formwork is lifted by a cable crane to the working area, and then lifted and installed on the segment surface.
3. The strength of the top layer of concrete in the bracket area should be $\geq$ 20MPa. After pouring to the upper layer, there should be an interval of 3days, and the internal and external support formwork can be removed.

4) Steel lining construction

According to the reserved concrete embedded parts, a steel rack is erected for the installation of the bottom steel lining, which is used to connect and reinforce the steel lining. During the pouring of concrete corresponding to the bottom steel lining, embedded parts are buried for the installation and reinforcement of the previous layer of steel lining.

Fig. 8. Installation of bottom formwork for supporting beams

4.3 Research on Direct Burial Construction Technology for Gate Slot Phase I

1) Technical principles

The concept of "using high lifting vehicles to achieve direct burial construction of gate slots in the first phase of deep hole inclined gate slots for high arch dams" connects the embedded parts into a whole through trusses in narrow spaces, relying on a joint force system to solve the problems of supporting and fixing the embedded parts. At the same time, through specialized lifting design, suspension construction and self climbing are achieved, releasing the space under the trolley. Simplify the construction process, accelerate the construction progress, improve the accuracy of gate slot installation, enhance safety, and improve the construction quality of deep hole inclined gate slots for high arch dams.

2) Operation points

(1) Installation and pouring of bottom sill

Before pouring concrete, pre-embed channel steel within the range of the bottom sill, weld the support bracket of the bottom sill with channel steel and angle steel, and then lift the bottom sill onto the bracket. Adjust it with a jack and flower basket bolt, with the main control points being the deviation of the centerline of the door slot and the centerline of the hole opening, the elevation and flatness of the bottom sill. Before installing the bottom sill, set up an independent sample rack and adjust the bottom sill based on the centerline of the layout.

(2) Main and reverse track installation measurement control

The installation of the main and reverse tracks is mainly controlled by the flatness and distortion of the working surface of the main track. The first section of the main and reverse rails is based on the permanent hole centerline and gate groove centerline set on the bottom sill, and the installation lines of the main and reverse rails on the left and right sides are drawn as the bottom installation reference. Then, the top installation position is determined by using a line cone and oil drum (deviating from a specific size). At the same time, independent sample racks are set up at the corresponding positions of each layer's main and reverse tracks, completely separated from the high lifting vehicle for door slot. The absolute coordinates set by measurement and layout are used as the size composite of the top installation reference.

(3) Door lintel installation measurement control

The installation of the lintel is mainly controlled by controlling the flatness of the water-stop surface and the misalignment between the lintel and the main rail water-stop surface. Based on the installed main rail water seal seat surface, adjust the water-stop surface of the door lintel, and set up an independent sample frame at the door lintel position. The sample frame should be completely separated from the high lifting vehicle for door slot, and the installation coordinates set by measurement and layout should be used as the composite size of the door lintel installation.

(4) Gate slot high lifting vehicle climbing

1. Before climbing the gate slot high lifting truck, the connecting components between the high lifting truck and the track need to be cut off. After removing the climbing obstacles, lift the high lifting truck as a whole by 100mm, stay for 20-30min, check or adjust locally, eliminate safety hazards, and then continue to lift.
2. After the high lifting truck climbs into place, it is welded and fixed to the poured door slot with channel steel. After the high lifting truck lifting is completed, use the top wall support and guide wheels to fasten the door slot high lifting truck, so as to safely fix the door slot high lifting truck in the already poured door slot.
3. The gate slot high lifting truck climbs and cycles until the installation of the gate slot embedded parts is completed. The gate slot high lifting truck is removed and the overall inspection and acceptance of the gate slot embedded parts are carried out.

4.4 Research on the Application of Hydraulic Self Climbing Formwork and Inverted Prefabricated Concrete Formwork for High Arch Dam Concrete Pouring

According to statistical analysis, the shortest impact time of strong winds above level 7 in winter on the Baihetan Hydropower Station from 2015 to 2016 was 23.6 days, and the longest was 44.9 days. In order to reduce the impact of strong winds on the concrete construction period of the dam, hydraulic self lifting formwork is used for the river bed dam section and cantilever formwork is used for the bank slope dam section.

4.4.1 Hydraulic Self Climbing Formwork

1) Ordinary single-sided cantilever formwork

The dam concrete adopts ordinary single-sided cantilever formwork, and internal tension rods need to be installed to resist the lateral pressure of the concrete. Its main disadvantages are as follows: firstly, due to the structural limitations of the formwork itself, a single formwork should not be too large, and there are many concrete joints, making it difficult to control the quality of the joints; Secondly, there are many safety hazards during the installation and dismantling process, and the de-moulding and lifting process is easily affected by strong winds. The swinging of the formwork can easily collide with concrete or equipment; Thirdly, it is necessary to set up internal pull rods and occupy lifting equipment, which increases the investment in labor, materials, and equipment costs; Fourthly, the construction process is easily constrained by the engineering structure, and complex structural shapes cannot be used; The fifth issue is that the dismantling process takes a long time and the construction efficiency is low.

2) Composition of single-sided hydraulic self climbing formwork

The single-sided self climbing cantilever formwork combines the advantages of single-sided cantilever formwork and hydraulic self-climbing formwork in other engineering fields (civil engineering, bridges), solving the problems of lifting and safety of single-sided cantilever formwork, greatly improving construction efficiency and safety construction guarantee coefficient.

The concrete pouring upstream and downstream of Baihetan Dam mainly uses cantilever formwork, and the transverse joints mainly use spherical keyway formwork. The upstream and downstream cantilever formworks are 3.0 × 3.5m (width × height) and made of $\delta = 6$mmsteel plate; The transverse seam formwork is 3.0 × 3.3m (width × height), and the panel is made of $\delta = 6$mm steel plate. The spherical keyway formwork is assembled from a cantilever formwork and a spherical keyway. The keyway is pressed from a 3mm thick steel plate and has a hemispherical surface with a spherical diameter of 80cm and a depth of 20cm. The hydraulic climbing platform includes embedded parts, supports, guide rails, load-bearing frame, platform and enclosure, and lifting mechanism assembly, which can be assembled into hydraulic self lifting cantilever formwork and hydraulic self lifting spherical keyway formwork, as shown in Figs. 9 and 10.

3) Application of hydraulic self climbing formwork

The climbing platform can be made to carry multiple sets of cantilever formworks, achieving simultaneous climbing of multiple sets. The single or multiple sets of templates are shown in Fig. 11.

Fig. 9. Downstream hydraulic climbing formwork

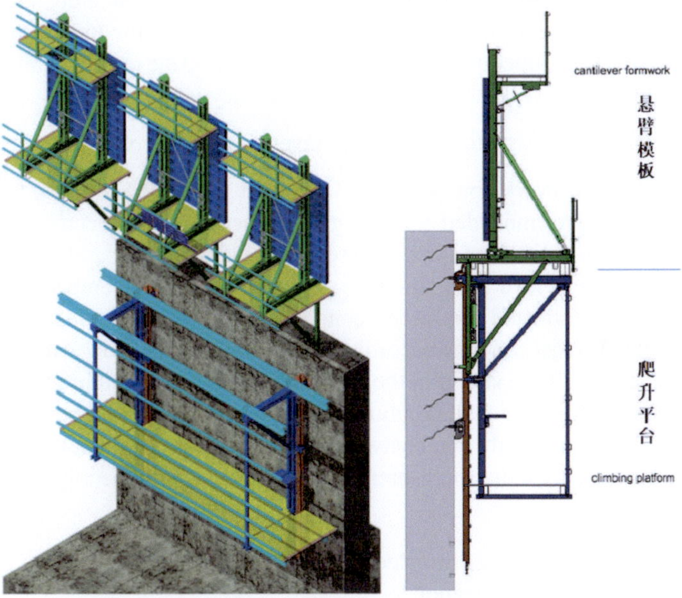

Fig. 10. Effect diagram of single-sidedhydraulic self climbing formwork

The self climbing formwork lifting mechanism alternately lifts the guide rail and climbing frame through hydraulic cylinders. It only takes 2-3people and 1–2 h to complete the climb for a 3m level.

1) Concrete prefabricated truss formwork

Large cantilever bracket structures are arranged upstream and downstream of the surface holes, deep holes, and diversion bottom holes of the dam body. Due to the high construction safety risks and difficulty in quality control of the formwork, steel bars, and

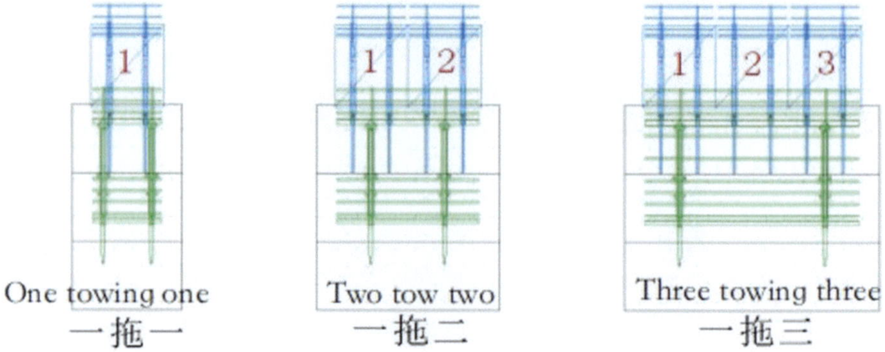

Fig. 11. Schematic diagram of single or multiple sets climbing simultaneously

concrete pouring in the inverted part of the dam, it is difficult to ensure the construction progress. The pre-cast concrete truss formwork support adopts an embedded structure, which avoids personnel from being exposed to high edges for long-term formwork installation and removal operations. At the same time, due to the modular and mechanized installation of the formwork, the construction efficiency is improved.

1) Composition and construction process of inverted precast concrete formwork

Prefabricated formwork consists of prefabricated reinforced concrete panels, internal support trusses, pre embedded anchoring components, and connectors, as detailed in Fig. 12. The prefabricated reinforced concrete panel is operated parallel to the internal support truss, and is welded to the internal support truss through embedded steel (I25a I-beam) in the prefabricated reinforced concrete panel; The embedded parts shall be embedded in the pre cast concrete according to the control precision. After the prefabricated parts reach the lifting strength, the formwork shall be hoisted in place as a whole, and then the formwork error shall be adjusted, and then the truss shall be welded with the support plate. To ensure a tight bond between the precast concrete panel and the concrete interface, it is necessary to use high-pressure water guns to roughen the interface at the prefabrication site.

2) Construction technology of inverted precast concrete formwork
(1) Formwork assembly verification

Prefabricated concrete formwork adopts fixed molds, and the mold design is an adjustable structure in the length direction. The side panels are opened in a translational manner; One end of the end plate is fixed, and the other end is hinged; One end can move in three sizes, positioned by holes on the side plate, and fixed by a reinforced positioning plate above to prevent loosening during concrete pouring.

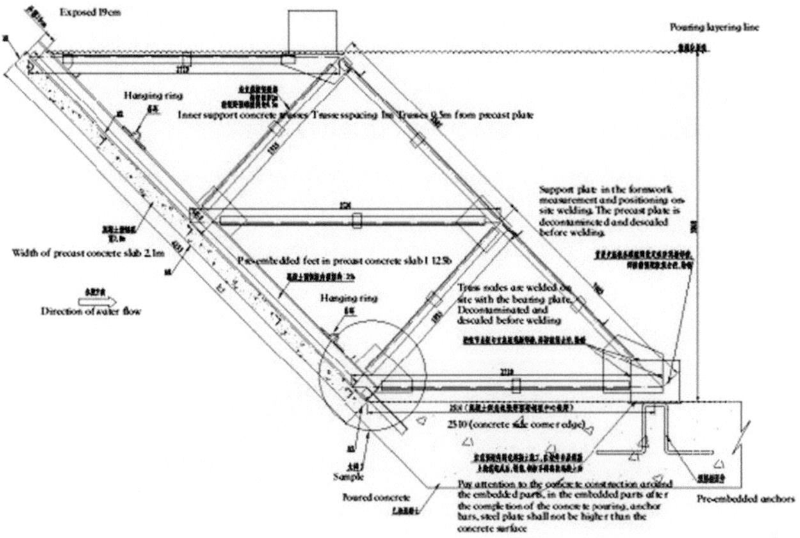

Fig. 12. Internal steel truss prefabricated formwork

After the formwork enters the site, trial assembly is carried out to ensure that the dimensions of each part of the formwork used for precast concrete components are accurate, the edges are straight, and the surface has good smoothness. Please refer to Figs. 13 and 14 for details. The manufacturing accuracy requirements for steel molds are shown in Table 1.

Table 1. Manufacturing accuracy requirements for steel molds

Steel mold width	± 5mm
Steel mold length	± 5mm
Height (thickness) of steel mold cavity	0~ + 2mm
Diagonal length difference of steel mold	<5mm

(2) Reinforcement installation

Before installing the steel bars and embedded parts, draw lines on the erection bars according to the spacing requirements, and manually bind or weld the processed steel bars and embedded partsaccording to the marked positions.After binding the steel bars and embedded parts, check the specifications, quantity, spacing, size, elevation, binding method, installation position, and protective layer thickness of the steel bars and embedded parts to ensure compliance with the design requirements, as shown in Fig. 15.

(3) Concrete pouring

1. Preparation before pouring

Fig. 13. Formwork assembly verification

Fig. 14. Formwork flatness verification

Fig. 15. Reinforcement in mold verification

Before pouring, a detailed safety technical briefing should be given to all operators, and a comprehensive inspection should be conducted on the stability of the formwork, steel bars, and embedded parts, as well as the machinery and equipment required for concrete mixing, transportation, and pouring.

After all the steel bars andembedded parts have been tied, installed accurately, and the formwork has been firmly installed, the team first conducts self inspection and then the quality inspection personnel conduct final inspection, fill out the steel bar and formwork quality inspection evaluation form, and report to the supervisor for acceptance; After obtaining approval from the supervisor and issuing the segment opening certificate.

2. Finish surface

After the concrete is collected, it is manually carried out with an iron trowel to finish the surface. The trowel should be smooth and even, without cracks or hollowing, and the surface of the concrete should be free of honeycombs, pockmarks, and other defects.

3. Demoulding, maintenance and protection

Requirements for demoulding: When dismantling the side formwork, remove each small formwork one by one, symmetrically on both sides, to avoid damage to the edges and corners of the prefabricated parts; When demolding prefabricated components, the concrete strength should reach at least 20Mpa. After demolding, the prefabricated components should be manually lifted to the curing area with the help of a crane for curing, and placed on square wooden pads.

After the concrete pouring is completed, watering and curing should be carried out in a timely manner. After the initial setting of the precast concrete components, rotary spraying curing should be carried out, and geotextile should be covered on the surface of the precast components to assist in moisture retention curing. The curing time is 28 days.

Concrete should be maintained and protected before final acceptance to prevent damage. Strengthen the protection of the edges and protruding parts of the pouring block. When the temperature drops sharply, the concrete is protected by surface insulation using 5cm thick polyethylene rolls as insulation material. Delay the demoulding time during low temperature seasons and periods of sudden temperature drops.

4. Internal support truss processing

The precision requirements for the production and installation of precast concrete formwork are high, and the size of precast concrete formwork is large, making on-site adjustment difficult. It is proposed to reinforce the inner side of the precast concrete formwork panel with a truss support structure, and only make minor adjustments during on-site precast formwork installation.

After the precast concrete panel is cured, the precast formwork support truss is welded and installed in the curing area.

5. On-site hoisting

Firstly, clean the upper openingof the prefabricated formwork, and then use a tower crane (cable crane) or 25t crane to lift the prefabricated concrete formwork to a height of about 5m above the installation position of the inverted bracket. Then, slowly lower it to the designated position, and the operator pulls the traction ropes pre-fixed on both sides of the prefabricated truss to position the prefabricated component. The on-site installation of prefabricated bracket formwork is shown in Fig. 16. After the prefabricated concrete formwork is initially in place, a 10t manual hoist is used to finely adjust the installation position of the prefabricated concrete formwork according to the measurement requirements. Afteradjustment, the node plate and support plate are welded and fixed.

Fig. 16. Installation of pre-fabricated bracket formwork on site

(3) Construction of T-shaped precast concrete formwork.

The damsurface hole beam adopts a gate pier joint, with a beam elevation of 820.0 ~ 834.0m, a length along the river of 15.291 ~ 18.133m, and a rectangular beam structure with a span across the river of 17.258 ~ 18.341m.

The surfaceholesupport hingebeam of Baihetan Damadopts an inverted "T"- shaped precast concrete beam as the bottom formwork, replacing the traditional load-bearing bottom formwork + support system. The installation and construction of precast beams are shown in Fig. 17, which can reduce the installation and production of cantilever support structures. Through off-site prefabrication and cable crane hoisting, it can greatly shorten the storage time and improve the safety of construction.

Fig. 17. Installation and construction of inverted "T" - shaped concrete prefabricated beam with table hole support and articulation

4.5 Research on Rapid Construction Technology ForBank Slope Dam Sections

1) Cause analysis

The shoulder groove of the high arch dam is steep, and the amount of excavation on the bedrock surface is large. The layout of construction equipment is difficult, and the area of the segment surface is small according to the 3m lifting layer, resulting in a small amount of work. As a result, the bank slope dam section cannot rise quickly, which affects the overall balanced rise of the dam. The concrete layering diagram of the left bank slope dam section is shown in Fig. 18.

2) Technical principles

When the distance between the bottom of the upstream dam surface and the upstream bedrock surface is less than3m, PKPM structural software is used to establish a model and conduct structural stress analysis. High rise construction is achieved by using internal and external support. Multiple segments are poured at once to reduce resource waste and time consumption caused by multiple segments.

3) Key points of construction technology

The bottom shape of the archdamprecursoris in the formofabroken line, and the formwork support below the broken line is supported by internal tension and internal support. The #12.6 I-beam diagonal support inside the segment ensures the stability of the formwork support below the broken line before pouring, and the #12.6 I-beam diagonal support outside the segment ensures the compressive stability of the concrete

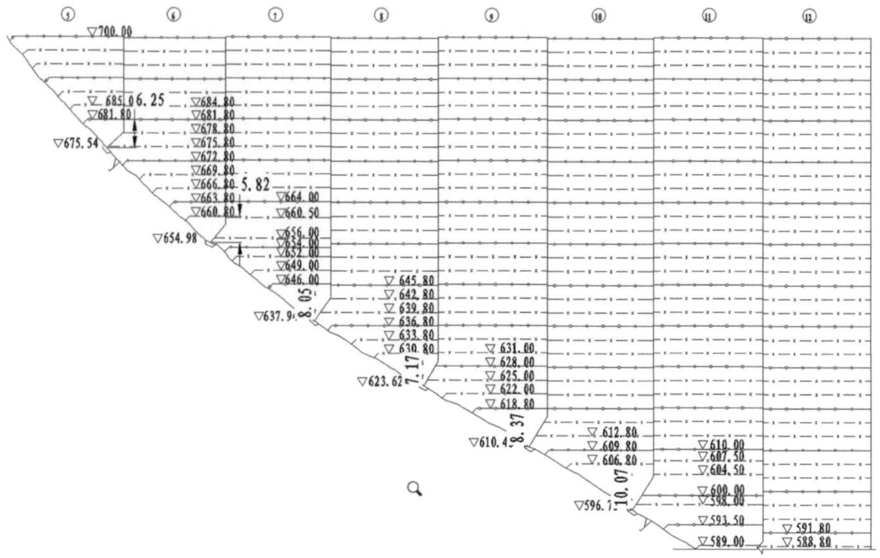

Fig. 18. Concrete layering diagram of left bank slope dam section

pouring stagesupport above the broken line. The vertical edges of the formwork are made of #16 channel steel, and the supports are welded together to achieve a height of 9.55m for the first segment and up to 11.5m for subsequent construction, achieving the goal of rapid rise of the bank slope dam section. The formwork facade support is shown in Fig. 19, and the formwork configuration is shown in Fig. 20.

5 Effects

Through the research on the construction of the concrete high archdam bank slope and orifice dam section of Baihetan Hydropower Station, the first segment of the bank slopedam section adopts high lift construction, the brackets adopt truss style prefabricated formwork, the riverbed dam section adopts hydraulic climbing formwork, the gate groove adoptsone-stage direct burial technology, the deep hole steel lining adopts pre-embedded steel lining installation track, the deep hole support hinge beam adopts the form of embedded steel truss + steel lining, and the surface hole support beam adopts inverted "T" beam and other processes, achieving the goal of balanced rise of the concrete high arch dam.

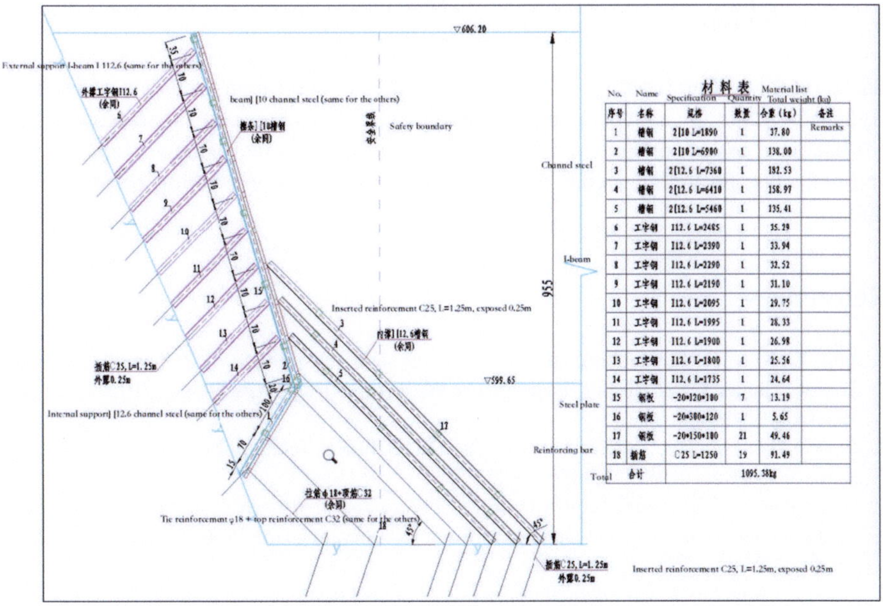

Fig. 19. Elevation view of concrete high rise construction support for bank slope dam section

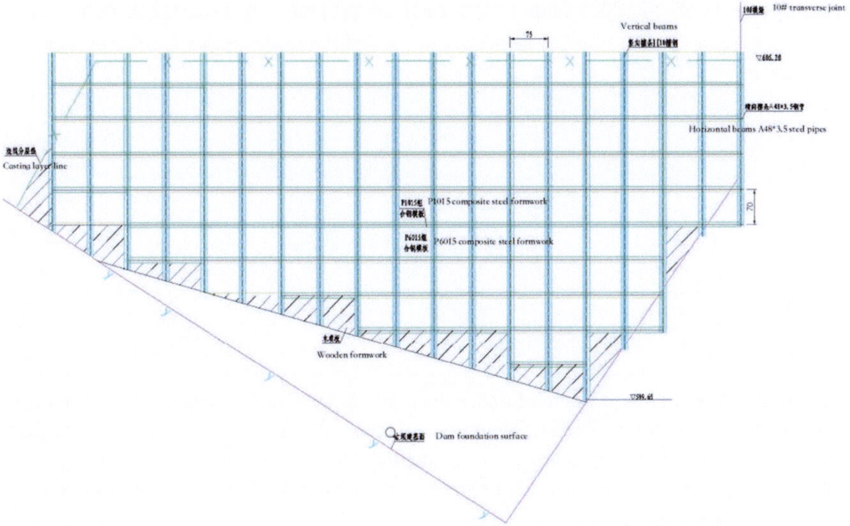

Fig. 20. Front view of formwork configuration for high rise concrete construction of the bank slope dam section

6 Conclusion

Based on the research on the balanced rise of the concrete high arch dam of Baihetan Hydropower Station, a set of technical, theoretical, and management support for the rapid and balanced rise of concrete high arch dams has been developed using new technologies and concepts, providing technical references for similar construction projects in the future.

References

1. Jing, Y., Xule, P.: Key technologies for high arch dam construction of Baihetan Hydropower Station. Hunan Water Resour. Hydropower **06**, 8–14 (2022). https://doi.org/10.16052/j.cnki.Hnslsd.2022.06.030
2. Jianxin, W., Huawei, J.: Discussion on the Direct Burial Technology of Gate Slot Phase I. Haihe Water Resour. **08**, 106–109 (2023)
3. Wenzhuan, Z.: Direct burial construction technology for the first phase of the intake gate of the right bank powerhouse of the Datongxia Water Conservancy Hub Project. Value Eng. **42**(22), 87–89 (2023)
4. Peng, Z.: Analysis and study on deformation of connection between gate slot and steel lining of Baihetan Dam. Yunnan Hydroelectr. Power **39**(12), 349–352 (2023)
5. Zhiyong, H., Shunhe, Z., Chaoxiu, L.: Rapid construction technology for deep hole support beam of Baihetan Hydropower Station arch dam flood discharge. Hunan Water Resour. Hydropower **6**, 49–52 (2022). https://doi.org/10.16052/j.cnki.Hnslsd.2022.06.006

Research on the Dispatching Decision Method of Cascade Hydropower Stations Based on the BVWS

Ma Haoyu[1,2](✉), Cao Hui[1,2], Liu Yaxin[1,2], Xu Yang[1,2], and Tian Rui[1,2]

[1] China Yangtze Power Co., Ltd, Yichang 443002, China
mahaoyu1014@163.com

[2] Hubei Key Laboratory of Intelligent Yangtze and Hydroelectric Science, Yichang 443002, China

Abstract. Compared with the operation of a single hydropower station, the difficulty of operation and management for cascade hydropower stations increases exponentially. Especially during the critical periods such as concentrated drawdown before flood season and concentrated storage during late flood season, the dispatching strategies of cascade stations are crucial for successfully realizing operation objectives and enhancing power generation benefits of the cascade system. Scientific methods should be used to formulate long-, medium-, and short-term operation strategies, and the sequence of drawdown or storage for cascade reservoirs should be arranged reasonably to maximize the system benefits. Current research typically utilizes reservoir dispatching diagrams or optimal dispatching models to develop scheduling plans. Reservoir dispatching diagrams can be used to quickly access the operation schemes for reservoirs, but it cannot guarantee the optimal power generation benefits within a given dispatching period. Optimal models typically utilize algorithms such as dynamic programming to solve and obtain corresponding scheduling solutions, achieving the maximization of reservoir dispatching objectives. However, the problem of dimension disaster often easily occurs. To balance the optimization of system benefits and the efficiency of formulating operation plan, this paper proposes a benefit evaluation index, namely Benefit Variation from Water Storage (BVWS), which simultaneously couples water head benefits and backwater jacking influence of hydropower stations.Based on this, the graphs of BVWS for cascade stations are drawn, which can assist decision-makers in quickly formulating dispatching strategies for different time scales. Taking a cascade system composed of four hydropower stations in the upper reaches of the Yangtze River as the research object, this paper compares the calculation results and efficiency of the proposed method with the progressive optimality algorithm (POA). The results indicate that the proposed method can remarkably reduce the time required to develop optimal operation schemes, enabling a rational allocation of water resources among the cascade stations, therefore proving the scientific rationality of this method.

Keywords: Hydropower optimization · Cascade hydropower stations · Head benefits · Backwater jacking · Benefit variation from water storage

S. Zheng et al. (Eds.): IHDC 2024, LNCE 487, pp. 137–145, 2025.
https://doi.org/10.1007/978-981-97-9184-2_12

1 Introduction

The upper reaches of the Yangtze River have a large head drop and abundant hydropower resources. The water conservancy planning and water resources development are usually carried out in stages, using the form of cascade hydropower stations to achieve efficient utilization of water resources. Cascade hydropower stations usually take on multiple tasks such as flood control, power generation, navigation, ecology, etc. Compared to the operation of a single hydropower station, the difficulty of managing cascade hydropower stations increases exponentially. Especially, the dispatching strategies during critical transition periods such as the drawdown period and the storage period are crucial for enhancing the annual power generation of the cascade system [1, 2]. It is essential to arrange the drawdown and storage sequences of cascade reservoirs rationally, and orderly connect the drawdown with flood control and flood control with water storage, in order to maximize the power generation benefits of the cascade system.

In recent years, many scholars have conducted extensive research on the dispatching and control strategies of cascade hydropower stations. Gong [3] proposed different drawdown methods for cascade reservoirs of lower Jinsha River and the Three Gorges Reservoir. Based on comprehensive benefits, they suggested the optimal drawdown sequence and schemes for cascade reservoirs. Yang [4] optimized the operation rules of Hanjiang cascade reservoirs based on the PA-DDS algorithm, and compared the calculation results with NSGA-II. Cheng [5] proposed the real-time optimization scheduling strategies based on "forecast optimization operation", focusing on the flood control operation of Wan'an Reservoir in the middle reaches of Ganjiang River. Using the forecast information of different magnitudes of floods as the discrimination criteria, Ma [6] constructed a self-identifying discharge control model to optimize the flood control of Longtan Reservoir.

The main task of cascade hydropower stations is to maximize the utilization of water resources for power generation. Therefore, this paper formulates dispatching strategies for key periods such as the drawdown period and the storage period with the goal of maximizing the power generation benefits. There are usually water level connections between the upstream and downstream reservoir. The downstream reservoir retains water to raise the water level, which can improve its own power generation efficiency. However, it may cause backwater jacking on the upstream reservoir, affecting the power generation efficiency of the upstream station, resulting in conflicts of interest between the upstream and downstream stations. To address this, this paper defines the index of Benefit Variation from Water Storage (BVWS), taking into account both the increased power generation from storage and the power generation loss caused by backwater jacking in each reservoir. The curves of BVWS for cascade stations are drawn, and based on this, we can formulate the operation strategies for cascade stations under different operating conditions.

2 Power Generation Increased from Water Storage

This paper first defines the index of Power Generation Increased from Water Storage (PGIWS), which is calculated as follows: Calculate the daily power generation of a hydropower station at the given flow rate and water level. Then, the reservoir is required to impound an additional unit volumn of water, and calculate the daily power generation under the same discharge. The difference in power generation before and after the water level rises is the PGIWS under the specified operating condition. The calculation formula is as follow.

$$\begin{aligned}
&PGIWS_i = E_{i,2} - E_{i,1} = (N_{i,2} - N_{i,1})\Delta t = (K_{i,2} H_{i,2} - K_{i,1} H_{i,1}) R_i \,\Delta t \\
&K_{i,1} = f_{i,HK}(H_{i,1}) \\
&K_{i,2} = f_{i,HK}(H_{i,2}) \\
&H_{i,1} = Z_i - Z_i^{down} - H_{i,loss} \\
&H_{i,2} = Z_i' - Z_i^{down} - H_{i,loss} = f_{i,VZ}\left[f_{i,ZV}(Z_i) + 1\right] - Z_i^{down} - H_{i,loss}
\end{aligned} \tag{1}$$

where, $PGIWS_i$ is the increased power generation of ith station due to water storage under the specified discharge Q_i and water level Z_i. $E_{i,1}$ and $E_{i,2}$ are the daily electricity generation of ith station before and after the storage. $N_{i,1}$ and $N_{i,2}$ are the output of ith station before and after the storage. Δt is the length of daily time period. $K_{i,1}$ and $K_{i,2}$ are the output coefficients of ith station before and after the storage. R_i is the power generation flow of ith station. $H_{i,1}$ and $H_{i,2}$ are the water head of ith station before and after the storage. $f_{i,HK}(\cdot)$ is the function relating the water head to output coefficient for ith station. Z_i and Z_i' are the upstream water level of ith station before and after the storage. *Zdown i* is the downstream water level of ith station. $H_{i,loss}$ is the head loss of ith station. $f_{i,ZV}(\cdot)$ and $f_{i,VZ}(\cdot)$ are the functions describing the relationship between the water level and reservoir capacity for ith station.

Taking the cascade system composed of four stations on the upper reaches of the Yangtze River as the research object, we further elaborates the indicators and decision-making methods proposed in this paper. The four hydropower stations are series reservoirs, which are station A, station B, station C and station D from upstream to downstream, all of which have strong regulation abilities.

Calculate the additional electricity generated at each reservoir due to an increase in the water head caused by impounding a unit volume of water (100 million m^3) under different discharge and water level. This is the PGIWS of each station, as shown in Fig. 1, where the legend represents different outflow (2000–8000 m^3/s). It can be seen that: (1) For each hydropower station, under the same outflow, the higher the current operating water level, the smaller the PGIWS due to storage. (2) When the outflow has reached full-load flow of the station under current water level, the PGIWS at and above such water level is 0. For example, when the outflow is 8000m^3/s, if the water level of station C is above 575m, the PGIWS of station C is 0. (3) The average PGIWS for four stations is 970 MWh, 520 MWh, 930 MWh and 790 MWh respectively.

3 Power Generation Loss from Water Storage

When the distance between the upstream and downstream dam sites is close, there are often complicated hydraulic connections between the adjacent upstream and downstream reservoirs. When water level of the downstream reservoir is high, it will generate back-water. If the tail water of the upstream reservoir is in the backwater region, the original stable relationship between the tail water level and outflow of the upstream reservoir will

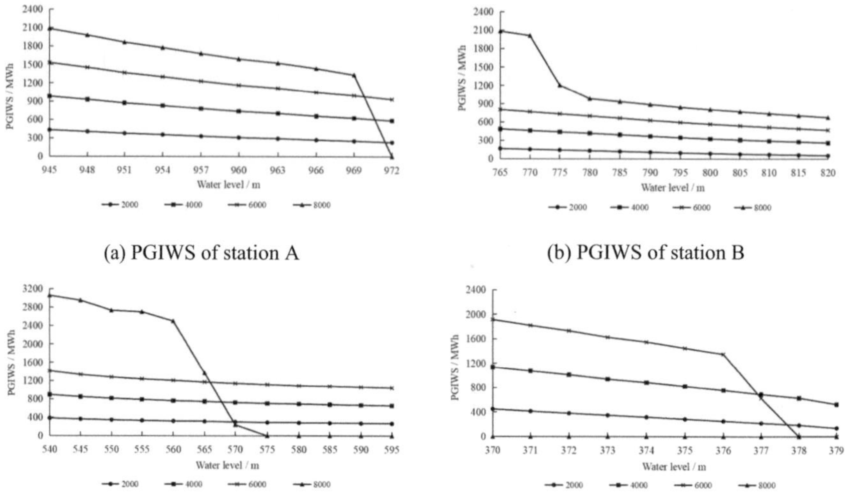

(a) PGIWS of station A

(b) PGIWS of station B

(c) PGIWS of station C

(d) PGIWS of station D

Fig. 1. PGIWS under different outflow and water level

be destroyed. The above phenomenon is called backwater jacking, which will reduce the power generation head and conversion efficiency of hydro-electric energy of the upstream station. Based on this, the paper defines the index of Power Generation Loss from Water Storage (PGLWS), which is calculated as follows: Calculate the daily power generation of the adjacent upstream hydropower station under the given outflow and water level. Then let the downstream reservoir retain an additional unit of water, and calculate daily power generation of the upstream station under the same outflow. The difference in electricity of the upstream station before and after the storage of downstream reservoir is the PGLWS under the specified outflow and water level.

$$
\begin{aligned}
& PGLWS_i = E_{i-1,1} - E_{i-1,2} = (N_{i-1,1} - N_{i-1,2})\Delta t = (K_{i-1,1} H_{i-1,1} - K_{i-1,2} H_{i-1,2}) R_{i-1} \Delta t \\
& K_{i-1,1} = f_{i-1,HK}(H_{i-1,1}) \\
& K_{i-1,2} = f_{i-1,HK}(H_{i-1,2}) \\
& H_{i-1,1} = Z_{i-1} - Z_{i-1,1}^{down} - H_{i-1,loss} = Z_{i-1} - f_{i-1,zdown}(Q_{i-1}, Z_i) - H_{i-1,loss} \\
& H_{i-1,2} = Z_{i-1} - Z_{i-1,2}^{down} - H_{i-1,loss} = Z_{i-1} - f_{i-1,zdown}(Q_{i-1}, Z_i') - H_{i-1,loss} \\
& Q_{i-1} = R_{i-1} + q_{i-1} = Q_i = R_i + q_i
\end{aligned}
\tag{2}
$$

where, $PGLWS_i$ is the electricity loss due to water storage of ith station under the specified outflow Q_i and water level Z_i. $E_{i\text{-}1,1}$ and $E_{i\text{-}1,2}$ are the daily power generation of the upstream (i-1)th station before and after water storage of ith station; $K_{i\text{-}1,1}$ and $K_{i\text{-}1,2}$ are the output coefficients of (i-1)th station before and after the impoundment. $R_{i\text{-}1}$ is the power generation flow of (i-1)th station. $H_{i\text{-}1,1}$ and $H_{i\text{-}1,2}$ are the power generation head of (i-1)th station before and after the impoundment. $f_{i\text{-}1,HK}(\cdot)$ is the function relating water head to output coefficient of (i-1)th station; *Zdown i-1,1* and *Zdown i-1,2* are the downstream water level of (i-1)th station before and after the storage; $f_{i\text{-}1,zdown}(\cdot)$ is the function for tail water level of (i-1)th station; $Q_{i\text{-}1}$ and Q_i are the outflow of (i-1)th station and ith station; $q_{i\text{-}1}$ and q_i are the abandoned flow of (i-1)th station and ith station.

Analyzing the electricity loss due to backwater by retaining a unit volume of water (100 million m^3) under different discharge and water level in three downstream stations, i.e. the the PGLWS of each station. Figure 2 shows the PGLWS of three downstream stations under different operating conditions, where the legend indicates different discharge flow. From Fig. 2, it can be seen that: (1) Under the same outflow, the higher the operating water level of hydropower station, the greater the PGLWS due to storage. (2) The average PGLWS for three stations is 100 MWh, 240 MWh and 470 MWh, respectively. Overall, station B has the lowest PGLWS, while station D has the highest.

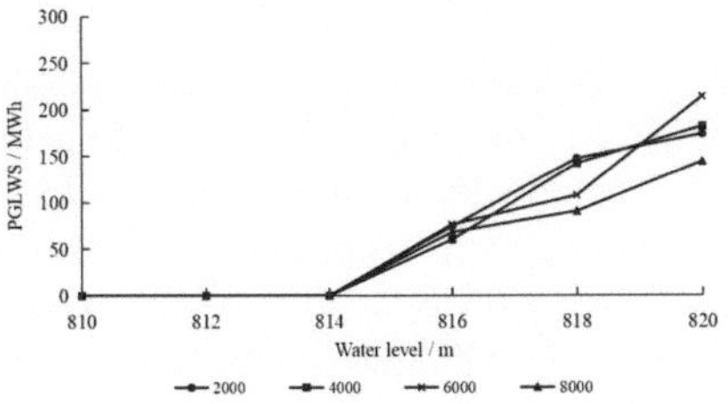

(a) PGLWS of station B

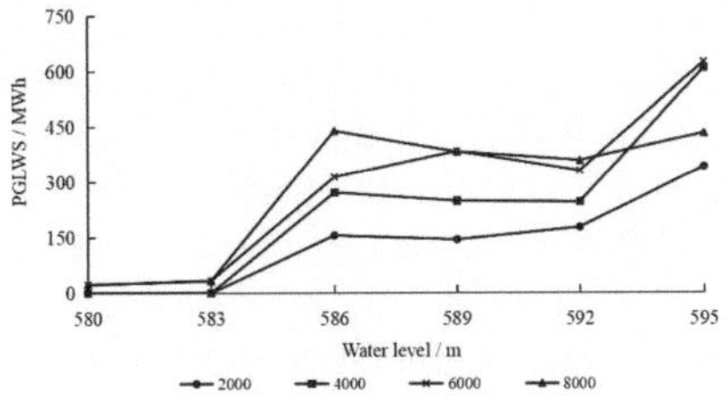

(b) PGLWS of station C

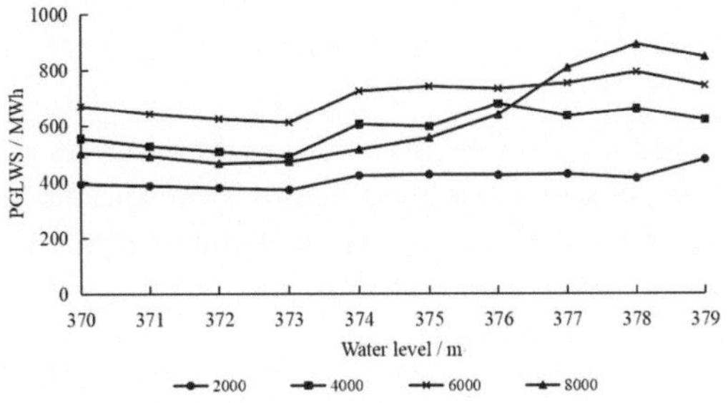

(c) PGLWS of station D

Fig. 2. PGLWS under different outflow and water level

4 Benefit Variation from Water Storage

4.1 Index Definition

Based on the two indicators defined above, this paper constructs a benefit evaluation index for cascade hydropower stations - BVWS. It is defined as follows: Calculate the PGIWS and PGLWS for the hydropower station under the given outflow and water level. The difference between the two indicators is the change in power generation of the cascade system due to water storage of this station, that is, the BVWS under the specified discharge and water level.

$$BVWS_i = PGIWS_i - PGLWS_i \tag{3}$$

where, $BVWS_i$ is the benefit variation due to water storage of ith station under the specified outflow Q_i and water level Z_i.

Normalize the water level of each hydropower station of the cascade system, plot the BVWS of each station under the same outflow on the graph, and finally obtain the graphs of BVWS for cascade stations under different outflow. The curves of BVWS can guide the dispatching of cascade stations to maximize the power generation benefits. Figure 3 shows the BVWS of cascade stations at different water level under various outflow, where the legend represents different hydropower stations, and the horizontal axis represents the normalized value of water level.

When the discharge of hydropower station is given, we can find the corresponding figure of BVWS for this station. Then, given the operating water level, we can get the value of BVWS for this station from the graph. The operation modes of reservoirs can be divided into three types: water storage, balance of inflow and outflow and drawdown. When storing water, the hydropower stations with higher BVWS will prioritize in retaining the water volume. When drawing down, the stations with lower BVWS will release water first. This strategy can be beneficial for improving the overall power generation of the cascade system.

4.2 Case Study

Based on the actual scheduling process of cascade hydropower stations, this paper sets up two cases under the scenarios of water storage and drawdown, and then uses the curves of BVWS and POA respectively to develop optimal operation plans. The calculation results and time consumption of two methods are compared to verify the scientific rationality of the method proposed in this paper.

For Case 1, the current water levels of the four stations are 950m, 785m, 555m and 373m, respectively. Based on the daily time scale, compile a scheduling plan for the cascade stations with a total storage capacity of 1.5 billion m^3 over 10 days. Table 1 shows the comparison of the calculation results between the method presented in this paper and the widely used POA. From Table 1, it can be seen that the scheduling strategy developed based on the BVWS graphs is basically consistent with POA. Station A and C respectively retain 0.9 billion m^3 and 0.6 billion m^3 of water. Additionally, the calculation time of the method proposed in this paper is significantly less, reducing by 95.5% compared to POA. The BVWS graphs can also be used to explain the calculation results of optimization algorithms. Under the given operating conditions in Case 1, the BVWS of station A and C is relatively high, both exceeding 1200 MWh. The BVWS of station B and D is relatively small, below 1200 MWh. Therefore, it is recommended to arrange station A and C for storage, which is more conducive to improving the overall power generation capacity of the cascade system.

For Case 2, the current water levels of the four stations are 965m, 820m, 590m and 378m, respectively. Using ten days as the calculation scale, develop an operation

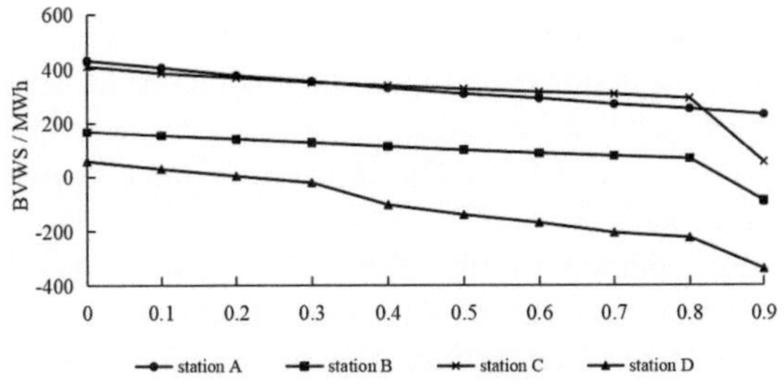

(a) BVWS of each station at different water level under the discharge of 2000 m^3/s

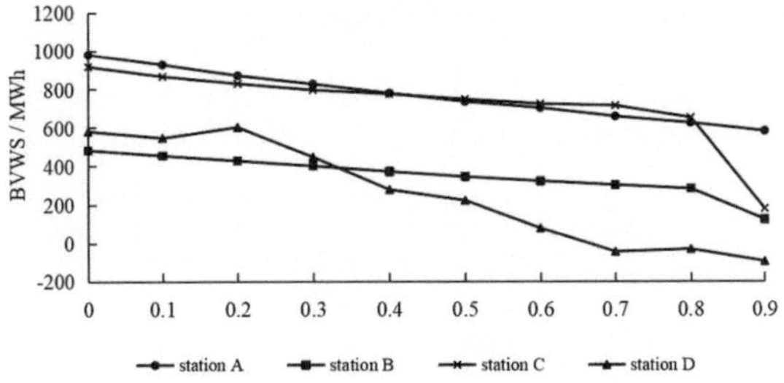

(b) BVWS of each station at different water level under the discharge of 4000 m^3/s

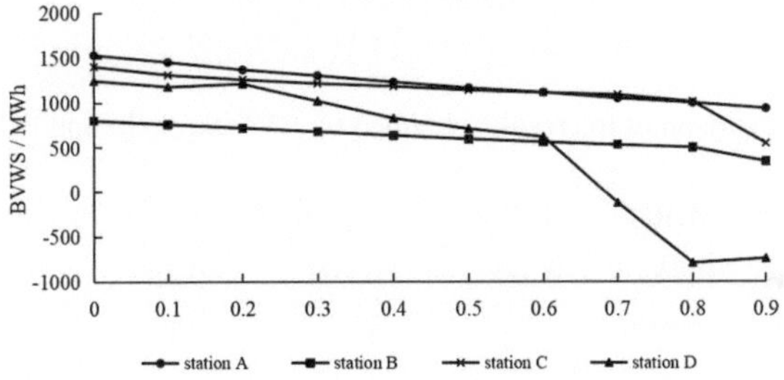

(c) BVWS of each station at different water level under the discharge of 6000 m^3/s

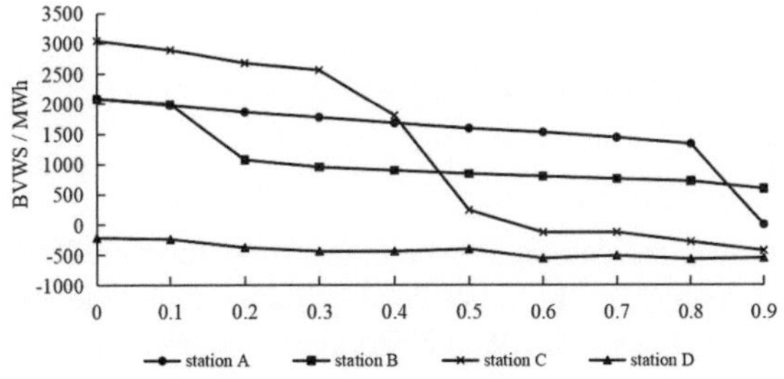

(d) BVWS of each station at different water level under the discharge of 8000 m^3/s

Fig. 3. BVWS under different outflow and water level

plan for the cascade stations that releases 8 billion m^3 of water over twelve ten-day periods. Table 2 shows the comparison of the calculation results between two methods. From Table 2, it can be seen that the scheduling strategy developed based on the BVWS graphs is basically consistent with POA. The drawdown is mainly carried out by station B, releasing 6 billion m^3 of water. Station D lowers the water level to around 370m, releasing 0.7 billion m^3 of water. Additionally, the calculation time of the method proposed in this paper is significantly less, reducing by 94.5% compared to POA. The BVWS graphs can also explain the results of the optimization algorithm. Under the given operating conditions, the BVWS of station B and D is relatively small when the water level is high, below 200 MWh. Therefore, it is recommended to arrange station B and D for drawdown, which will improve the overall benefits of the cascade system.

Table 1. Comparison of the results obtained by BVWS graphs and POA on Case 1

Indicator	Method	Station A	Station B	Station C	Station D	Total
Water storage (billion m^3)	POA	0.953	0	0.547	0	1.5
	BVWS	0.9	0	0.6	0	1.5
Calculation time (s)	POA	–	–	–	–	112
	BVWS	–	–	–	–	5

Table 2. Comparison of the results obtained by BVWS graphs and POA on Case 2

Indicator	Method	Station A	Station B	Station C	Station D	Total
Water storage (billion m^3)	POA	0.868	6.265	0.187	0.681	8
	BVWS	0.6	6	0.7	0.7	8
Calculation time (s)	POA	–	–	–	–	128
	BVWS	–	–	–	–	7

5 Conclusion

This paper defines two indicators: Power Generation Increased from Water Storage (PGIWS), which characterizes the water head benefits of cascade hydropower stations, and Power Generation Loss from Water Storage (PGLWS), which characterizes the impact of backwater jacking between cascade stations. Based on this, an evaluation index for the power generation benefits of cascade stations, namely Benefit Variation from Water Storage (BVWS), is constructed. This indicator can intuitively display the impact of a unit water volume retained by each station on the power generation efficiency of the cascade system.

The study then focuses on the cascade hydropower stations in the upper reaches of the Yangtze River. The curves of the BVWS for the cascade stations are drawn, which can clearly show the BVWS of each station under different conditions and the relative magnitude of the indicators among the stations in a simple and clear manner. Compared with an existing scheduling decision-making method in two scenarios, it is proven that this method can quickly and efficiently obtain the dispatching schemes for cascade stations at different time scales. It maximizes the power generation of the cascade system during the operation period while avoiding the complex solving process of optimization algorithms.

Acknowledgements. This study was financially supported by National Key Research and Development Program of China (Grant No. 2022YFC3202805). The authors are grateful to the reviewers for their comments and valuable suggestions.

References

1. TURGEON A.: Stochastic optimization of multireservoir operation: The optimal reservoir trajectory approach. Water Resour. Res. **43**, W05420 (2007)
2. Zhou, Y., Guo, S., CHANG, F.J., et al.: Methodology that improves water utilization and hydropower generation without increasing flood risk in mega cascade reservoirs. Energy **143**, 785–796 (2018)
3. Gong, W.T., Li, S., Hu, T. et al.: Research on drawdown sequence for cascade reservoirs of downstream Jinsha River and Three Gorges Reservoirs. Yangtze River **53**(09), 187–194 (2022)
4. Yang, G., Guo, S.L., Liu, P. et al.: Multiobjective cascade reservoir operation rules and uncertainty analysis based on PA-DDS algorithm. J. Water Resour. Plan. Manag. **143**(7), 04017025 (2017)
5. Cheng, Y.X., He, Z.Z., Chen, J.T. et al.: Research on the forecasting-optimization operation strategy of real-time flood control operation of Wan'an Reservoir in post-flood seasons. China Rural Water and Hydropower, 66–70 (2022)
6. Ma, Z. P., WANG, S., Li, S. Z. et al.: Research on the classified flood control rules of Longtan Reservoir. China Rural Water and Hydropower, 116–120 (2018)

Assimilating FY-4A AGRI Data Based on a WRF-GSI NWP System and Its Impact on Precipitation Forecasts

Chen Jian, Yang Dengyu(✉), Wang Jianping, Cao Nianhong, and Tang Zhaokang

NARI Water Resources and Hydropower Technology Company Limited, Jiangsu 210003, China
yangdengyu@sgepri.sgcc.com.cn

Abstract. Hydropower and renewable energy prediction require accurate precipitation forecast as fundamental, which is computed from Numerical Weather Forecast (NWP) models. Assimilating satellite data has become an important method to improve NWP model results, especially geostationary meteorological satellites (GMS) that can provide continuous observation information of weather systems. The Advanced Geostationary Radiance Imager (AGRI) onboard China's new generation GMS, FY-4A, can provide observations of temperature and water vapor covering the land and surrounding sea areas. Using a newly established WRF-GSI NWP system, this paper analyzes the impact of AGRI data assimilation on several precipitation case forecasts in China. First, channel selection of AGRI data, along with data thinning and observation error adjustment are performed. Then, four experiments of precipitation case forecasts are conducted and compared with each other. The results show that the influence of assimilating conventional data is mainly located in the inland area of China, while AGRI data can provide more observation information in the offshore ocean area. AGRI data assimilation can significantly improve the accuracy of atmospheric temperature and water vapor at a height of 500 hPa in the coastal area of the model, which further enhanced precipitation forecasts. The improvement of ETS score for heavy precipitation is most obvious, the biggest improvement can be up to 60% in the 15 July 2021 case, indicating the addition of AGRI data can ensure the improvement of the ETS scores of heavy precipitation that larger than 50 mm. The results prove that the AGRI data has added value to precipitation forecasts in the WRF-GSI NWP system, the improved forecast accuracy has great potential on hydrological forecasts, supporting the development of related hydropower applications.

Keywords: Data assimilation · Geostationary satellite · Rainfall · Numerical weather prediction

1 Introduction

Hydropower and renewable energy prediction depend on accurate weather forecasts, which typically relies on the numerical weather forecast (NWP) models. The initial condition of NWP models is the most crucial part, and the data assimilation (DA) technique is developed to produce better initial conditions.

S. Zheng et al. (Eds.): IHDC 2024, LNCE 487, pp. 146–154, 2025.
https://doi.org/10.1007/978-981-97-9184-2_13

Abundant observation information is a prerequisite for DA methods to improve the initial field of NWP model. Because of the limitations of the spatial and temporal distribution of conventional observations, remote sensing data from satellite platforms have become the most important source of observations. At present, satellite radiance data that can be effectively assimilated into NWP can be divided into two categories, the one is from polar-orbiting satellites and the other one is from geostationary meteorological satellites (GMS). GMS have a fixed observation position that allows which to continuously observe weather systems over a large area, providing adequate observation information for data assimilation. Especially, with the development of new generations of GMS (e.g., the U.S. GOES-R series (Schmit et al., 2017), China's FY-4 satellites (Yang et al., 2017), and Japan's Himawari-8 satellites (Bessho et al., 2016), etc.), the high temporal and spatial resolution of GMS data have greatly improved the capacity of monitoring the Earth's environment and forecasting of various catastrophic weather.

The research on direct assimilation of GMS data have been developed for years. The imager data of the U.S. GOES-R satellite and the Japanese Himawari-8 satellite have been released earlier, and many researchers have been devoted to analysing and evaluating the impact of assimilation of data from these satellites (Zou and Zhuge, 2016; Ma et al., 2017; Qin et al., 2017; Honda et al., 2018; Wang et al., 2018; Xu et al., 2023). China's FY-4A satellite also carries a new generation of GMS imager named AGRI (Advanced Geosynchronous Radiation Imager). The number of channels of this new imager has been increased from the original 5 channels to 14 channels, with higher spatial and temporal resolution, and capacity of finer temperature and high-level water vapor detection. However, there are relatively few studies on the assimilation of AGRI data, especially for operational circumstances.

This paper studies on the assimilation of AGRI data on a newly built WRF-GSI NWP system, the system, data and methodology are described in order, and the added value of assimilating AGRI data is evaluated and analysed. The research results can provide reference for the operational and scientific applications of AGRI data assimilation.

2 Description of NWP System and AGRI Data

2.1 The WRF-GSI NWP System

The WRF-GSI NWP system is a preliminary operational forecasting system built by the author's group, which mainly includes three parts: data preprocessing module, assimilation module, and regional forecasting module. The regional forecasting module adopts the WRF-ARW model. The model has a triple nested forecast area of 9km-3km-1km, which are shown in Fig. 1. The model provides gradually refined forecast results from the whole country to the southwest region, and then to Chongqing city, covering key river basins of concern to the water conservancy and hydropower industry. In terms of parameterization, the Kain Fritsch cumulus scheme was applied to the outermost domain, the Thompson microphysical scheme was applied to the nested domain. Other parameterization schemes include the Yonsei PBL scheme and so on.

The data assimilation module is based on the Gridpoint Statistical Interpolation (GSI) system. The GSI system is a three-dimensional variational data assimilation system (Shao et al., 2016) developed by the National Centers for Environmental Prediction (NCEP) of the U.S. National Oceanic and Atmospheric Administration (NOAA). GSI is capable of assimilating most of satellite, radar, and conventional observations worldwide, and can run on different hardware platforms. It has good expansion ability in terms of new observations, new quality control methods, new assimilation control variables etc.

The observation data source of the WRF-GSI NWP system are mainly the sounding observation data and the AGRI infrared imager data, and the assimilation is carried out in the outermost domain with 9km grid resolution.

2.2 The AGRI Infrared Imager Data

The AGRI data is a new generation of infrared imager data carried by China's FY-4A GMS. The specific parameter information of the AGRI data is given in Table 1 (Yang et al, 2017), which shows major wavelengths, observation errors and spatial resolution information of the seven channels. The AGRI data includes visible, near-infrared, and infrared wavelength bands, but at present, it is mainly the infrared wavelength bands, i.e., channels 8–14, that can be used for assimilation.

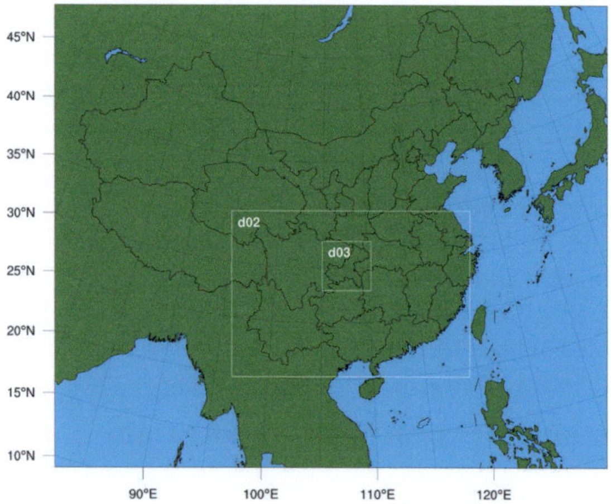

Fig. 1. The nested domain configuration of WRF-GSI operational NWP system

2.3 Configurations of Assimilating AGRI Data

The study directly utilized the corresponding cloud product data that matches the AGRI data for cloud detection during the assimilation process. Currently, only clear-sky observations are assimilated in the assimilation system while the cloudy-area observations are not.

Through the comparison of preliminary tests, all the channels are assimilated in the oceanic region, but in the terrestrial region, considering the lack of accuracy of the numerical model in the simulation of the land surface temperature, as well as the variable surface emissivity which is difficult to be determined accurately, the channels 8, 11–14 in

Table 1. Parameters of AGRI instrument

Channel No	Central wavelength (μm)	Observation error (K)	Spatial resolution (km)
8	3.75	0.2	4
9	6.25	0.3	4
10	7.1	0.3	4
11	8.5	0.2	4
12	10.8	0.2	4
13	12.0	0.2	4
14	13.5	0.5	4

the terrestrial region are excluded and only the data of channels 9 and 10 are assimilated. The assimilation of data from other channels on land requires further study.

To eliminate the correlation of observation errors as much as possible, through analysis and comparison, the original data is thinned to 60km, that is, to select one observation every 60 km from the 4 km resolution observations, and the selected one is closest to the model grid point.

Although the observation errors of each channel are given in the parameters of the AGRI instrument, these errors are only the characteristics of the instrument itself. Considering the simulation errors of the radiative transfer modes and the forecast errors of the numerical modes, the observation errors for channels 8–14 are adjusted to 0.8, 0.8, 0.8, 1.2, 1.3, 1.2, and 1.3 K respectively, according to the results of preliminary tests.

The bias correction of the observations is carried out in two steps: firstly, the static bias-revision, and then the application of the air mass bias-correction method to further eliminate the systematic bias of the observations. For each channel, four bias correction operators are used in the air mass bias-correction. The first operator $p_{1,i}$ is equal to 1.0; the second operator $p_{2,i}$ is calculated as $(\frac{1}{\cos(\alpha)} - 1)^2$, where α is the zenith angle of the satellite; the third operator $p_{3,i}$ and the fourth operator $p_{4,i}$ are related to the attenuation rate of the transmittance:

$$\tau_i^{\tau} = \sum_{k=2}^{lev-1} (\tau_i^{k+1} - \tau_i^{k}) \times (T^{k-1} - T^{k+1}) \tag{1}$$

where *lev* is the number of mode layers; T^k is the Kelvin temperature on the mode layer; on the lowest layer, the temperature on the layer is replaced by the ground temperature; and the fourth operator $p_{4,i}$, is $(\tau_i^{\tau} - \overline{\tau}_i^{\tau})^2$, which stands for mean transmittance.

3 Experiments and Results

3.1 Experiment Design

In order to evaluate the effect of AGRI data assimilation on the 24-h model forecasts, four experiments are designed in the study: the first one is the control experiment (CTRL), which directly uses the FNL data as the initial field for the 24-h forecast; the second one is the CONV experiment, which assimilates only the conventional data; the third one is the SAT experiment, which assimilates only the AGRI data; the fourth one is the BOTH experiment, where both of conventional and the AGRI data is assimilated. The FNL reanalysis data are used as the initial field, after a 6-h forecast for initialization, the obtained forecast results are used as the assimilation background field. Considering the computational wall-time requirements in operational scenarios, data assimilation is performed only once, following a 24-h numerical forecast at present.

3.2 Results of Analysis Increments

The study utilizes a precipitation forecast case experiment on July 15, 2021 to analyze the improvement of the assimilation system on the results of numerical forecasting. Figure 2 presents the spatial distribution of conventional observations (Fig. 2a) and AGRI satellite data (Fig. 2b) utilized by the assimilation system on July 15, 2021 at 0000 UTC. The assimilated conventional observations are mainly sounding data, and the stations are evenly distributed in the eastern part of China. Satellite data, on the other hand, are able to cover the whole model area, but only observations of clear-sky area are assimilated in the system at present because it is difficult to effectively estimate the influence of clouds on satellite data. The small black dots in the figure below represent the satellite observations rejected by cloud detection and quality control, while the colors of the other dots represent the AGRI data that passed the quality control.

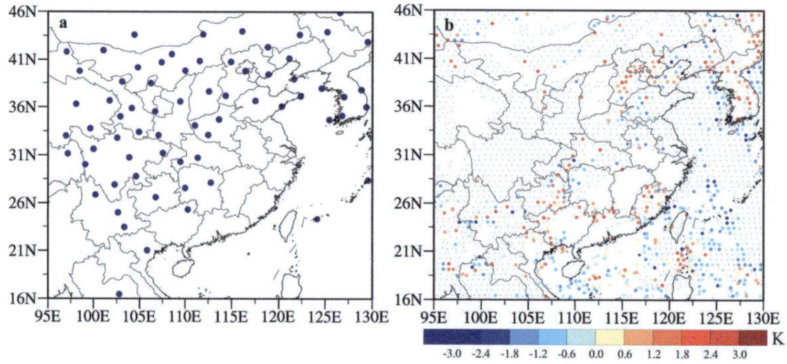

Fig. 2. Spatial distributions of assimilated (a) conventional observation and (b) AGRI satellite data at 0000UTC on 15 July 2021

The spatial distribution of 500 hPa specific humidity (contours in g/kg) and temperature (contours in °C) and their corresponding analysis increments (shaded) at 0000 UTC on July 15, 2021 are given in Fig. 3. Firstly, all available observations are assimilated (the BOTH experiment) and the results are shown in Fig. 3a and Fig. 3b. It can be seen that through data assimilation, the water vapor in central China has been significantly reduced, while the temperature has been significantly increased. From the subsequent precipitation results, it can be seen that these changes not only favors the concentration of the central precipitation area, but also more conducive to the enhancement of the precipitation in the eastern part of the country. Then, in order to clarify the impact of the assimilation of AGRI data, the study further analyzes the improvement effect of the assimilation of AGRI data through sensitivity tests, and the results of specific humidity for CONV and SAT experiment are shown in Fig. 3c and Fig. 3d respectively. Comparing these two analysis increment maps with BOTH experiment, it can be seen that the land area is dominated by the analysis increment of the conventional data, especially in the central part of China, the water vapor is significantly reduced, but the coastal area is dominated by the analysis increment of the AGRI data, and by assimilating the AGRI data, it can significantly increase the water vapor in the coastal area of east China. In addition, for Beijing and Fujian, the assimilation of AGRI data can also significantly reduce the specific humidity of these two regions.

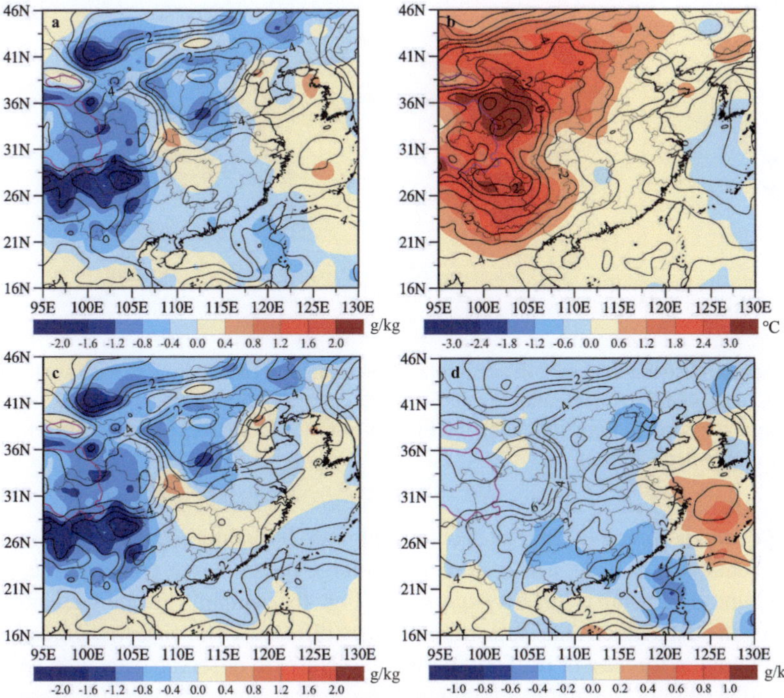

Fig. 3. Spatial distributions of 500hPa specific humidity (a,c,d contours, unit: g/kg) and temperature (b, contours, unit: °C) with corresponding analysis increments (shaded) at 0000UTC on 15 July 2021. Figure 3a and Fig. 3b are for BOTH experiment while Fig. 3c and Fig. 3d are for CONV experiment and SAT experiment respectively

4 Results of Precipitation Forecast

The study focuses on the impact of data assimilation on precipitation forecasting. Figure 4 shows the spatial distribution of the observed and forecast 24-h cumulative precipitation (in mm) during 0000 UTC July 15, 2021 to 0000 UTC July 16, 2021. Figure 4a is for the observation provided by the China Meteorological Administration (CMA) (Shen et al, 2014) and Fig. 4b-e are for CTRL, BOTH, CONV and SAT experiment respectively.

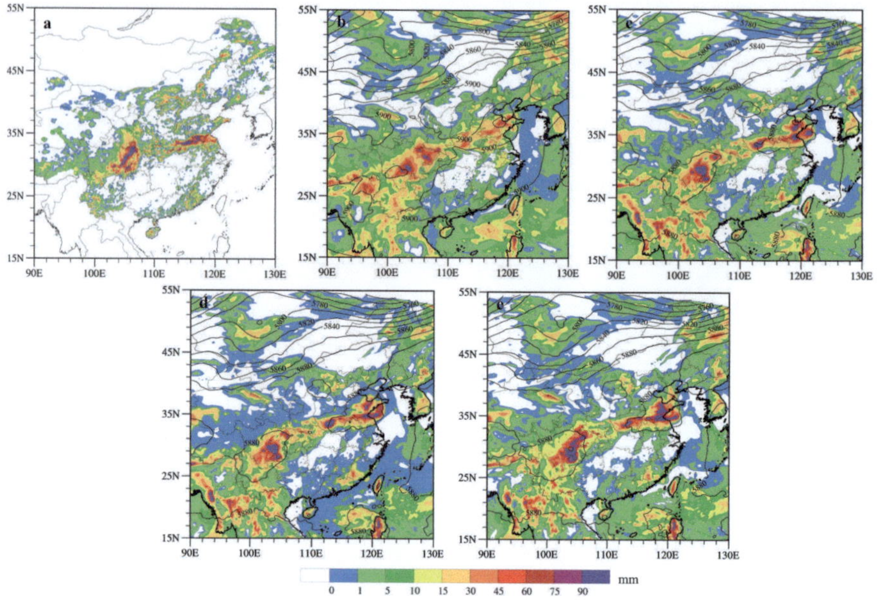

Fig. 4. Spatial distributions of 24h accumulated precipitation (shaded, unit: mm) within 0000-2400UTC and 500hPa geopotential height (contours, unit: gpm) at 0000UTC on 15 July 2021. Figure 4a is for observed precipitation, and Fig. 4b-e are for forecasts of CTRL, BOTH, CONV and SAT experiment, respectively

The observed precipitation, as shown in Fig. 4a, is mainly located in the northern part of Anhui and Jiangsu, and there is also a large precipitation center in the central part of China. The result of CTRL experiment (Fig. 4b) is mainly a southwest-northeast oriented rain band, which distribution is obviously different from observed precipitation. The precipitation forecast of BOTH experiment is closer to observed data, with two large precipitation centers, and the precipitation in the central part of the country is obviously more concentrated. The benefit of assimilation at eastern area is more obvious, the precipitation forecast of CTRL is mainly located in Shandong, and the rain band have an obvious southward shift after assimilation, basically located in the northern part of Anhui and Jiangsu.

Figure 4d and Fig. 4e are the results of CONV and SAT experiments respectively, these two sensitivity experiments can also overcome the northward bias of rain bands that yield in the CTRL. In comparison, the assimilation of conventional data improves the precipitation in the inland areas of China obviously, but the satellite data mainly improves the precipitation forecast in the eastern part of China, especially the eastern coastal areas. The conventional data have a better effect on the calibration of the erroneous precipitation located in Shandong, which is consistent with the fact that the conventional data assimilation can significantly reduce the specific humidity in the northern part of

China. The precipitation forecast in Fujian region is significantly stronger by the model, as the assimilation of AGRI data can better reduce the specific humidity in this region, it has a better effect on overcoming the overestimate phenomenon.

In order to quantify the impact of AGRI data assimilation, the ETS scores of 24-h cumulative precipitation for different threshold scenarios are given in Fig. 5a. The black line is the CTRL experiment that use FNL data as initial field, the red line is the result of CONV experiment, and the blue line is the result of BOTH experiment. It can be seen that the addition of AGRI data further improved the model's prediction of 24-h cumulative precipitation, with the most obvious improvement for precipitation more than 30 mm. In addition, both assimilation experiments can improve the precipitation forecasts by about 20%. While the assimilation of AGRI data slightly improves the forecasts for 10 mm precipitation, the biggest improvement is in the area of rainstorms where precipitation larger than 50 mm. The addition of AGRI data can ensure the improvement of the ETS scores of heavy precipitation, and it can be seen in Fig. 5b that the biggest percentage of improvement can be up to 60%.

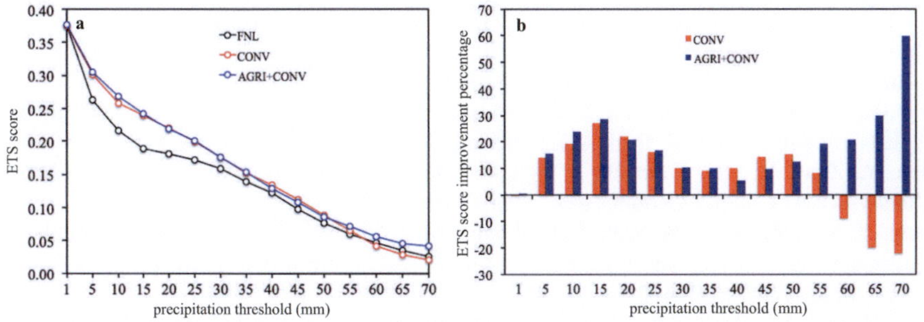

Fig. 5. ETS scores of 24h accumulated precipitation under different thresholds (a) and percentages of ETS score improvement of 24h accumulated precipitation under different thresholds (b)

5 Conclusions

This study focuses on the assimilation of conventional sounding data and the AGRI data of China's FY-4A satellite in a newly built WRF-GSI NWP system. Based on the case of precipitation on July 15, 2021, the results showed that the assimilation of AGRI data can improve the accuracy of the NWP system, especially for the prediction of heavy precipitation, and the AGRI data can provide a wide range of observation information of China's offshore, which complements the conventional observation data that basically located on the land. The sensitivity test also proves that the assimilation of the AGRI data can improve the skills of precipitation prediction for China's eastern coastal area.

The improved precipitation forecast has great potential on hydrological forecasts and renewable energy predictions, the accurate NWP results can further support the development of related hydropower applications. As a newly built NWP system, it is necessary to refine the research on the key technologies of AGRI data assimilation in the future. Meanwhile, the introduction of more satellite data is also the next step in the development of this NWP system.

Acknowledgements. This article is funded by the NARI Group Corporation (State Grid Electric Power Research Institute) project: Research and System Development of Key Technologies for New Energy Power Prediction Based on High Precision Numerical Meteorological Forecasting (5246C5220027).

References

Bessho, K., Date, K., Hayashi, M., et al.: An introduction to Himawari-8/9—Japan's new-generation geostationary meteorological satellites. J. Meteorol. Soc. Jpn. **94**(2), 151–183 (2016)

Honda, T., Miyoshi, T., Lien, G.Y., et al.: Assimilating all-sky Himawari-8 infrared radiances: A case of Typhoon Soudelor. Mon. Weather Rev. **146**(1), 213–229 (2018)

Ma, Z., Maddy, E.S., Zhang, B., et al.: Impact Assessment of Himawari-8 AHI Data Assimilation in NCEP GDAS/GFS with GSI. J Atmos Ocean Tech **34**(4), 797–815 (2017)

Qin, Z., Zou, X., Weng, F.: Impacts of assimilating all or GOES-like AHI infrared channels radiances on QPFs over Eastern China. Tellus A **69**(1), 1345265 (2017)

Schmit, T.J., Griffith, P., Gunshor, M.M., et al.: A closer look at the ABI on the GOES-R series. Bull. Am. Meteorol. Soc. **98**(4), 681–698 (2017)

Shao, H., Derber, J., Huang, X.Y., et al.: Bridging research to operations transitions: status and plans of community GSI. Bull. Am. Meteorol. Soc. **97**(8), 1427–1440 (2016)

Shen, Y., Zhao, P., Pan, Y., et al.: A high spatiotemporal gauge-satellite merged precipitation analysis over China. J Geophys Res: Atmos **119**(6), 3063–3075 (2014)

Wang, Y., Liu, Z., Yang, S. et al.: Added value of assimilating Himawari-8 AHI water vapor radiances on analyses and forecasts for "7.19" severe storm over north China. J Geophys Res-Atmos **123**(7): 3374–3394 (2018)

Xu, L., Cheng, W., Deng, Z.R., et al.: Assimilation of the FY-4A AGRI Clear-Sky Radiance Data in a Regional Numerical Model and Its Impact on the Forecast of the "21·7" Henan Extremely Persistent Heavy Rainfall. Adv. Atmos. Sci. **40**(5), 920–936 (2023)

Yang, J., Zhang, Z., Wei, C., et al.: Introducing the new generation of Chinese geostationary weather satellites, Fengyun-4. Bull. Am. Meteorol. Soc. **98**(8), 1637–1658 (2017)

Zou, X., Zhuge, X., Weng, F.: Characterization of bias of advanced Himawari Imager observations from NWP background simulations using CRTM and RTTOV. J Atmos Ocean Tech **33**(12), 2553–2567 (2016)

Improving Dam Safety Using Optical Fiber Seismic Sensing

Cicero Martelli[1(✉)], Xinjian Chen[3], Jean Carlos Cardozo da Silva[1], Uilian José Dreyer[1], João Paulo Bazzo[1], Daniel Rodrigues Pipa[1], Sidnei Helder Cardoso Teixeira[2], Gustavo Macioski[1], Alessandra de Barros[2], Silva Bongiolo[2], Beatriz Brusamarello[1], Larissa Wierzynski Kulik[1], Gilson Antônio Brunetto[3], Luis Fernando Pedrozo Melegari[3], Huiyi Zhang[3], Alexandre Frescki de Oliveira[4], and Marcelo Henrique Bernardy[5]

[1] Federal University of Technology – Paraná, Curitiba, Brazil
huiyipallas@163.com
[2] Federal University of Paraná (LPGA/UFPR), Curitiba, Brazil
[3] CPFL Energia, Campinas, Brazil
[4] Baesa Energetica, Florianópolis, Brazil
[5] CSC Energia, Florianópolis, Brazil

Abstract. This paper presents seismic measurements at a large hydroelectric power plant dam in the south of Brazil. The seismic measurements were realized using optical fibre distributed sensor system technology and aim at diagnosing possible structural problems inside of the dam. Studies carried out at a laboratory scale dam using geophones as well as optical fiber sensors are also presented. Results are very promising and indicate the potential of the technology using both active and passive seismic techniques to solve major monitoring problems that cannot be addressed by today's technology.

Keywords: Dam safety · Optical fibre sensing · Seismic sensing

1 Introduction

There are thousands of dams around the world for various purposes across distinct industries such as mining, electricity generation, residue processing and storing, water storage among many others. Dams are normally very large infrastructures with hundreds of meters in length and height composed of enormous volumes of materials. They can have very simple designs and structures, but they can also be complex in their structure using several layers of materials of different sizes and densities. There is, however, one common issue to all dams in the world today, the current structural monitoring techniques are not just good enough. They lack spatial coverage and resolution, and they cannot detect what is happening inside of the whole body of the dam. The technique proposed here based on optical fibre sensing for dynamic seismic reconstruction of the dam structure has the potential to solve all this issues at once.

S. Zheng et al. (Eds.): IHDC 2024, LNCE 487, pp. 155–164, 2025.
https://doi.org/10.1007/978-981-97-9184-2_14

Seismic monitoring techniques applied to dam monitoring is not something new. There have been several reports of such technique used in dams with interesting results (Goldswain and Wesseloo 2020).

Similarly optical fiber sensors have also been used to monitor dams as deformation and temperature point sensors as well as distributed sensor systems to detect leakage through the dam walls (Woschitz et al. 2015; Johansson and Sjödahl 2004).

The motivations for the work presented here are:

- There are thousands of dams in Brazil and many of them have structural problems.
- Dams are very large structures and are in remote places and they are subjected to extremely high loads of water that can generate defects and wearing.
- The standard monitoring technologies are very poor in spatial resolution, sensitivity and connectivity.

Optical fiber seismic sensing has unique advantages over any other solution and can allow high resolution and real time information of full body of the dam, increasing to unprecedent levels the safety of the dam. The seismic technique used in this work is called MASW (multichannel analysis of surface waves) and it is application with geophones is also demonstrated as a reference tool.

1.1 Distributed Seismic Optical Fiber Sensing

The sensor technology employed in this work consists of the Distributed Acoustic Sensor System (DAS). This technique has found several applications in the oil and gas industry and in the security sector. It is based on the capacity of reconstructing dynamic deformations along the entire length of an optical fibre by detecting the Rayleigh scattering that happens everytime light is transmitted by the optical fibre. The optical phase of the scattered light brings valuable information of vibration which can be used to generate seismic data, for instance. The sensing fibre can be as long as hundreds of kilometres and the spatial resolution of the measurements as small as 1m with the current technology.

The system used in the work presented here was developed by the authors and operates under the phase detection mode. The majority of the measurements specifications this system are just as good as any commercially available equipment.

2 Seismic Signal Reconstruction - MASW

The multichannel analysis of surface waves (MASW) method uses the dispersive properties of surface waves, commonly Rayleigh waves, propagating through a heterogeneous medium. The shear wave velocity profile is subsequently obtained by back calculation of the dispersion data by assuming a layered soil model (Olafsdottir et al. 2018). Compared to other available methods, surface wave analysis methods are low cost, as well as being non-invasive and environmentally friendly since they neither require heavy machinery nor leave lasting marks on the surface of the test site. These characteristics are favorable for application in already built dams, for example. Other advantages are that the method is compatible with various types of locations, very fine and coarse grained, granular gravel, soft rock, and others with limited penetration, and the combination of multiple dispersion curves (Multichannel) improves resolution and reliability (Yust et al. 2022).

An application of MASW includes three steps: field measurements, dispersion analysis and inversion analysis. Surface waves are generated by an active or passive seismic source and the wave propagation is recorded by multiple sensor points that are evenly spaced along the survey line.

Techniques with active source present greater precision for nearby layers, but are limited to depths of up to 30 m. Passive source techniques use environmental vibrations, such as street traffic or vibrations caused by equipment installations, as in the case of hydroelectric generators. Passive techniques can reach depths of up to 500 m, and can even be combined with active techniques to increase the precision of the layers (Park et al. 2007).

Each multichannel surface wave record is transformed into a dispersion image and the corresponding fundamental mode dispersion curve is identified. The elementary dispersion curves are subsequently combined into a single experimental curve and the uncertainty associated with the combined mean curve evaluated. Finally, the shear wave velocity profile is obtained by inversion of the combined mean dispersion curve by assuming a plane-layered elastic earth model. The parameters required to describe the properties of each layer are shear wave velocity (V_S), compressional wave velocity (V_P), mass density (ρ) and layer thickness (h).

In the work presented here we merge such powerful technique with the measurement capabilities of the DAS systems to promote the seismic monitoring of dams.

3 Laboratory Scale Dam

A prototype dam was developed for the study of the optical fiber sensing application in dams. Figure 1 show images of the dam and fiber cables that are placed at different locations inside the dam structure. This is the first and an unique laboratory dam in Brazil and has the potential to allow the development of several monitoring techniques.

There are over 1200 m of optical fibers, including fiber Bragg grating sensors, installed and six different cable designs to grant a comprehensive study of optical fiber sensing in dams. The dam has 12m in length, 5.5m in its base and 2m in height.

4 Seismic Analysis and Reconstruction – Field Trials and Preliminary Results

In the following sessions, the application of the MASW technique with geophones and the DAS system is presented. The results correspond to measurements that were carried out on a flat field in the university campus, and they have the objective of demonstrating some of the potential of the DAS as a tool for seismic analysis. The results with the DAS system are still preliminary as some numerical corrections are currently being investigated but they are enough to show that the DAS system has the capacity to measure just as good as the geophones.

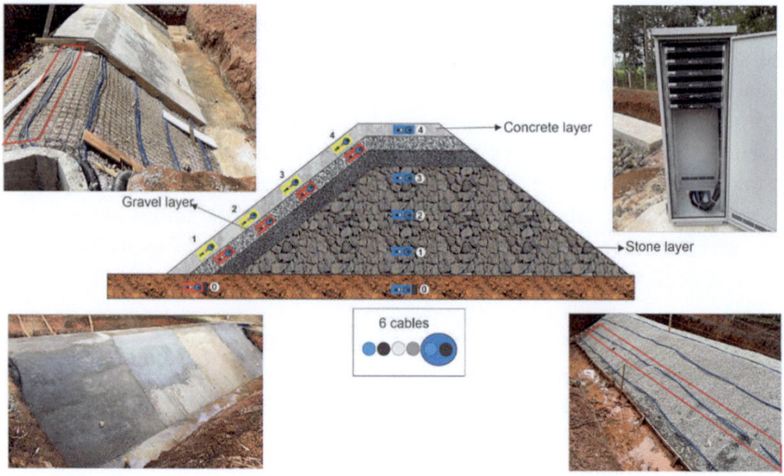

Fig. 1 Laboratory dam built to emulate the real dam where over 1200 m of optical fibers, including fiber Bragg grating sensors, were installed and six different cable designs to allow a comprehensive study of optical fiber sensing in dams. The dam has 12m in length, 5.5m in its base and 2m in height.

4.1 Geophones Measurements

For the survey with geophones, the technique based on refraction seismic was used (D18 COMMITTEE 2023). The equipment used for seismic acquisition comprised the following items:

- Geometrics GEODE seismographs;
- Geometrics 13–105-017D vertical coil geophones, with a fundamental frequency of 14Hz;
- Sercel SC3243 horizontal coil geophones, with a fundamental frequency of 14 Hz;
- 5 kg sledgehammer with impact trigger attached;
- 6 cm thick aluminum plate;
- Acquisition software 1 Seismodule Controller Software (SCS), developed by the equipment manufacturer.

The data was processed with the following programs:

- Visualization of files in the field: Geo2view2;
- Refraction Seismic: Refrapy, with inversion by the seismic tomography method.

The experimental arrangement used is represented in Fig. 2, where 24 geophones were placed with a spacing of 4 m, along a 100 m length in the field. In each point of execution, three hammer blows were made, which were accumulated (stacked) on the seismographs before being transferred to the computer. For the refraction test, excitations were carried out at relative positions of -18 m, 14 m, 44 m, 74 m, 106 m, with an acquisition rate of 0.125 ms during 2.048 s.

The result generated by software analysis after processing the files generated in the test is shown in Fig. 3.

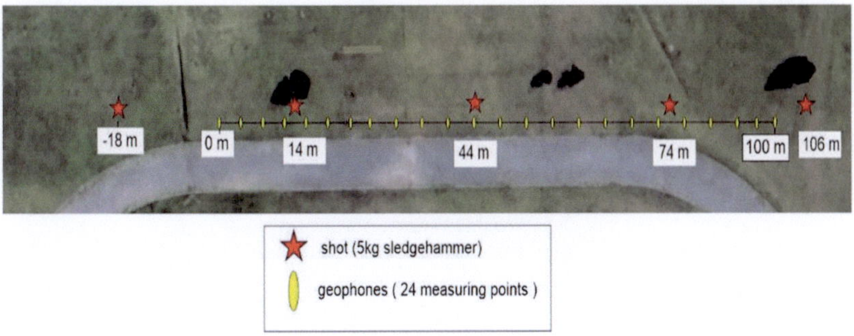

Fig. 2 Experimental setup used in the geophone tests for seismic tomography reconstruction. 24 geophones spaced at every 4 m along 100 m of length of the field were employed as detailed in the image by the yellow marks. The red stars consist of the place where the seismic waves were generated by the impact of a 5kg sledgehammer to the ground.

Table 1 summarizes the results of each layer identified through the refraction tomography technique considering S-waves. These results are used as a reference for the experiment using the DAS system, in the same 100 m region of the field.

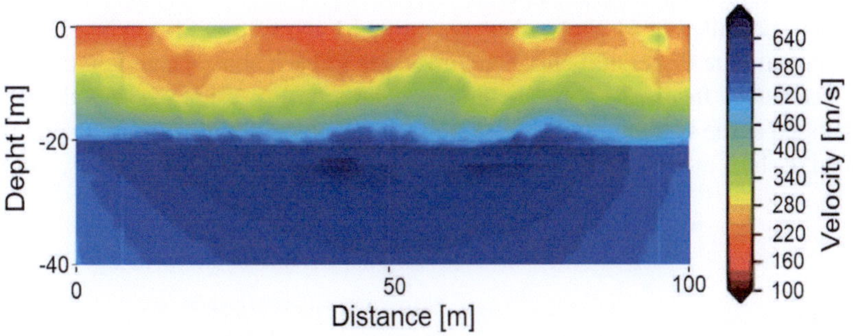

Fig. 3 Seismic MASW tomography result using geophones. The image shows a 3D reconstruction of the velocity profile for the S-waves as a result the experiment carried described in this figure. In the color map it is possible to observe distinct regions which are accounted for the different materials of different densities that lead to different acoustic wave propagation velocities.

4.2 Optical Fiber Sensors Measurements

For comparison with the results already obtained in soil assessment with geophones the MASW method with a "roll-along" type data set and fixed array, proposed by Park 2005 was used (Park 2005). The overview of the test is presented in Fig. 4.

In the test, a single-mode optical fiber, telecommunication standard model Furukawa OpticLAN, was used, buried 10 cm underneath the ground, with a length of 100 m along the field, in the same position as the geophones shown in Fig. 2.

The DAS interrogator equipment was configured with an acquisition rate of 1 kHz, pulse width of 50 ns, gauge length of 1.5 m, and acquisition time of 5 s for each test. The

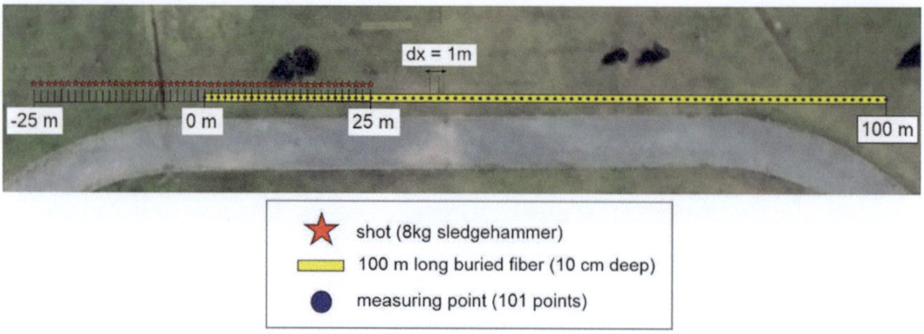

Fig. 4 Experimental setup for the distributed seismic optical fiber sensing system measurement. One hundred meters of fiber was used as informed in the image, the spatial resolution of the measurements was of 1.5 m and the cable was buried 10cm deep into the ground. 50 exciting shocks were used to generate the acoustic waves as presented by the red stars.

samples were considered for each meter of fiber, resulting in 100 sensor points, spaced every 1 m (dx = 1 m).

Based on the MASW method, an 8 kg sledgehammer was used as the active source, starting the shots at the −25 m position (before the buried fiber), advancing every 1 m until the position 25 m from the fiber.

This test resulted in 51 shots, also generating 51 acquisition files to be processed in order to obtain the dispersion curve and inversion of each shot, where you can observe the 3 basic steps for the MASW method, acquisition, dispersion curve and inversion, as discussed at the beginning of this section. As an example, Fig. 5 presents the results obtained in shots at −25 m. This procedure was carried out for 51 files shots, from −25 to 25 m.

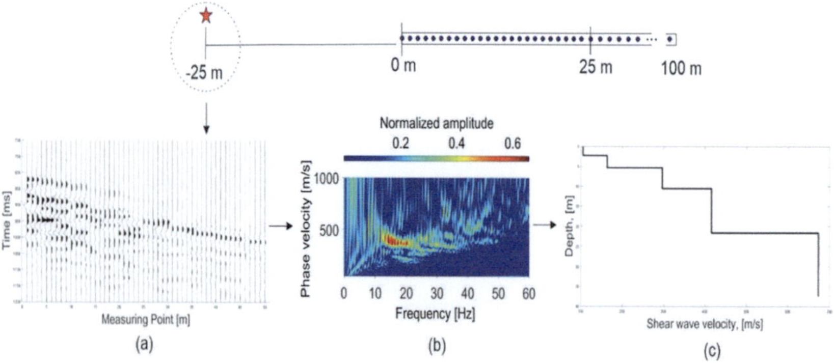

Fig. 5 Seismic signal resulting from the measurement of a shot at the −25 m position. (a) Wiggle plot of the response of the 101 measurement points in DataStrain/Time (ms). (b) Calculated dispersion curve - phase velocity (m/s) /frequency (Hz). (c) Inversion obtained through the dispersion curve and soil model, with depth (m) / velocity (m/s).

The results of each 51 files are interpolated to generate a 2D of velocities, where it is possible to check the variations of the layers along the entire length of the sensor, 100 m, as shown in the result in Fig. 6. As can be seen in the map obtained, 5 layers limited to 30 m depth were identified. The average characteristics of each layer are summarized in Table 1. This result, although preliminary, demonstrates the application capacity of the DAS system, being compatible with traditional geophone techniques, with advantages such as ease of installation, real-time monitoring, greater number of sensor points and greater monitoring distance with a single installation.

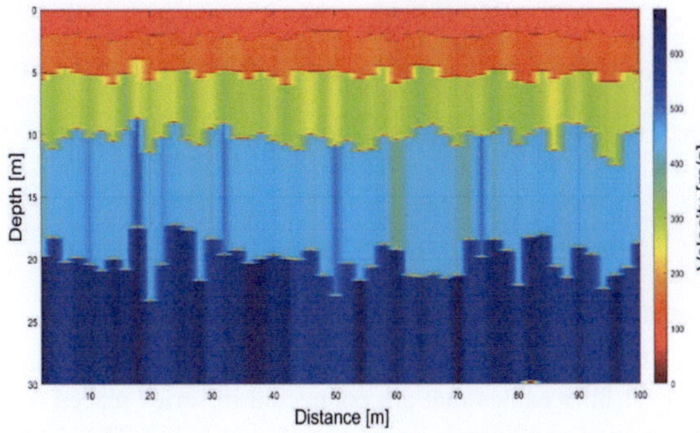

Fig. 6 Velocity map obtained with the MASW technique using the optical fiber seismic sensing system. The map is generated through the interpolation of the 51 files acquired in the test.

Table 1. Average velocities of each layer identified through MASW technique using the optical fiber seismic sensing system. It is possible to compare with Table 1 and see that the optical fiber system was able to reconstruct the exact same ground profile as the geophones and the average velocities of each layer identified through seismic tomography using geophones. Five different layers are found and they can be directly correlated to geology knowledge of the area.

Layer (m)		Type	Vs (m/s)	
Geophone	Optical fiber	Type	Geophone	Optical fiber
0–2	0–2	Vegetable Soil	100	96
2–5	2–5	Sandy Soil	160	165
5–10	5–10	Altered Mudstone	300	323
10–20	10–20	Compact Mudstone	460	440
25–35	20–30	Compact Mudstone	650	620

4.3 Seismic Signals at BAESA Hydroelectric Power Plant

A first field trial was carried out at the BAESA hydroelectric power plant with the objective of identify natural vibration signals originated from the high power electric generators as well as to define parameters of a installation with focus on the seismic analysis and the MASW method. Figure 7 shows a satellite photograph image of the BAESA dam. The yellow line represents the two optical fiber cables that are connected in series placed on top of the dam as indicated in the other detailed photograph image at right hand side. The bottom graph shows the seismic signal measured by the cable as a car goes over the dam. It is possible to see the two inclined lines indicating the displacement of the vehicle over time.

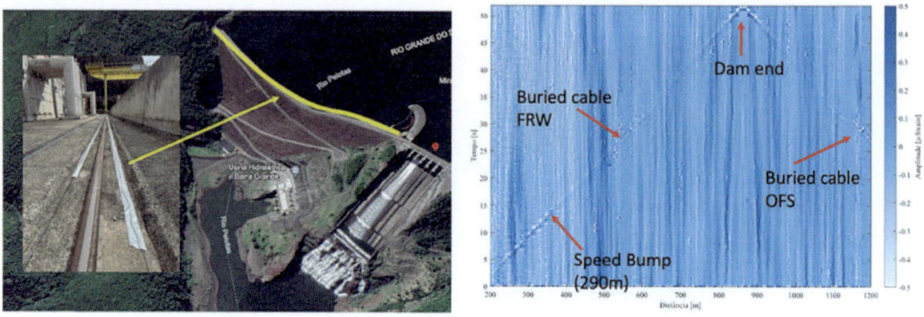

Fig. 7 The top image shows a satellite photograph image of the BAESA dam. The yellow line represents the two optical fiber cables that are connected in series placed on top of the dam as indicated in the other detailed photograph image at right hand side. The bottom graph shows the seismic signal measured by the cableas a car goes over the dam. It is possible to see the two inclined lines indicating the displacement of the vehicle over time.

Figure 8 shows the natural vibration signals at two locations, first nearby the electric generators, that are the sources of such vibration at approximately 16Hz, and secondly at the top of the dam showing that the vibration signal propagates the entire dam and can, consequently, be use as a seismic source for passive seismic.

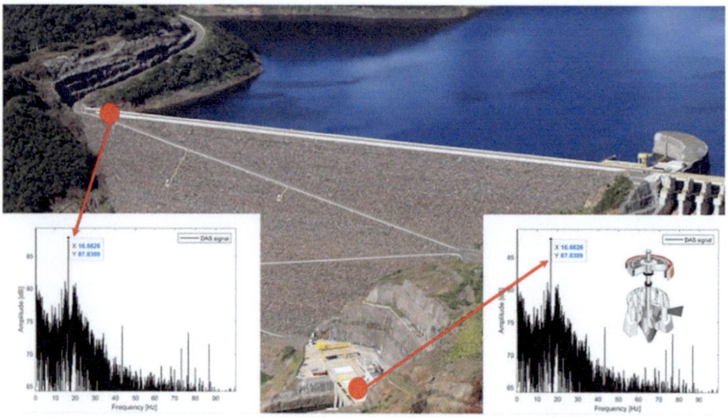

Fig. 8 Power plant natural vibration signals at two locations, first nearby the electric generators, that are the sources of such vibration at approximately 16Hz, and secondly at the top of the dam.

Considering the possible adjustments to identify forced variations in the soil, such as the region of wells, infiltration and small and medium-sized drillings, it is also intended to apply the same MASW method to the crest of the Barra Grande dam, for preliminary calibration using an active source technique, as already demonstrated in controlled environment laboratory tests. And after this calibration, there will be the application of passive MASW (Park et al. 2007), where the source of excitation will be the vibration of the generators, enabling online monitoring and allowing assessment with depth up to 180 m, the approximate height of the dam.

5 Conclusions and Future Work

In conclusion, method for seismic analysis of dams was proposed and partially validated and the first results are very promising. The prototype dam, unique in Brazil, was built for studies of optical fiber instrumentation and sensing and will bring a lot of opportunities on the development of novel sensing techniques as well as methods for processing the generated signals. First results for both controlled environment and the real dam (Barra Grande) demonstrate the huge potential of this technique that can be applied to virtually any dam in. Moreover, finally, is our expectation that by the end the project we will have demonstrated and developed the full solution that will allow the detection of possible defects inside the dam structure. There is also the potential of combining the seismic technique with machine learning methods to measure the flow of liquids inside the dams.

The authors would like to thank CPFL, Baesa, Enercan, and Ceran for technical and financial support through the Research and Development project PD-03936-2201/2022 – “Sistema de Monitoramento Sísmico de Alta Resolução de Barragens” with resources from ANEEL’s R&D program. We also especially thank the financial support provided by CNPq, CAPES, and FINEP. The laboratory LPGA/UFPR is thanked for the supply the geophones.

References

1. D18 COMMITTEE: Guide for using the seismic refraction method for subsurface investigation. [s.l.] ASTM International, 2018. https://doi.org/10.1520/D5777-18. Available in: http://www.astm.org/cgibin/resolver.cgi?D5777-18. Access 6 dez. 2023 (2023)
2. Olafsdottira, E.A., et al.: Combination of dispersion curves from MASW measurements. Soil Dyn. Earthq. Eng. (2018). https://doi.org/10.1016/j.soildyn.2018.05.025
3. Park, C.B.: MASW — Horizontal Resolution in 2D Shear-Velocity (Vs) Mapping. KGS Open-File Report 2005–4, February 17, 2005 (2005)
4. Yust, M.B., et al.: DAS for 2D MASW imaging: a case study on the benefits of flexible sub-array processing (2022). https://doi.org/10.48550/arXiv.2210.14261
5. Park, C.B., et al.: Multichannel analysis of surface waves (MASW)—active and passive methods. **26**(1), 1–112 (2007). ISSN (print): 1070-485X ISSN (online): 1938-3789 (2007). https://doi.org/10.1190/1.2431832
6. Johansson, S., Sjödahl, P.: Downstream seepage detection using temperature measurements and visual inspection—monitoring experiences from Røsvatn field test dam and large embankment dams in Sweden. In: Proceedings of international seminar on stability and breaching of embankment dams, vol. 21 (2004)

7. Woschitz, H., Klug, F., Lienhart, W.: Design and calibration of a fiber-optic monitoring system for the determination of segment joint movements inside a hydro power dam. J. Lightwave Technol. **33**(12), 2652–2657 (2015)
8. Goldswain, G., Wesseloo, J.: Advances in seismic monitoring technologies. In: Proceedings of the second international conference on underground mining technology, pp 173–188. Australian Centre for Geomechanics, Perth (2020)

Research on Real-Time Intelligent Control Technology for Runoff Cascade Hydropower Station Group

Bian Lijuan(✉), Yi Zhang, and Li Shuming

Guodian Nanjing Automation Co., Ltd, No. 8 Xinghuo Road, Pukou High Tech Development Zone, Nanjing, Jiangsu, China
849565465@qq.com

Abstract. An in-depth research was conducted on the real-time scheduling issues involved in the joint operation of runoff cascade hydropower station under high-intensity peak-load regulation and frequency regulation requirements, and proposed an intelligent control method for real-time load regulation of runoff cascade hydropower stations. Based on the characteristics of real-time power generation scheduling of cascade hydropower stations, ensuring the effectiveness and flexibility of scheduling strategies, taking into account the constraints of reservoir operation, grid safety, unit operation, and the impact of different water heads on unit power generation efficiency, aiming to achieve stable water level control of runoff type hydropower stations and rapid response of cascade total load control. Real time load Control is divided into two categories: dispatching mode and non dispatching mode. In the dispatching mode, constructed water power determination, stable water level, less load regulation, load balance economic dispatching models, to predict in advance whether the day ahead load planning curve needs to be adjusted and automatically track the planning curve. In non dispatch mode, automatically match the abnormal water level control model based on PID regulation and the stable water level model that meets the requirements of efficient power generation. At the same time, proposed a water level control solution method based on successive approximation and multi-objective dynamic programming, as well as a mixed integer programming model for reducing the difficulty of solving real-time hydropower scheduling for load allocation. The application of the proposed model in the cascade hydropower stations in the Shaxi River Basin shows that it can achieve high-precision automatic load adjustment, effectively reduce the number of regulation times by 6%, and only retain one dispatcher per shift, greatly reducing the work intensity of operators on duty. This research effectively improves the centralized control capability and economic operation level of the watershed, and has rich theoretical and practical significance for promoting the intelligent and intelligent construction of real-time load scheduling for cascade power stations.

Keywords: Cascade hydropower stations · Optimize scheduling · Real time load · Intelligent regulation

S. Zheng et al. (Eds.): IHDC 2024, LNCE 487, pp. 165–175, 2025.
https://doi.org/10.1007/978-981-97-9184-2_15

1 Introduction

The Shaxi River Basin is located in the central western part of Fujian Province and is one of the important tributaries of the Minjiang River. The basin covers an area of 11793 square kilometers and has a slightly feather shaped shape, with abundant hydraulic resources. The total storage capacity of Longtou Ansha Hydropower Station is 740 million m^3, with an installed capacity of 115000 kW. The reservoir has incomplete seasonal regulation performance, and its scheduling situation has a significant impact on downstream hydropower stations at all levels. The Shaxi Water Diversion Center is responsible for the economic optimization and flood control coordination of Ansha Reservoir, as well as six power stations including Fenghai, Ximen, Gongchuan, Shaxian Chengguan, and Gaosha. Except for Ansha Power Station, all other power stations are daily regulating runoff type, and the water level is greatly affected by the hydrological characteristics of the upstream and basin. The load and water level safety control of the power station are very sensitive to water level information. Water level fluctuations exceeding 10cm may lead to overtopping and dam accidents. Moreover, for every two personnel on duty in the centralized water control and regulation system, the load adjustment of the runoff power plant is very frequent and the workload is large, which seriously affects other flood prevention and dispatch work. It is urgent to achieve automation, digitization, and intelligence in load adjustment.

This paper mainly studies the joint operation problem of reservoirs in the Shaxi River Basin, except for the Ansha Hydropower Station. It grasps the operation characteristics of each cascade power station, research on water level control of runoff type daily regulation power stations, realizes load optimization distribution, ensures safe operation of power stations, improves the water energy utilization rate and comprehensive benefits of cascade power stations in the basin, and has important practical significance in reducing flood control pressure, reducing flood losses, reducing labor costs, and improving water energy utilization rate.

2 Current Situation and Technical Difficulties

Cascade hydropower stations are usually composed of multiple hydropower stations, each of which has complex water flow, energy conversion, and water level correlation relationships. The real-time scheduling of cascade hydropower stations needs to consider the changing water conditions and coordinated operation between various hydropower stations. Some hydropower stations are limited by reservoir capacity and real-time changes in water level on the dam [1]. Real time scheduling of cascade hydropower stations requires consideration of multiple functions such as power generation, water supply, and flood control, and requires balancing and balancing different objectives. At the same time, real-time scheduling of cascade hydropower stations needs to fully consider the safe and stable operation of hydropower equipment, to avoid equipment damage or accidents caused by frequent operations.

In terms of real-time automatic scheduling control, existing joint optimization scheduling technologies mostly achieve scheduling optimization through optimizing the preparation of power generation plans. However, due to the deviation between the actual load of the power system and the planned load, the method of simply optimizing power generation plans cannot meet the frequency stability requirements of the power system. The Gongchuan, Chengguan, and Gaosha power stations in the Shaxi River Basin allow for the modification of the scheduling plan curve 15 min before the set time to meet the scheduling needs of the central control center and respond more timely to changing electricity demand and water conditions. Therefore, it is necessary to calculate a scheduling control scheme with an appropriate length of time. The intelligent scheduling model of hydropower stations is generally nonlinear and multi constrained, and the solution of the hydropower scheduling model is very important for timeliness. The successive approximation dynamic programming method [2] is generally applied to solve the short-term scheduling model of cascade reservoirs. Many experts and scholars have conducted research in this area and achieved certain results [3].

3 Overall Technical Framework

3.1 Overall Strategy

The operation personnel of Shaxi Water Diversion Center issue real-time power generation load instructions to the cascade hydropower station group based on the system load demand or frequency changes. Real time load regulation of cascade power stations in the Shaxi River Basin can be divided into two categories: non dispatching mode and dispatching mode. The Fenghai and Ximen power stations adopt a non dispatching mode of water to electricity regulation, while the Gongchuan power station, Shaxian Chengguan power station, and Gaosha power station adopt a dispatching mode of market dispatch. These five power station reservoirs have very small storage capacity and basically do not have any regulating ability. The water level of the reservoirs is prone to steep rise and fall under the influence of the load of the higher-level power station.

In order to effectively control the changes in reservoir water level and avoid unnecessary water abandonment or reservoir depletion, fine monitoring and control of reservoir water level can be carried out by dividing the water level operation zone into high water level operation zone, normal operation zone, and dead water level operation zone. Set a water level control range $Z_{S,down} \sim Z_{S,up}$ between the dead water level $Z_{S,dead}$ and the normal storage level $Z_{S,store}$. If the real-time water level $Z_{S,t}$ on the dam meets the requirements $Z_{S,up} \leq Z_{S,t} \leq Z_{S,store}$, it is considered to have entered the high water level operation zone. If the real-time water level $Z_{S,t}$ on the dam meets the requirements $Z_{S,dead} \leq Z_{S,t} \leq Z_{S,down}$, it is considered to have entered the dead water level operation zone; If the real-time water level $Z_{S,t}$ meets $Z_{S,down} \leq Z_{S,t} \leq Z_{S,up}$, it is considered to be in the normal water level operating zone.

In non dispatch mode, take Ximen Power Station as an example, when the water level enters the operating range of high or dead water level and there is no trend of returning to the operational zone, the abnormal water level model is automatically matched to redistribute the power station load, so that the abnormal water level can return to its operational zone as soon as possible. When the water level is within the operational range, real-time parameters such as upstream flow rate and current water level are integrated to match the water level stationary model and maintain the water level in a stable state (Fig. 1).

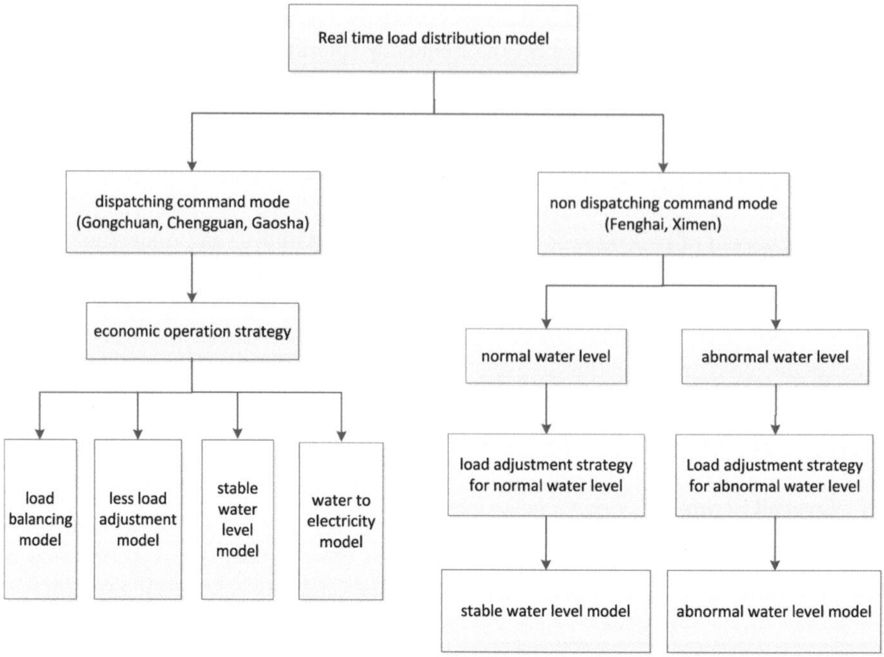

Fig. 1. Composition of real time load distribution model

The short-term power generation optimization scheduling of the Shaxi cascade hydropower station group generally takes 15 min as a scheduling period, and one or several days as the scheduling period. In the actual scheduling process, the total load of the hydropower station group needs to be adjusted in real-time according to the load indicators issued by the scheduling center, and it needs to meet the scheduling requirements while meeting the assessment requirements. In the dispatch mode, it is necessary to calculate the load for the next period 15 min in advance, and the system will automatically match load balancing model, stabile water level model, less load adjustment model, and water to electricity model of the economic operation strategy Model composition.

3.1.1 Constraints

1. Water balance constraint

$$V_{n,t+1} = V_{n,t} + (Q_{n,t}^{in} - Q_{n,t}^{out})\Delta t + V_{n,t}^{af} \tag{1}$$

where,

$V_{n,t}$ final storage capacities of the hydropower station during the time period [m^3].

$V_{n,t+1}$ initial storage capacities of the hydropower station during the time period [m^3].

$Q_{n,t}^{in}$ average inflow of the hydropower station during the time period [m^3].

$Q_{n,t}^{out}$ average outflow of the hydropower station during the time period [m^3].

$V_{n,t}^{af}$ changes in storage capacity caused by other factors of the hydropower station during the time period can be ignored [m^3].

2. Hydraulic connection between steps

$$Q_{n,t}^{in} = Q_{n-1,t-\tau}^{out} + Q_{n,t}^{mid}, n \geq 2 \tag{2}$$

where,

$Q_{n,t}^{in}$ average inflow of the power station during the time period [m^3/s].

$Q_{n-1,t-\tau}^{out}$ average outbound flow rate of the previous power station during the time period [m^3/s].

τ the time required for the outflow flow from the previous power station to reach the hydropower station [h].

$Q_{n-1,t-\tau}^{out}$ the interval flow between the previous power station and the hydropower station during the time period [m^3/s].

3. Reservoir water level constraint

$$Z_n^{\min} \leq Z_{n,t} \leq Z_n^{\max} \tag{3}$$

where,

$Z_{n,t}$ Real time water level on the dam for the power station during the time period [m].

$Z_n^{\min}$ The dead water level of the power station during the time period [m].

$Z_n^{\max}$ The storage level of the power station during the time period [m].

4. Power plant output range limitations

$$P_{n,t}^{\min} \leq P_{n,t} \leq P_{n,t}^{\max} \tag{4}$$

wheres,

$P_{n,t}^{\min}$ represents the minimum output of the power station during the time period [MW].

$P_{n,t}^{\max}$ represents the maximum output of the power station during the time period [MW].

5. Water level and storage capacity relationship

$$Z_{n,t} = f_n^{ZV}(V_{n,t}) \tag{5}$$

$Z_{n,t}$ the real-time water level on the dam of the power station at the beginning of the time period [m].

f_n^{ZV} the water level storage capacity relationship function of a hydropower station.

6. Tail water level outflow flow relationship

$$Z_{n,t}^{d} = f_n^{ZQ}(Q_{n,t}^{out}) \tag{6}$$

wheres,

$Z_{n,t}^{d}$ the tail water level of the power station during the time period [m].

f_n^{ZQ} the tail water of the hydropower station is a function of the outflow flow rate.

Power plant dynamic characteristics - NHQ curve function

$$P_{n,t} = f_n^{phq}(h_{i,t}Q_{n,t}^P) \tag{7}$$

whereses,

f_n^{phq} the dynamic characteristic function of the hydropower station.

3.1.2 Mathematical Models

1. Water to electricity model

On the premise of meeting the water level control requirements of various water conservancy comprehensive utilization departments and cascade power stations, calculate the discharge volume of each level of power station, and strive to generate electricity with the highest efficiency and increase power generation, enhancing the safety, stability, and economic operation of the hydropower system. Objective function:

$$E = \max \sum_{i=1}^{N} \sum_{i=1}^{T} k_i Q_{i,t} H_{i,t} \tag{8}$$

where,

E Generate electricity for cascade power stations;

k_i the power plant output coefficient;

$Q_{i,t}$ the power generation flow rate of the power station during the specified time period [m^3/s].

$H_{i,t}$ The average net head of power generation for the power station during the first period [m].

T the number of hours in the time period.

N the total number of cascade power stations.

2. Stable water level model

When the water level of the reservoir is within the control range or in the high water level operating area, but the inflow flow is less than the outflow flow, or when the water level is in the dead water level operating area, but the inflow flow is greater than the outflow flow, under the premise of meeting various safety constraints, with the minimum change in water level on the dam as the control objective, the inter plant load distribution is carried out according to flow balance, so as to achieve the matching of load and flow, and achieve the goal of stabilizing the water level of the reservoir on the dam as much as possible. Objective function:

$$\Delta H = \min \sum_{i=1}^{N} \sum_{j=1}^{T} \Delta H_{i,j} \tag{9}$$

ΔH the amount of water level change during the time period [m].

$\Delta H_{i,j}$ the change in water level of the power station during the time period [m].

3. Water level anomaly model

For power plants with small storage capacity and poor regulation capacity, the water level is greatly affected by upstream water. When the reservoir water level enters the high water level operating area or the dead water level operating area and there is no trend of returning to the operational area, the water level on the dam of the power plant is redistributed by the power plant load to be as close as possible to the middle value of its operational area after a certain period of execution according to the allocation results, achieving the goal of returning to the operational area.

PID controller has proportional, integral, differential and other links, and has become one of the main technologies in industrial control due to its simple structure, good stability, reliable operation, and convenient adjustment. By taking the outflow flow in front of the dam, the water level on the dam, and the difference between the calculated and actual load values as inputs, the real-time operating load can be flexibly increased or decreased to enhance the robustness of the water level control effect in front of the dam.

Set the input deviation $e(t)$ to and control the output to $u(t)$; When using PID control, there are:

$$u(t) = K_p e(t) + K_i \int_o^t e(t)dt + K_d \frac{de(t)}{dt} \tag{10}$$

wheres,

K_p proportional coefficient

K_i integral coefficient

K_d differential coefficient

Due to the different water level differences that deviate from the operational benchmark water level when the water level is abnormal, the adjustment of the unit load will also have different sizes of adjustment processes. Using the same set of PID adjustment parameters in the abnormal water level range is difficult to meet the adjustment requirements of different dam water levels, which may lead to slow adjustment speed, large overshoot, and multiple oscillations. Determine the corresponding relationship between water level difference and load based on the NHQ curve, determine the benchmark coefficient of the PID model, and take into account the amplitude of the current outflow flow rate change of the superior power station, and make a correction to this coefficient. It can be represented by the following equation.

$$K = K\prime \times K_b \tag{11}$$

wheres,

K the adjusted PID coefficient

$K\prime$ Model benchmark coefficient

K_b adjustment coefficient

4. Less load adjustment model

In actual operation, it is necessary to minimize the load fluctuations of certain power plants to ensure that the load allocated to the power plant has the smallest change relative to the current actual output. This requires inter plant load allocation to achieve the goal of some power plants adjusting their load more and some power plants adjusting their load less. On the basis of inter station economic operation scheduling, for specific regulation objects, it is necessary to calculate the number and magnitude of load changes, and introduce penalty factors separately. The ultimate optimization goal is to minimize the comprehensive penalty amount, thereby achieving the optimal regulation effect. Objective function:

$$\max F = \max \sum_{i=1}^{N} \left(\sum_{t=1}^{T} k_i \left| \Delta N_{i,t} \right| \times \sum_{i=1}^{N} l_i \left| \Delta h_{i,t} \right| \right) \tag{12}$$

wheres,

k_i penalty coefficient for power plant load adjustment.
l_i penalty coefficient for power plant load adjustment frequency.
$N_{i,t}$ real time load during power station period [MW].

3.2 Solving Algorithm

The water level changes of power plants are influenced by various factors, such as upstream outflow flow, weather changes, sensor measurement noise, etc. Therefore, water level changes are nonlinear, multi constrained, and time delayed. By adjusting the load to cause changes in the water level on the dam, the water level on the dam is self adjusted based on feedback and differential error states, making it closer to the operational water level range. By gradually approaching the optimal water level operating range through finite steps, the influence of time delay and nonlinear links on the control process can be overcome. The regulating system has adaptive ability and can cope with the characteristic changes between the water level on the dam and the load.

For runoff hydropower stations, water level control is divided into two processes. Firstly, it is necessary to control the real-time change of water level on the dam towards the benchmark water level. Then, when the actual water level reaches the target water level, the water level is kept stable, which is actually a multi-objective scheduling problem. In response to the "curse of dimensionality" problem in multi-objective dynamic programming, Xiwei Li proposed an iterative algorithm for multi-objective dynamic programming in reference [4], which can effectively obtain the non inferior solution set of multi-objective problems and has achieved good application results. Shushan Li, Shushan Li, and others proposed a cascade hydropower peak shaving method based on complex constraint normalization processing strategy in reference [5], and constructed a mixed integer programming peak shaving model based on MILP. The algorithm process is shown in Fig. 2.

3.3 Results and Analysis

The controlled water level range and vibration zone information in the Shaxi River Basin are shown in Table 1.

Fenghai and Ximen power stations do not need to operate according to the superior dispatch curve. Taking the flow rate change of Ansha Power Station exceeding 100 m^3/s

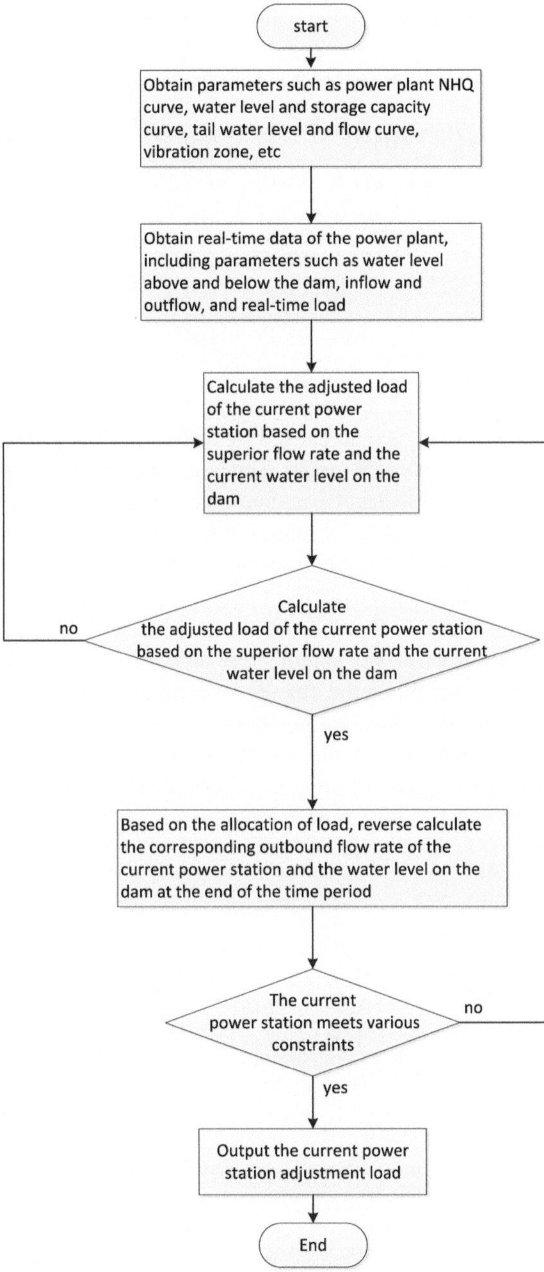

Fig. 2. Real time load distribution solution process

as an example, when the real-time water level on the dam in Fenghai and Ximen is within the benchmark operating range, the water level stable operation model is preferred, and the inter plant load distribution is based on flow balance. If it is not within the benchmark

Table 1. Power station parameters

station	High water level operating zone (m)	Benchmark operating zone (m)	Dead water level operating zone (m)	Single machine vibration zone (MW)
Fenghai	157.48	157.4–157.45	157.35	4–15
Ximen	166	165.9–165.95	165.87	3–15
Gongchuan	157.5	157.4–157.45	157.35	3–21.5
Chengguan	114.55	114.45–114.5	114.4	5–16
Gaosha	103.1	103–103.05	102.95	7–12.5

operating range, the water level change caused by upstream flow change is considered on the basis of the water level anomaly model. The Gongchuan, Chengguan, and Gaosha power stations adopt a water to electricity model, which is later transformed into models such as stable water level and low load regulation to ensure that each power station can achieve the goals of low load regulation and stable water level. The target load adjustment for each station is shown in the table for different changes in outbound flow of Ansha. From this Table 2, it can be seen that when the change in the flow rate of Ansha is small, all levels of power stations can achieve less or no load adjustment through reasonable allocation; When the change is significant, the load of each level of power station can respond and adjust in the shortest time possible. At the same time, through secondary load distribution, it ensures that each level of power station operates in a more economical and reasonable load range, reducing the load adjustment of power stations while achieving maximum economic efficiency.

Table 2. Load distribution adjustment results

Ansha flow changes (m^3/s)	Initial load (MW)	10	50	150
		Adjusting load (MW)	Adjusting load (MW)	Adjusting load (MW)
Fenghai	8	0	5.6	14.6
Ximen	10	0	3.6	10.9
Gongchuan	12	0	4.3	13
Chengguan	14	0	4	11.8
Gaosha	12	0	5	13.5

4 Conclusion

In the cascade power station group in the Shaxi River Basin, the runoff type power station aims to achieve water level safety control and rapid response to the total load regulation of the cascade. It constructs economic dispatch models such as water fixed electricity, abnormal water level, stable water level, less load regulation, and load balance, and establishes model clusters for both regulation and non regulation modes to achieve intelligent load regulation of high-intensity peak shaving and frequency regulation needs in the power system. Taking 2023 as an example, the proposed model can achieve high-precision automatic load adjustment, effectively reducing the number of adjustments by 6%, and retaining only one dispatcher per shift. This study effectively enhances the centralized control capability and economic operation level of the watershed, and has rich theoretical and practical significance for promoting the intelligent construction of real-time load scheduling for cascade power stations.

References

1. Jiekang, W., Jianquan, Z.: A new strategy for short-term scheduling optimization of cascade hydro plants based on chance-constrained programming. Proc. CSEE. **28**(1), 41–46 (2008)
2. Gao, H.: Run-of-river hydropower plant power generation optimize and short-term load capacity forecast. School of Electrical and Electronic Engineering (2017)
3. Wang, B., Li, H.-G., Ai, X.-R.: Optimal operation of cascade reservoirs based on dynamic feasible region (2022). https://doi.org/10.20040/j.cnki,1000-7709.2022.20212599
4. Li, J.: Research on multi-objective optimal operation and decision methods for cascaded reserviors. School of Economics and Management (2014)
5. Shushan, L., Li, C., Wu, H.: Peak shaving method of cascade hydropower stations based on normalization processing strategy with complex constraints. Power Syst. Technol. **47**(9), 3576–3585 (2023)

Application of Tower Type Cyclone Stabilizing Cylinder Concentration Technology in Wastewater Treatment of Sand and Gravel Processing System of Batang Hydropower Station

Wei Zhang[1,2(✉)], Xingyu Li[3], and Dong Zhang[4]

[1] Zhejiang Huadong Engineering Consulting Co., Ltd., Hangzhou, China
707014588@qq.com
[2] Huadong Engineering Corporation Limited, Hangzhou, China
[3] Sichuan Duowei Filtration Equipment Co., Ltd., Deyang, China
573581624@qq.com
[4] Guiyang Engineering Corporation Limited, Guiyang, China
759439326@qq.com

Abstract. The artificial sand and stone processing plant has always been a major pollutant discharge household in hydropower station construction, and the problems of dust and waste water are more prominent. In order to solve the problem of waste water, the artificial sand and stone processing system often uses the wet production process, combined with the sewage treatment facilities to treat the waste water produced by the system, and the pollution particles are precipitated and dehydrated again. The proper measures can realize the zero discharge of waste water. The treatment method of flocculation sedimentation and mechanical dehydration is often used in domestic sand and stone processing system wastewater. The key of the treatment process is flocculation sedimentation, and the control of sludge floc formation speed and separation effect. This paper introduces a new flocculation and sedimentation technology "tower type cyclone stabilizing cylinder concentration technology". This technology absorbs the advantages of tower type sedimentation technology and cyclone type sedimentation tank technology. By setting up mixed flow tank, stabilizing cylinder and other structures, and combining with field flocculation test and technical fine-tuning, the optimal dosage of flocculant is determined, so as to control the formation speed and separation effect of sludge flocs in a relatively ideal range Status. The process has been successfully applied in the no wastewater treatment of the sand and gravel processing system of Batang hydropower station, and good results have been achieved, which is worthy of reference.

Keywords: artificial sand and stone processing plant · flocculation sedimentation · tower sedimentation · cyclone sedimentation · sludge thickening · chamber filter press

S. Zheng et al. (Eds.): IHDC 2024, LNCE 487, pp. 176–185, 2025.
https://doi.org/10.1007/978-981-97-9184-2_16

1 Preface

With the development of China's economy and society, the problem of environmental pollution has become increasingly prominent, people's awareness of environmental protection has gradually increased, and green, efficient and sustainable development has become the trend of current social development. A large amount of sand and gravel aggregates are required in the construction process of hydropower projects, and artificial sand and gravel processing systems are often set up in the work area in order to solve the aggregate supply. The artificial sand and gravel processing system has always been a major polluter in the construction of hydropower stations, and the problems of dust and wastewater are more prominent. The processing technology of artificial sand and gravel processing system is divided into two types: dry and wet, and the wet production process is often used in engineering construction, with the wastewater generated by the sewage treatment facility treatment system, and the pollution particles are precipitated and dehydrated, and the measures are appropriate to achieve zero discharge of wastewater.

The wastewater of domestic sand and gravel processing system often adopts the treatment method of flocculation sedimentation + mechanical dehydration, the flocculation and sedimentation methods mainly include advection sedimentation, radiation precipitation and cyclone precipitation, and the mechanical dehydration methods mainly include belt filter press, disc vacuum filter and chamber filter press [1–3]. At present, the commonly used mechanical dehydration method is the chamber filter press, which has the characteristics of good filtration effect, low equipment cost, low operating cost, good equipment stability, easy maintenance and overhaul, etc., and has a wide range of applications in sand and gravel processing systems. The selection of flocculation and sedimentation mode is generally determined according to the production intensity of the sand and gravel processing system and the content of waste cement sand. Several wastewater flocculation and sedimentation treatment methods have the characteristics of large footprint, limited sludge collection volume and large operation workload, and the sludge cannot be removed and transported in time, and often cause the sludge discharge facilities to be easily blocked and damaged [4]. In this paper, a new flocculation and sedimentation technology "tower cyclone stabilizer cylinder concentration technology" is introduced, which has been successfully applied in the wastewater treatment of the sand and gravel processing system of Batang Hydropower Station, and has achieved good results, which is worthy of reference.

2 Project Overview

Batang Hydropower Station is located on the main stream of the Jinsha River at the junction of Sichuan and Tibet, surrounded by lofty mountains and mountains, the right bank of the dam site is Mangkang County, Qamdo, Tibet, and the left bank is Batang County, Ganzi, Sichuan, and the dam site is 9 km south of Batang County. The power station adopts the hub layout pattern of asphalt concrete core wall dam rockfill dam, left bank spillway, open pipe water diversion and ground plant, mainly for power generation, and is the 9th power station among the 13 cascade power stations in the approved hydropower plan in the upper reaches of the Jinsha River, with a total installed capacity of 750 MW.

The sand and gravel processing system of Batang Hydropower Station needs to undertake the production and supply of sand and gravel aggregates of 1,309,200 m^3 of concrete (including about 1,270,100 m^3 of normal concrete and about 39,100 m^3 of sprayed concrete), 214,800 m^3 of dam transition material and 61,900 m^3 of dam filter material of Batang Power Station, and the production capacity of finished aggregate is not less than 700 t/h and the wool processing capacity is not less than 850 t/h. The water equipment in the system is the second screening workshop, large stone, medium stone, small stone washing screen, rod mill, coarse crushing, medium and fine crushing workshop spray dust removal, stone powder washing and other equipment, as well as road washing. Combined with the process design of the sand and gravel processing system, the water consumption of the system is calculated to be 512.14 m^3/h, and the wastewater is generated to be 451.2 m^3/h, and the process design and configuration of the wastewater treatment system is carried out according to 500 m^3/h in order to ensure the normal and stable production and operation of the wastewater treatment system.

Fig. 1. Geology of the source of concrete wool mining

The raw material of the sand and gravel processing system is weakly weathered and slightly weathered material on the left bank slope, and the main component is biotite quartz schist. The geological structure of the wool reclaiming area is relatively complex, mostly strong weathering, fault joints are relatively developed, and a large amount of mud is mixed in the fractures, which is poor in overall quality as wool and has a large concentration of pollutants in the production of waste cement slurry. See Figs. 1 and 2 for details. On the basis of a comprehensive comparison of the layout area and sewage treatment capacity of different sewage treatment process systems, the technical staff selected the tower swirl stabilizer wastewater concentration and purification process. The process has the characteristics of simple structure, flexible size, low cost, small footprint, good sludge concentration effect, high treatment efficiency, high circulating

Fig. 2. Wastewater from sand and gravel processing system

water quality, low failure rate and easy operation, which can achieve all sewage recycling and realize "zero" discharge of production sewage.

3 Tower Cyclone Stabilized Cylinder Sewage Wastewater Concentration and Purification Process

3.1 Process Principle

The core of the wastewater concentration and purification process of the tower cyclone stabilized cylinder lies in the sludge thickening technology. First of all, an appropriate amount of flocculant is added to the wastewater for mixing reaction, and then the sludge floc and clarified water with a higher concentration are separated from the wastewater through the tower cyclone stabilizer thickening tank (hereinafter referred to as the high-level concentration tank) dedicated to this process. The clarified water overflows directly into the clear water pool, and the sludge floc with higher concentration is transported to the box filter press through the sludge pump and sludge pipeline, and is dewatered and separated into sludge cake and clean water [4–8].

3.2 Process Flow Diagram

See Fig. 3.

3.3 Schematic Diagram of the Overall Sewage Treatment Process

See Fig. 4.

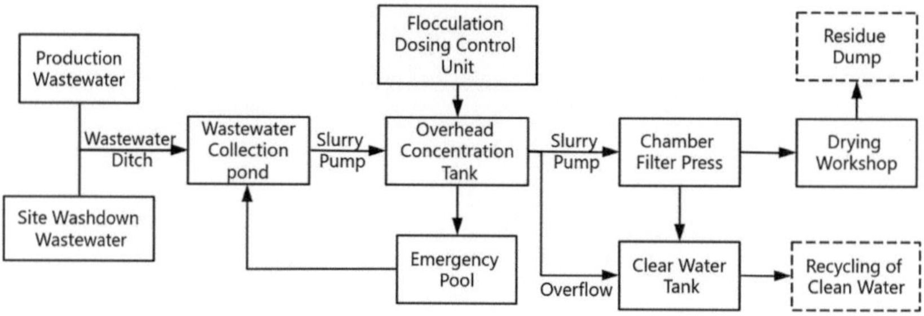

Fig. 3. Sewage treatment process flow diagram

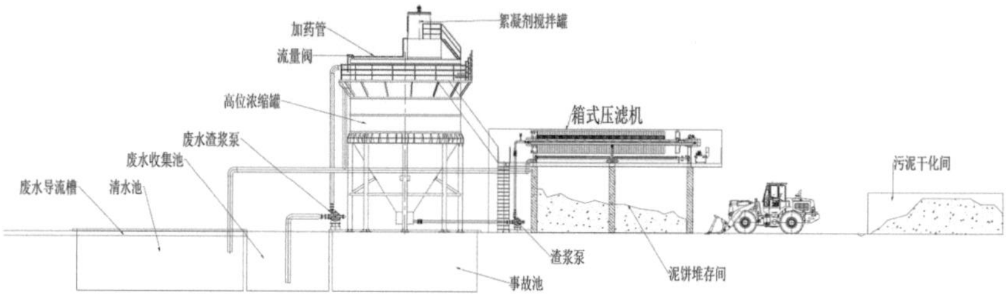

Fig. 4. Schematic diagram of sewage treatment process

3.4 Wastewater Collection

The wastewater generated in the process of sand and gravel aggregate processing and site flushing wastewater are introduced into the wastewater collection tank through the wastewater diversion tank, and the wastewater is transported to the mixed flow tank of the high-level concentration tank through the slurry pump and conveying pipe set up.

3.5 High-Level Concentrator Tank

The high-level concentration tank is a solid-liquid separation equipment specially designed based on the principle of gravity sedimentation, which is composed of wastewater feed pipeline, mixed flow tank, steady flow cylinder, concentration tank body, overflow tank, mud transportation pipeline and other components.

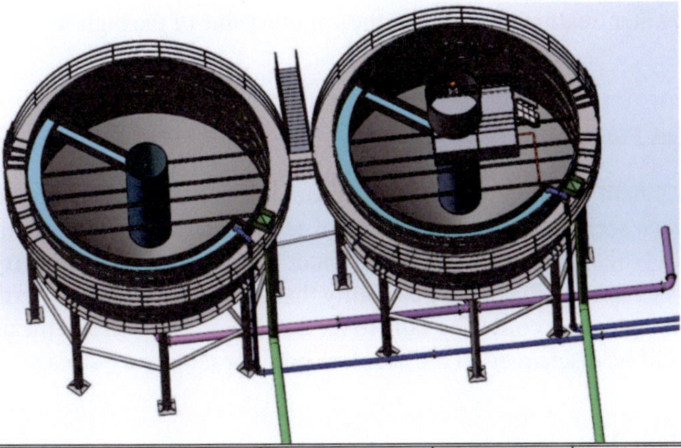

Wastewater	Flocculation agents	Wastewater after flocculation	Clarified water	Concentrated slurry

The mixed flow tank is a mixing device of waste water and flocculant, the wastewater in the collection tank is transported to the top of the mixed flow tank through the slurry pump, the flocculation device injects the flocculant into the sewage, and the wastewater and the flocculant are mixed evenly in the mixed flow tank under the action of gravity, and are injected into the steady flow barrel through the guide tank.

The steady flow barrel is vertically arranged in the center of the tank, and the evenly mixed sewage is injected into the high-level concentration tank through the steady flow barrel, and the component mainly plays the role of stabilizing the water flow of the tank.

The upper part of the thickening tank is cylindrical, and the lower part is conical, which is the reaction vessel of wastewater and flocculant. Wastewater and flocculant are mixed and reacted in the mixed flow tank and then flow into the steady flow cylinder of the high-level concentration tank, and the flocculation reaction occurs in the concentration tank body, and the flocculation reaction occurs in the concentration tank body, and the flocculent matter with large specific gravity is separated from the water and deposited downward to the bottom of the high-level concentration tank, and the clarified water after separation gradually rises and flows into the clear water pool from the tank overflow tank.

The bottom of the lower conical tank is provided with a slurry outlet, which is connected with the chamber filter press through the pipeline, and a slurry pump is arranged on the pipeline to transport the high-concentration mud at the bottom of the tank to the filter press. The sludge deposited at the bottom of the tank enters the filter press through the sludge output pipeline and the filter press feed pump for solid-liquid separation.

The practice shows that the precipitation effect of wastewater staying in the high-level concentration tank for more than 40 min is better. Therefore, the structure design of the high-level tank steady flow cylinder should ensure that the effective volume of the tank is not less than 40 min sewage production, and generally take 1–1.5 times of the hourly wastewater treatment volume as the volume value of the high-level concentration tank.

3.6 Accident Pool

There is an accident pool below the sludge thickening tank, which is a temporary wastewater storage tank set up to prevent the failure of each operating component in the wastewater treatment process and affect the wastewater treatment. When the slurry pump fails and needs to be replaced, there is a possibility that the wastewater will overflow from the cone-bottom wastewater collection tank to the outside of the pool, and the tank overflow wastewater will be discharged to the accident pool for temporary storage.

3.7 Flocculant Test

The dosage of flocculant is one of the core keys of the control of this process method, and the appropriate amount of flocculant can accelerate the precipitation of sludge particles in sewage, and at the same time can prevent the separation of sludge and water too quickly, resulting in the blockage of the sludge pipeline. The amount of flocculant should ensure that most of the suspended flocculent particles in the wastewater can be separated and precipitated to the lower part of the tank within 15–30 min under the action of gravity.

The dosage of flocculant needs to be determined in the indoor flocculant precipitation test before formal production, and the method steps are as follows:

1. Take 3 1000 mL graduated cylinders and fill each with 1000 mL of wastewater.
2. Different doses of flocculant were mixed into three graduated cylinders, the sedimentation of particulate matter in the water was observed, and the position of the sedimentation interface was recorded every 1 min.
3. Draw the settlement curve based on the recorded data.
4. After standing for 10 min, the supernatant was precipitated and tested, and the SS should be less than 100 mg/L.

Considering that the water in the high-level concentrated tank is constantly stirring during the actual production, the sedimentation time of the wastewater should be controlled at 10–15 min in the indoor flocculation and sedimentation test. According to the flocculation and precipitation test, the amount of flocculant with a sedimentation completion time of 10–15 min was selected as the production control dosage. If the precipitation time of the three groups of tests cannot meet the requirements, the amount of flocculant will be adjusted and then another round of tests will be carried out until the flocculant output that meets the requirements is selected.

3.8 Flocculation Dosing Control Device

The flocculation dosing control device is arranged at the top of the high-level concentration tank, and is composed of a drug filling device, a stirring tank device and a drug storage box. The main function is to dissolve the flocculant and control the measurement of the flocculant added to the wastewater. The step is to first dilute the flocculant in the agent storage box, then further dilute the diluted agent in the stirring tank device, and finally adjust the flow control valve on the screw push rod of the drug addition device and fill it into the wastewater of the mixed flow tank.

The mixed flow tank is a circular shape, and the flow direction of the wastewater and the flocculant is constantly changing while flowing in the mixed flow tank, and the hydraulic turbulence of the wastewater flow makes the flocculant and the wastewater fully mixed and stirred, and the wastewater and the flocculant react with the flocculant. Observe the alum flower produced after the reaction between the wastewater and the flocculant in the mixed flow tank, and appropriately adjust the flow control valve on the automatic agent filling device to ensure that the wastewater and the flocculant react quickly to form large flocculents to achieve the purpose of sludge thickening.

3.9 Filter Press

The chamber filter press is a kind of intermittent pressure filtration equipment, and its main function is the dewatering of sludge, so that the mud cake with low moisture content formed after the filter press is convenient for transportation [7].

The main structure of the chamber filter press is that the filter plates are arranged to form a filter chamber, the two sides of the filter plate are recessed, and every two filter plates are combined to form a compartment-shaped filter chamber, and the filter cloth is embedded. There are through holes in the center and corners of the filter plate, which after assembly form a complete channel, which can pass through the suspension, wash water, and lead out the filtrate. There are handles on both sides of the filter plate to support the cross beam, and the filter plate is pressed by the pressing device. The filter cloth between the filter plates acts as a seal.

In wastewater treatment, under the pressure of the conveying pump, the wastewater that needs to be filtered is sent to each filter chamber, and the solid and liquid are separated through the filter cloth. Filter residue is formed on the filter cloth until it fills the filter chamber to form a filter cake. The filtrate passes through the filter cloth and flows along the filter plate groove to the lower outlet hole channel for centralized discharge. After filtration, open the chamber filter press to remove the filter cake.

In order to ensure that the sewage treatment work of the filter press does not stagnate when the filter cake is discharged, one unit is generally added as a backup on the basis of the theoretical configuration number.

By means of a chamber filter press, most of the free water in the large floc sludge deposited at the bottom of the high-level thickening tank can be separated from the sludge.

3.10 Clear Water Pool

The clear water tank is used to collect the overflow clarified supernatant of the high-level concentration tank and the clarified filtrate filtered by the chamber filter press, and the clean water in the clear water tank is used for the production water of the artificial aggregate production site to achieve zero discharge of wastewater. The clear water pool is made of concrete and arranged around the perimeter of the high-level concentration tank.

3.11 Sludge Drying Tank

The sludge drying tank is mainly used to store the sludge cake mixed with the filter press for drying and dewatering. When the sludge in the sludge drying tank reaches a certain amount, it is cleaned and transported to the slag yard by dump truck for storage.

4 Running Effects

The wastewater treatment system of the sand and gravel processing system of Batang Hydropower Station was officially put into operation in May 2020, and it was controlled in strict accordance with the "zero discharge" of production wastewater. According to statistics: as of January 10, 2021, sand and gravel processing has produced a total of 765,173 tons of various aggregates and 1,007,300 tons of production wastewater, all of which have been recycled by the concentration technology of the lap cyclone stabilizer cylinder, with a wastewater recovery rate of 100% and about 100,000 tons of waste residue.

After testing, the SS of solid suspended solids in sewage produced before treatment was about 10000 mg/L; After being treated by the tower cyclone stabilizer cylinder concentration technology, the SS of the sewage solid suspended solids is not higher than 100 mg/L, and the removal rate of the main pollutants reaches more than 99.5%, which meets the requirements of production water, realizes the "zero" discharge of sewage, effectively protects the water quality of local rivers, and achieves good environmental protection effects.

5 Summary

The key to the sand and gravel processing wastewater treatment process is flocculation and sedimentation, and the key to the flocculation and sedimentation treatment process is the speed of sludge floc generation and separation effect. The formation of sludge floc is slow, and the solid-liquid separation effect is poor, which will lead to poor removal rate of contaminants from the overflow supernatant, and the system has to increase the secondary sedimentation facility, which increases the treatment cost [5]. The formation of sludge floc is too fast, the solid-liquid separation effect is good, it is easy to cause the blockage of the sludge pipeline, and the system failure maintenance rate is high [8, 10].

The new flocculation and sedimentation technology "tower cyclone stabilizer cylinder concentration technology" introduced in this paper can better solve the problems of sludge floc generation speed and separation effect. This technology draws on the advantages of tower sedimentation technology and cyclone sedimentation tank technology [4, 5, 10]. By setting up the mixed flow tank, stabilizer cylinder and other structures, and combining the on-site flocculation test and technical fine-tuning to determine the optimal dosage of flocculant, the formation rate and separation effect of sludge floc are controlled in a relatively ideal state.

Although the construction investment and operating cost are relatively high, the sewage treatment effect is good, and the large water resources can be saved, and it has a wide range of popularization and use value in areas where environmental protection requirements are relatively high, the terrain is narrow or water resources are scarce.

References

1. Liu, W., Tu, M.: Preliminary discussion on wastewater treatment process and equipment selection of artificial gravel processing system. Sichuan Hydropower **27**(006), 43–45,67 (2008)
2. Shi, X., tie, C., Liu, J., et al.: Research and application of energy conservation, emission reduction and environmental protection technology of artificial gravel system. China Three Gorges **6**, 81–85 (2013)
3. Zhu, C., Lin, C.: Wastewater treatment design of Mayanpo sand and gravel processing system of Xiangjiaba Hydropower Station. In: Proceedings of the second sand and gravel production technology exchange conference of China water resources and hydropower projects (2008)
4. Shi, Z., Sun, G.: Application of tower sedimentation technology in sewage treatment of gravel aggregate processing system of hydropower project. Water Resour. Hydropower Constr. 000(004), 37–40 (2018)
5. Wang, F.: Study and application of two-stage series cyclone flocculation instead of ordinary concentration and sedimentation in the treatment of beneficiation wastewater. Environ. Eng. **014**(001), 32–33 (1999)
6. Liu, Z., Wang, H., Liu, K., et al.: Preliminary study of wastewater pre-treatment design key points in hydropower project aggregate processing system. Zhejiang Water Conservancy Sci. Technol. **43**(006), 39–42 (2015)
7. Sun, J.: Experimental study on sediment vacuum dehydration of aggregate processing wastewater. Chongqing University (2011)
8. Yang, Q.: The designation of stone processing sewage purification and tower separation recycling system. Shandong University (2011)
9. Wang, T., Sun, J., Lang, J., et al.: Technological exploration for wastewater treatment of the aggregate processing system in the hydropower station. Water Treat. Technol. **37**(5), 66–69 (2011)
10. Luan, C.: Optimization design on CFD of grit chamber with rotational flow in sandstone wastewater treatment of hydropower engineering. Tianjin University (2009)

Seismic Safety Evaluation of a High Arch Dam-Foundation Coupling System

Chunli Yan, Jin Tu(✉), Hui Liang, Shengshan Guo, and Deyu Li

China Institute of Water Resources and Hydropower Research, Beijing 100038, China
tujin@iwhr.com

Abstract. The seismic safety evaluation of the dam is performed based on either dam strength failure or dam abutment instability failure according to the traditional deterministic methods and concepts in the current code. However, these two failure modes are interactive and inseparable, considering only one failure mode under a strong earthquake fails to fully reflect the actual seismic performance of high arch dams. To develop a more realistic seismic safety evaluation framework for high arch dams, this paper constructs a finite element analysis model that considers the coupling of dam strength failure and dam abutment instability failure. The nonlinear dynamic response analysis of the high arch dam-foundation coupling system is conducted using the seismic overload analysis method. Different performance evaluation indexes, such as damage depth-thickness ratio, sliding area ratio, and the residual displacement of the dam crest relative to the dam bottom in the stream direction, are proposed. The performance evaluation criteria and ultimate seismic capacity are defined and quantitatively assessed. The results indicate that 2.0 times the horizontal PGA can be considered the ultimate seismic capacity of the high arch dam-foundation coupling system, providing a reliable scientific basis for seismic safety evaluation of high arch dams.

Keywords: high arch dam · strength failure · instability failure · seismic safety evaluation

1 Introduction

Seismic response and failure mechanisms of concrete dams are highly complex under strong earthquakes. Overstressing, joint opening and sliding, and other failure modes may simultaneously occur under strong earthquakes, according to the existing cases of concrete high dams suffering from strong earthquakes [2, 3, 6]. These different failure modes are key concerns in the seismic design and safety evaluation of high dams.

Fruitful research on dam strength failure and dam abutment instability failure has been achieved. Omidi et al. [16, 17] employ a plastic-damage model to simulate the stiffness degradation and permanent deformation of dam concrete. Alembagheri et al. [1] conduct damage assessment of arch dams by nonlinear incremental dynamic analysis and investigate the damage propagation through the dam body. Guo et al. [8] investigate seismic damage distribution and maximum joint opening of arch dams under different

S. Zheng et al. (Eds.): IHDC 2024, LNCE 487, pp. 186–196, 2025.
https://doi.org/10.1007/978-981-97-9184-2_17

water levels, taking into account contraction joints and concrete damage cracking. Zenz et al. [19] suggest that while the rigid body method may be conservative for analyzing abutment stability under strong earthquakes, the finite element method is highly suitable for such analysis. Liang et al. [12] establish a 3D finite element model of an arch dam-block-foundation system considering contraction joints, dam-foundation interface, and potential sliding blocks of dam abutment. The influence of uplift pressure on the stability of dam abutment is studied [15]. However, all of these existing studies only consider dam strength failure or sliding instability failure of dam abutment, making it difficult to truly reflect the seismic failure process and mechanism of high arch dam-foundation systems. At present, there is no relevant research on seismic analysis of high arch dam foundation systems considering the coupling of two failure modes.

Moreover, seismic analysis and safety evaluation of high arch dams are quite complicated problems that are affected by many complex factors. Plenty of researchers have been devoted to finding appropriate evaluation indexes and evaluation criteria. Hariri-Ardebili et al. [9, 10] use demand-capacity ratio, cumulative inelastic duration, and extension of overstressed areas on upstream and downstream faces as performance indexes to emphasize the necessity of nonlinear analysis. Liang et al. [13] select sliding displacement and sliding area ratio as an index for seismic sliding stability evaluation of high dams. While there are no universal performance indexes and safety evaluation criteria for concrete dams, a single evaluation standard can no longer meet the requirements [20]. Therefore, it is necessary to perform seismic safety evaluation of high arch dams from various perspectives, using a comprehensive finite element model that truly reflects the seismic response of high arch dam-foundation systems.

Hence, a high arch dam-foundation coupling model is established, considering the material nonlinearity of dam concrete and contact nonlinearities of the dam-foundation interface, contraction joints, and sliding surfaces of potential dam abutment blocks. Nonlinear seismic response analyses are performed under different overload coefficients. Then, the seismic failure process and mechanism are investigated using evaluation indexes such as the damage depth-thickness ratio, sliding area ratio, and residual displacement of the dam crest relative to the dam bottom in the stream direction. Moreover, the ultimate seismic capacity is comprehensively evaluated, which provides a reliable scientific basis for the seismic safety evaluation of high dams.

2 Computational Model and Condition

2.1 Finite Element Model

Taking an arch dam as an example, a three-dimensional finite element model of a high arch dam-foundation coupling system is first constructed, incorporating the dam, potential sliding blocks at the left and right dam abutments, and the foundation, as shown in Fig. 1. The finite element model of the dam body is depicted in Fig. 2. Figure 3 shows potential sliding blocks at left and right dam abutments. The large-scale finite element model comprises 3548298 nodes and 3329118 elements, resulting in a total of 10.64 million degrees of freedom. The dam is separated into 31 adjacent monoliths by 30 vertical contraction joints according to the actual situation.

Fig. 1. The finite element model of high arch dam-foundation coupling system

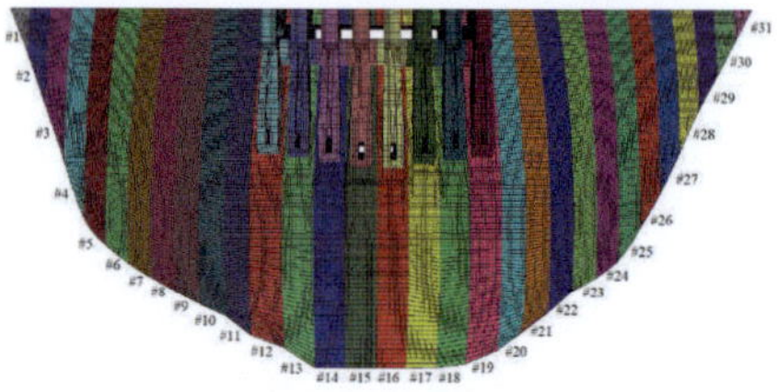

Fig. 2. The finite element model of the dam

Fig. 3. The left and right sliding blocks and its supporting surfaces

2.2 Computational Condition

Material nonlinearity of the dam and contact nonlinearity of contraction joints, dam-foundation interface, and potential sliding blocks at the dam abutment are comprehensively considered. The material nonlinearity of the dam concrete is simulated using the concrete damage model, as described in references [4, 5, 11]. The damage evolution curve of dam concrete is shown in Fig. 4. The contact nonlinearity of various joints is simulated using the dynamic contact model, detailed in reference [7]. The shear parameters of different joint interfaces are shown in Table 1.

The static and dynamic load: the self-gravity of the dam, water pressure, sediment pressure, uplift pressure, and seismic load. The upstream normal water level is 600 m. Upstream sediment elevation is 490 m. The bulk density of sediment is 5 kN/m3 and the internal friction angle is 0°. Dynamic interaction between the dam and reservoir is modeled using Westergaard's added mass model [18]. The horizontal peak ground acceleration (PGA) is 0.432 g, and the vertical PGA is 2/3 of it. The time histories of three components of artificial earthquakes are shown in Fig. 5.

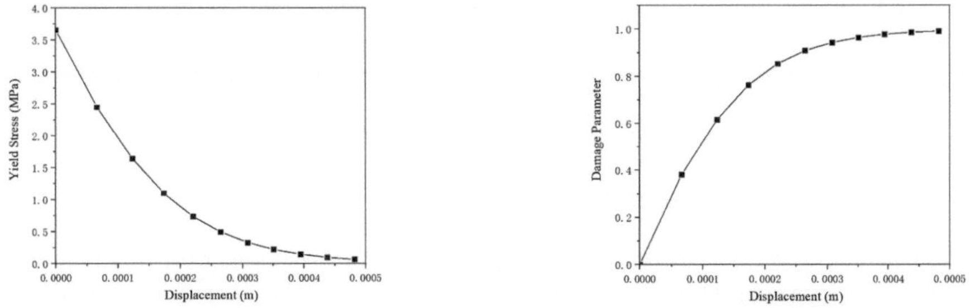

(a) Softening curve (b) Damage curve

Fig. 4. The curve of dynamic damage evolution of dam concrete

Table 1. Shear parameters of different joint interfaces

Joint interface	Friction coefficient	Cohesion (MPa)
Contraction joints	0.8	0.0
Dam-foundation interface	1.1	1.1
Left side sliding surface	1.2	1.8
Left bottom sliding surface	0.435	0.09
Left cracking surface	0.0	0.0
Right side sliding surface	1.2	1.8
Right bottom sliding surface	0.5	0.17
Right cracking surface	0.0	0.0

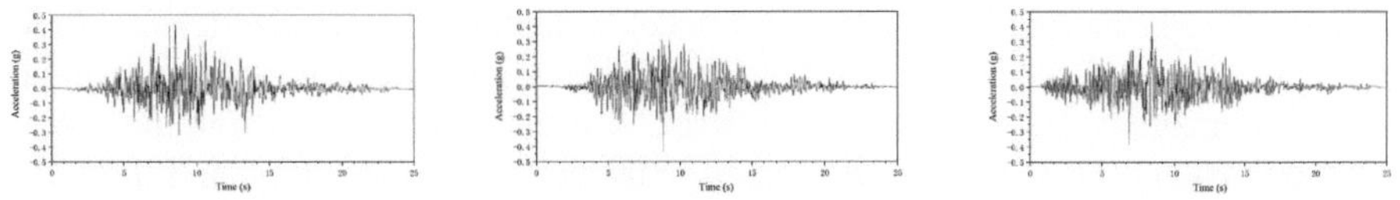

(a) Cross-stream direction (b) Stream direction (c) Vertical direction

Fig. 5. Artificial earthquake

3 Results and Discussions

The nonlinear dynamic response analysis of the high arch dam-foundation coupling system is performed using the seismic overload analysis method. Different seismic overload coefficients of 0.2, 0.4, 0.6, 0.8, 1.0, 1.2, 1.3, 1.4, 1.5, 1.6, 1.7, 1.8, 2.0, 2.2, and 2.4 are employed. Different indexes are proposed to qualitatively and quantitatively evaluate the seismic performance of dams, including the ratio of maximum damage cracking depth Ldc to dam thickness T at the same elevation (maximum damage depth-thickness ratio, $\alpha_{LT} = \mathrm{L_{dc}}/\mathrm{T}$), the ratio of effective sliding area Ars to total area A of the bottom sliding surface (sliding area ratio, $\beta_{rs} = A_{rs}/A$), and the residual displacement of the dam crest relative to the dam bottom in the stream direction. Additionally, performance criteria are defined.

3.1 Strength Failure Mode-Based

Figure 6 shows damage distributions of the dam cantilever section corresponding to the maximum damage depth-thickness ratio under different seismic overload coefficients. Figure 7 shows the maximum damage depth-thickness ratio under different seismic overload coefficients. From Fig. 6 and Fig. 7, slight damage occurred at the dam heel, and no macroscopic cracking (generally corresponding to a damage factor greater than 0.8) occurred at the dam head when earthquake overload coefficient ranges from 0.0 to 0.6. Slight damage occurred at the upstream and downstream surfaces, with the maximum damage depth-thickness ratio ranging from 2.5% to 49% when the earthquake overload coefficient ranges from 0.8 to 1.6. As the overload coefficient continues to increase, damage cracking rapidly increases and gradually extends to the interior of the dam. Moreover, severe damage occurs at upstream and downstream surfaces. When the overload coefficient is 1.8, the maximum damage depth-thickness ratio reaches 100%, indicating the occurrence of a penetrating crack.

Thus, if the occurrence of a penetrating crack is taken as the failure criterion, the ultimate seismic capacity of the dam based on the damage indexes is 1.7 times the horizontal PGA, i.e. 0.7344 g.

3.2 Instability Failure Mode-Based

Figure 8 shows the residual sliding displacement contours of left and right bottom sliding surfaces under different overload coefficients. Figure 9 shows the sliding area ratio of left and right bottom sliding surfaces under different overload coefficients. From Fig. 8 and Fig. 9 can be observed that the residual sliding displacement and sliding area ratio gradually increase with the increase of overload coefficient, and the overall variation trend of the corresponding curves on the bottom surface of left and right dam abutment blocks are similar. When the overload coefficient is 0.2, the bottom sliding surfaces have slid, but the sliding area is relatively small and the arch dam remains stable. Then, the sliding area expands rapidly with the increase of seismic overload coefficients. When the overload coefficient is 1.5, the left sliding area ratio reaches 100%, indicating that the left bottom sliding surface has overall sliding, while the right sliding area ratio is 80%. Subsequently, the right sliding area ratio reaches 100% when the seismic overload coefficient is 2.4, indicating the occurrence of overall instability of the high arch dam-foundation system.

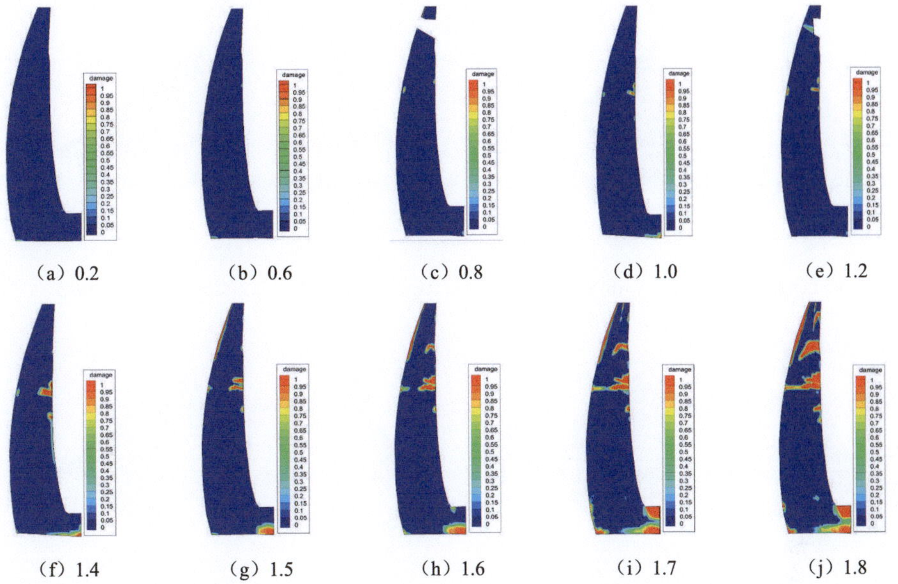

Fig. 6. Damage distribution of the dam cantilever section corresponding to the maximum damage depth-thickness ratio under different seismic overload coefficients

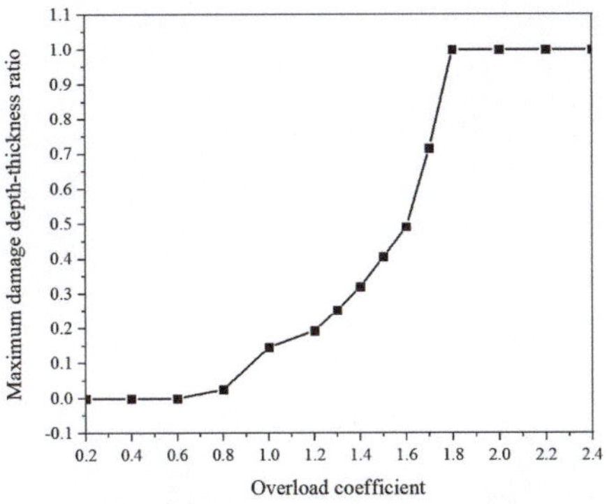

Fig. 7. The maximum damage depth-thickness ratio under different seismic overload coefficients

Thus, if the overall sliding of the left or right bottom sliding surface is taken as a failure criterion, the ultimate seismic capacity of the dam is 1.4 times the horizontal PGA, i.e. 0.6048 g. Compared with the strength failure mode-based index, the instability failure mode-based index is used to evaluate the ultimate seismic capacity of a high arch dam is safer.

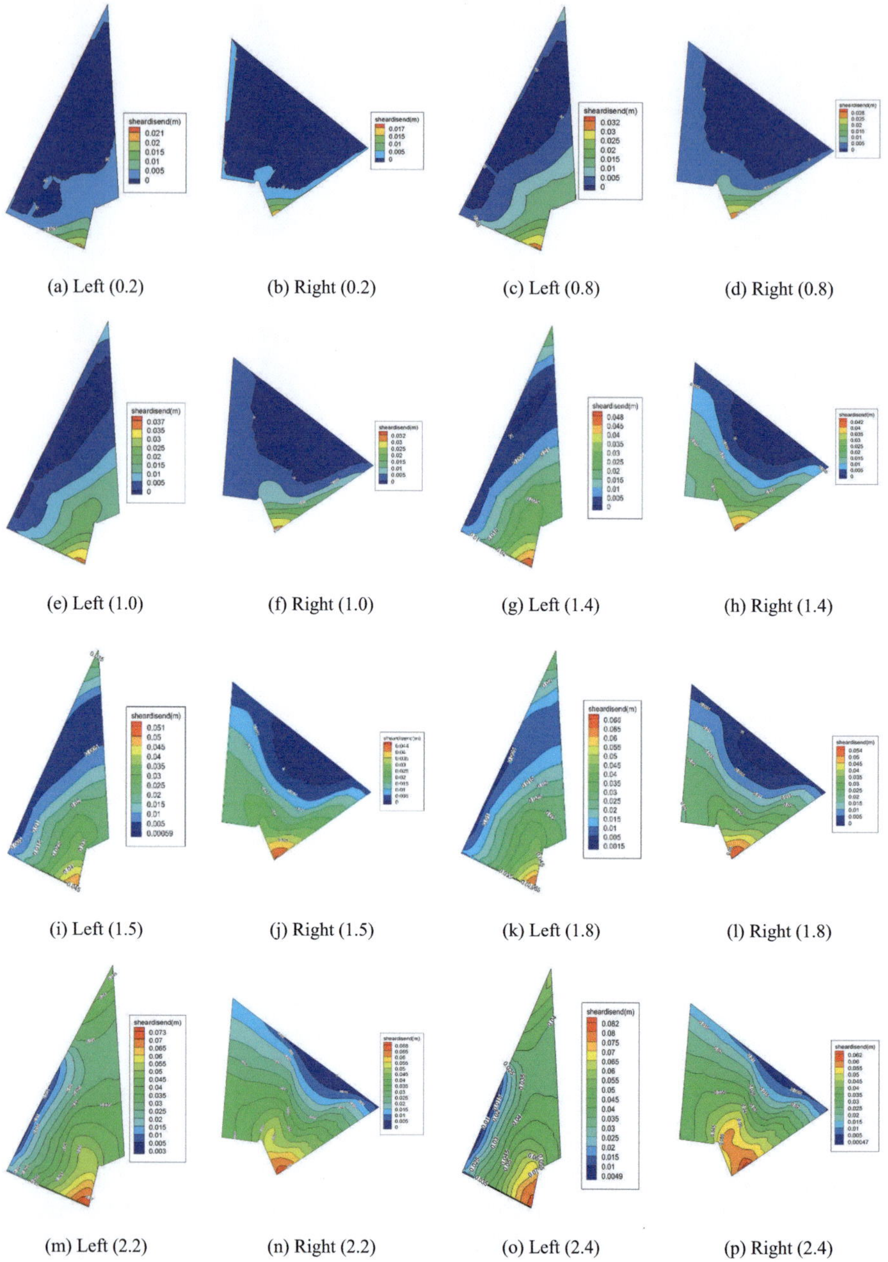

Fig. 8. Residual sliding displacement contours of left and right bottom sliding surfaces under different overload coefficients

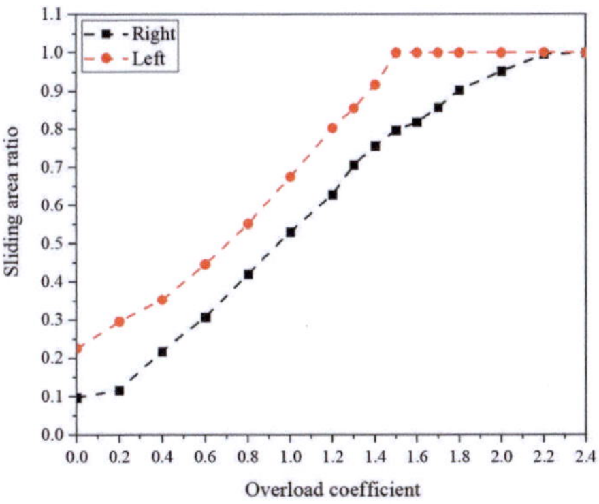

Fig. 9. Sliding area ratio of left and right bottom sliding surfaces under different overload coefficients

3.3 Coupling of Strength Failure and Instability Failure Modes

Based on the above results, the maximum damage depth-thickness ratio, representing the dam strength failure, and the sliding area ratio, representing dam abutment blocks instability failure, are taken as evaluation indexes. However, these indexes are specific to individual failure modes and do not fully characterize the coupling of the two failure modes. Therefore, the residual displacement of the dam crest relative to the dam bottom in the stream direction is proposed as a comprehensive performance evaluation index, capable of representing the coupling of these two failure modes.

Figure 10 shows the residual displacement of the dam crest relative to the dam bottom in the stream direction under different overload coefficients. From Fig. 10, the residual displacement gradually increases with the change of the overload coefficient. A slight change occurs when the overload coefficient is 1.0, which may be caused by the sudden development of the dam damage cracking (Fig. 7). When the overload coefficient is 2.0, there is a sudden change and rapid growth of the residual displacement curve.

Thus, if a sudden change and rapid growth of the residual displacement curve is taken as a failure criterion, the ultimate seismic capacity of the dam is 2.0 times the horizontal PGA, i.e. 0.864 g.

It should be noted that the quantitative evaluation indexes of the ultimate seismic capacity are very complex and are still in the exploratory stage. The current standard [14] recommends using the turning point on curves of deformation at typical locations on the dam body, which vary with the increase in input acceleration, to evaluate the safety of the dam-foundation system. This approach is based on the principle that the change in the working performance of the dam-foundation system will increase the risk of dam failure. The occurrence of penetrating cracks or overall sliding of sliding surfaces does not imply immediate loss of bearing capacity and failure of the dam-foundation system due to the reciprocal nature of the seismic load. Therefore, it is suggested that 2.0 times the horizontal PGA can be taken as the ultimate seismic capacity of the dam.

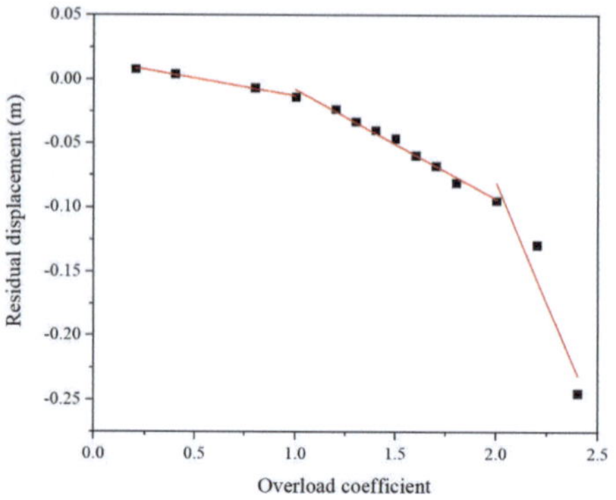

Fig. 10. Residual displacement of the dam crest relative to the dam bottom in stream direction under different overload coefficients

4 Conclusion

A high arch dam-foundation coupling system is established, which for the first time considers the material nonlinearity of dam concrete and contact nonlinearities of the contact surfaces with a total degree of freedom of 10.64 million. The seismic failure mode and process of the proposed model are discussed using different indexes including damage depth-thickness ratio, sliding area ratio, and the residual displacement of the dam crest relative to the dam bottom in the stream direction. Meanwhile, evaluation criteria and ultimate seismic capacity are defined and quantitatively assessed. Based on the present numerical analyses, the following conclusions can be obtained:

(1) If the occurrence of a penetrating crack is taken as the failure criterion, the ultimate seismic capacity of the dam based on the damage indexes is 1.7 times the horizontal PGA, i.e. 0.7344 g.
(2) If the overall sliding of the left or right bottom sliding surface is taken as a failure criterion, the ultimate seismic capacity of the dam is 1.4 times the horizontal PGA, i.e. 0.6048 g. Compared with the strength failure mode-based index, the instability failure mode-based index is used to evaluate the ultimate seismic capacity of a high arch dam is safer.
(3) If a sudden change and rapid growth of the residual displacement curve is taken as a failure criterion, the ultimate seismic capacity of the dam is 2.0 times the horizontal PGA, i.e. 0.864 g.
(4) The occurrence of penetrating cracks or overall sliding of sliding surfaces does not imply immediate loss of bearing capacity and failure of the dam-foundation system due to the reciprocal nature of the seismic load. Therefore, it is suggested that 2.0 times the horizontal PGA can be taken as the ultimate seismic capacity of the dam.

Acknowledgments. This study is supported by China Huaneng Group Co., Ltd. (Grant No. HNKJ22-H108) and Power Construction Corporation of China (Grant No. DJ-ZDXM-2021-03). All authors are grateful for this support.

References

1. Alembagheri, M., Ghaemian, M.: Damage assessment of a concrete arch dam through nonlinear incremental dynamic analysis. Soil Dyn. Earthq. Eng. **44**(1), 127–137 (2013)
2. Chen, H.Q.: Challenge confronted in seismic design of high concrete dams. Hydropower and Pumped Storage **3**(2), 1–13 (2017)
3. Houqun, C.: Seismic safety analysis of tall concrete dams, investigation and insights on critical challenges. Earthq. Eng. Eng. Vib. **19**(3), 533–539 (2020). https://doi.org/10.1007/s11803-020-0578-6
4. Chen, H.Q., Guo, S.S.: Seismic damage analysis of high concrete dam-foundation system. J. Hydraul. Eng. **43**(Supp. 1), 2–7 (2012)
5. Chen, H.Q., Li, D.Y., Guo, S.S.: Damage-rupture process of concrete dams under strong earthquakes. Int. J. Struct. Stab. Dyn. **14**(7), 1–21 (2014)
6. Ghanaat, Y.: Failure modes approach to safety evaluation of dams. In: Proceedings of the 13th World Conference on Earthquake Engineering, Vancouver, Canada (2004)
7. Guo, S.S.: Study on seismic damage process mechanism and quantitative evaluation criteria of concrete dam-foundation system based on parallel computation. China Institute of Water Resources and Hydropower Research, Beijing (2013)
8. Guo, S.S., Liang, H., Wu, S., et al.: Seismic damage investigation of arch dams under different water levels based on massively parallel computation. Soil Dyn. Earthq. Eng. **129**, 1–8 (2020)
9. Hariri-Ardebili, M.A., Mirzabozorg, H.: Seismic performance evaluation and analysis of major arch dams considering material and joint nonlinearity effects. Int. Sch. Res. Netw. **2012**, 1–10 (2012)
10. Hariri-Ardebili, M.A., Mirzabozorg, H., Ghasemi, A.: Strain-based seismic failure evaluation of coupled dam-reservoir-foundation system. Coupled Syst. Mech. **2**(1), 85–109 (2013)
11. Lee, J., Fenves, L.G.: A plastic-damage concrete model for earthquake analysis of dams. J. Earthq. Eng. Struct. Dyn. **27**, 937–956 (1998)
12. Liang, H., Guo, S.S., Tu, J., Li, D.Y.: Seismic stability sensitivity and uncertainty analysis of a high arch dam-foundation system. Int. J. Struct. Stab. Dyn. **19**(6), 1–26 (2019)
13. Liang, H., Guo, S.S., Tian, Y.F., et al.: Probabilistic seismic analysis of the deep sliding stability of a concrete gravity dam-foundation system. Adv. Civ. Eng. **2020**, 1–10 (2020)
14. Ministry of Water Resources of the People's Republic of China: GB 51247–2018 Standard for Seismic Design of Hydraulic Structures. Beijing China Planning Publishing House, Beijing (2018)
15. Mostafaei, H., Sohrabi Gilani, M., Ghaemian, M.: Stability analysis of arch dam abutments due to seismic loading. Sci. Iran. Trans. A: Civ. Eng. **24**, 467–475 (2017)
16. Omidi, O., Lotfi, V.: Earthquake response of concrete arch dams: a plastic-damage approach. Earthquake Eng. Struct. Dynam. **42**(14), 1–21 (2013)
17. Omidi, O., Lotfi, V.: Seismic plastic-damage analysis of mass concrete blocks in arch dams including contraction and peripheral joints. Soil Dyn. Earthq. Eng. **95**, 118–137 (2017)
18. Westergaard, H.M.: Water pressure on dams during earthquakes. Trans. ASCE **98**, 418–432 (1933)
19. Zenz, G., Glodgruber, M., Feldbacher, R.: Seismic stability of a rock wedge in the abutment of an arch dam. Geomech. Tunn. **2**, 186–194 (2012)
20. Zhang, S.R., Wang, G.H., Wang, C.: Study on ultimate aseismic capacity evaluation of concrete gravity dam. J. Hydroelectr. Eng. **32**(3), 168–175 (2013)

Research on the Sediment Flushing Scheme Under the Layout of "Reservoir Replacing Pool + Bypass Flushing" Based on 2D Flow and Sediment Model

Shuangchao Yang(✉), Jinyang Liu, and Yu Liao

Guangdong Zhurong Architecture and Engineering Design Co., Ltd, Guangzhou 510610, Guangdong, China
30684910@qq.com

Abstract. A two-dimensional flow and sediment model was used to study the sedimentation problem of a high head and high sand content hydroelectric power station in Nepal. Under the layout scheme of "reservoir replacing pool + bypass flushing", a total of 12 sediment discharge operation schemes were proposed for the reservoir area under the combination of 3 flow limits and 4 sand discharge water levels. The results show that: After 5 years of operation of the project, there is a difference in the elevation of the sedimentation surface in front of the water intake. Different sand flushing operation methods have a significant impact on the elevation of the sedimentation surface in front of the water intake during the period of 7–17 years. Under various schemes, the trend of the elevation change of the sedimentation surface in front of the water intake after 17 years of operation of the head hub is basically consistent, and the difference is not significant. The elevation of the sedimentation surface in front of the water intake can be controlled at around 2505 m under each scheme; When operating at a sand discharge water level of 2520 m, the elevation of the sedimentation surface in front of the sand discharge tunnel can be controlled at around 2510 m, and the probability of sedimentation in the bypass sand discharge tunnel is not high; The project operates for about 10–15 years and is basically in a balanced state; Under the condition of 2520 m sediment discharge water level, the remaining effective storage capacity of the hub is larger after 20 years of operation under each scheme; This project follows the 2520 m sediment discharge water level scheme. When operating at a sediment discharge flow limit of 120 m^3/s, the effective storage capacity of the reservoir meets the requirements of the designed daily regulating storage capacity. Therefore, it is recommended that this project operate according to the 2520 m sediment discharge water level scheme.

Keywords: Hydropower stations · Sedimentation · Reservoir replacing pool · Sediment flushing scheme

S. Zheng et al. (Eds.): IHDC 2024, LNCE 487, pp. 197–214, 2025.
https://doi.org/10.1007/978-981-97-9184-2_18

1 Introduction

The sedimentation and sediment discharge of reservoirs are the key factors affecting the normal operation of hydropower stations. Excessive sediment content and particle size can cause severe wear and tear of water turbines, even leading to their cessation of operation.

Many domestic and foreign scholars have conducted extensive research on the problem of sediment interception and discharge at hydropower stations on sediment-laden river. Zheng Hexin et al. [1] focused on the high sediment concentration problem at the Kohala Hydropower Station in Pakistan. By analyzing the water and sediment characteristics and combining them with mathematical simulation calculations, a reservoir instead of a pool scheme was adopted to solve the sediment problem; Wang Xiaofeng et al. [2] used a mathematical model to calculate the sand settling effect of a hydropower station in Pakistan, simulated the efficiency of the sand settling tank, and studied the replacement of traditional sand settling tanks with a reservoir instead of a tank scheme; Wang Xinhong et al. [3] used numerical simulation methods to study the sediment erosion and sedimentation process and water sediment regulation process of Wangyao Reservoir on high sediment content rivers, and determined the operation mode of regular open discharge and sediment discharge during flood season; Xiao Jun et al. [4] studied the sediment discharge operation plan of the Karabe Reservoir in Xinjiang through physical model experiments on a moving bed; Gao Donghong et al. [5] used numerical simulation methods to study the sedimentation and flushing effects of the Nasuwakari Hydropower Station in Nepal; Dong Dian [6] established a physical model to conduct experimental research on the sedimentation and flushing effects of the sedimentation facilities at the Shangma Xiangdi A hydropower station in Nepal; Zhou H et al. [7] studied the measures for preventing and reducing sedimentation of old hydropower stations built on high sediment content rivers through experiments; Richter et al. [8] studied the sedimentation and sediment discharge effects of a newly designed sedimentation device for high sediment content river hydropower stations through physical model experiments and CFD simulations.

The above research mainly adopts the sedimentation tank method and the "reservoir replacing tank" method for flushing and sand discharge. The sedimentation tank scheme is adopted, and the operating water level during the flood season determines the elevation of the water intake. If the operating water level is too low, the distance between the water intake and the riverbed elevation is low, the power generation efficiency is small, and the sediment content entering the sedimentation tank is high. The sedimentation tank has a large scale and requires a large investment; If the operating water level is high, it is easy to cause sedimentation in the reservoir, and the sediment discharge effect is not significant. The "reservoir replacing pool" plan often encounters two situations that fail to achieve the expected design results: firstly, during the flood season, the operating water level is high, and sediment accumulates at the end of the reservoir prematurely, causing a sharp decrease in storage capacity. Under the condition of open discharge and flushing, the reservoir cannot be washed away, resulting in premature loss of benefits. Or, in order to maintain the regulation of storage capacity and the effectiveness of reservoir sediment discharge, a longer period of open discharge and flushing of the reservoir is required, which affects the power generation efficiency; Secondly, the operating water level during the flood season is relatively low, and the sedimentation effect of the reservoir is limited. The reduction of sediment concentration through the turbine is relatively small, and the particle size or content of sediment passing through the turbine cannot meet the design requirements.

This study focuses on the problem of "high head and high sediment concentration" faced by a hydroelectric power station project in Nepal. Based on the "reservoir replacing pool" scheme, a bypass sand discharge tunnel and a sand blocking dam are set up in the upstream reservoir, dividing the reservoir into upper and lower storage areas. During the flood season, most of the sediment is stored in the upper reservoir area using a sand dam, and then directly discharged downstream through a bypass sand discharge tunnel to reduce the incoming sediment in the lower reservoir area. The lower reservoir area forms a natural sedimentation tank, ensuring that the sediment content and particle size entering the water intake meet the requirements of the water turbine. This article establishes a two-dimensional mathematical model of water and sediment, studies the characteristics of reservoir sedimentation changes under different sediment discharge operation schemes under the "reservoir replacing pool + bypass sediment discharge" mode, and comprehensively considers the engineering effect and economic benefits, proposing a reasonable sediment discharge operation mode.

2 Overview and Engineering Layout of Reservoir

A hydropower station is located on a river in central northern Nepal. Its main task is to generate electricity, with a normal water level of 2530 m and a corresponding storage capacity of about 30.2 million cubic meters. The lowest water level is 2511 m, and the engineering regulation storage capacity is 26.2 million cubic meters. The installed capacity is 163 MW, and it is a runoff type power station. According to domestic regulations, this project is classified as a Class III medium-sized project.

The river where the power station is located is a major tributary of the Shapta Gandaki River system, with a catchment area of 3474 km^2 at the dam site and an average annual flow rate of 42.45 m^3/s. The incoming water is mainly snowy mountain meltwater. The median particle size of riverbed bed load sand is $D_{50} = 3.3$ cm, with a maximum of about 20 cm of pebbles. The average annual bed load sediment transport is 5.37 million tons. The median particle size D_{50} of suspended sediment in the river is about 0.035 mm, with an average annual suspended sediment transport of 10.73 million tons and an average sediment concentration of over 5 kg/m^3. The reservoir capacity to sediment ratio of this hydropower station is less than 5, making it a hydropower station with serious sediment problems. The design head of the reservoir reaches 602 m, which is a typical "high head and high sediment concentration" hydropower station. Without effective sediment discharge measures, the reservoir can be filled with sediment within 3–5 years.

The head hub water retaining dam consists of the left bank earth-rock dam section, overflow dam, flushing bottom hole, and right bank gravity dam section from left to right. The water inlet of the hydropower station is located 35 m in front of the right bank dam, with a bottom elevation of 2505.5 m; In order to solve the problem of sediment accumulation, a sediment blocking dam is set up 1.3 km upstream of the head hub dam site, from left to right, which are earth-rock dams, spillway gates, and side channels in sequence; The bypass sand discharge tunnel is located on the right bank of the sand dam, with a total length of 1980 m, an inlet elevation of 2507 m, an outlet elevation of 2486.45 m, and a longitudinal slope of 1.0%. This arrangement of sand blocking dams distinguishes the reservoir into upper and lower storage areas. The sand blocking dam intercepts sediment in the upper storage area and discharges it into the river downstream of the retaining dam through the sand discharge hole located upstream of the discharge gate, greatly reducing the amount of sediment entering the lower storage area (Fig. 1).

Fig. 1. Engineering layout diagram

3 Two-Dimensional Flow-Sediment Mathematical Model

3.1 Basic Equations

The two-dimensional flow-sediment mathematical model employed in this study is primarily based on the two-dimensional turbid water continuity equation and turbid water motion equation for natural open channel water flow, which are derived from the laws of conservation of mass and momentum.

(1) Flow Control Equations

$$\frac{\partial H}{\partial t} = -\frac{\partial HU_j}{\partial x_j}, j = 1, 2 \tag{1}$$

$$\frac{\partial HU_i}{\partial t} + \frac{\partial HU_j}{\partial x_j} + gH\frac{\partial \zeta}{\partial x_i} + \frac{gn^2 U_i\sqrt{U_j^2}}{H^{1/3}} - \frac{\partial^2 HU_i}{\partial x_j^2} = 0, i, j = 1, 2 \tag{2}$$

$$\frac{\partial HS}{\partial t} + \frac{\partial HSU_j}{\partial x_j} + \rho\prime\frac{\partial \bar{z}_b}{\partial t} - \frac{\partial}{\partial x_j}\left(D_s\frac{\partial HS}{\partial x_j}\right) = 0, i, j = 1, 2 \tag{3}$$

where H is the water depth, ζ is the water level, U_j is the vertical average velocity, S is the vertical average sediment concentration, and D_s is the sediment turbulent diffusion coefficient.

(2) Suspended Sediment Convection-Diffusion Equation

The sediment continuity equation is integrated horizontally to obtain a two-dimensional continuity equation along the water depth, which is also the suspended sediment convection-diffusion equation:

$$\begin{aligned}&\frac{\partial(hS)}{\partial t}+\frac{\partial(huS)}{\partial x}+\frac{\partial(hvS)}{\partial y}\\&=\frac{\partial}{\partial x}\left(\nu_{sx}\left(\frac{\partial(hS)}{\partial x}\right)\right)+\frac{\partial}{\partial y}\left(\nu_{sy}\left(\frac{\partial(hS)}{\partial y}\right)\right)+Q_LS_L-F_S\end{aligned}\tag{4}$$

where S is the vertical average sediment concentration, ν_{sx}, ν_{sy} are the turbulent diffusion coefficients, F_S is the erosion and deposition term, Q_L, S_L is the horizontal unit area source quantity and source sediment concentration, and other symbols are defined similarly as in the hydrodynamic model.

(3) River Bed Deformation Equation

The flux of suspended sediment on the water surface is generally 0, while the unit flux of suspended sediment on the riverbed bottom is equal to the deformation of the suspended sediment riverbed. The erosion and deposition term can be represented by erosion and sedimentation functions, which are represented by sediment carrying capacity and bed shear stress.

(1) Erosion-Deposition Function Based on Sediment Carrying Capacity

Introducing the relationship coefficients between the vertical average sediment concentration, the vertical average sediment carrying capacity, the bed sediment concentration, and the sediment carrying capacity, the bed deformation equation for suspended sediment can be expressed as:

$$F_s=\rho_s\frac{\partial b}{\partial t}=\alpha_3\omega(\alpha_1S-\alpha_2S_*)\tag{5}$$

where α_1 and α_2 are the coefficients of sediment concentration and saturation recovery of sediment carrying capacity, respectively, α_3 is the probability of sediment settling, ω is the sediment settling velocity; $S*$ is the sediment concentration under the saturated state (i.e. the sediment carrying capacity), ρ_s is the sediment density (2650 kg/m^3).

(2) Erosion-Deposition Function Represented by Bed Shear Stress

The erosion and deposition function represented by the bed surface shear stress is:

$$F_s=\begin{cases}S_D,\ \tau_b\le\tau_{cd}\\0,\ \tau_{ce}>\tau_b>\tau_{cd}\\-S_E,\ \tau_b>\tau_{ce}\end{cases}\tag{6}$$

where S_D is the sedimentation rate of sediment, S_E is the rate of sediment erosion, τ_b is the bed shear stress(N/m^2), τ_{cd} is the critical bed shear stress during sedimentation (N/m^2), τ_{ce} is the critical bed shear force during erosion or resuspension (N/m^2).

(3) Bedload Equation

Currently, there are two types of bedload transport and the resulting bed deformation: one type derives the bedload transport imbalance equation from the conservation principle, and the other directly uses the bedload transport rate formula to represent the bedload transport equation.

For bedload, the mass conservation resulting from bed elevation changes due to bedload inflow and outflow is also considered. Thus, the bed deformation equation caused by bedload is:

$$\frac{\partial T}{\partial t} = -F_s = -\rho_s \frac{\partial b}{\partial t} = -\left(\frac{\partial g_{bx}}{\partial x} + \frac{\partial g_{by}}{\partial y}\right) \tag{7}$$

where T is the displacement mass per unit area (kg/m^2), g_{bx} and g_{by} represent the bed load sediment flux in the x—and y—directions, respectively.

Using explicit format to process the total deformation equation of the riverbed:

$$b_i^{t+\Delta t} = b_i^t + \frac{1}{\rho_s}\frac{\Delta t}{\Omega_i}\left\{\sum_{n=1}^{N}\left(\sum_{k=1}^{4}\left(g_{i,k}^{\mathrm{bx}} n_{i,k}^x + g_{i,k}^{\mathrm{by}} n_{i,k}^y\right) l_{i,k} + \alpha_3 \omega (\alpha_1 S - \alpha_2 S_*)\right)\Big|_i^n\right\} \tag{8}$$

where b is the bottom elevation of the unit center, N is the number of sediment size classes, n is the index of the sediment size class.

3.2 Model Establishment

The upstream boundary of the two-dimensional model is located approximately 11 km upstream of the reservoir dam site, and the downstream boundary is at the hydropower station dam site. The total number of grid cells in the model is 38,974, with grid sizes ranging from 5 to 60 m. Larger grid scales are used in areas with higher or flatter terrain, while denser grids are applied in key areas such as river channels, dam sites, and sediment control dams. The grid scale around important hydraulic structures is refined to about 5 m. The reservoir topography and model computational grid are shown in Fig. 2 and Fig. 3.

3.3 Model Validation

Due to the lack of measured erosion and deposition data of the reservoir, according to relevant specifications, the model validation in this calculation is carried out by using the 7-year riverbed longitudinal profile erosion and deposition measurement data of the natural river channel in the section where this project is located.

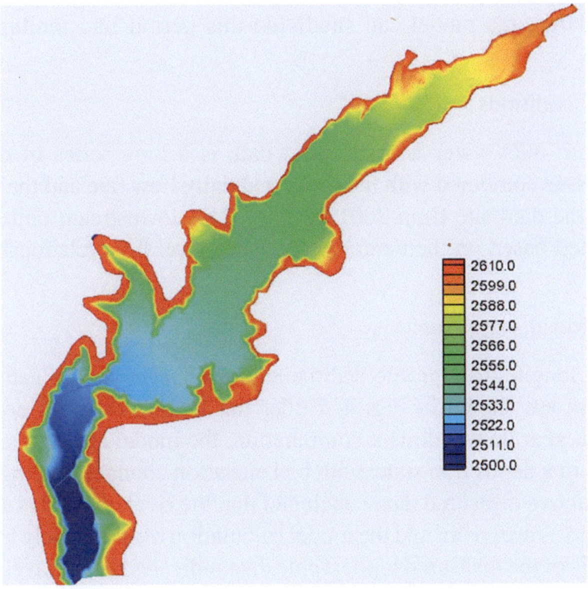

Fig. 2. Topography of the reservoir area

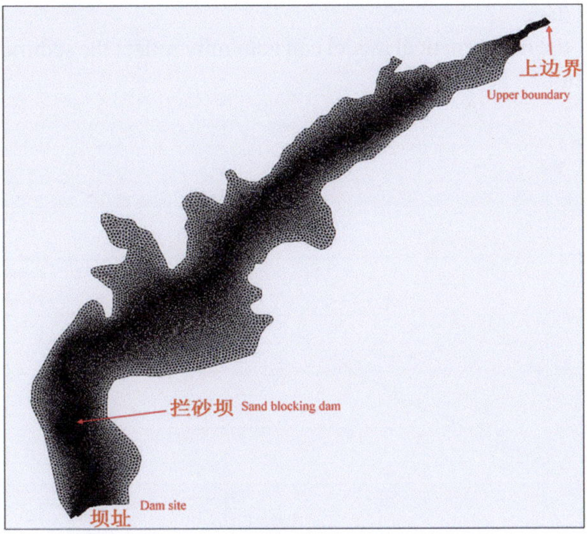

Fig. 3 2D model calculation grid

The validation river section is the interval from about 1.2 km upstream of the hydropower station dam site to the dam site. The validation period series is from 2001 to 2007, a total of 7 years. The calculation period is divided into daily intervals during the flood season (June to October) and ten-day intervals during the non-flood season (November to May). If the erosion and deposition amount in a particular period is excessively large, the model can subdivide this period into multiple intervals for calculation.

(1) Boundary Conditions

The upstream inlet water and sediment data is a long series of daily water and sediment processes combined with the measured daily flow rate and the water sediment relationship at the dam site from 2001 to 2007. The downstream outlet control water level is calculated based on the natural river water level flow relationship curve at the dam site.

(2) Validation Results and Analysis

The 7-year longitudinal profile sedimentation changes in the natural river channel over 7 years are shown in Fig. 4. Under the long-term series conditions of the designed typical year flow-sediment combination, the model calculates that the natural river channel is in a deposition state, with bed elevation changes ranging from −0.02 m to 1.10 m. The above measured data concluded that the river channel is an accumulative depositional sand-gravel river, and the model calculation results closely match this actual pattern. Analysis of the measured topographic data indicates that the average annual sedimentation volume in the reservoir area is approximately 700,000 m^3, translating to a sedimentation intensity of about 0.29 m/a. The calculated average sedimentation intensity in the reservoir area under natural river conditions is about 0.38 m/a, which is close to the measured data.

In summary, the mathematical model can generally reflect the sedimentation patterns of this river section.

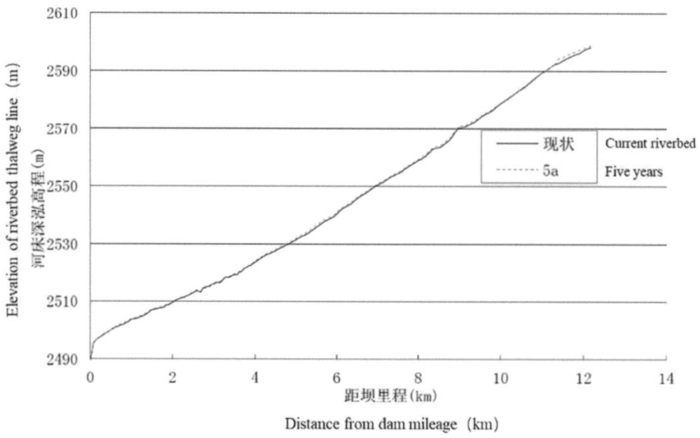

Fig. 4. Longitudinal profile sedimentation changes in the natural river channel over 7 years

4 Simulation Calculation and Analysis of Long-Term Sediment Discharge Operation in Hub

4.1 Sediment Discharge Scheme Formulation

Since the inflow water and sediment are concentrated during the flood season, characterized by the typical "large water inflow and large sediment inflow" pattern, the focus of sediment management is on the flood season.

Based on the statistical analysis of the number of days with various flow rates in long-term series, and considering the contradiction between reservoir erosion and power generation efficiency of the hydropower station, the preliminary selected inflow flow rate thresholds are 200 m^3/s, 150 m^3/s, and 120 m^3/s. The elevation of the sand barrier in front of the water intake is 2505 m, and the bottom elevation of the sediment discharge tunnel at the dam front is 2499 m, thus determining the sediment control elevation at 2498.5 m. In the upper reservoir area, the bottom elevation control for bypass sediment discharge to discharge bedload is set at 2508 m. The proposed flood season sediment discharge operation control water levels are 2518 m, 2520 m, 2522 m, and 2525 m.

Combining the development objectives of this hydropower station, different sediment discharge operation schemes under various flow thresholds and control water levels in the reservoir area are proposed, as shown in Table 1.

4.2 Calculation Boundary Conditions

(1) Upstream Water and Sediment Boundary Conditions

According to the sediment design requirements and available data, a representative series of water and sediment is selected for this design. The annual mean values of the runoff series from 2001 to 2007 at the dam site show little difference, but since the series is less than 30 years, it lacks representativeness. The representative series needs to choose the natural continuous water and sediment series. Therefore, the runoff series from 2001 to 2007 is used as a basic series unit and recycled four times to form a 28-year long series. The corresponding sediment series is derived using the most adverse water-sediment relationship.

(2) Downstream Boundary Conditions

The downstream boundary condition is the operating water level in front of the dam in the first hub reservoir area of the hydropower station. During the non-flood season, when the hub is in normal storage operation, the water level at the front of the retaining dam is controlled at 2530 m. During flood season, when the hub is operated according to the flood limit water level, the water level in front of the retaining dam is controlled at 2520 m. During the flood and sediment discharge period of the dam, the water level in front of the hub dam during a certain period is calculated from the inflow flow and discharge curve of that period. During the flushing period of the bypass sand discharge tunnel in the upper reservoir area, the water level in front of the upstream sand dam during a certain period is calculated from the inflow flow rate and discharge curve of that period.

Table 1. Explanation of various calculation schemes

Scheme Number	Reservoir Operation Mode	
	Flood season (June–October)	November to May of the following year
Scheme 1-1	Flow rate greater than 200 m^3/s for open discharge and sand discharge, with a water level of 2518 m for sand discharge	Normal water storage level 2530 m
Scheme 1-2	Flow rate greater than 200 m^3/s for open discharge and sand discharge, with a water level of 2520 m for sand discharge	
Scheme 1-3	Flow rate greater than 200 m^3/s for open discharge and sand discharge, with a water level of 2522 m for sand discharge	
Scheme 1-4	Flow rate greater than 200 m^3/s for open discharge and sand discharge, with a water level of 2525 m for sand discharge	
Scheme 2-1	Flow rate greater than 150 m^3/s for open discharge and sand discharge, with a water level of 2518 m for sand discharge	
Scheme 2-2	Flow rate greater than 150 m^3/s for open discharge and sand discharge, with a water level of 2520 m for sand discharge	
Scheme 2-3	Flow rate greater than 150 m^3/s for open discharge and sand discharge, with a water level of 2522 m for sand discharge	
Scheme 2-4	Flow rate greater than 150 m^3/s for open discharge and sand discharge, with a water level of 2525 m for sand discharge	
Scheme 3-1	Flow rate greater than 120 m^3/s for open discharge and sand discharge, with a water level of 2518 m for sand discharge	

(*continued*)

Table 1. (*continued*)

Scheme Number	Reservoir Operation Mode	
	Flood season (June–October)	November to May of the following year
Scheme 3-2	Flow rate greater than 120 m^3/s for open discharge and sand discharge, with a water level of 2520 m for sand discharge	
Scheme 3-3	Flow rate greater than 120 m^3/s for open discharge and sand discharge, with a water level of 2522 m for sand discharge	
Scheme 3-4	Flow rate greater than 120 m^3/s for open discharge and sand discharge, with a water level of 2525 m for sand discharge	

4.3 Calculation Results and Analysis

(1) The Elevation Change of the Sedimentation Surface in front of the Water Intake of The Hydropower Station

Using the validated numerical model of water and sand to calculate different working conditions, the change of the siltation surface elevation in front of the hydropower station intake with different operating years is shown in Fig. 5. It can be seen that the change of the siltation surface elevation in front of the intake is basically the same in the first 5 years of operation of the first hub under various schemes. After 5 years of operation of the hub, there is a difference in the elevation of the sedimentation surface in front of the water intake. Different sand flushing operation methods have a significant impact on the elevation in front of the water intake between 7 and 17 years. The reason for this is that although the inflow process is consistent, but due to the differences of flow limits and sediment discharge water level, the time and magnitude of sand flushing in each scheme for the open discharge of the low valley show differences. After 17 years of operation under various schemes, the trend of changes in the elevation of the sedimentation surface in front of the water intake of the first hub is basically consistent, and the difference is not significant, indicating that the flushing operation mode in the later stage of hub operation has little impact on the elevation in front of the water intake. The siltation surface elevation in front of the intake can be controlled at about 2505 m under each scheme.

Under the same flow limit, the higher the water level for sediment discharge, the higher the elevation of the upper reservoir sediment dam, the higher the efficiency of sediment interception, the longer it takes for the siltation surface in front of the intake to reach the elevation of 2505 m. Under the same sediment discharge water level and different flow limits, the larger the flow limit, the longer it takes for the sedimentation surface in front of the water intake to reach an elevation of 2505 m.

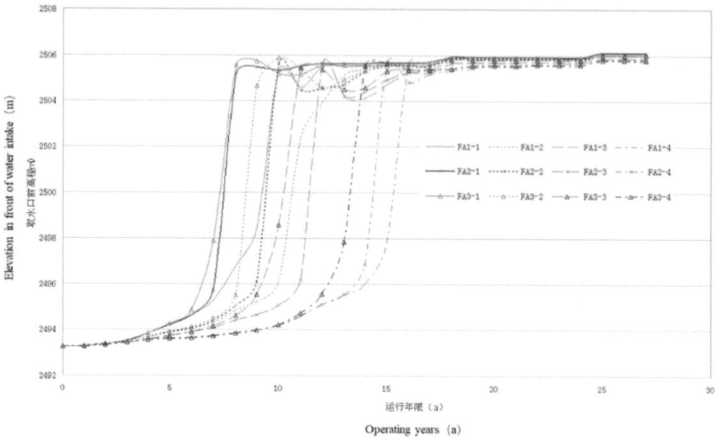

Fig. 5. Changes in the elevation of the siltation surface in front of the water intake with different operating years

(2) Changes in the Elevation of the Siltation Surface in front of The Bypass Sand Discharge Tunnel

From Fig. 6, it can be seen that the elevation changes of the sedimentation surface in front of the bypass sand discharge tunnel are consistent for the first 5 years of operation of the head hub under different schemes. Although the inflow process of each operation scheme is consistent, there may be differences in the bypass flushing time and flushing amplitude due to the difference in the water level for sediment discharge. After 5 years of operation of the hub, there is a difference in the elevation of the sedimentation surface in front of the bypass sand discharge hole.

The elevation of the siltation surface in front of the sand-discharge hole in four sand discharge water level schemes is the lowest with a sand discharge water level of 2518 m, which can be controlled around 2508 m. In the sand discharge water level of 2520 m and 2512 schemes, the elevation of the sedimentation surface in front of the bypass sand discharge tunnel can be controlled around 2510 m and 2512 m, respectively. In the scheme of a sand discharge water level of 2525 m, the elevation of siltation surface in front of sand discharge hole can be controlled to 2514 m, and the risk of siltation in sand discharge hole will be increased greatly due to the elevation of the top of the hole is 2515 m.

At the same water level for sediment discharge, under different flow limit schemes, the elevation of the sedimentation surface in front of the bypass sediment discharge tunnel is not sensitive to the flow limit. At high sand discharge water levels, the elevation of the sedimentation surface in front of the bypass sand discharge tunnel under each flow limit eventually tends to be consistent. At low sand discharge level, the elevation of siltation surface in front of bypass sand discharge holes under each flow limit is slightly different, among which, the smaller the flow limit, the lower the elevation of siltation surface in front of bypass sand discharge holes.

From the perspective of the elevation change process of the sedimentation surface in front of the bypass sand discharge hole, the higher the water level of the sand discharge, the higher the elevation of the sedimentation surface in front of the bypass sand discharge hole, the higher the probability of siltation, and the higher the risk of pushing the quality of the dam. When operating at the recommended sand discharge water level of 2520 m according to the pre feasibility study, the elevation of the sedimentation surface in front of the bypass sand discharge hole can be controlled around 2510 m. The depth of siltation surface in front of the hole is 3 m, and the hole height is 8 m, which accounts for only 42.5% of the hole height, therefore, the siltation probability of the bypass sand discharge hole is not large.

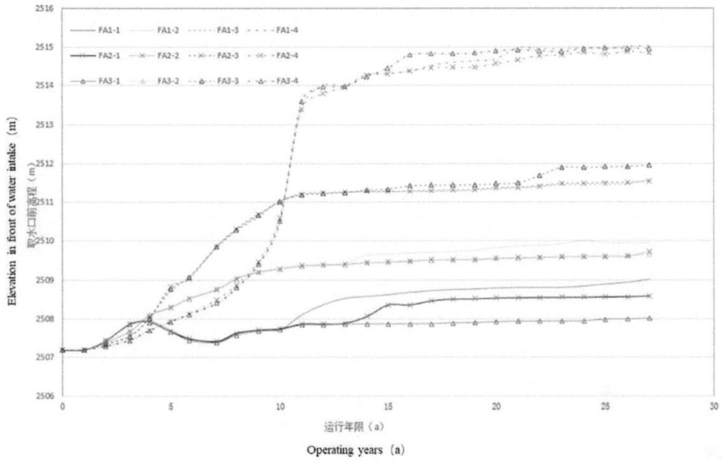

Fig. 6. Changes in the elevation of the sedimentation surface in front of the bypass sand discharge tunnel with different operating years

(3) Longitudinal Profile Erosion and Sedimentation Morphology

Figure 7 and Fig. 8 show the longitudinal profiles of sediment deposition in the reservoir area after 10 and 15 years of operation under different schemes. Under the same sediment discharge water level and different flushing flow limits, the longitudinal section of the reservoir area is the lowest under the 120 m^3/s flow limit scheme.

Under the same limit of sand flushing flow rate and different sand discharge water levels, the longitudinal section of the reservoir area is the lowest under the 2508 m sand discharge water level scheme.

(4) Changes in Storage Capacity Curve

From Table 2, it can be seen that under the same sediment discharge water level and different sediment discharge flow limits, the effective storage capacity remaining in

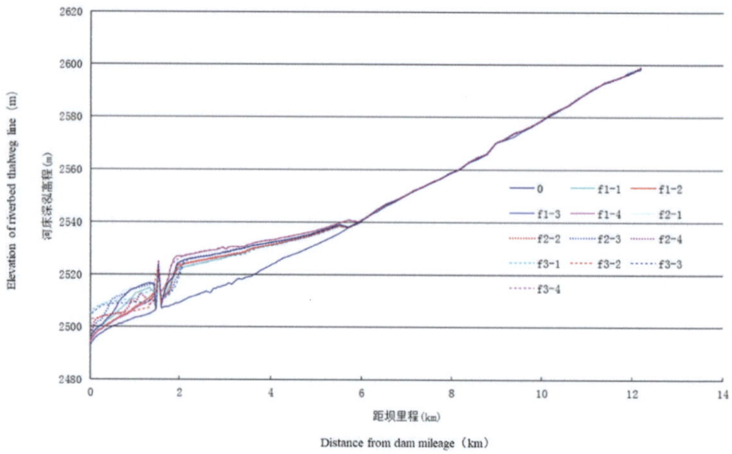

Fig. 7. Longitudinal section of sedimentation in the reservoir area after 10 years of operation of each plan

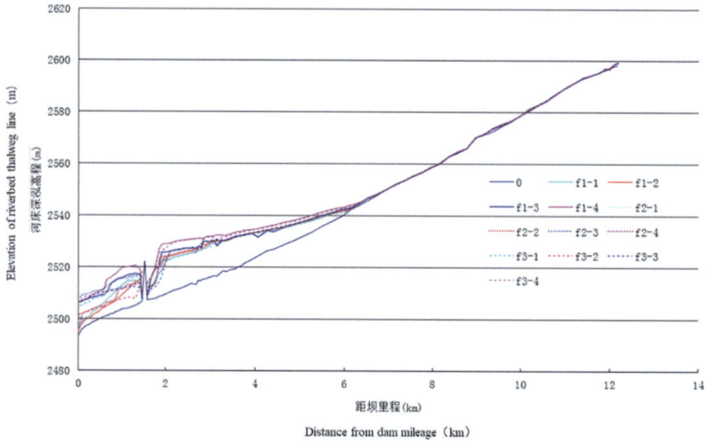

Fig. 8. Longitudinal section of sedimentation in the reservoir area after 15 years of operation of each plan

the reservoir area is the highest under the 120 m^3/s flow limit scheme. Under the same limit of sand flushing flow rate and different sand discharge water levels, the effective storage capacity of the mud storage area in the reservoir area is the highest under the 2508 m sand discharge water level scheme. With the increase of operation years, the effective storage capacity decreases, but the magnitude is not significant, indicating that the reservoir area has basically reached a balance of erosion and sedimentation after 15 years of operation.

From this, it can be concluded that under 2520 m sediment discharge water level, the remaining effective storage capacity of the hub is larger after 20 years of operation under each scheme; When the flow limit is 120 m^3/s, the effective storage capacity is maximum.

At a sediment discharge water level of 2520 m, the remaining effective storage capacity of the hub storage area after 10 years of operation under different flow limit schemes 1-2, 2-2, 3-2 is 11.36 million cubic meters, 12.75 million cubic meters, and 14.7 million cubic meters (corresponding to a normal storage capacity of 2530 m). On the basis of previous sedimentation, with the operation of open discharge and sand flushing, the effective storage capacity has been restored to a certain extent. After 20 years of operation in the three schemes, the remaining effective storage capacity in the hub storage area is 12.61 million meters, 13.44 million meters, and 15.13 million meters, respectively, in the 1-2, 2-2, and 3-2 schemes.

(5) Recommended Operating Mode

The main task of this project is to generate electricity, so the long-term operation of the hub needs to meet the designed daily regulating storage capacity. The regulating capacity, the water intake before the sedimentation surface elevation, bypass sand discharge hole before the sedimentation surface elevation and other related control factors for comparison and analysis, to determine the reasonable operation of the project reservoir for the flood season to limit the reservoir level and control the flow rate level to limit the reservoir level of sand discharge application, a reasonable sediment discharge period is from June to September, and when the reservoir water level is controlled at 2520 m during the flood season and the flow limit is 120 m^3/s, open discharge and sand flushing are carried out, which is Scheme 3-2. At a sediment discharge water level of 2520 m, the remaining effective storage capacity of the hub is larger after 20 years of operation under various scheme conditions, and the maximum effective storage capacity is achieved when the flow limit is 120 m^3/s.

5 Conclusion

A two-dimensional water and sediment model was used to study the sediment deposition problem of a high head and high sand content hydroelectric power station in Nepal. Under the layout plan of "reservoir replacing pool + bypass sediment discharge", a total of 12 sediment discharge operation schemes for the reservoir area were combined with 3 flow limits and 4 sediment discharge water levels. The study concludes that:

(1) After 5 years of operation of the hub, there is a difference in the elevation of the sedimentation surface in front of the water intake. Different sand flushing operation methods have a significant impact on the elevation in front of the water intake between 7 and 17 years. After 17 years of operation under various schemes, the trend of the elevation change of the sedimentation surface in front of the water intake is basically consistent, and the difference is not significant. Under each scheme, the elevation of the sedimentation surface in front of the water intake can be controlled at around 2505 m.
(2) When operating at a sand discharge water level of 2520 m, the elevation of the sedimentation surface in front of the sand discharge hole can be controlled at around 2510 m, and the probability of sedimentation in the bypass sand discharge hole is not high.

Table 2. Storage capacity curves for different operation years of each scheme

Elevation		Storage capacity curve for different operation years (10000 m^3)						
(m)		Original	5a	10a	15a	20a	25a	28a
Scheme 1-1	2511	380	199.8	8	11.7	14.6	10.3	52.4
	2520	1200	591.1	207.9	297.7	323.9	326.6	356.4
	2530	3000	1883.7	1350.9	1401	1371.5	1334.3	1323.1
Scheme 1-2	2511	380	276.2	11.9	12.4	15.4	11.5	17.7
	2520	1200	761.4	173	291.8	346.9	320.6	346.9
	2530	3000	1992	1148	1271.4	1276.3	1200.1	1192.6
Scheme 1-3	2511	380	291.9	106.7	20.6	17.6	12.3	92.6
	2520	1200	800.5	367.7	402.9	382.4	350.6	419.2
	2530	3000	1927.5	1221.1	1340.8	1256.9	1154.6	1151.3
Scheme 1-4	2511	380	323.7	227.9	4.9	15	12.4	59.3
	2520	1200	946.1	677.9	180.2	347.8	325.8	362.1
	2530	3000	2088.7	1417.8	717	859.1	770.5	774
Scheme 2-1	2511	380	199.1	7.8	12.9	14.7	10.7	50.1
	2520	1200	594.3	290.1	342	379.7	350.1	381.7
	2530	3000	1888.7	1437.6	1457.2	1435.5	1372.4	1380.4
Scheme 2-2	2511	380	276.3	8.6	12	14.6	11.7	13
	2520	1200	761.8	289.1	354.3	382.6	383.3	389
	2530	3000	1992.5	1284	1382.9	1358.7	1309.6	1274.8
Scheme 2-3	2511	380	287.9	52.4	19.7	53.7	13.6	65.1
	2520	1200	791	331.8	352.8	388	352.9	382.9
	2530	3000	1906	1188.5	1300.2	1294.4	1174	1151.1
Scheme 2-4	2511	380	323.8	226.4	5.1	57.9	54.3	15.7
	2520	1200	941.6	680.6	283.1	320.6	318.2	315
	2530	3000	2067.2	1426.6	867.9	842.6	868.8	829
Scheme 3-1	2511	380	174.2	5.1	16.6	12.6	11.2	14.2
	2520	1200	590.5	349.5	411.2	451.5	423.6	413.6
	2530	3000	1883.9	1475.3	1558.7	1544.8	1480.6	1462.8
Scheme 3-2	2511	380	277.7	14.4	15.4	14.9	11.8	55
	2520	1200	765.8	385.4	444.4	469.3	455.6	474.8
	2530	3000	2102	1484.9	1562	1527.9	1469.6	1458.2

(continued)

Table 2. (*continued*)

Elevation		Storage capacity curve for different operation years (10000 m^3)						
Scheme 3-3	2511	380	291	12.5	16.1	13.1	10.9	49.8
	2520	1200	800.4	422.1	451.7	461.8	455.2	484.8
	2530	3000	1927.1	1287.5	1368.5	1328.8	1258.2	1243.2
Scheme 3-4	2511	380	324.4	227.4	5	44	14.3	52.1
	2520	1200	946.3	684.2	377.7	482.5	435.8	444.3
	2530	3000	2083.8	1416.4	1020.3	1127.7	1060.2	1046

(3) The hub operates for about 10–15 years and is basically in a state of balance between erosion and sedimentation; At a sediment discharge water level of 2520 m, the remaining effective storage capacity of the hub is larger after 20 years of operation under various scheme conditions.

(4) This hydropower station operates with a sediment discharge water level of 2520 m and a sediment discharge flow limit of 120 m^3/s, the effective storage capacity of the reservoir meets the requirements of the designed daily regulating storage capacity. Therefore, it is recommended that the hydropower station operate according to Scheme 3-2.

References

1. Zheng, H., Wang, X., Sun, T.: Research on replacing pools with reservoirs in the Kohala hydropower station project in Pakistan. Water Resour. Hydropower Eng. Des. **35**(2), 36–56 (2016)
2. Wang, X., Xu, Z.: Research on a reservoir replacing reservoir scheme for a muddy sand river power station in Pakistan. Guangdong Water Resources and Hydropower (09): 4–6+25 (2015)
3. Wang, X., Gong, L., Wu, W., et al.: Study on the operation mode of high sand content river water supply reservoirs—taking Wangyao reservoir as an example. Sediment Res. **43**(2), 33–39 (2018)
4. Xiao, J., Hu, J.: Experimental study on the sediment discharge effect of Karabe reservoir in Xinjiang. People's Yellow River **45**(6), 18–23 (2023)
5. Gao, D., Zou, H.: Study on the sedimentation and flushing effects of the sedimentation tank at the Nasuwakali hydropower station. People's Yangtze River **53**(3), 121–126 (2022)
6. Dong, D.: Research and application of sedimentation facilities at Shangma Xiangdi a hydropower station in Nepal. Fujian Hydroelectric Power **1**, 3–5+37 (2017)
7. Zhou, H., et al.: Study on measures of prevention and reducing silting of old hydropower stations in high sedimentation content reserve area. IOP Conference Series Earth and Environmental Science **768**(1), 012004 (2021). https://doi.org/10.1088/1755-1315/768/1/012004
8. Richter, W., et al.: Retrofitting of pressurized sand traps in hydropower plants. Water **13**, 2515 (2021). https://doi.org/10.3390/w13182515

Research on Load Distribution Method of Cascade Hydropower Station with Maximum Energy Storage at the End of Dispatching Period

Tianqing Li(✉), Peng Lu, Pengcheng Zhou, Bing Han, Zijun Yang, and Kaibin Yang

Kunming Engineering Corporation Limited, Renmin East Road. 115, Kunming, China
1074761688@qq.com

Abstract. The paper focuses on how to rationally distribute the load of cascade hydropower station in the short term economic operation to meet the grid requirements and improve the water energy efficiency of cascade hydropower stations. In this paper, a calculation method of energy storage for cascade hydropower station is presented, the change of cascade storage caused by power generation of different hydropower stations is studied, and the influence of reservoir capacity characteristics on load distribution of cascade hydropower stations is analyzed. According to the short term economic operation and dispatching requirements of cascade hydropower station, the load distribution model of cascade hydropower station based on the maximum storage capacity at the end of the term is constructed. Taking the short-term load distribution of cascade hydropower stations in the Nam Ou River Basin of Laos as an example, four scheduling schemes with different boundary conditions are calculated. The results show that the load distribution results of cascade hydropower stations in the Nam Ou River based on the maximum energy storage the end of the period are as follows: Nam Ou4 and Nam Ou1 in the downstream with larger changes in water level per unit storage capacity are preferentially stored, Nam Ou7, Nam Ou6 and Nam Ou2 in the upstream with larger storage capacity are preferentially supplied and generating power. The research results can provide guidance for the short-term economic operation and dispatch of cascade hydropower stations, and help to improve the water energy utilization efficiency of cascade hydropower stations.

Keywords: Cascade hydropower stations · energy storage · load distribution · Reservoir capacity characteristics

1 Introduction

Under the influence of climate change, electricity energy is transforming from dependence on fossil energy to dependence on renewable energy. To construct a new power system with renewable energy as the main body is the development trend of power energy system [1–3]. Hydropower is an important energy component in the new power system, the reason is that hydropower station is not only a renewable clean energy, but also has the ability to adjust the wind and wind power generation fluctuations, improve

S. Zheng et al. (Eds.): IHDC 2024, LNCE 487, pp. 215–227, 2025.
https://doi.org/10.1007/978-981-97-9184-2_19

the stability of the power system [4–6]. In short-term operation scheduling, the main goal of cascade hydropower stations operation scheduling is not to pursue the maximum of its own power generation, but to make full use of its regulating capacity and cooperate with the stable load output process of the power grid. In this case, reducing the energy consumption of cascade hydropower station and realizing the economic operation of cascade hydropower station is the goal of short-term operation and dispatching of cascade hydropower station [7, 8]. Many scholars have carried out a lot of research in this field.

From the Angle of water utilization, some scholars take average water consumption rate or total water consumption as the target to optimize the load distribution of cascade hydropower stations [9–13]. This method equates water consumption with energy consumption, which is correct when there is only one hydropower station, the more water there is, the more water energy there is, and the more electricity the station can produce. However, when there are multiple cascaded hydropower stations, the amount of water and energy in cascaded reservoirs cannot be completely equivalent. The amount of electricity that can be generated by cascade hydropower station is related not only to the total water storage capacity of cascade hydropower station, but also to the distribution of these water storage capacity among hydropower stations.

Therefore, some scholars also try to study from the perspective of energy. Kai Zheng developed a load distribution model for the daily planned cascade hydropower stations with the objective of minimizing the potential water level [14]. Qian Cheng proposed a load distribution method aimed at minimizing the cumulative energy consumption of cascade hydropower stations during their operational period [15]. Bai T proposed the objective function of maximizing the efficiency of cascade hydropower stations [16].

The energy storage of cascade hydropower station is related to the relationship between upstream and downstream of hydropower station and the characteristics of hydropower station. It is an extremely complex concept, which can not be directly expressed in terms of the total water volume of a cascade hydropower station or the gravitational potential energy of the total water volume. The existing calculation methods do not deeply study the energy storage of cascade hydropower stations, and rarely analyze the key factors affecting the energy storage. In this paper, an energy storage calculation method of cascade hydropower station is proposed, the influencing factors related to the storage of cascade hydropower station are analyzed, and the load distribution method of cascade hydropower station based on the maximum storage capacity at the end of the term is constructed. The method has been applied and verified in cascade hydropower stations in the Nam Ou River.

2 Definition and Calculation of Energy Storage of Cascade Hydropower Stations

The energy storage of cascade hydropower stations is defined as: Without considering the future local inflow, based on the current water level, each hydropower station successively reduces the reservoir water level to the dead water level from upstream to downstream, and the total electricity capacity of all hydropower stations.The total storage energy of cascade hydropower stations is equal to the sum of the storage energy of each hydropower station. The storage energy of each hydropower station includes self-storage energy and the storage capacity provided by the water level of the downstream hydropower stations. For convenience of writing, self-storage energy is called storage energy A, and the storage energy provided by the water level of the downstream hydropower stations is called storage energy B. Energy storage is calculated as follows.

- Total energy storage:

$$EN = \sum_{i=1}^{N} EN_i \tag{1.1}$$

where EN is the total energy storage of cascade hydropower stations. N is the number of hydropower station. EN_i is the energy storage of ith station.

- The energy storage of ith station:

$$EN_i = EN_{i,0} + \sum_{j \in \theta_i} EN_{i,j} \tag{1.2}$$

where $EN_{i,0}$ is The energy storage A of ith station. $EN_{i,j}$ is the storage capacity of ith station provided by the water level of the downstream jth station. θ_i is the collection of stations those are located downstream of ith station.

- The energy storage A of ith station:

$$EN_{i,0} = k_i \cdot \Delta V_i \big(\overline{Z}_i - Z_{Tailwater,i}(Q_i) - H_{Loss,i}(Q_i)\big) \tag{1.3}$$

where k_i is output coefficient of ith station. ΔV_i is available water supply of ith station. $\overline{Z}_i$ is the water level corresponding to the storage capacity that is average value of current storage capacity and dead storage capacity. Q_i is full discharge of the unit. $Z_{Tailwater,i}$ is the relation function between tailwater level and discharge. $H_{Loss,i}$ is the relation function between head loss and generating flow.

- The energy storage B of ith station:

$$EN_{ij} = k_j \cdot \Delta V_i \big(Z_j - Z_{Tailwater,j}(Q_i) - H_{Loss,j}(Q_i)\big) \tag{1.4}$$

where Z_j is the current water level of the reservoir of ith station.

According to the calculation method of energy storage, for one station, the greater the amount of water stored, the greater the energy storage. For the cascade stations, the size of the total energy storage is related to the distribution of water among each station. The head effect of the downstream hydropower station can be utilized by the upstream hydropower station, and the head effect of the most downstream hydropower station can be utilized by all the upstream hydropower stations. The higher the water level of the downstream hydropower station, the larger the energy storage B of the upstream hydropower station. When the water storage capacity of upstream reservoir is large, the head effect of downstream hydropower station is more obvious. Therefore, generally speaking, raising the water level of the downstream power station is conducive to the increase of total energy storage of the cascade hydropower stations, but it needs to be analyzed in detail.

The task of load distribution of cascade hydropower station is to determine which power stations give priority to water supply and generate more power, and which power stations give priority to water storage and generate less power. According to the different inflow conditions, the load distribution between plants can be divided into two working conditions. he first scenario is the water supply condition, where the available power from the current range of the river is less than the load requirement of the power grid, and to meet the load requirement of the power grid, the cascade power station needs to provide water; the second scenario is the water storage condition, where the available power from the current range of the river is greater than the load requirement of the power grid, and in order to meet the load requirement of the power grid, the cascade power station needs to store the remaining water in the reservoir. The differences between the water supply and water storage of the upstream station or the downstream station are analyzed below.

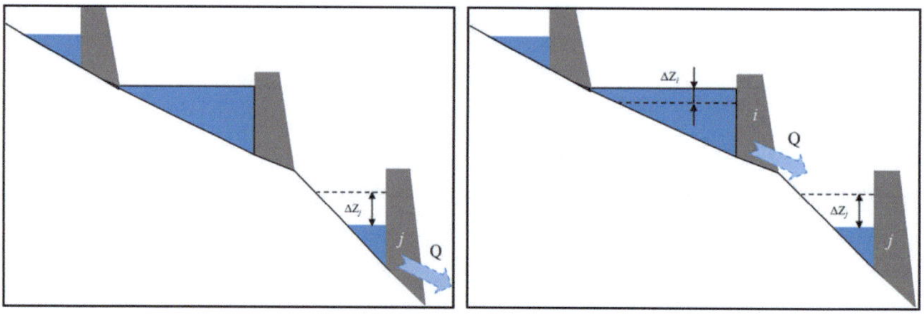

Fig. 1. Cascade hydropower station releases water. (a) Power Station j, located at the bottom of the cascade, supplies water; (b) Power Station i, located in the middle of the cascade, supplies water. ΔZ_j or ΔZ_i is the water level corresponding to the water consumed by the power station to meet the load requirements of the grid.

In Fig. 1(a), the energy storage of *j*th station decreases due to the decrease of water level. At the same time, because the water level of *j*th station is reduced, the energy storage B of all stations located upstream of *j*th station is reduced, so the total energy storage of the cascade power stations is reduced.

In Fig. 1(b), *i*th station supplies water and *j*th station stores water. In this scenario, the energy storage A of *i*th station decreases, and the change of energy storage B depends on the influence between the reduction of water quantity of *i*th station and the elevation of water level of *j*th station. Energy storage A and B of *j*th station are both increased. Energy storage A and B of the power station located downstream of *j*th station remain unchanged. The energy storage A of the power station located upstream of *i*th station is unchanged, and the energy storage B is changed, because the water level of *i*th station and *j*th station is changed. The specific changes are as follows.

- If $\Delta Z_i < \Delta Z_j$, the energy storage B of the station located upstream of *i*th station increases;
- If $\Delta Z_i = \Delta Z_j$, the energy storage B of the station located upstream of *i*th station unchanged;

- If $\Delta Z_i > \Delta Z_j$, the energy storage B of the station located upstream of ith station decreases.

Due to the short time scale of short-term dispatching, which is generally only 24 h, the change of energy storage A has little impact, and the change of energy storage B will change greatly due to the change of water level of downstream hydropower station. The variation of energy storage B of the station located upstream of the ith station is determined by the relationship between the sizes of ΔZ_i and ΔZ_j. The value of ΔZ depends on the capacity characteristics of the reservoir. Reservoir capacity characteristic is the variation of water level per unit storage capacity. As shown in Fig. 3, if the variation of water level per unit storage capacity of ith station is smaller than the variation of water level per unit storage capacity of jth station, it exists $\Delta Z_i < \Delta Z_j$. Therefore, the energy storage B of the power station upstream of ith station increases, and the total energy storage of the cascade hydropower station may increase. On the contrary, the total energy storage of cascade hydropower stations may be reduced (Fig. 2).

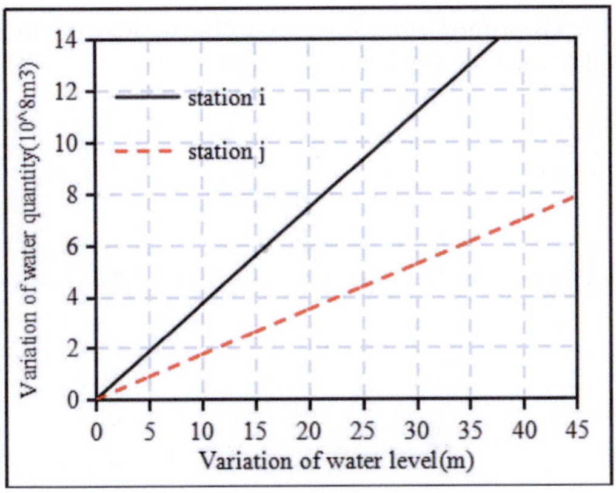

Fig. 2. The relationship between the variation of reservoir water level and the variation of reservoir capacity

In the following, this paper establishes the load distribution model of cascade hydropower station based on the maximum energy storage at the end of the term, and analyzes it with specific cases.

2.1 Load Distribution Model of Cascade Hydropower Station

The objective function of the model is the maximum cascade energy storage at the end of the scheduling period, and the expression is as follows.

$$max\, EN = \sum\nolimits_{i=1}^{N} EN_i \tag{1.5}$$

The constraints of the model are as follows.

- Load balance constraint

$$P_t = \sum_{n=1}^{N} P_{i,t} \quad (1.6)$$

where P_t is the tth load of the grid. $P_{i,t}$ is the tth output of ith station.

- Water balance constraint

$$V_{i,t+1} = V_{i,t} + I_{i,t} \cdot \Delta t - O_{i,t} \cdot \Delta t \quad (1.7)$$

$$V_{i,t}^{min} \le V_{i,t} \le V_{i,t}^{max} \quad (1.8)$$

$$O_{i,t}^{min} \le O_{i,t} \le O_{i,t}^{max} \quad (1.9)$$

where $V_{i,t}$ is the tth reservoir capacity of ith station, $I_{i,t}$ is the tth inflow flow of ith station. $O_{i,t}$ is the tth outflow flow of ith station. $V_{i,t}^{min}$ is the tth minimum allowable storage capacity of ith station. $V_{i,t}^{max}$ is the tth maximum allowable storage capacity of ith station. $O_{i,t}^{min}$ is the tth minimum allowable outflow flow of ith station. $O_{i,t}^{max}$ is the tth maximum allowable outflow flow of ith station.

- Outflow constraint

$$O_{i,t} = Q_{i,t} + R_{i,t} \quad (1.10)$$

where $Q_{i,t}$ is the tth power generation flow of ith station. $R_{i,t}$ is the tth flood discharge of ith station.

- Head constraint

$$H_{i,t} = \frac{Z_{i,t} + Z_{t+1}}{2} - Z_{Tailwater}(O_{i,t}) - H_{loss}(Q_{i,t}) \quad (1.11)$$

$$H_i^{min} \le H_{i,t} \le H_i^{max} \quad (1.12)$$

where $H_{i,t}$ is the tth water purification head of ith station. $Z_{i,t}$ is the tth reservoir level of ith station. H_i^{min} is the tth minimum head of ith station. H_i^{max} is the tth maximum head of ith station.

- Output constraint

$$P_{i,t} = P(Q_{i,t}, H_{i,t}) \quad (1.13)$$

where $P_{i,t}$ is the tth output of ith station under $Q_{i,t}$ and $H_{i,t}$.

- Steady output constraint

$$P_{i,t} \in \left(P_{i,1}^{low}, P_{i,1}^{up}\right) \cup \left(P_{i,2}^{low}, P_{i,2}^{up}\right) \cup (\ldots) \quad (1.14)$$

3 Model Application

3.1 Basin Introduction

Nam Ou river the largest tributary of the Mekong River on the left bank of Laos. It originates in the border mountains between Jiangcheng County, Yunnan Province, China and Phongsari Province, Laos. The river flows from north to south, with a basin area of 256,34km^2 and a river length of 475km.There are seven cascaded hydropower stations on the Nam Ou rive, from upstream to downstream, respectively, Nam Ou7, Nam Ou6, Nam Ou5, Nam Ou4, Nam Ou3, Nam Ou2 and Nam Ou1.Nam Ou7 station is the leading reservoir of hydropower development planning, which has the annual regulation capacity; The regulatory capacity of Nam Ou6 and Nam Ou5 stations is seasonal. The regulatory capacity of other stations is daily regulatory capacity. The characteristic parameters of cascade hydropower station are shown in Table 1.

Table 1. Table of characteristic parameters of station

Station name	Normal storage level/m	Dead water level/m	Regulating storage/10^8m^3	Installed capacity/MW
Nam Ou7	635	590	12.45	210
Nam Ou6	510	490	2.46	180
Nam Ou5	441	430	1.42	240
Nam Ou4	386	384	0.98	132
Nam Ou3	360	358	0.24	210
Nam Ou2	325	323	0.25	120
Nam Ou1	307	305	0.22	180

The reservoir capacity characteristic curve is as follows.

The regulated storage capacity of Nam Ou7, Nam Ou6 and Nam Ou5 hydropower stations accounts for 91% of the total regulated storage capacity, of which Nam Ou7 hydropower stations account for 69%..Therefore, the water storage capacity of Nam Ou7, Nam Ou6 and Nam Ou5 hydropower stations has a great influence on the total energy storage of the cascade hydropower stations. Nam Ou4, Nam Ou3, Nam Ou2 and Nam Ou1 hydropower stations mainly provide head effect for cascade energy storage, raising the water level of this four hydropower stations downstream, the energy storage B of Nam Ou7, Nam Ou6 and Nam Ou5 will increase significantly, and the total energy storage of cascade will also increase significantly. In terms of the characteristics of reservoir water level, the variation of water level of the same amount of water in Nam Ou4 is greater than that in Nam Ou1, the Nam Ou3 and the Nam Ou2.For the four downstream cascade hydropower stations, Nam Ou4 and Nam Ou2 give priority to water storage, which is more beneficial to increase cascade energy storage.

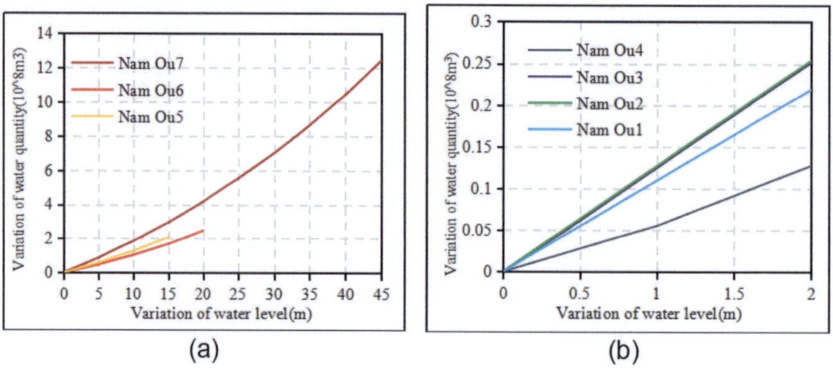

Fig. 3. The relationship between the variation of reservoir water level and the variation of reservoir capacity. (a) is the curve of three hydropower stations located upstream. (b) is the curve of four hydropower stations located downstream

3.2 Case Result Analysis

Take intraday hourly short-term economic scheduling as an example. It is assumed that the 24-h hourly load of the grid is 600MW.Two local inflow conditions are considered: one is that there is more local inflow in the power station section, and the generating capacity of local inflow in the section is greater than the given load of the power grid; the other is that there is no local inflow in the section, which depends on the power station water supply to meet the load demand of the power grid. Three initial water level states of Nam Ou7, Nam Ou6 and Nam Ou5 are considered. The first is that the water level of the three power stations is lower and the water volume of the reservoirs is less. The second is that the water level of the three power station reservoirs is in the middle, and the water volume of the reservoirs is in the middle. The third is that the water level of the three power stations is higher and the water volume of the reservoirs is more. Four scheduling schemes are set according to different local inflow and initial water levels. The scheme information is shown in Table 2.

Table 2. Table of Scheme information

Scheme number	Initial water level/m							Inflow condition
	Nam Ou7	Nam Ou6	Nam Ou5	Nam Ou4	Nam Ou3	Nam Ou2	Nam Ou1	
1	612	500	435	385	359	324	306	Meet load requirements
2	595	495	432	385	359	324	306	=0
3	612	500	435	385	359	324	306	
4	634	509	440	385	359	324	306	

The statistical table of cascade energy storage of the four schemes is shown in Table 3. The total cascade energy storage is increased in Scheme 1, and reduced in scheme 2, Scheme 3 and Scheme 4. The load distribution and reservoir water level changes of the specific four schemes are shown in Figs. 4, 5, 6, 7 (Table 4).

Table 3. Table of Scheme information

Scheme number	Energy storage before dispatch/10^8kWh	Energy storage after dispatch/10^8kWh	variation/10^8kWh
1	3.877	4.062	0.185
2	0.871	0.741	−0.130
3	3.877	3.771	−0.106
4	10.530	10.461	−0.068

Table 4. Table of average load distribution. The unit is MW.

Scheme number	Nam Ou7	Nam Ou6	Nam Ou5	Nam Ou4	Nam Ou3	Nam Ou2	Nam Ou1
1	16.77	13.64	9.96	5.01	8.18	3.26	3.17
2	13.83	14.26	12.75	4.90	9.33	3.35	1.59
3	19.12	16.15	13.34	4.67	4.75	1.98	0.00
4	18.17	13.75	17.32	5.33	3.46	1.96	0.00

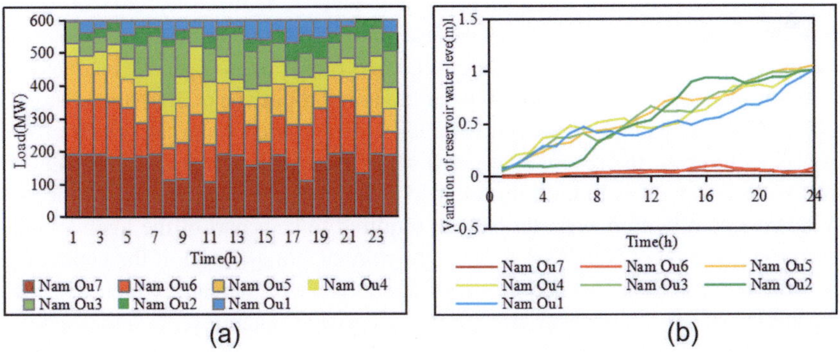

Fig. 4. The scheduling result diagram of scheme 1. (a) is bar chart of load distribution. (b) is the chart of water level change compared to initial condition.

In Scheme 1, all stations are impounded because there is more local inflow. The reservoir water level of Nam Ou4, Nam Ou3, Nam Ou2 and Nam Ou1 hydropower stations reaches the normal storage level at the end of operation period. The reservoir water level of Nam Ou7, Nam Ou6, Nam Ou5 is raised by 0.03 m, 0.07 m and 1.045 m respectively. The loads allocated to Nam Ou7, Nam Ou6 and Nam Ou5 accounted for 27.9%, 22.7% and 16.6% of the total load respectively. Nam Ou7 hydropower station generates more power, the water level of Nam Ou6 and Nam Ou5 rises, the energy storage B of Nam Ou7 and Nam Ou6 increases, and then the total energy storage of the cascade increases. Nam Ou5 generate less power than Nam Ou6, the reason is that the reservoir water level of Nam Ou4, Nam Ou3, Nam Ou2 and Nam Ou1 has reached the normal storage level, reducing Nam Ou5 power generation, can reduce the waste water of the downstream power plant.

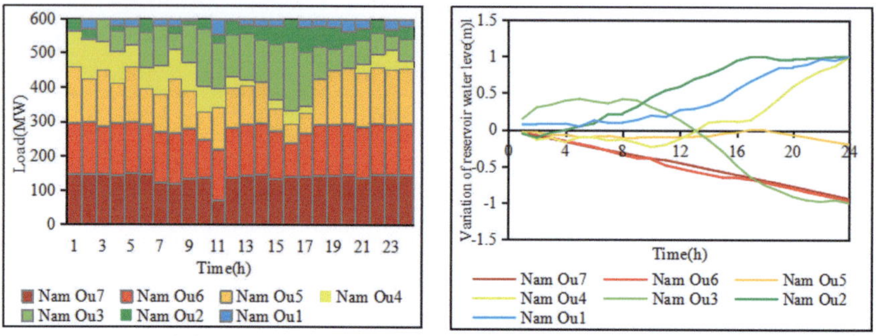

(a) (b)

Fig. 5. The scheduling result diagram of scheme 2. (a) is bar chart of load distribution. (b) is the chart of water level change compared to initial condition.

In Scheme 2, the hydropower stations need to use the water stored in the reservoir to generate electricity. Nam Ou7, Nam Ou6, Nam Ou5 and Nam Ou3 consume water to generate electricity, and the reservoir water level is reduced by 0.94 m, 0.97 m, 0.19 m, and 1.0 m, respectively. The load distribution of Nam Ou7, Nam Ou6, Nam Ou5 accounts for 68% of the total load of the cascade stations. According to the characteristics of reservoir water level, it is reasonable for the load proportion of Nam Ou7, Nam Ou6 and Nam Ou5 to be large. However, Nam Ou3 also provide water and electricity. The reason is as follows: the reservoir water level of Nam Ou7, Nam Ou6 and Nam Ou5 is low, so their output is limited. In order to meet the load requirements of the power grid, downstream hydropower stations are required to bear part of the load. While the reservoir water level of Nam Ou1 reaches the normal storage level, if the generation of Nam Ou2 increased, the discharge flow of Nam Ou1 will increase, which will lead to the reduction of the total energy storage of the cascade. Therefore, it is reasonable for Nam Ou3 to bear more load.

In Scheme 3, the hydropower station also needs to use the water stored in the reservoir to generate electricity. The load distribution of Nam Ou7, Nam Ou6, Nam Ou5 accounts for 81% of the total load of the cascade stations, and the reservoir water level is reduced by 0.67 m, 0.73 m, and 0.04m, respectively. The reservoir water level of Nam Ou4, Nam Ou3, Nam Ou2 and Nam Ou1 is raised by 1.00 m, 0.14 m, 0.38 m, and 0.99m, respectively. In Scheme 3, the same as in Scheme 2, the output of Nam Ou7, Nam Ou6

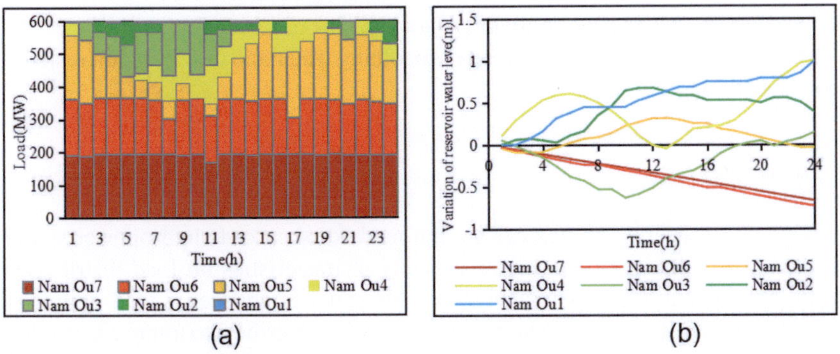

Fig. 6. The scheduling result diagram of scheme 3. (a) is bar chart of load distribution. (b) is the chart of water level change compared to initial condition.

and Nam Ou5 is also limited. Therefore, in this condition, it is reasonable for Nam Ou3 to bear more load.

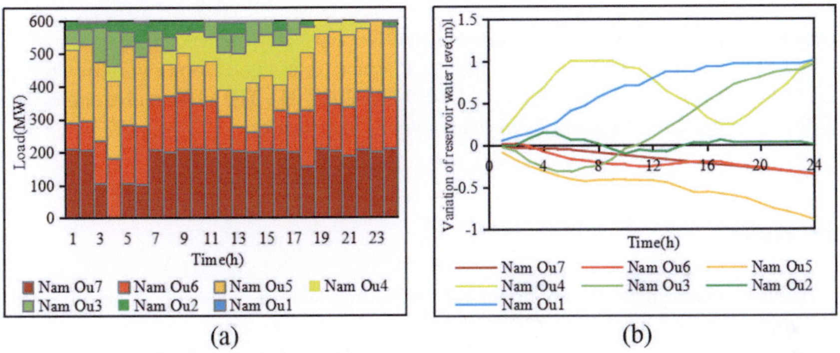

Fig. 7. The scheduling result diagram of scheme 4. (a) is bar chart of load distribution. (b) is the chart of water level change compared to initial condition.

In Scheme 4, the hydropower station also needs to use the water stored in the reservoir to generate electricity. The load distribution of Nam Ou7, Nam Ou6, Nam Ou5 accounts for 82% of the total load of the cascade stations, and the reservoir water level is reduced by 0.35 m, 0.35 m, and 0.89m, respectively. The reservoir water level of Nam Ou4, Nam Ou3, Nam Ou2 and Nam Ou1 is raised by 1.00 m, 0.96 m, 0.00 m, and 1.00m, respectively. The reservoir water level of Nam Ou4, Nam Ou3 and Nam Ou1 is the normal water level. The result accords with the reservoir capacity characteristics.

Based on the above analysis, the load distribution scheme of Nam Ou river hydropower stations is obtained, which is based on the maximum energy storage at the end of dispatching period. The order of cascade hydropower stations using reservoir water to generate power is as follows: the first priority is Nam Ou7, Nam Ou6, Nam Ou5 hydropower stations, the second priority is Nam Ou3 and Nam Ou2 hydropower stations, and the last is Nam Ou4 and Nam Ou1 hydropower station. In the actual load distribution, it is necessary to consider not only the characteristics of reservoir water level, but also the characteristics of load, the situation of incoming water and the operation restrictions of power station.

4 Summary

In this paper, an energy storage calculation method of cascade hydropower station is presented. The influence of reservoir capacity characteristics on load distribution of cascade hydropower stations is analyzed. This paper establishes a load optimization distribution model of cascade hydropower stations based on the maximum energy storage at the end of the term, and verifies the effectiveness of the model in the Nam Ou river. The results show that the optimal load distribution mode of cascade hydropower station based on the maximum energy storage at the end of the term is as follows: the hydropower station located in the lower reaches of the basin and the hydropower station with large variation of unit storage water level gives priority to water storage and raising water level, and the hydropower station located in the upper reaches of the basin gives priority to water supply.

Acknowledgement. This work is supported by the National Natural Science Foundation of China (No. U2243232) and the Technology Project of Power Construction Corporation of China, Ltd (DJ-ZDXM-2022-10, DJ-HXGG-2022-01, DJ-HXGG-2021-04). The authors also appreciate the insightful comments and suggestions from anonymous reviewers.

References

1. Zhang, B.: Recognize the important role of hydropower in achieving "Carbon Peak and Neutrality" goal. J. Eng. Stud. (2024)
2. Liang, Z., Kong, L., Pan, H.: Challenges and Key Technologies of the New Power System. Power Energy (2024)
3. Kong, L., Pei, W., Rao, J., et al.: Build new power system to promote carbon neutrality. Bull. Chin. Acad. Sci. (2022)
4. Chen, G., Wang, X.: Global hydropower development trend and China's role in context of carbon neutrality. J. Hydroelectric Eng. (2024)
5. Zhou, J., Du, X., Zhou, X.: Study on hydropower development strategy for new power systems. J. Hydroelectric Eng. (2022)
6. Zhou, J., Li, S., Gao, J.: Technical and economic analysis of water energy storage to promote new energy development. J. Hydroelectric Eng. (2022)
7. Chen, J., Wu, Y., Hu, B., et al.: Improved mathematical model of minimum energy consumption for load distribution of cascade hydropower plants. Autom. Electric Power Syst. **41**, 155–160 (2017)
8. Yu, P., Tang, J.Y., Pu, Y., et al.: Research on the method of short-term load plan distribution in cascade hydropower stations. China Rural Water Hydropower (2015)
9. Li, C., Li, S., Tang, H., et al.: Short-term optimal scheduling method of caseade hydropower stations based on segmented load construction strategy. Pearl River (2023)
10. Yu, P., Tang, J., et al.: Study on the method of short-term load plan distribution in cascade hydropower stations. China Rural Water and Hydropower (2015)

11. Wei, Q., Chen, S., Huang, W., et al (2020). Optimal load distribution of cascade hydropower stations based on memetic algorithm. Water Power
12. Xu, G. (2012). Study on load allocation algorithm for cascade hydropower stations. J. Hydroelectric Eng.
13. Xiao-Qing, R., Ting-Hong, Z., Xu-Fen, W.: Research on the Optimization Method for Load Distribution of Cascade Hydropower Station Faced to Units. Value Eng. (2015)
14. Zheng, K., Jiang, Z., Liu, M., et al.: Study on short-term joint optimal operation of cascaded hydropower stations. Electric Power Technol. Environ Prot. **39**(4), 305–313 (2023)
15. Cheng, Q., Liu, P., Ming, B.: Short-term economic operation of cascade hydropower station based on improved two-layer nested optimization algorithm. Eng. J. Wuhan Univ. (2022)
16. Bai, T., Chang, J.X., Huang, Q., et al.: Short-term optimal operation of xiaolangdi and xixiayuan cascade hydropower station. Energy Educ. Sci. Technol. (2012)

Exploring the Untapped Potential of Existing Hydropower Resources in the Context of New Energy Development: A Case Study of the Liyuan-Ahai Hybrid Pumped Storage Power Station

Hanmo Chen(✉), Chuting Miao, and Peng Lu

PowerChina Kunming Engineering Corporation Limited, Kunmign, China
chenhanmo1217@qq.com

Abstract. In recent years, countries and regions worldwide have set goals to increase the proportion of new energy source in their energy transition plans. However, the intermittent nature of new energy sources, represented by wind power and solar photovoltaics, necessitates the support of flexible resources like pumped storage and hydropower. This study takes the established Liyuan and Ahai Hydropower Stations along the Jinsha River as typical cases, thoroughly exploring the potential benefits of utilizing the reservoirs of these two stations to construct a Liyuan-Ahai hybrid pumped-storage power station. Through comprehensive analysis, we propose an installed capacity scheme that aims to maximize the benefits of the three power stations. This scheme not only provides a feasible reference method for the design of similar engineering projects, but also holds significant importance in promoting the efficient utilization and sustainable development of hydroelectric energy. We hope that through this research, we can provide valuable reference and inspiration for experts and scholars in the field of hydropower engineering.

Keywords: Hybrid Pumped Storage Power Station · New Energy · Installed Capacity · Existing Hydropower Station

1 Research Background

In recent years, countries and regions worldwide have set goals to increase the proportion of new energy source in their energy transition plans. China is also focused on developing new energy and constructing a new power system. However, the intermittent nature of new energy sources, represented by wind power and solar photovoltaics, necessitates the support of flexible resources like pumped storage and hydropower.

Yunnan Province stands as a prominent hydroelectric powerhouse in China, with a preliminary assessment indicating a potential economic hydropower capacity of around 130,000 MW across the province. However, most of the remaining resources are situated in ecologically sensitive areas, rendering the development conditions relatively immature.

S. Zheng et al. (Eds.): IHDC 2024, LNCE 487, pp. 228–235, 2025.
https://doi.org/10.1007/978-981-97-9184-2_20

Considering the potential, constraints, and implementation timeline for wind and solar energy development in Yunnan Province, the provincial power grid has devised plans for the addition of new energy sources during the "14th Five-Year Plan," "15th Five-Year Plan," and "16th Five-Year Plan" periods. These strategic plans entail the targeted addition of approximately 50,000 MW, 30,000 MW, and 20,000 MW of new energy capacity respectively. By 2035, it is projected that the share of new energy installed capacity will surpass 50% of the total power capacity.

Hybrid pumped storage hydropower plants combine the functions of pumped storage and traditional hydropower plants, offering peak load shifting, backup power supply, and other benefits. They also have the advantages of relatively short construction cycles and the ability to increase power generation during flood seasons.

In the case of existing hydropower plants in Yunnan Province that have insufficient installed capacity or reservoir capacity exceeding the demand for regulation, the addition of reversible units can be considered. By integrating these reversible units with the operation of new energy sources, these plants can be developed into hybrid pumped storage hydropower plants. This approach not only complements the operation of new energy sources but also enhances the overall efficiency of the power plants.

2 Research Object

Due to the absence of the Longpan Reservoir, a crucial control reservoir on the mainstream of the middle reaches of the Jinsha River, downstream hydropower stations face significant water discharge during the flood season. As the first-level power station built in the middle section of the Jinsha River, the Liyuan Hydropower Station operates with a normal storage level of 1618 m, a dead storage level of 1605 m, a regulated storage capacity of 173 million m^3, and an installed capacity of 2400 MW. Downstream, the Ahai Hydropower Station operates a normal storage level of 1504 m, a dead storage level of 1492 m, and a regulated storage capacity of 238 million m^3.

Both the Liyuan and Ahai Hydropower Stations possess significant untapped storage capacity for regulating water flow, with daily required storage capacities for regulation at a mere 30 million m^3 and 40 million m^3 respectively, well below their regulated storage capacities. This advantageous situation allows for the construction of hybrid pumped storage hydropower plants, capitalizing on the existing infrastructure without compromising their primary regulatory functions.

In this study, a comprehensive analysis of the power system demand and market potential in Yunnan Province was conducted. The research focused on evaluating the regulation capabilities of the Liyuan and Ahai hydropower stations, and delved into the assessment of the overall benefits of the Liyuan-Ahai hybrid pumped storage hydropower plant. Based on these findings, an appropriate installed capacity for the Liyuan-Ahai hybrid pumped storage hydropower plant was determined, ensuring optimal utilization of the available resources.

3 Research Method

Liyuan-Ahai hybrid pumped storage hydropower plant demonstrates a strong correlation between its comprehensive benefits, installed capacity, and project investment. The installed capacity serves as a crucial indicator influencing the power generation efficiency, peak load regulation benefits, and its impact on the power generation of both the Liyuan and Ahai hydropower stations. Key factors affecting the installed capacity of the Liyuan-Ahai hybrid pumped storage hydropower plant include determining the appropriate unit capacity, available reservoir capacity, and considering the engineering construction conditions.

To determine the suitable unit capacity, data from existing and under-design pumped storage hydropower plants with similar head conditions were collected and analyzed. The stability of both slope sides, along with Yunnan Province's comprehensive load curve incorporating the influence of large-scale hydropower sources and eastward electricity transmission, was considered when calculating the available reservoir capacity for the Liyuan-Ahai hybrid pumped storage hydropower plant. Furthermore, the engineering layout was examined to analyze the number of units that can be accommodated.

By conducting a comprehensive analysis of the Liyuan-Ahai hybrid pumped storage hydropower plant, the Liyuan Hydropower Station, and the Ahai Hydropower Station, the study determined the appropriate installed capacity for the Liyuan-Ahai hybrid pumped storage hydropower plant, optimizing its overall efficiency and benefits.

4 Main Achievements

4.1 Appropriate Unit Capacity

The Liyuan-Ahai hybrid pumped storage hydropower plant operates within a head range of 90 m to 130 m. Currently, the unit capacity of pumped storage hydropower plants designed and operated within this head range is typically between 50 and 200 MW. Unit capacities below 100 MW are only found in the Panjiakou and Shahe pumped storage hydropower plants, while unit capacities exceeding 150 MW are seen in the Zagorsk and Kanelovsk pumped storage hydropower plants. Pumped storage hydropower plants with a unit capacity of 150 MW have been constructed, such as the Baishan and Langyashan plants, while projects are underway for the construction of the Weijiachong, Pankou, and Wuxi River pumped storage hydropower plants. There are no completed pumped storage hydropower plants with unit capacities between 100 and 150 MW, but the Zhongxiang pumped storage hydropower plant is currently under construction. Considering the manufacturing aspect of the units, the appropriate unit capacity for the Liyuan-Ahai hybrid pumped storage hydropower plant falls within the range of 100 MW to 150 MW.

4.2 Available Reservoir Capacity

4.2.1 Scheduling and Operational Mode

Before the low electricity demand during nighttime and the peak generation from solar power during midday, the Liyuan Hydropower Station operates by appropriately reducing the water level in the reservoir and then operating at a high water level after pumping is completed. On the other hand, the Ahai Hydropower Station maintains a high water level in the reservoir and operates at a low water level after pumping is completed. During the evening peak hours, the Liyuan Hydropower Station, the Ahai Hydropower Station, and the Liyuan-Ahai hybrid pumped storage hydropower plant increase their output according to system requirements for peak load regulation. When the Liyuan Hydropower Station has excess water during the flood season, the Liyuan-Ahai hybrid pumped storage hydropower plant directly utilizes the excess water for additional power generation without pumping. The typical daily operational process can be seen in Figs. 1 and 2.

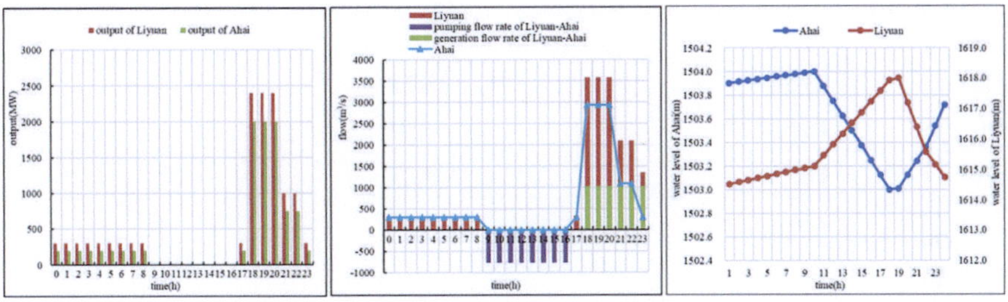

Fig. 1 The Typical Daily Operational Process (Dry Season)

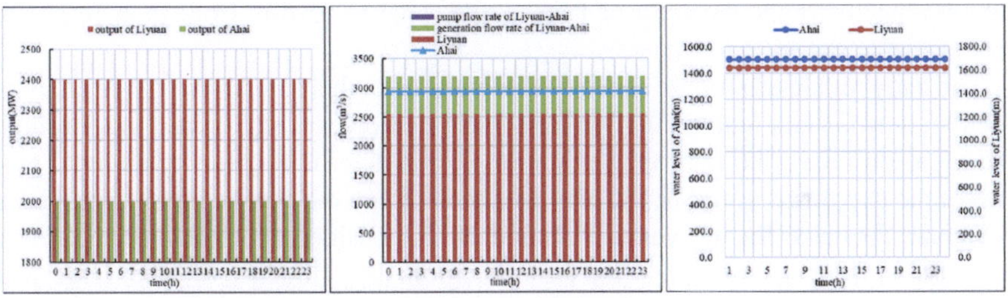

Fig. 2 The Typical Daily Operational Process (Flood Season)

4.2.2 Maximum Available Capacity

The Liyuan Hydropower Station reservoir's bank area is prone to the accumulation of Nianshengken deposits, Xiazari deposits, as well as the risk of Longmen steep slope, Xingpeidang and Caokedu landslide. According to the assessment and analysis of bank stability, it is concluded that only the overall instability of the Caokedu landslide in the Liyuan Hydropower Station reservoir bank may result in secondary hazards caused by surging waves. By analyzing the seepage and slope stability under different rates of reservoir water level decline, the correlation between landslide stability and reservoir water level decline rate under natural and heavy rainfall conditions is established as shown in Table 1.

In general, the stability state K of landslide bodies should be greater than 1. However, considering a certain safety margin, this study recommends controlling the daily water level decline of Liyuan Reservoir at 5.5 m/day. For Ahai Reservoir, considering the daily operational data, the recommended control is a water level variation of 3.5 m/day. The daily available reservoir capacity for Liyuan Hydropower Station ranges from 62.1 million m^3 to 67.98 million m^3. For Ahai Hydropower Station, the daily available reservoir capacity ranges from 57.06 million m^3 to 68.62 million m^3.

Table 1 Relationship between critical reservoir water level decline rate of landslides and stability state K-value

K	natural conditions (m/d)	Heavy rainfall conditions(m/d)
1.05	10.38	3.93
1.025	13.50	5.71
1.00	16.63	7.50
0.975	19.75	9.29

4.2.3 Available Capacity of Liyuan-Ahai Hybrid Pumped Storage Power Station

The usable capacity for the Liyuan-Ahai hybrid plant is determined by subtracting the daily regulated storage capacity of Liyuan and Ahai reservoirs from their respective daily available storage capacities.

The daily regulated storage capacity of Liyuan and Ahai reservoirs will be analyzed based on independent load regulation for Liyuan hydropower station and synchronized load regulation for Ahai hydropower station. When determining the operational positions of the two power stations, the load will be based on the comprehensive load curve of Yunnan Province for the years 2030 and 2035, considering the addition of power transmission from the western to the eastern region. The influence of other large-scale power sources in the province, including Longpan Reservoir in the middle reaches of the Jinsha River and Xiaowan, Manwan, Dachaoshan, and Nuozhadu hydropower stations in the lower reaches of the Lancang River, will also be considered. As the downstream Xiaowan, Manwan, Dachaoshan, and Nuozhadu hydropower stations have strong regulation capabilities, this study will analyze the residual load curve after deducting the output of hydro, thermal, and new energy sources within the system, prioritizing load regulation by existing large-scale hydropower stations.

The calculated results indicate that, before the operation of the Longpan hydropower station, the maximum daily regulated storage capacity for the Liyuan hydropower station is 29.96 million m^3, and for the Ahai hydropower station, it is 5.86 million m^3. After the operation of the Longpan hydropower station, the maximum daily regulated storage capacity for the Liyuan hydropower station is 39.28 million m^3, and for the Ahai hydropower station, it is 14.95 million m^3. Therefore, the available capacity for the Liyuan-Ahai hybrid pumped storage plant is primarily influenced by the capacity of the Liyuan reservoir after the operation of the Longpan hydropower station. The usable capacity for the Liyuan-Ahai hybrid plant is 24.03 million m^3. Considering the predominance of solar energy in Yunnan Province, with a typical continuous pumping and full discharge duration of 6 h, the maximum installed capacity of the Liyuan-Ahai hybrid pumped storage plant is 976 MW.

4.3 Engineering Construction Conditions

Due to the presence of the Nianshengken at the inlet/outlet of the upper reservoir and the utilization of the existing Liyuan reservoir discharge tunnel for the outlet of the lower reservoir, if the number of units for the Liyuan-Ahai hybrid pumped storage power station exceeds 6, it could have adverse effects on slope stability and the stability of newly excavated chambers. Therefore, this study suggests that the number of units for the Liyuan-Ahai hybrid pumped storage power station should not exceed 6.

4.4 Engineering Benefit Comparison

Based on the above study, under the premise of a feasible engineering plan, the following analysis will be conducted for the Liyuan-Ahai hybrid pumped storage power station with different installed capacities: 600 MW (4×150 MW), 750 MW (5×150 MW), and 900 MW (6×150 MW).

4.4.1 Impact on the Liyuan Hydropower Station

According to the calculations, before the operation of the Longpan hydropower station, different installed capacity scenarios for this project result in a net increase in the Liyuan hydropower station's average annual electricity generation by approximately 535 million kWh to 679 million kWh. Among these, the expansion of capacity leads to an additional electricity generation during the flood season of about 563 million kWh to 719 million kWh, while the decrease in the average water level of the power station results in a reduction in the electricity generation of the original conventional units at the Liyuan hydropower station by about 28 million kWh to 40 million kWh.

After the operation of the Longpan hydropower station, different installed capacity scenarios for this project result in a net increase in the Liyuan hydropower station's average annual electricity generation by approximately 51 million kWh to 67 million kWh. Among these, the expansion of capacity leads to an additional electricity generation during the flood season of about 121 million kWh to 157 million kWh, while the decrease in the average water level of the power station results in a reduction in the electricity generation of the original conventional units at the Liyuan hydropower station by about 69 million kWh to 90 million kWh.

4.4.2 Impact on the Ahai Hydropower Station

According to the calculations, before the operation of the Longpan hydropower station, different installed capacity scenarios for this project have an impact on the Ahai hydropower station's average annual electricity generation of approximately 9 million kWh to 20 million kWh.

After the operation of the Longpan hydropower station, different installed capacity scenarios for this project have an impact on the Ahai hydropower station's average annual electricity generation of approximately 10 million kWh to 18 million kWh.

4.4.3 Economic Comparison

The different installed capacity scenarios for the Liyuan-Ahai hybrid pumped storage power station are evaluated based on the system-designed level of annual electricity generation, aiming to meet the system's electricity demand to an equal extent. The capacity and energy benefits of each scenario are calculated, and the total present value of costs for each scenario is determined to evaluate the economic feasibility of the comparative options.

According to the calculations, the present value of costs for the different installed capacity scenarios of the Liyuan-Ahai hybrid pumped storage power station are 6,562.34 million yuan, 5,675.51 million yuan, and 4,835.54 million yuan, respectively. It is observed that the higher installed capacity scenarios have lower present value of costs.

5 Conclusion

When utilizing an existing reservoir for the construction of a pumped storage power station, it is necessary to analyze the surrounding renewable energy resources and development potential. The operational mode of the project should be determined, taking into account factors such as topography, geological conditions, and existing infrastructure. It is important to assess whether the site is suitable for operating as a conventional power station. The benefits of the newly constructed pumped storage power station and its adverse impacts on the existing reservoir should be calculated. After considering various factors, the installed capacity should be determined to maximize the benefits of both the newly constructed and existing power stations.

Acknowledgements. The National Natural Science Foundation China (U2243232), Science and Technology projects of POWERCHINA (DJ-ZDXM-2022-10, DJ-HXGG-2022 -01) funded this research.

References

1. Luo, B., Chen, Y., Miao, S., Liu, Z., Liu, X.: A joint short-term operation model for wind power and hybrid pumped-storage hydropower plant. J. Hydraulic Eng. (2023)
2. Guo, A., Chang, J., Wang, Y., Wang, X., Sun, X.: Research on capacity computation of cascade on-stream integral pumped storage hydropower plants II: multi-scale nesting model for capacity computation and specific example. J. Hydraulic Eng. (2024)

Research on Identification of Deep Leakage Channels in Karst Pumped Storage Reservoirs Based on Multi Field Data Fusion

Zheng Kexun[1,2], Gan Feifei[1,2(✉)], Zhao Daiyao[1,2], Chen Xiao[1,2], Liu Xianggang[1,2], and Zhang Ning[1,2]

[1] Powerchina Guiyang Engineering Corporation Limited, Guiyang 550081, Guizhou, China
2294153304@qq.com

[2] Geotechnical Engineering Corporation Limited, Sinohydro Guiyang Survey and Design Corporation Limited, Guiyang 550081, Guizhou, China

Abstract. The most prominent engineering geological problem of pumped storage power station reservoirs in karst areas is karst leakage, the development of karst leakage channels has a significant impact on the selection of reservoir locations, layout of engineering buildings, design of anti-seepage measures, and engineering costs. Therefore, the survey and evaluation of reservoir leakage channels are the foundation for the construction of pumped storage power station reservoirs in karst areas. The field analysis method plays an important role in karst leakage survey. Traditional karst groundwater field analysis methods, based on the representative indicators of each field measured and determined by experience, fail to fully reflect the temporal and spatial change information of each field indicator, and the data cannot be fully utilized and compared for verification. The multi field data fusion analysis method for karst groundwater proposed in this article comprehensively considers the relationship between measured field indicators and leakage sources, natural conditions, adjacent spaces, and different time field indicators, and obtains the characteristic values of the tracer index, background index, gradient index, and time series index of each field, and overlay calculation of single field comprehensive eigenvalues and multi field composite eigenvalues, which can realize the fusion of multiple fields and multiple indicators, amplify the abnormal location signal of seepage, and delineate the location of centralized seepage, so as to quantitatively determine the location information of the seepage channel of karst groundwater. This method is applied to the survey of karst leakage in the lower reservoir of a pumped storage power station in Guizhou Province, field data fusion analysis shows that there is an abnormal seepage field in the anti-seepage curtain line of the site, and there is good evidence for the temperature and conductivity field data. There is a deep karst leakage channel in the reservoir; The burial depth of the channel is more than 170 m below the normal water level. The research results can provide support for subsequent anti-seepage methods and engineering treatments, as well as relevant engineering experience for other projects.

Keywords: groundwater field analysis · normalized treatment · tracer indicators · background indicators · gradient indicators · timing indicators · the eigenvalue

S. Zheng et al. (Eds.): IHDC 2024, LNCE 487, pp. 236–246, 2025.
https://doi.org/10.1007/978-981-97-9184-2_21

1 Preface

China has abundant karst groundwater resources, and the karst aquifer system exhibits strong heterogeneity due to the uneven distribution and development of pipelines and cracks, which seriously restricts the exploration and evaluation of karst water resources [1]. The multi index field data of karst groundwater system can not only be used to analyze the spatiotemporal changes in groundwater quality, but also provide information on groundwater occurrence conditions, seepage pathways, circulation depth, prevention and remediation of groundwater pollution, and water resource development and utilization [2, 3]. The multi index field analysis method has become a commonly used research method in exploring engineering and environmental research such as reservoir leakage, groundwater pollution, and underground cave water inrush investigation and treatment [4–6].

Field refers to the distribution of objects in space and is a special form of material existence characterized by spatial position functions [7]. The variable information related to groundwater undergoes a certain period of special geological processes in the geological medium space, and the characteristic parameters and changes of groundwater seepage, temperature, conductivity, chemical composition, and isotopes reflect certain characteristics and laws objectively existing in the groundwater system [8, 9]. The main methods for analyzing karst groundwater systems include four field analysis methods: groundwater seepage, hydrochemical field, water temperature field, and isotope field [10]. The general application of field analysis method is to measure the representative indicators of each field, and then draw the corresponding indicator hole depth distribution curve [11], indicator profile line (axis) distribution curve [12] or cloud map [13]. Then, based on the shape of the curve and the shape of the cloud map, the abnormal areas of the indicators in space are delineated based on experience, and combined with the characteristics of buildings and geology, the existence of seepage channels is determined [14–20]. This process of data processing, analysis, and judgment fails to fully reflect the temporal and spatial changes of a single indicator, resulting in the omission of a significant amount of information and insufficient utilization of the data; The judgment of seepage information is still mainly qualitative, and the influence of experience is significant; The numerical differences between different field indicators are significant, and comparative verification between indicators has not been achieved.

This article normalizes and standardizes the indicators of temperature field and conductivity field according to some factors, including the relationship between field indicators and water sources, the relationship with natural field background values of geological media, variation of indicators in adjacent spaces, and changes in indicators at different times. After processing, the characteristic values of the field indicator's tracer indicator, background indicator, gradient indicator, and temporal indicator are obtained. Each characteristic value can be stacked with different weights based on the reliability of the data and the correlation with the water source to obtain a unified field comprehensive indicator characteristic value. Overlay calculation of single field comprehensive eigenvalues and multi field composite eigenvalues to achieve multi field and multi index fusion, amplify the signal of abnormal seepage location, and delineate the concentrated seepage location. By quantifying and systematizing the field analysis method for karst groundwater systems, this method can effectively process field analysis data and more significantly analyze possible groundwater seepage location information.

2 Data Normalization Preprocessing Session

Firstly, the water level, temperature, hydrochemistry, conductivity, and isotope field data of the water source in the research area are measured, and the measured values of each field of drilling holes, caverns, or groundwater dew points are obtained. Based on the variation law of water source indicators, multiple repeated measurements can be carried out at different times. The space of the field can be divided into one-dimensional lines (drilling or profile lines), two-dimensional surfaces (profiles), and three-dimensional geological spaces. Perform necessary data interpolation on one-dimensional data such as boreholes or profile lines based on measured values, so that the field data is uniformly distributed in elevation or horizontal space, and obtain one-dimensional distribution data of the field; Obtain two-dimensional distribution data of the field through plane interpolation based on a certain field data of groundwater seepage points from different boreholes and underground caverns; Interpolate a certain field data from different profiles to obtain the three-dimensional distribution data of the field.

3 Normalization Processing of Two Data Sets

3.1 Calculation of Characteristic Values of Field Indicators

Normalize the field data with adjacent spatial field data, leakage water source field data, different time field data, and natural field data to obtain the gradient index characteristic values, tracer index characteristic values, temporal index characteristic values, and background index characteristic values of the field.

The one-dimensional, two-dimensional, and three-dimensional distribution data of the field only reflect the spatial differences of the field, and the calculation of indicator characteristic values is similar. Represent the one-dimensional, two-dimensional, or three-dimensional distribution data V obtained from actual measurements and interpolation as a matrix:

$$V = \left(v_{ij}\right)_{m\times n} \tag{1}$$

In the formula, i represents a certain field, which can be defined as 1, 2, 3, 4, etc. for temperature field, seepage field, chemical field, isotope field, etc.; j represents the position corresponding to the measured points in the research area, one-dimensional data can represent the drilling elevation, and two-dimensional data can represent the profile position; m represents the total number of measured fields; n represents the total number of measured data points in each field; v_{ij} is the representative value of the measured point data.

3.1.1 Gradient Index

The gradient index reflects the degree of spatial variation between the measured representative value of a certain field at a certain location and the measured value of the corresponding field index at nearby locations. The expression is:

$$v^1{}_{ij} = \frac{|v_{ij} - v_{i(j-1)}|}{\max\limits_{1 \le j \le n} |v_{ij} - v_{i(j-1)}|} \tag{2}$$

In the formula, $v^3{}_{ij}$ is the characteristic value of the gradient index corresponding to a certain field in the study area, and the calculated result is 0–1, dimensionless. 0 represents that the measured field indicators of the borehole have no changes near the elevation, indicating no anomalies and a low possibility of seepage; 1 represents the maximum variation in the measured field indicators of the borehole near the elevation, with anomalies and a high possibility of seepage.

3.1.2 Tracer Indicators

The tracer index reflects the degree of closeness between the measured representative value of a certain field at a certain location and the corresponding representative value of the water source field, expressed as:

$$v^2{}_{ij} = 1 - \frac{|v_{ij} - R|}{\max\limits_{1 \le j \le n} |v_{ij} - R|} \tag{3}$$

In the formula: $v^1{}_{ij}$ is the characteristic value of the tracer indicator corresponding to a certain field in the study area, and the calculated result is 0–1, dimensionless; 0 represents the maximum difference between the measured field indicators of the borehole and the water source, and the possibility of seepage is small; 1 represents that the measured field indicators of the borehole are the same as the water source, indicating a high possibility of seepage; *R* is the representative value of a certain field indicator of the water source, measured at the water source.

3.1.3 Time Series Indicators

The time series indicator reflects the degree of temporal variation between the measured representative value of a certain field at a certain location and the measured values of corresponding field indicators at other time periods, expressed as

$$v^3{}_{ij} = \frac{|v_{ij} - t_{ij}|}{\max\limits_{1 \le j \le n} |v_{ij} - t_{ij}|} \tag{4}$$

In the formula, $v^4{}_{ij}$ is the characteristic value of the temporal index corresponding to a certain field in the study area, and the calculated result is 0–1, dimensionless. 0 represents that the measured field indicators of the borehole have not changed compared to a certain period in the past, that is, there is no abnormality, and the possibility of seepage is small; 1 represents the maximum variation in the measured field indicators of the borehole compared to a certain period in the past, with anomalies and a high possibility of seepage; t_{ij} is the representative value of a certain field corresponding to a certain time period and location in the research area; The selection of time periods generally considers the time when the indicators of the water source field have undergone significant changes.

3.1.4 Background Indicators

The background indicator reflects the degree of anomaly between the measured representative value of a certain field and the background value of the corresponding field indicator under general conditions, expressed as:

$$v^4{}_{ij} = \frac{|v_{ij} - b_{ij}|}{\max\limits_{1 \leq j \leq n} |v_{ij} - b_{ij}|} \tag{5}$$

In the formula, $v^2{}_{ij}$ is the characteristic value of the background indicator corresponding to a certain field in the study area, and the calculated result is 0–1, dimensionless. 0 represents that the measured field indicators of the borehole are the same as the background values under normal conditions, indicating no abnormalities and a low possibility of seepage; 1 represents the maximum difference between the measured field indicators of the borehole and the background values, indicating anomalies and a high possibility of seepage; b_{ij} is the background value corresponding to a certain field in the research area, which can be obtained through theoretical analysis or empirical values.

3.2 Superposition of the Characteristic Values of the Field Indicators

3.2.1 The Characteristic Value of the Comprehensive Index of Each Field

According to the reliability of the data and the correlation between the data and the seepage, different weights are superimposed to obtain the unified comprehensive index eigenvalues of a certain field. The calculation expression is:

$$f_{ij} = \alpha_{i1} v^1{}_{ij} + \alpha_{i2} v^2{}_{ij} + \alpha_{i3} v^3{}_{ij} + \alpha_{i4} v^4{}_{ij} \tag{6}$$

$$\alpha_{i1} + \alpha_{i2} + \alpha_{i3} + \alpha_{i4} = 1 \tag{7}$$

where: fij is the characteristic value of the comprehensive index corresponding to a field somewhere in the study area, and the calculated result is 0–1, dimensionless. 0 means that there is no abnormality in the field index of the borehole, and the possibility of seepage is small; 1. It means that the field index of the borehole is abnormal, and the possibility of seepage is high; α_{i1}, α_{i2}, α_{i3}andα_{i4} are the weights of the eigenvalues of a field tracer index, the eigenvalue of the background index, the eigenvalue of the gradient index and the eigenvalue of the time series index, respectively, according to the reliability of the data and the correlation with the seepage, they are selected between 0–1, and if the correlation between the eigenvalues of each index and the seepage is comparable, 0.25 can be taken on the whole.

3.2.2 Eigenvalues of Multi-Field Composite Indicators

According to the reliability of the data of each field, different weights are selected to superimpose the calculated eigenvalues of each field composite index to obtain the eigenvalues of multiple field composite indicators. The calculation expression is as follows:

$$F_j = \beta_1 f_{1j} + \beta_2 f_{2j} + \beta_3 f_{3j} + \cdots + \beta_n f_{nj} \tag{8}$$

$$\beta_1 + \beta_2 + \beta_3 + \cdots + \beta_n = 1 \tag{9}$$

In the formula, Fj is the characteristic value of the multi-field composite index in the study area, and the calculated result is 0–1, indicating that the seepage possibility is from small to large. β_1–βn is the weight of each field, which is selected between 0–1 according to the reliability and correlation of the field, and if the characteristic value of the comprehensive index of each field is equivalent to the correlation of seepage, it can be the same.

3.3 Analysis of Eigenvalues of Field Indicators

For one-dimensional distribution data, for example, for a single borehole data, the relationship curve between the multi-field composite eigenvalue f_{ij} and the elevation at different elevations in the borehole is plotted, and the seepage may be divided into large, medium and small hole segments according to the size of the multi-field composite eigenvalue.

For two-dimensional distribution data, for example, for profile data, the multi-field composite eigenvalues f_{ij} and profile positions at different positions of the profile are plotted, and the seepage may be large, medium and small areas are quantitatively divided according to the size of the multi-field composite eigenvalues.

For the three-dimensional distribution data, the relationship between the multi-field composite eigenvalues f_{ij} and the location of the geological body at different locations was plotted, and the seepage may be large, medium and small seepage according to the magnitude of the multi-field composite eigenvalues.

4 Engineering Case Analysis

Taking the lower reservoir of a pumped storage power station in Guizhou Province as an example, the karst leakage problem of the anti-seepage curtain line is analyzed by multi-field data fusion. Drill holes were arranged at the anti-seepage curtain line, and the temperature field and conductivity field data of each borehole were measured in April, and the actual measurements were carried out according to the 2m spacing. The elevation of the orifice of the hole of the ZK8 borehole is 958.8m, the elevation of the groundwater level is 858.8m, and the field data of the borehole elevation of 748.8–858.8m are measured.

For the gradient index, the difference between the boreholes and the measured points was calculated sequentially with a distance of 2m, and the maximum difference between the temperature field was 0.3 °C and the conductivity field was 0.0068. The characteristic values of the gradient index $v^1{}_{1j}$ and $v^1{}_{2j}$of the borehole temperature field and conductivity field can be calculated according to Eq. (2).

For the tracer index, the temperature and conductivity of the leakage source were taken as the average value of the temperature and conductivity of the water outlet point of the reservoir measured on the same day, that is, R was 20.2 °C and 0.4286. The characteristic values $v^2{}_{1j}$ and $v^2{}_{2j}$ of the tracer index of the temperature field and conductivity field of the borehole can be calculated according to Eq. (3).

For the time series index, the temperature field and conductivity field data of the ZK8 borehole were measured one month apart, and the maximum temperature difference between the two corresponding elevations was 0.79 °C, and the maximum conductivity difference was 0.0487. The characteristic values $v^3{}_{1j}$ and $v^3{}_{2j}$ of the time series index of the temperature field and conductivity field of the borehole can be calculated according to Eq. (4).

For the background index, the lowest temperature in all boreholes is located at the elevation of 756 m and the temperature is 19.59 m in hole ZK3, which can be regarded as the natural groundwater temperature not affected by leakage, and the geothermal gradient from the elevation of 756 m to the bottom of each borehole is calculated, with a minimum of 2.06 ºC /100 m, which is used as the natural geothermal gradient of the survey area, so that the natural temperature field without the influence of leakage can be restored, and the background value of natural groundwater temperature at different elevations b1j can be obtained. The lowest conductivity in all boreholes is located at the elevation of 805.3 m and the conductivity is 0.519m in ZK2 hole, which can be regarded as the natural groundwater conductivity that is not affected by seepage, and the geothermal gradient from the elevation of 805.3 m to the bottom of each borehole is calculated, with a minimum of 0.1429/100 m, which is used as the natural geothermal gradient of the survey area, so that the natural conductivity without the influence of leakage can be restored, and the background value of natural groundwater conductivity at different elevations b2j is obtained. The characteristic values $v^4{}_{1j}$ and$v^4{}_{2j}$ of the background index of the temperature field and conductivity field of the borehole can be calculated according to Eq. (5).

Due to the short interval between borehole measurements, the difference is small, which can interfere with the analysis and judgment. Therefore, the weights corresponding to the eigenvalues of the time series index are 0.1, and the weights of the other three indexes are 0.3, respectively, and the comprehensive index eigenvalues f_{1j} and f_{2j} of the borehole temperature field and conductivity are calculated according to Eq. (6). The relationship curves between the characteristic values of each index of temperature field and conductivity field and the characteristic values of comprehensive indexes and elevation were made (see Figs. 1, 2, 3, and 4).

The correlation between temperature field and conductivity field is equivalent, with weights of 0.5. According to Eq. (8), the composite index characteristic values of the two fields are calculated, as shown in Fig. 5, $F_{1j} \geq 0.4$ is the area with a high possibility of leakage, in the elevation range of 780–854 m, and karst leakage treatment is required, $F_{1j} < 0.4$ is a small area with a possibility of leakage, at an elevation of 748–780 m. The rock mass is intact and does not require leakage investigation and treatment.

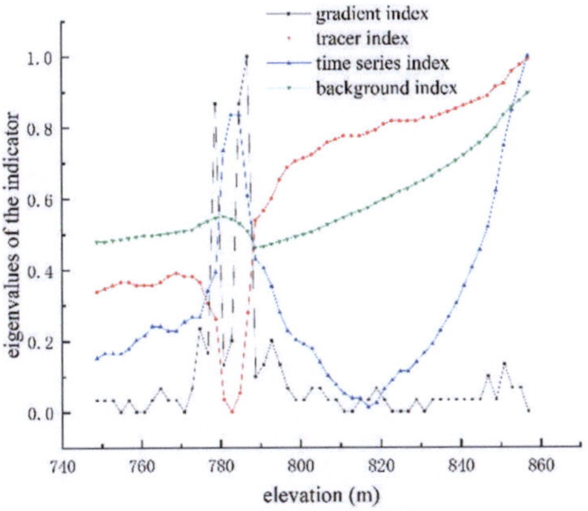

Fig. 1. Curve of the relationship between characteristic values of various indicators in ZK8 temperature field and elevation

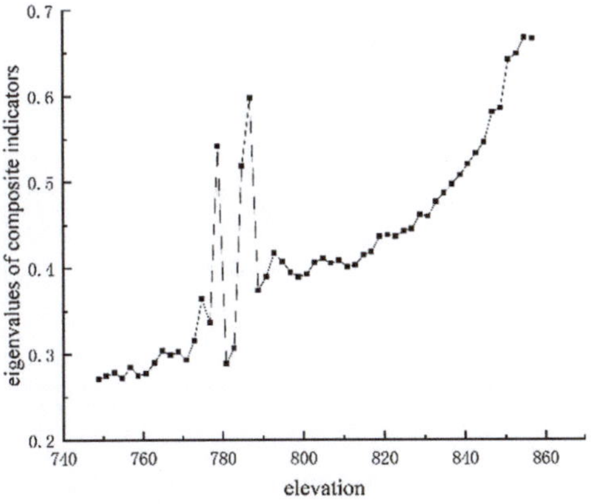

Fig. 2. Curve of the relationship between the characteristic values of ZK8 temperature field comprehensive indicators and elevations

5 Conclusion

The multi field data normalization analysis method studied in this article can be applied in engineering and environmental research such as reservoir leakage, groundwater pollution, and underground cave water inrush investigation and treatment, making the multi index field analysis method of karst groundwater quantitative and systematic. This method can effectively process the field analysis data, and more significantly analyze the possible location information of groundwater seepage. Taking the leakage study of a pumped storage power station in Guizhou as an example, the elevation of the leakage interval of the anti-seepage curtain line is analyzed, providing a theoretical basis for later engineering measures.

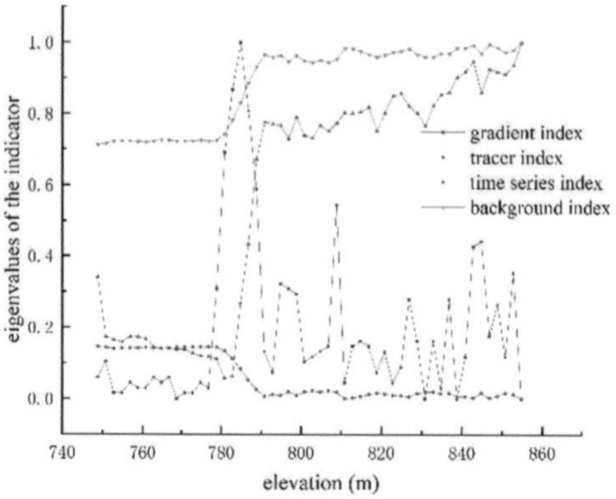

Fig. 3. Curve of the relationship between characteristic values and elevations of ZK8 conductivity indicators

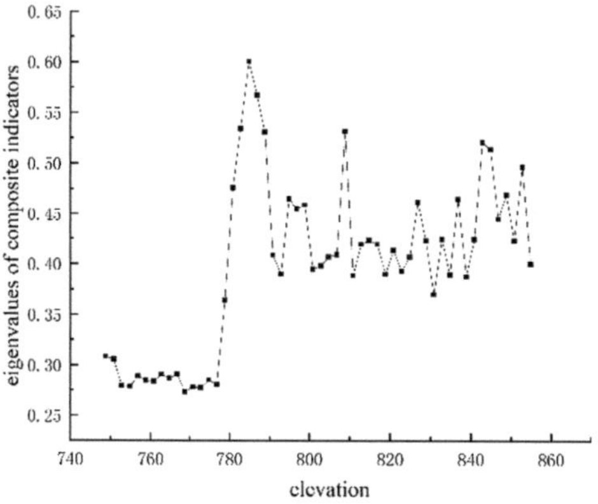

Fig. 4. Characteristic values and elevation relationship curve of ZK8 conductivity comprehensive index

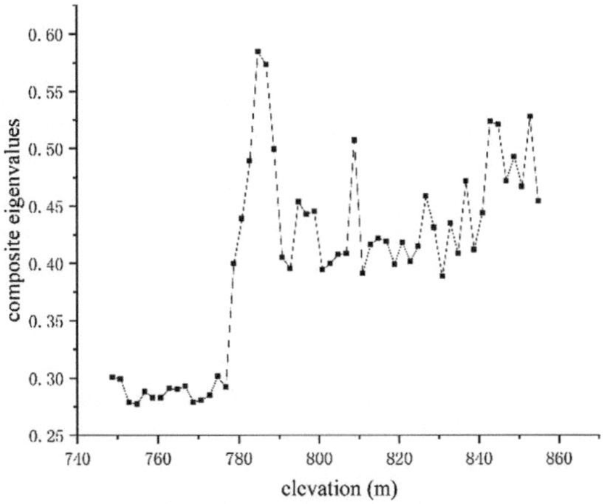

Fig. 5. Curve of the relationship between characteristic values and elevations of ZK8 field fusion indicators

References

1. Yang, Y., Zhao, L.J., Pang, X.D., et al.: A comparative study of groundwater resources assessment methods in Karst mountainous areas of Southwest China: A case study of Zhaidi Underground River Basin. Carsologica Sinica **41**(1), 111–123 (2022)
2. Luo, M.M., Cheng, J., Ji, H.S., et al.: Research progress of solute exchange between karst pipes and fractured media. Earth Sci. 1–17 (2022)
3. Luo, M.M., Ji, H.S.: Mechanism of solute transient storage between karst conduit and fissures. Adv. Water Sci. **33**(1), 145–152 (2022)
4. Guo, Q.H., Wang, Y.X.: Indicative significance of hydrogeochemical information to the characteristics of karst groundwater flow system: A case study of Shentou Spring Field, Shanxi Province. Bull. Geolog. Sci. Technol. **3**, 85–88 (2006)
5. Shi, Y., Liao, J.Q., Cheng, Y., et al.: Study on comprehensive prevention and control Model of Groundwater Pollution in Karst Area of southern Sichuan —— Taking Shau Jitang Karst Spring pollution control in Maoxi Town, Gulin County, Sichuan Province as an example. Coal Geol. China **34**(5), 50–54 (2022)
6. Zheng, K.X., Zhang, G.J.: Investigation on Karst seepage pipeline of left bank anti-seepage line of Xhaixiangkou Hydropower Station. Guizhou Water Power **26**(2), 10–15 (2012)
7. Zou, C.J.: Water conservancy and hydropower Karst engineering geology. Water Resources and Electric Power Press, Beijing (1994)
8. Zhang, Q.F.: Study on evolution law and transformation characteristics of complex seepage field. Hohai University, Nanjing (2002)
9. Wu, Z.W., Song, H.Z.: Research progress on theory and method of groundwater temperature tracing. Adv. Water Sci. **22**(05), 733–740 (2011)
10. Zeng, X.W.: Study on Karst Development Characteristics and Karst water system in dam site of Qizong Hydropower Station. Chengdu University of Technology, Chengdu (2009)
11. Fan, Z.C.: Theoretical research and engineering application of comprehensive tracer method for dam leakage. Hohai University, Nanjing (2006)
12. Dong, H.Z., Cheng, J.S.: Study on natural tracing method of seepage at left abutment of Xiaolangdi Reservoir. Yellow River **26**(004), 43–45 (2004)
13. Yang, D.H., Duan, Y.X., Wang, J.J., et al.: Study on mechanical behavior of loess water tunnel considering local seepage. J. Water Res. Arch. **20**(3), 8–15 (2022)

14. Li, X.: Study on inversion calculation of seepage field and its application in reservoir leakage control. Jilin University, Changchun (2004)
15. Wang, P.J., Guo, H.Y., Luo, H., et al.: Seepage study of Pushihe Pumped Storage Power Station based on hydration method. South-to-North Wat. Trans. Wat. Sci. Technol. **14**(A01), 108–111 (2016)
16. Cheng, J.S., Liu, J.G., Dong, H.Z., et al.: Environmental isotope tracing method for seepage flow around the right abutment of Xin 'an River. Strat. Stud. CAE **6**(1), 57–63 (2004)
17. Zou, C.J.: Study on geothermal field and karst seepage in karst area. Carsologica Sinica **2**, 58–65 (1989)
18. Altuğ, S.: Leakage study of the west side of the oymapinar reservoir, Turkey. Bull. Int. Assoc. Eng. Geol. **13**(1), 147–152 (1976)
19. Gangopadhyay, S.: An approach to estimation of leakage from a karstic limestone reservoir. Bull. Int. Assoc. Eng. Geol. - Bulletin de l'Association Internationale de Géologie de l'Ingénieur **20**(1), 189–191 (1979)
20. Wang, M., Chai, J., Xu, Z., et al.: Leakage safety analysis of anti-seepage measures in reservoir basins: A case study of the Okinawa seawater pumped storage system in Japan. Arab. J. Geosci. **14**(5) (2021)

Analysis of Dynamic Response Characteristics of Towering Intake Towers Under the Action of Main-Aftershock Sequences

Zhiyu Song[1,2], Yafei Zhai[1,2](✉), and Guangkun Liu[1,2]

[1] Yellow River Engineering Consulting Co., Ltd, Zhengzhou 450003, China
yfzhai@hhu.edu.cn

[2] Key Laboratory of Water Management and Water Security for Yellow River Basin of Ministry of Water Resources, Zhengzhou 450003, China

Abstract. After a strong earthquake occurs, it can cause a certain degree of damage to the structure, and the strong aftershock effect can cause secondary damage to the already damaged structure. In this study, taking a actual project of intake tower in the district of western strong earthquake as an example, the acoustic unit is used to simulate the dynamic effect of reservoir water on the tower body, and the overall nonlinear model of the water-intake tower-foundation is established. Combined with the site and seismic motion characteristics of the engineering area, the seismic motion sequence of the main-aftershocks was constructed based on the statistical relationship between the main shocks and strong aftershocks, as well as the NGA seismic motion attenuation model. The effects of main shock, aftershock, and main-aftershock on the structural damage evolution of intake towers in strong earthquake zones were investigated separately. The results show that after considering the aftershock effect, the damage and failure of the intake tower structure intensify, and its dissipation energy and residual displacement increase by about 20%~25%. Compared with the main shock, the aftershock alone causes less damage to the tower structure. However, for the intake tower structure that is damaged after the main shock, the aftershock can cause larger secondary residual deformation of the tower.

Keywords: Intake tower · Main aftershock sequence · Structural safety · NGA model · Damage evolution

1 Introduction

Seismic statistical data show that a large number of aftershocks typically occur after a main earthquake. Due to the short interval between the main shock and its subsequent aftershocks, structures damaged by the main shock often do not receive timely repairs and are further subjected to strong aftershocks (Yabe and Ide 2018; Du et al. 2023). This cumulative damage effect undoubtedly weakens the seismic resistance of structures, implying that designs considering only the impact of a single main shock may not meet the demands imposed by both the main shock and the aftershocks (Xu

S. Zheng et al. (Eds.): IHDC 2024, LNCE 487, pp. 247–257, 2025.
https://doi.org/10.1007/978-981-97-9184-2_22

et al. 2022; Yu et al. 2022). As a crucial component of hydroelectric power projects, the seismic safety of intake towers under earthquake conditions is of great significance to the entire project. Current seismic research and design codes, both domestically and internationally, generally consider only the effects of the main shock, overlooking the secondary damage caused to intake tower structures by strong aftershocks following the main shock. This oversight may result in inadequate seismic resistance of the structures under the combined effects of the main shock and subsequent aftershocks.

Regarding the seismic safety of intake towers, some scholars have conducted relevant research and made significant progress. Liu et al. (2015) used time-history analysis to simulate seismic damage to the tower body and upper frame of the Shapai Hydropower Station intake tower. The simulation results showed good agreement with experimental data. Liu et al. (2016) studied the seismic response of tall intake towers using a viscoelastic artificial boundary. The results indicated that, compared to a massless foundation, considering the far-field radiation damping effect of the foundation led to a reduction in the stress and displacement of the tower body. Zhang et al. (2018) investigated the impact of backfill height behind the tower on the seismic performance of towering intake towers. The results showed that a reasonable backfill height could effectively reduce the dynamic response of the tower body. Zhao et al. (2019b) investigated the seismic performance of intake towers using the endurance time-history method. The results indicated the effectiveness of this method in analyzing the seismic response of structures. Yang et al. (2019) investigated the impact of different concrete pouring methods on the dynamic response of intake towers. The results indicate that employing the method of casting the tower body and backfilling concrete together can effectively reduce the structural dynamic response of intake towers. Currently, in the study of the dynamic response of intake towers, the focus has largely been on the effects of individual main seismic events, while the nonlinear dynamic response of intake towers under the combined action of main and aftershocks remains to be investigated.

The attenuation pattern of seismic parameters is a crucial aspect of earthquake prediction, as it broadly reflects the influence of factors such as seismic sources, propagation pathways, and site conditions on seismic characteristics (Wang et al. 2019; Shi 2022). In studying the impact of main and aftershocks on the damage to concrete structures, it's essential to consider the characteristics of the structure site and seismic activity to determine corresponding main and aftershock seismic parameters. When existing seismic data is insufficient, selecting appropriate seismic attenuation relationships becomes necessary to address this issue (Zhao et al. 2019a; Zhai et al. 2020). Considering the similarities in tectonics, crustal composition, and seismic activity characteristics between mainland China and North America, this study opts for the "Next Generation Attenuation (NGA) relationships" developed in the United States to determine seismic parameters associated with site-specific seismic geological conditions.

To investigate the effects of main and aftershock sequences on the nonlinear dynamic response of intake towers, this study employs an actual engineering project of intake towers located in a strong seismic zone in the western region as an example. It employs acoustic unit simulation to model the hydrodynamic water pressure, establishes a comprehensive nonlinear model of water-tower body-foundation system, and combines statistical relationships between main shocks and strong aftershocks with NGA seismic attenuation models to construct main and aftershock seismic sequences. Under the influence of individual main shocks, aftershocks, and the combined effect of main and aftershocks, comparative studies are conducted on the tower's damage areas, displacement responses, and energy dissipation characteristics. The aim is to quantify the cumulative damage effects of aftershocks on intake towers, in order to reveal the evolution of damage in towering intake towers under the action of main and aftershock sequences.

2 Attenuation Relation of Ground Motion in NGA-WEST2-BSSA13 Model

The Pacific Earthquake Engineering Research Center Lifeline Program (PEER-LL), the U.S. Geological Survey Institute (USGS), and the Southern California Research Center (SCEC) jointly launched the NGA Program (Next Generation of Ground Motion Attenuation) in 2003 Models), five independent and collaborative working groups obtained five different sets of ground motion attenuation relationships by using the ground motion record database developed by the NGA program. The BSSA13 attenuation model selected in this study is the research achievement made by Boore et al. (2013), one of the five teams. Compared with other models, this model is simple in form, clear in physical meaning, and has a relatively wide range of adaptation.

The main parameters of the BSSA13 model include moment magnitude M, Joyner-Boore fault distance R_{JB}, and average shear wave velocity V_{S30}, 30 m below the ground. Secondary parameters include fault fracture depth Z_{or} and basin depth Z_1. In addition, the model takes into account the influence of fault type, which can be distinguished by strike-slip fault, normal fault, reverse fault or uncertain fault type. The basic equation of ground motion attenuation relation of BSSA13 model is expressed as follows:

$$\begin{aligned}\ln Y = {} & F_E(M, mesh) + F_{P,B}(M, mesh) + F_{S,B}(V_{S30}, R_{JB}, M) \\ & + \delta_n \psi(M, V_{S30}, R_{JB})\end{aligned} \tag{1}$$

where, Y is the ground motion parameter to be predicted; F_E, $F_{P,B}$, $F_{S,B}$ are earthquake magnitude terms, path terms and site conditions terms respectively. mesh is the seismic fault type. ψ is the population standard deviation of the model. δ_n is the ln$Y$ standard deviation ratio of the forecast average. For example, when $\delta_n = 0$, it means that no accidental uncertainty is considered, that is, the predicted ground motion parameter is the average value.

The population standard deviation σ consists of the internal term $\phi(M, V_{S30}, R_{JB})$ and the mutual term $\tau(M)$, which can be expressed as:

$$\psi(M, V_{S30}, R_{JB}) = \sqrt{\varphi^2(M, V_{S30}, R_{JB}) + \tau^2(M)} \tag{2}$$

where, ϕ is a polynomial function related to M, V_{S30}, and R_{JB}; τ is a linear function related to M; The specific parameters in the formula are shown in reference.

3 Finite Element Model

The intake tower of the hydropower station in western China is chosen as the focus of this study. According to engineering data, the top width of the intake tower measures 17 m, while the bottom width is 14.0 m. The tower stands at a height of 86.0 m, with the reservoir water reaching a depth of 78.0 m. Based on the ABAQUS finite element software, a finite element model was established. The tower body and foundation were meshed using reduced integration elements C3D8R. Considering the compressibility of reservoir water, the acoustic element AC3D8 is used to simulate the dynamic interaction between the reservoir water and the tower. An absorption boundary is set at the cutoff boundary of the reservoir water to simulate the radiation damping effect. The coupling between the acoustic medium and the structure is simulated using a tie constraint on the contact surfaces between the intake tower and water. The finite element model of the water-intake tower-foundation system is shown in Fig. 1. The model consists of a total of 151,432 elements, of which 39,731 are water acoustic elements.

According to engineering data, the tower concrete is divided into C30 and C25 materials (Fig. 1). The concrete plastic damage model is used to simulate the nonlinear properties of the tower concrete materials. The elastic modulus of C30 concrete is 30.0 GPa, Poisson's ratio is 0.15, and the density is 2500 kg/m^3. The elastic modulus of C25 concrete is 28.0 GPa, Poisson's ratio is 0.2, and the density is 2450 kg/m^3. In the calculation, the dynamic elastic modulus of concrete is 1.5 times of its static elastic modulus according to the Seismic Design Standard for Hydraulic Buildings (2019). The bedrock is simulated by elastic materials, the elastic modulus is 14.0 GPa, Poisson's ratio is 0.167, and the density is 2850 kg/m^3. The density of water is 1000 kg/m^3, and the volume modulus is 2.0 GPa.

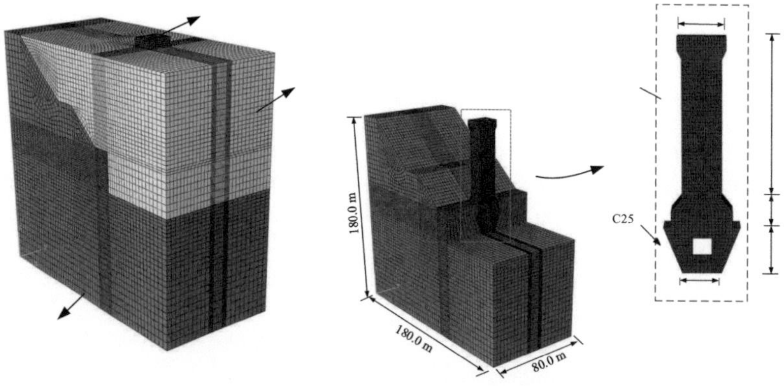

Fig. 1 Finite element model of reservoir water-intake tower-foundation system

4 Synthesis of Main and Aftershock Sequences

Taking the above intake tower project as an example, the project site is selected to represent a strong main earthquake magnitude of 7.2, and the synthesis of the main aftershock sequence seismic structure is performed. It is assumed that a strong aftershock occurs after the main earthquake, with an aftershock magnitude of 6.0 according to Bath's law (1965). Based on the characteristic parameters of the bedrock at the project site, the average shear wave velocity (V_{S30}) at a depth of 30 m below the surface is 1100 m/s. Using the ground motion attenuation relationship of the BSSA13 model, the accidental uncertainty option is not considered (δ_n value is 0), and the fault distance is selected as 10 km, 15 km, 20 km, 30 km, and 50 km, respectively. The corresponding acceleration response spectrum (PSA) calculated is depicted in Fig. 2. It is evident from the figure that the peak acceleration of the earthquake is inversely proportional to the fault distance.

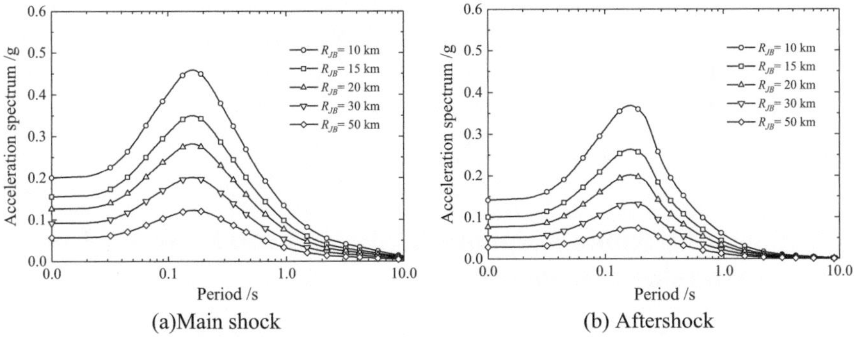

Fig. 2 Response spectra of BSSA13 for different R_{JB}

To ensure safety redundancy, this study adopts a fault distance of 10 km. Both the main shock and aftershock are considered as homologous earthquakes, with peak ground accelerations (PGA) of 0.20 g and 0.14 g, respectively. In determining the duration of ground motion, this study follows the research findings of Huo (1991). Based on the seismic source parameters (magnitude M and epicenter distance R), the duration of the main shock ground motion is determined to be 23 s, while the aftershock ground motion is set to 17 s. According to the obtained acceleration response spectrum curve (Fig. 2), artificial seismic waves are separately fitted for the main shock and aftershock. Each seismic wave is treated independently, with a 10 s interval between the main shock and the aftershock. The seismic waves are then combined to obtain the main-aftershock sequence ground motion. Figure 3 presents the constructed acceleration time history curves for the main and aftershock sequences, with the peak acceleration for vertical ground motion taken as 2/3 of the horizontal ground motion.

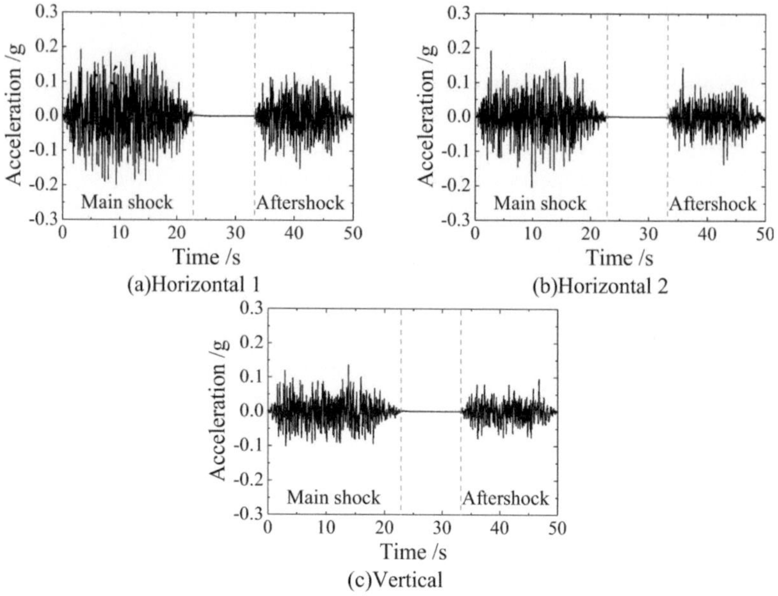

Fig. 3 Acceleration time history curves for the main-aftershock sequence

5 Nonlinear Dynamic Response Analysis of Intake Tower Under Main-Aftershock Sequence

In finite element analysis, the load calculation considers both static and dynamic loads. The dynamic interaction between reservoir water and the tower body is simulated using acoustic elements, while a viscoelastic artificial boundary is employed to simulate the dissipation effect of seismic energy propagating to the far field (Zhai et al. 2022). The synthesized main and aftershock ground motions are used as seismic inputs. The nonlinear dynamic responses of the intake tower under the single main shock, aftershock, and combined main-aftershock actions are analyzed separately.

5.1 Structural Damage Distribution

Figure 4 illustrates the damage distribution of the intake tower body under various operating conditions. It can be observed that the damage distribution of the tower body remains consistent across different operating conditions, primarily concentrated at the connection between the lower part of the tower body and the tower shaft. Under the sole action of aftershocks, the damage area of the tower structure is limited. However, under the single action of the main shock, the damage area of the tower body significantly expands. Due to the unique characteristics of the intake tower structure, seismic actions tend to concentrate damage within the tower body. When subjected to aftershocks following the main shock, the increase in the damaged area is not significant. However, this does not imply that the influence of aftershocks can be ignored. Therefore, further analysis is necessary from the perspectives of structural energy dissipation characteristics and dynamic response.

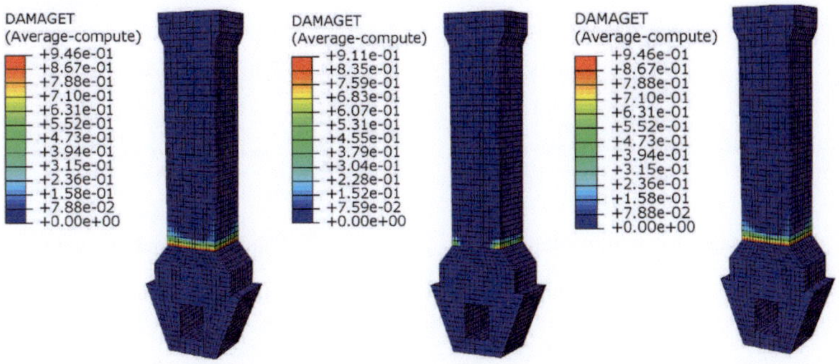

Fig. 4 Damage zone of tower under different seismic sequences

5.2 Analysis of Energy Dissipation Characteristics

Under seismic action, the damage and failure of concrete structures in the intake tower accumulate in the form of energy dissipation (Zhai et al. 2022). This study analyzes the damage energy dissipation and plastic energy dissipation as two indicators. The energy dissipation of the intake tower structural system under different conditions is depicted in Fig. 5. It can be observed that following the main shock, the curves of damage dissipation energy and plastic dissipation energy for the tower body's concrete structure exhibit a secondary growth process under the influence of aftershocks. After the single action of the main shock ends, the structural damage dissipation energy and plastic dissipation energy are 14.0 kN·m and 63.3 kN·m, respectively. Under the combined action of the main shock and aftershocks, the final structural damage dissipation energy and plastic dissipation energy are 17.0 kN·m and 75.5 kN·m, respectively. Compared to the sole action of the main shock, the structural damage dissipation energy and plastic dissipation energy of the system under the combined action of the main shock and aftershocks increase by 21.4% and 19.3%, respectively.

Under the single action of aftershocks, the structural damage and plastic dissipation energy of the tower body are 1.8 kN·m and 6.8 kN·m, respectively. However, under the combined action of the main shock and aftershocks, the structural damage and plastic dissipation energy induced by aftershocks are 3.0 kN·m and 12.2 kN·m, respectively. A comparison between the two reveals a significant increase in structural damage and plastic dissipation energy caused by aftershocks following the main shock, with increases of 67.7% and 79.4%, respectively. This indicates that aftershocks can cause greater structural damage and failure for the already damaged intake tower structure.

5.3 Structural Displacement Responses

Under the combined action of the main shock and aftershocks, the time history curves of relative horizontal displacement at the tower's top point are shown in Figs. 6 and 7.

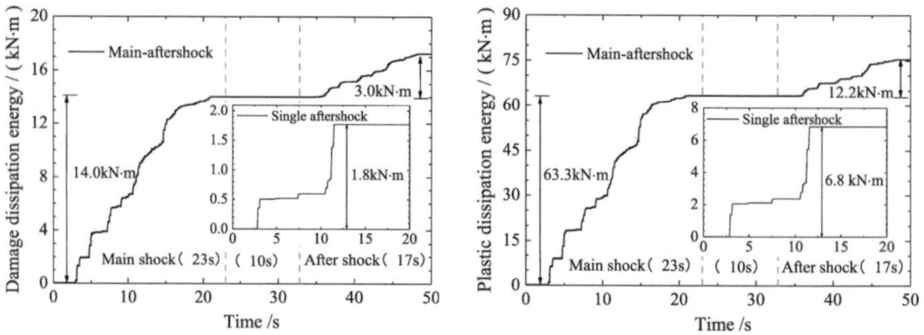

(a)Damage dissipation energy (b)Plastic dissipation energy

Fig. 5 Curve of structural energy dissipation process

From the tower's top displacement response caused by aftershocks, it can be observed that, compared to the sole action of aftershocks, the time history curves of relative displacement of the tower body induced by the combined action of the main shock and aftershocks exhibit similar patterns. Under the Single action of aftershocks, when the dam body damage is minimal, the relative displacement curve at the tower's top point exhibits repetitive motion, and no significant residual displacement occurs after the seismic action ends. However, under the combined action of the main shock and aftershocks, aftershocks continue to develop on the basis of residual deformation caused by the main shock, resulting in a noticeable increase in residual deformation compared to the single action of the main shock.

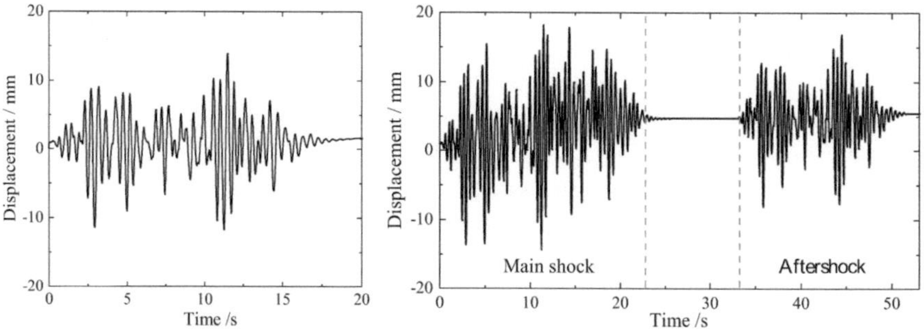

(a)Single aftershock (b)Main-aftershock

Fig. 6 The time history curves of relative horizontal X displacement

Table 1 presents the residual displacement values at the tower's top point under different conditions. It can be observed that, compared to the action of a single main shock, the horizontal residual displacement at the tower top increases by 15.9% and 24.6% under the combined action of the main shock and aftershocks. Under the sole action of aftershocks, the residual displacements at the tower top are 0.83 mm and 0.64 mm, respectively. However, the residual displacement increments at the tower top due to aftershocks following the main shock are 1.51 mm and 1.97 mm, respectively. This

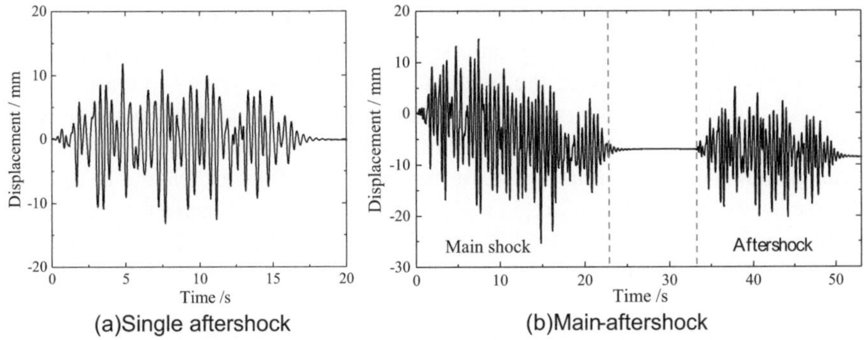

Fig. 7 The time history curves of relative horizontal Y displacement

indicates that although the residual displacement caused by aftershocks alone might be small, for a tower body already damaged by the main shock, aftershocks can still result in significant secondary residual displacements.

Table 1 the residual displacement at the tower's top point

Direction	Residual displacements /mm			Increase due to aftershocks /%
	Main shock	Aftershock	Main-aftershock	
Horizontal X	4.73	0.83	5.48	15.9
Horizontal Y	7.01	0.64	8.73	24.6

6 Conclusions

Combining the NGA seismic attenuation model with considerations of magnitude, main-aftershock parameter relationships, and site characteristics, this study proposes a method for constructing main-aftershock ground motions when seismic parameters are insufficient. Using an actual intake tower project in a seismically active western region as a case study, acoustic elements were used to simulate the dynamic interaction between reservoir water and the tower. A nonlinear model of the reservoir water-tower-foundation system was established to investigate the damage response of the intake tower under single main shocks, single aftershocks, and main-aftershock sequences. The study yielded the following insights:

After the main shock, the subsequent aftershocks have a significant cumulative effect on the damage and plastic energy dissipation of the tower structure. When aftershocks are considered, the damage to the intake tower intensifies, with energy dissipation and residual displacement increasing by approximately 20%~25%. Compared to the effects of aftershocks alone, the aftershocks following a main shock result in a 67.7% increase in structural damage and a 79.4% increase in plastic energy dissipation. This indicates that for an already damaged intake tower structure, aftershocks can cause significantly greater structural damage. Although aftershocks alone cause relatively minor residual displacements at key points of the intake tower, for intake tower structures already damaged by a main shock, aftershocks can induce significantly larger secondary residual displacements. The impact of aftershocks on intake tower structures post-main shock is crucial and cannot be overlooked, as it may be a critical factor leading to structural failure or even collapse. Therefore, in seismic safety analyses of intake towers in strong seismic zones, it is essential to consider the cumulative damage and deformation effects induced by aftershocks.

References

Bath, M.: Lateral inhomogeneities in the upper mantle. Tectonophysics **2**, 483–514 (1965)

Boore, D.M., Stewart, J.P., Seyhan, E., et al.: NGA-WEST2 equations for predicting response spectral accelerations for shallow crustal earthquake. California: Pacific Earthquake Engineering Research Center (2013)

Du, M., Zhang, S., Wang, C., et al.: Seismic fragility assessment of aqueduct bent structures subjected to main-aftershock sequences. Engineering Structures (Oct.1), 292 (2023)

Huo, J., Hu, Y., Feng, Q.: Research on time-history intensity envelope function of ground motion. Earthquake Eng. Eng. Vibration **1**, 1–12 (1991)

Liu, Y., Zhao, L., Qian, W.: Simulation analysis on seismic damage of intake tower of Shapai Hydropower Station. Water Resour. Hydropower Eng. **46**(01), 30–33 (2015)

Liu, Y., Zheng, X., Zhang, X.: Seismic dynamic response analysis of tall water intake tower structure considering viscoelastic artificial boundary. J. Xian Univ. Technol. **32**(02), 134–141 (2016)

Ministry of Housing and Urban-Rural Development of the People's Republic of China: Seismic design standard for hydraulic buildings: GB 51247–2018, pp. 55–65. China Planning Press, Beijing (2018)

Shi, Y., Rui, J., Huang, X., et al.: Study on PGA attenuation relationship of aftershock ground motion based on main aftershock theory. J. Qinghai Univ. **40**(02), 78–85 (2022)

Wang, S., Lu, Y., Shi, Y., et al.: Application of ground motion attenuation model in the rapid assessment of the affected field of Madom_W7.3 earthquake. World Earthquake Eng. **38**(03), 192–202 (2019)

Xu, Q., Zhang, T., Chen, J., et al.: A new endurance time analysis method for damage evaluation of high arch dams under the oblique incidence of mainshock-aftershock seismic sequences by wavelet decomposition. Soil Dyn. Earthquake Eng. (2022)

Yabe, S., Ide, S.: Why do aftershocks occur within the rupture area of a large earthquake? Geophys. Res. Lett. (2018)

Yang, G., Li, S., Li, L., et al.: Dynamic characteristics of tall water intake tower structures under different concrete casting and forming methods. J. Hydropower Energy **38**(06), 92–95 (2019)

Yu, X., Zhou, Z., Lu, D.: Damage-state-dependent aftershock fragility analysis for reinforced concrete frame structures. J. Building Struct. **43**(04), 8–16 (2022)

Zhai, Y., Bi, Z., Tang, Y., et al.: Study on damage and failure of gravity dam under main aftershock sequence based on NGA Model. Shui Li Xue Bao **51**(02), 152–157+168 (2020)

Zhai, Y., Zhang, L., Bi, Z., et al.: Seismic performance evaluation of AAR-affected concrete gravity dams under main aftershock sequence. Soil Dyn. Earthquake Eng. **157**, 107258 (2022)

Zhang, Y., Li, S., Xia, K., et al.: Research on seismic resistance of tower backfill height in tall water intake tower. Water Resour. Hydropower Technol. **49**(11), 62–67 (2018)

Zhao, J., Yang, J., Li, X.: Seismic performance analysis of water intake tower structure based on time-history method. J. Earthquake Eng. **43**(05), 1244–1250 (2019a)

Zhao, X., Wen, Z.P., Xie, J., et al.: Applicability of NGA-West2 ground motion prediction model to various components of velocity pulse ground motion. Acta Seismologica Sinica **45**(02), 356–372 (2019b)

Research on Deformation Monitoring and Early Warning and Safety Control of Hydraulic Tunnel in Extremely Fractured Rock Mass

Bin Duan[1](✉), Haisheng Wang[1], Deqiang Feng[1], Shihe Qin[1], Zhen Li[2], and Haoyu Mao[3]

[1] CHN Energy Dadu River Jinchuan Hydropower Project Construction Co., Ltd, Aba Sichuan 624100, China
iamduanbin@163.com
[2] China Anneng Group First Engineering Bureau Co., Ltd., Nanning 530028, Guangxi, China
[3] National Key Laboratory of Mountain River Protection and Management, Sichuan University, Chengdu 610065, Sichuan, China

Abstract. Taking the underground cavern group of Dadu River Jinchuan Hydropower Station as the research object, the key technology of deformation monitoring, early warning and safety control of hydraulic tunnels in extremely fractured rock mass is systematically studied by theoretical research, on-site monitoring and numerical computation, the characteristics of the surrounding rock of hydraulic tunnels and stability of the import and export slopes are investigated, and the safety coefficient of the side slopes and danger of slope rockfall are evaluated under multiple conditions, and the stress-strain characteristics of the surrounding rock during excavation of the tunnels and the sides of the side slopes under the condition of extremely fractured rock mass are simulated; On this basis, a monitoring system for hydraulic tunnels and slopes in extremely fractured rock mass is constructed, and an index system for judging the risk of surrounding rock stability is established; finally, an intelligent control system for underground cavern group construction based on BIM is constructed, and a platform for the safety and intelligent control of hydraulic tunnels is set up. The results show that the system improves the accuracy of construction risk identification of hydraulic tunnels in extremely fractured rock mass, realizes comprehensive and efficient real-time monitoring of project status, and provides a scientific basis for safety control of tunnels and slopes.

Keywords: extremely fractured rock mass · hydraulic tunnel · deformation monitoring · early warning · safety control

1 Introduction

As China's socio-economic development continues to drive the escalating demand for clean energy, a vast array of large-scale hydropower projects have been constructed, are under construction, or are in planning stages in the southwestern region 12 (Chen 2024;

S. Zheng et al. (Eds.): IHDC 2024, LNCE 487, pp. 258–271, 2025.
https://doi.org/10.1007/978-981-97-9184-2_23

Xu 2017). Situated amidst the mountainous canyons on the eastern edge of the Qinghai-Tibet Plateau, these projects often adopt subterranean powerhouse structures to meet the exigencies of hub layout and construction in deep valley conditions, thereby giving rise to numerous subterranean hydraulic tunnel structures (Mao 2020). Hydraulic tunnels play a pivotal role in hydropower engineering, undertaking crucial functions such as regulating water flow and facilitating hydroelectric power generation. Nevertheless, owing to the complex and mutable geological conditions, a plethora of challenges typically confront these hydraulic tunnels during their construction phase, including deformation of surrounding rock (Hou 2023; Liu 2016) and instability of inlet/outlet slopes (Zhang 1983; Zhu 2006). Notably, at the Jinchuan Hydropower Station, the inlet/outlet slopes of hydraulic tunnels are characterized by steep gradients and intense unloading, with the surrounding rock comprising metamorphic fine sandstone interlaced with carbonaceous shale, thus exhibiting poor stability. These challenges are particularly pronounced. Consequently, elucidating the deformation mechanisms of surrounding rock and inlet/outlet slopes in hydraulic tunnels within extremely fractured rock masses, and establishing a comprehensive suite of technologies for deformation control from preliminary design to post-construction, represents a pressing and substantial engineering conundrum and technical challenge.

In recent years, leveraging the cavernous structures of spillway discharge tunnels and diversion tunnels at the Dadu River Jinchuan Hydropower Station, a comprehensive approach integrating theoretical research, on-site monitoring, and numerical computation has been adopted (Duan 2022a). Through continuous innovation in research and engineering applications, a series of original achievements have been made in the monitoring, early warning, and safety control of deformation and failure in hydraulic tunnels under extremely fractured rock mass conditions. These accomplishments play a crucial role in vigorously advancing research related to the monitoring, early warning, and safety control of deformation in hydraulic tunnels within extremely fractured rock masses.

2 Project Overview

The Jinchuan Hydropower Station, the sixth cascade station in the mainstream regulation plan of the Dadu River, is located within the administrative boundaries of Jinchuan County and Barkam City in Aba Prefecture, Sichuan Province. Classified as a Grade II large (2) project, its primary function is power generation, with a designed installed capacity of 860 MW and an annual average power generation of 34.857 billion kWh.

The key components of the Jinchuan Hydropower Station's hub project include a concrete-faced rockfill dam, a left bank water diversion power generation system, a right bank spillway, ecological spillway, and a discharge tunnel. The dam, serving as the main water-retaining structure, stands at an elevation of 2258 m with a width of 10 m and a maximum height of 112 m. Its toe is placed on a covering layer approximately 65 m thick. The spillway structures consist of a two-bay open-sided overflow channel on the right bank and a pressured discharge tunnel. The overflow channel, open on the bank side and adjoining the concrete-faced rockfill dam, comprises intake channel section, weir gate section, spillway section, and jet flow nose section, with a total length of about 425 m and a maximum discharge capacity of 6589 m^3/s. The discharge tunnel is located on the right side of the overflow channel, primarily used for flood discharge during permanent operation, with a total length of 938.5 m and a maximum discharge capacity of 1612 m^3/s. The pressured section of the discharge tunnel consists of inlet transition section, circular pressured section, and outlet transition section, transitioning from a rectangular section of 7m × 12m to a circular hole with a diameter of 10 m at the inlet, and from a circular section of 10 m to a rectangular section of 8m × 7m at the outlet. The unpressurized tunnel section has a total length of 456.0 m, with a cross-section of 8m × 14m in the shape of a circular arch straight wall. The water diversion power generation system is situated within the left bank mountain, employing shore-tower intake, single-machine single-tunnel water diversion, and underground powerhouse housing four units, with a tailrace tunnel arranged in a "two units one tunnel" configuration. The diversion method involves the use of a cofferdam for a one-time shutdown of the entire tunnel for year-round diversion. The diversion tunnel is located on the right side of the discharge tunnel, with a diameter of 12.5m × 14.5m. The overall layout of the underground cavern complex of the Jinchuan Hydropower Station is illustrated in Fig. 1.

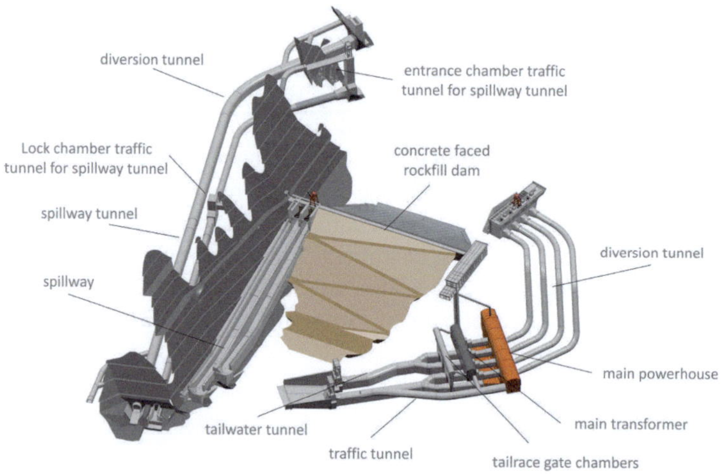

Fig. 1. The layout of the underground cavern complex at Jinchuan Hydropower Station

3 Topographical and Geological Conditions

The dam site area is situated in a "V"-shaped valley, presenting an inverse "S" shape in plan view, with predominantly exposed bedrock on both banks forming lateral valleys. The valley slopes are steep, with a substantial mountain mass on the left bank. Downstream on the right bank, a major tributary, Xinza Gully, has been incised by the Dadu River, resulting in the formation of a relatively thin ridge. Numerous small gullies are distributed on both sides of the dam site area, with Xinza Gully forming a fan-shaped distribution of first-order alluvial terraces at its mouth. The exposed geological formations in the dam site area consist of the Upper Triassic Zhaga Nao Group (T3z2), characterized by thin to thick layers of metamorphic fine sandstone interbedded with carbonaceous shale. Based on lithology, bed thickness, and composition characteristics, a total of eight engineering geological lithological units have been delineated, arranged from old to young. The structural framework of the dam site area and its periphery is relatively simple, primarily comprising a series of roughly parallel NW-trending linear tight anticlines of varying scales. In summary, the geological conditions in the hub project area of the Jinchuan Hydropower Station exhibit the following characteristics:

1. Uncommonly complex geological conditions at the dam site area: The steep and heavily loaded rock masses on both banks are characterized by weathering and fragmentation, predominantly consisting of metamorphic fine sandstone interbedded with carbonaceous shale, containing three sets of tilted deformations and multiple sets of faults and fractures.
2. Complex topography intertwined with geological hazards: The reservoir area features high mountain canyon landforms, with steep mountain slopes conducive to the development of slope-related geological hazards. The narrow and deeply incised valleys with steep slopes contribute to the dense development of geological hazards (Qin 2023).

4 Research Work and Technological Innovation

4.1 Characteristics of Surrounding Rock and Stability of Inlet/Outlet Slopes of Hydraulic Tunnels

1. The inlet and outlet slopes of the Jinchuan Hydropower Station's hydraulic tunnels are steep and heavily loaded, with the surrounding rock consisting of metamorphic fine sandstone interbedded with carbonaceous shale, exhibiting poor stability. In response to these extremely fractured rock conditions, a geological exploration borehole positioning device has been proposed to enhance the accuracy of geological exploration in fractured rock masses. By precisely analyzing geological and geophysical data, combined with on-site reconnaissance, an in-depth investigation into the characteristics of surrounding rock and the stability of inlet/outlet slopes of hydraulic tunnels has been conducted. A database of tunnel and slope fractures has been established, and a probability analysis method has been employed to calculate the stability of structural plane combinations under different conditions (natural conditions, heavy rainfall, seismic loads), resulting in the determination of stability coefficients for favorable orientations of rigid structural plane combinations. Geological and geophysical data analysis for diversion tunnels and spillway discharge tunnels is illustrated in Fig. 2.

2. The stability and deformation control of steep slopes during excavation unloading and throughout the operation of the hydropower station are particularly critical. Addressing the full-cycle stability issues of steep slopes in hydraulic tunnel construction and operation, based on limit equilibrium theory and finite element method, a comprehensive assessment has been conducted on the safety factors and deformation stress characteristics of slopes under more than ten different conditions, including natural conditions, excavation unloading, reinforcement, heavy rainfall, reservoir filling,

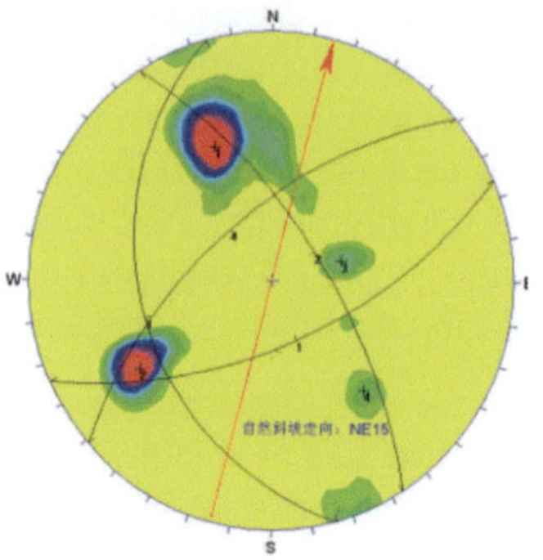

inlet slope fissure isometric map of tunnel spillway

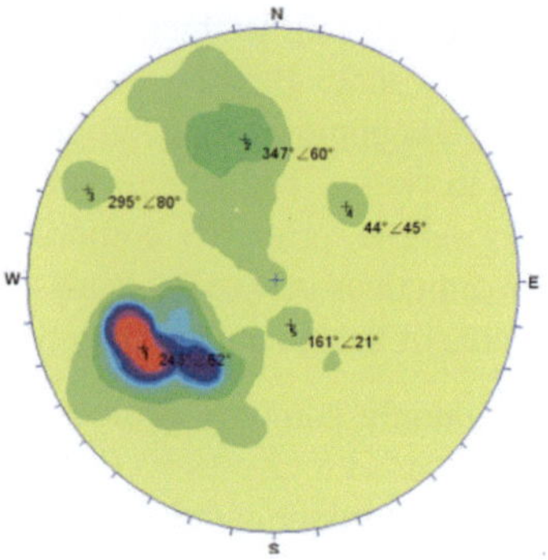

inlet slope fissure isometric map of diversion tunnel

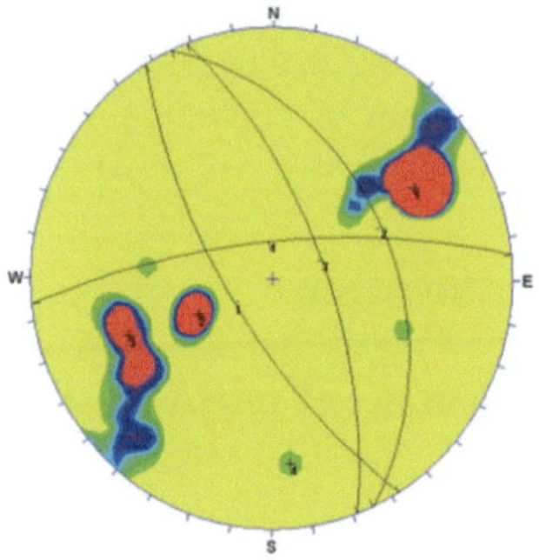

major structural plane isometric map of tunnel spillway and diversion tunnel outlet slopes

block combination mode

Fig. 2. Geological data and geophysical data analysis

reservoir filling during heavy rainfall, design earthquake during reservoir filling, seismic check during reservoir filling during heavy rainfall, and sudden water level drop, ensuring slope stability. The three-dimensional finite element slope calculation results for slope excavation are shown in Fig. 3. A shear strength parameter inversion analysis method based on equivalent soil pressure has been proposed to eliminate the influence of internal sliding failure of slopes in the covering layer, making the inverted parameters of slope shear strength more accurate and reasonable. A hydraulic landslide model experimental device has been developed to provide guidance for risk prevention and emergency response measures. During slope construction, rockfall is also a major challenge affecting construction safety and support design. Through dynamic and statistical simulations of slope rockfall, the trajectories and motion characteristic parameters of rockfall on inlet and outlet slopes have been analyzed, providing efficient and accurate support for engineering design and construction optimization.

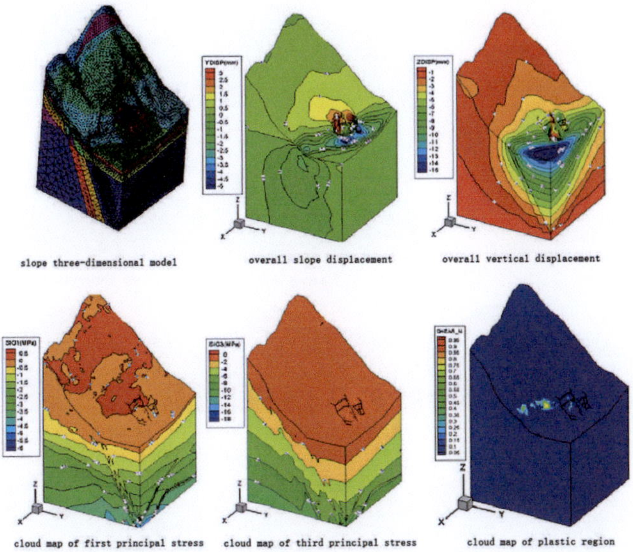

Fig. 3. 3D finite element calculation results of slope excavation

3. At the Jinchuan Hydropower Station, the surrounding rock is characterized by simultaneous fragmentation and the development of weak structural planes. Blasting excavation unloading increases the disturbance to the surrounding rock of hydraulic tunnels, which is detrimental to the stability of the surrounding rock. The unclear deformation characteristics and mechanisms of surrounding rock during excavation pose challenges. Utilizing finite difference software, high-precision numerical models of the spillway discharge tunnel and diversion tunnel have been established to simulate and reproduce the stress-strain characteristics of surrounding rock during excavation in extremely fractured rock mass conditions. This approach delves into the deformation mechanisms of surrounding rock and proposes a two-dimensional numerical calculation method suitable for underground cavern excavation, providing a scientific and systematic theoretical basis for the safe construction of tunnel projects. The evolution of surrounding rock deformation during excavation unloading is depicted in Fig. 4.

4.2 Safety Monitoring of Hydraulic Tunnels and Slopes in Extremely Fractured Rock Masses

1. Conventional safety monitoring of hydraulic tunnels primarily adopts a "poin-tline" monitoring approach, which cannot provide robust safety support for the entire tunnel section. Addressing the spatial limitations of conventional monitoring methods, a multidimensional monitoring system incorporating three-dimensional laser scanning and microseismic monitoring has been introduced. This system constructs a multidimensional monitoring framework encompassing "points, lines, surfaces, and volumes" for

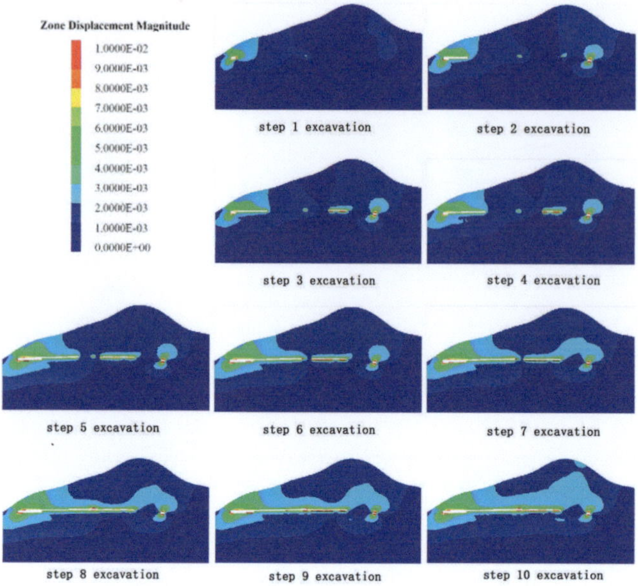

Fig. 4. Deformation evolution characteristics of surrounding rock during excavation unloading process

extremely fractured rock mass hydraulic tunnels. It monitors the stress, strain, internal defects, acoustic wave velocity changes, apparent deformation of surrounding rock, and microfracture fields of the surrounding rock, enabling comprehensive real-time monitoring of the engineering status and providing solid technical support for tunnel safety and stability. Additionally, to address the issue of background noise affecting the accuracy of microseismic monitoring localization, a denoising method based on CEEMD-CS-ST (Complementary Ensemble Empirical Mode Decomposition - Compressed Sensing - Soft Thresholding) is proposed. Compared with denoising methods such as empirical mode decomposition and wavelet transform, the CEEMD-CS-ST method exhibits a signal-to-noise ratio improvement of 28.121 dB, minimal signal standard deviation, and maximum correlation coefficient, fully preserving the transient non-stationary characteristics of microseismic signals. The CEEMD-CS-ST denoising method is depicted in Fig. 5.

2. To address the limitations of monitoring scope imposed by terrain, high equipment installation and maintenance costs, and the potential for overlooking monitoring points in the complex surface deformation monitoring of large-scale hydropower engineering slopes, a comprehensive slope monitoring system has been developed. This system integrates various monitoring methods such as Ground-Based Synthetic Aperture Radar (GB-InSAR) interferometric measurement technology and anchor stress gauges (Duan 2022b). It enables real-time and accurate assessment of slope stability, enhancing the effectiveness and specificity of engineering safety precautions. For areas with significant deformation, optimization of the layout of GB-InSAR

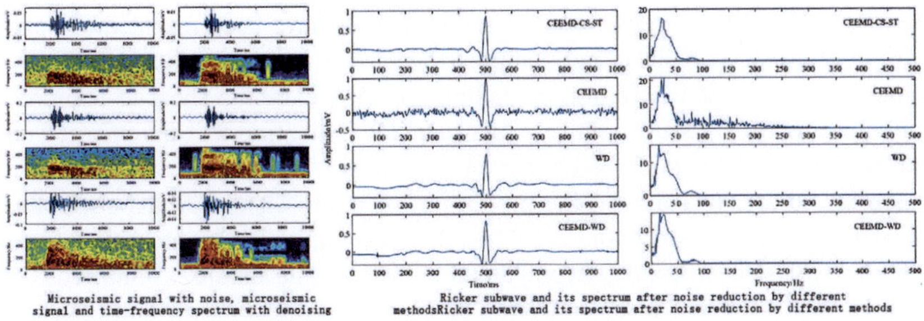

Fig. 5. CEEMD-CS-ST denoising method

and stress gauge measurement points has been conducted to improve the utilization value of data and provide precise guidance for on-site construction. An improved radar data denoising model based on three-dimensional laser scanning technology has been constructed, allowing for intelligent judgment and screening of abnormal on-site data, thereby enhancing the visibility, effectiveness, and reliability of monitoring data. Slope ground-based synthetic aperture radar monitoring is illustrated in Fig. 6.

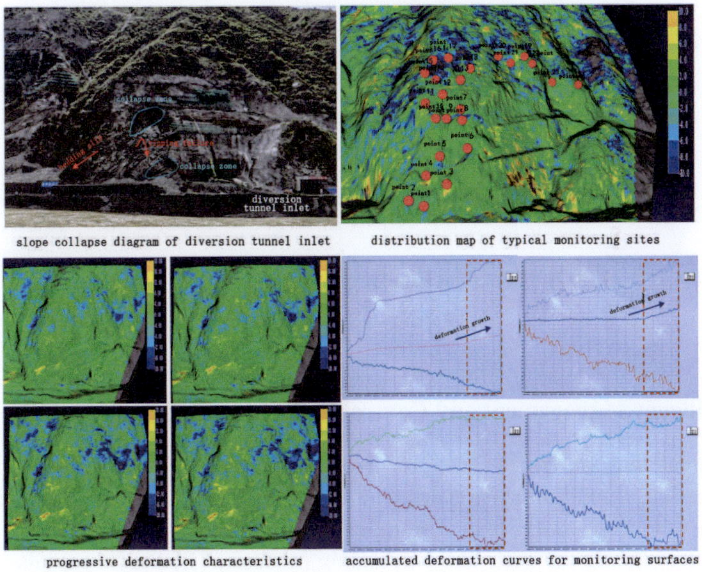

Fig. 6. GB-InSAR monitoring of slope foundation

3. The inherent relationship between surrounding rock deformation and failure, numerical simulation, and multidimensional monitoring information has been thoroughly analyzed. The mechanism and evolution characteristics of rock mass deformation

in extremely fractured rock mass hydraulic tunnels have been revealed through the ratio of microseismic shear and compressional wave energies. Microseismic events influenced by strong excavation unloading typically exhibit a shear-to-compression wave ratio (ES/EP) almost entirely less than 10, indicating that the primary mechanism of rupture for microseismic events induced by strong excavation unloading is tensile failure. In contrast, the proportion of microseismic events controlled by adverse structural planes with ES/EP less than 3 is minimal. Therefore, their primary rupture type is shear failure accompanied by less tensile failure, as numerous shear failures occur due to fractures developing along the direction of fault planes. This understanding of rupture mechanisms provides a theoretical basis for deformation early warning and control during tunnel construction. Numerical simulation reveals the on-site failure conditions as depicted in Fig. 7, while microseismic monitoring unveils the rock mass deformation mechanism as illustrated in Fig. 8.

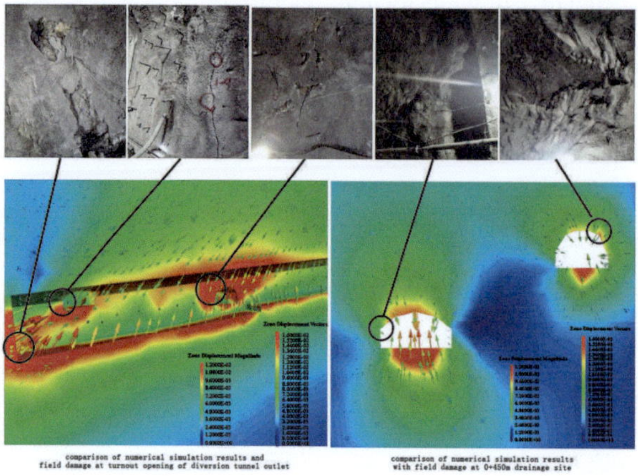

Fig. 7. Numerical simulation revealing on-site damage

4.3 Intelligent Management and Control of Surrounding Rock Stability Risks

1. To address the issue of unclear precursor information regarding surrounding rock deformation, a multidimensional monitoring information risk discrimination index system has been established. Through the organization and analysis of multiple parameters during the excavation unloading process of hydraulic tunnels, the variation patterns of monitoring data during tunnel construction have been revealed. This approach reduces the risk of surrounding rock deformation and failure in hydraulic tunnels caused by complex geological conditions, thereby enhancing the accuracy and practicality of engineering risk identification. The risk discrimination index based on multidimensional monitoring information is illustrated in Fig. 9.

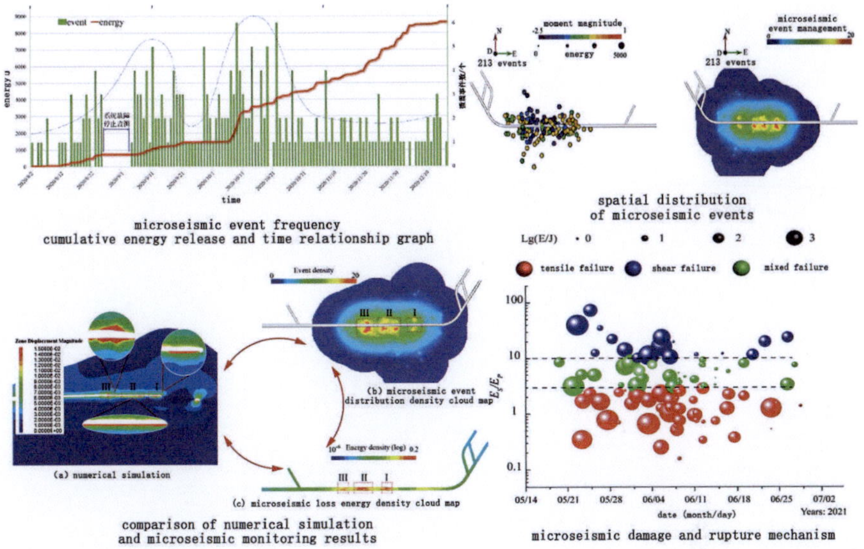

Fig. 8. Microseismic monitoring revealing rock mass deformation mechanism

2. To achieve integration, intelligence, visualization, and refinement of safety management business information, and to enhance the level of intelligent safety control, a smart management platform for construction safety of hydraulic tunnels in extremely fractured rock masses has been constructed. This platform is based on multidimensional monitoring information and Ultra-Wideband (UWB) positioning technology. It integrates multi-parameter information during the process of rock mass instability and failure, enabling data collection, risk assessment, intelligent warning, and corrective action throughout the construction process. This approach facilitates the scientific and standardized construction of hydraulic tunnels, improves the level of engineering safety management, and provides a new intelligent and smart mode for safety management in hydropower engineering construction (Wang 2021). The smart control platform is illustrated in Fig. 10.

3. Addressing the challenges of control arising from surrounding rock fragmentation in hydraulic tunnels, a construction safety management system based on Building Information Modeling (BIM) technology has been developed. This system introduces multi-source data management techniques and integrates numerical simulation with multidimensional safety monitoring methods. It achieves integration of heterogeneous data from multiple sources based on BIM technology and establishes a digital twin engineering system, ensuring the security, reliability, and efficiency of data throughout the entire process of engineering construction and management. Intelligent geological forecasting is illustrated in Fig. 11, while the integration of multi-source data is depicted in Fig. 12.

early-warning index warning level	red alert level 1	orange alert level 2	yellow alert level 3	normal	warning index classification
steady state	instability	substable	basically stable	stable	/
microseismic index W (microseismic monitoring)	at least three simultaneous exceptions	two exceptions	an exception	all normal	main control index 1
longitudinal wave velocity change rate η (%) (monitoring of rock loose circle)	>15	10~15	5~10	<5	main control index 2
deformation rate v(mm/d) (multi-point displacement meter)	>0.8	0.5~0.8	0.3~0.5	<0.3	secondary control index 1
particle vibration rate v(cm/s) (blasting vibration monitoring)	>15	10~15	5~10	<5	secondary control index 2
stress change rate of anchor (Mpa/d) (anchor stress meter)	>1	0.2~1	0.1~0.2	<0.1	validation index 1
three-dimensional laser scanning first judgment surrounding rock category	IV~V	III~IV	II~III	I~II	validation index 2
Microseismic index includes: microseismic event activity frequency, moment magnitude, microseismic b value, microseismic energy release, microseismic signal frequency.					

Fig. 9. Risk discrimination index based on multivariate monitoring information

4.4 Application of Novel Construction Methods

Combining on-site geological conditions with multidimensional monitoring information, novel construction methods for hydraulic tunnel sections with adverse geological conditions have been proposed. These include the use of nano-concrete for shotcrete anchorage support and the preservation of core soil-layered excavation in the upper

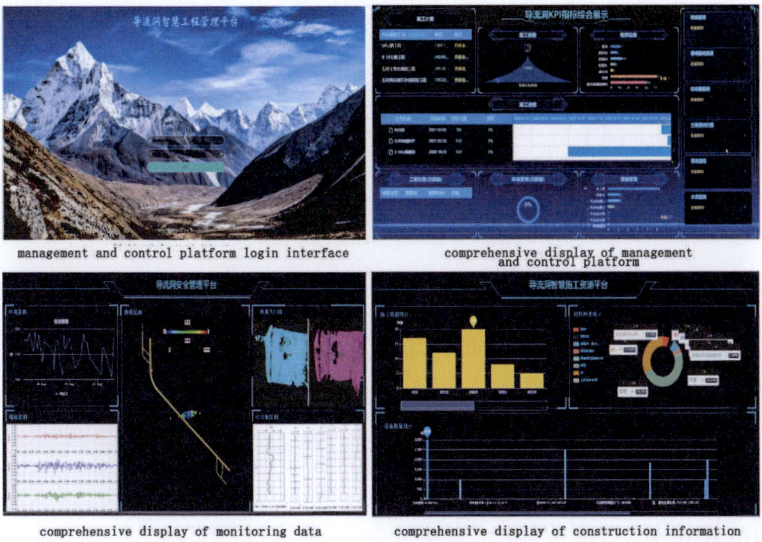

Fig. 10. Intelligent control platform

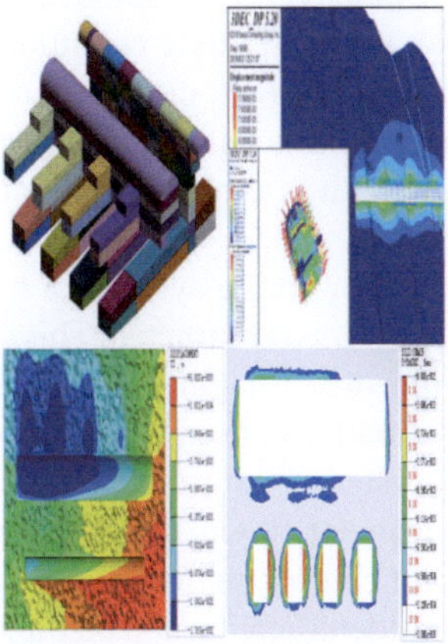

Fig. 11. Intelligent geological forecast

half-tunnel. Innovative support methods such as the combination grouting anchor head have been introduced, overcoming the limitations of traditional construction deformation control techniques under complex conditions. This breakthrough has bridged the gap between economical construction and safety coordination, enabling macroscopic control of multiple risk levels during construction. It provides significant insights for integrated disaster prevention and control in hydraulic tunnels under fractured rock mass conditions.

Fig. 12. Multi-Source data fusion

5 Conclusion

1. The characteristics of surrounding rock in hydraulic tunnels and the stability of inlet and outlet slopes were investigated. A comprehensive evaluation of slope safety coefficients and rockfall risk under multiple conditions was conducted. The stress-strain characteristics of surrounding rock during tunnel and slope excavation under extremely fractured rock mass conditions were accurately simulated. The mechanism of surrounding rock deformation was explored, providing a scientific basis for safety control during tunnel excavation.
2. A multi-level and multi-scale monitoring system for hydraulic tunnels and slopes in extremely fractured rock masses was constructed. This system enables comprehensive and efficient real-time monitoring of engineering status, providing strong technical support and effective prevention for the safety and stability of tunnels and slopes.
3. A risk discrimination index system for surrounding rock stability was established. A BIM-based intelligent management system for underground cavern construction was developed, along with a smart control platform for hydraulic tunnel safety. This has improved the accuracy of risk identification and facilitated innovation in macroscopic control and deformation control techniques.

References

1. Chen, G.F., Wang, X.H.: The global hydropower development trend and China's role in the context of carbon neutrality. J. Hydroelectr. Eng. **43**(4), 1–11 (2024)
2. Duan, B., Zhou, X.: Intelligent engineering planning and practice of large hydropower station. Hydropower and Pumped Storage **8**(03), 29–34 (2022)
3. Duan, B., He, J.P.: Surface deformation monitoring of high slope in hydropower project based on GB-InSAR technology. China Saf. Sci. J. **32**(S2), 64–69 (2022)
4. Hou, Q.L., Zhang, Y.T.: Mechanical parameter inversion and stability evaluation of weakly cemented soft rock in hydraulic tunnels. Yangtze River **54**(S2), 202–207 (2023)

5. Liu, T.X., Bai, X.J.: Study on prediction of surrounding rock deformation in soft rock tunnel TBM construction. Yangtze River **47**(03), 93–97 (2016)
6. Mao, H.Y., Xu, N.W.: Stability analysis of an underground powerhouse on the left bank of the Baihetan hydropower station based on discrete element simulation and microseismic monitoring. Rock and Soil Mechanics **41**(7), 2470–2484 (2020)
7. Qin, S.H., Duan, B.: Study on the characteristics and distribution of geological hazards in Jinchuan section of Dadu River. Journal of Yangtze River Scientific Research Institute 1–10 (2023)
8. Wang, J.S., Duan, B.: Study on construction scheme of intelligent safety management and control system for hydropower projects. China Saf. Sci. J. **31**(S1), 96–102 (2021)
9. Xu, N.W., Li, T.: Stability analysis on the left bank slope of Baihetan hydropower station based on discrete element simulation and microseismic monitoring. Rock and Soil Mechanics **38**(08), 2358–2367 (2017)
10. Zhu, Y., Tian, D.S.: Study on the treatment of the entrance and exit slopes of the right bank tunnel at Xiangjia-ba hydropower station. Hunan Water Resources and Hydropower **02**, 24–25 (2006)
11. Zhang, J.Y.: Slope instability and mitigation measures at Tarbela hydropower station. Hunan Water Resources and Hydropower **05**, 62–64 (1983)

Study on Adaptive Heads for Flip Bucket with Small Slope of Aeration Facilities in High-Flow and Slow-Bottom-Slope Flood Discharging Tunnel

Chuang Liu[1], Anzhe Cui[2], Ming Yin[3,4], Luchen Zhang[1(✉)], and Shaoze Luo[1]

[1] Nanjing Hydraulic Research Institute, Nanjing, China
110455898@qq.com
[2] China Gezhouba Group Co., Ltd, Wuhan, Hubei, China
[3] Design and Research Co, Changjiang Survey, PlanningWuhan, Hubei, China
[4] National Dam Safety Research Center, Nanjing, China

Abstract. The use of flip bucket with small slope in aeration facilities along high-velocity flood discharging tunnels can effectively improve the flow conditions within tunnels, but it is difficult to guarantee the aeration effect under large variations of head. Based on the Kashi Hydropower Station, this paper adopts a hydraulic model test to study the hydraulic characteristics of the aeration facilities in flood discharging tunnel under various heads, and analyzes the adaptability of the flip bucket with small slope configuration. The study shows that when the head is 70m or above, the cavity of the flip bucket with small slope without backwater, and has good aeration effect; when the water head is between 25m and 70m, there is varying degrees of backwater in the air cavity, and the aeration effect slightly decreases with the decrease of head. The average reduction of effective air cavity length, ventilation hole airflow, and aeration concentration are 20.4%, 13.4%, and 12.1% respectively, indicating that the aeration facilities with flip bucket with small slope have a wide range of water head adaptability; when the head is below 25m, the aeration effect is significantly reduced, and even the entire cavity is filled with backwater, without aeration mitigation cavitation effect.

Keywords: High-speed flow · Aeration mitigation cavitation · Flip bucket with small slope · Large variation of head · Hydraulic model test

1 Introduction

With the continuous development of water conservancy technology, dams under construction or planned are becoming taller, and their operating heads for flood discharging tunnels have generally exceeded the medium-to-high head of 80m. This change has led to increasingly prominent issues of cavitation damage under high flow velocities, posing significant threats to the safe operation of water release structures. Cavitation damage not only reduces water discharge efficiency but also harms the structural integrity of buildings and significantly increases maintenance costs.

S. Zheng et al. (Eds.): IHDC 2024, LNCE 487, pp. 272–281, 2025.
https://doi.org/10.1007/978-981-97-9184-2_24

To address this issue, aeration technology is widely used in engineering practice as an effective means to prevent cavitation damage on the flow wall. By properly arranging aeration facilities, air can be forcibly mixed into the water flow, thereby avoiding cavitation damage.

The main types of aeration facilities include spillway, drop, and aeration troughs, as well as combinations of these basic forms. Among them, the spillway is widely used due to its simple structure, ease of layout, and ability to easily form a stable cavity. However, due to differences in the adaptability of the spillway design to various projects, unfavorable flow phenomena such as unstable flow patterns, severe backwater in the cavity, and unsatisfactory aeration effects often occur in practical applications.

In response to these issues, many experts have adopted various design solutions in different projects to improve hydraulic conditions. Nonetheless, there are still certain limitations in the applicability of different shaped water release structures, especially regarding the adaptability of flip bucket with small slope aeration facilities under large water head variations. Therefore, it is necessary to further explore the aeration effect of flip bucket with small slopes under large water head variations to improve relevant engineering design theories and ensure the safe and stable operation of water conservancy projects.

2 Project Overview and Model Design

The Yulongkashi Hydro Project is a critical water control project in the mountainous section of the Yulongkashi River. The flood discharging tunnel, located on the right dam abutment, adopts a combined pressurized and non-pressurized discharge method. It mainly consists of an approach channel section, a pressurized tunnel section, a gate well section, a non-pressurized tunnel section, a chute section, a flip bucket section, and an apron section. The bottom slope of the tunnel section is $i = 0.08$, and the bottom slope of the chute section is $i = 0.4$ (Fig. 1).

The experiment adopts a normal hydraulic model designed according to the similarity criteria of gravity and water flow movement. The physical model's pressurized inlet section, gate chamber section, and non-pressurized flood discharging tunnel section are all scaled down, with a model scale of 1:50. This model experiment mainly observes the cavity development and aeration behind the aeration ridge. To facilitate observing the water flow pattern, all water release structures are made of transparent plexiglass. The shape of the aeration ridge is shown in Fig. 2. Five aeration ridges are arranged along the non-pressurized tunnel section, located at stake numbers 0 + 115.000m, 0 + 215.000m, 0 + 315.000m, 0 + 415.000m, and 0 + 505.000m. All adopt a small flip bucket type, with a length of 10m, a height of 0.5m, and a vent size of 1m × 1m. The range of test head variation is from 25m to 70m, and there are a total of five operating conditions classified according to the outlet head, with the gate fully open, as shown in Table 1.

In this experiment, 11 measurement points and sections are arranged along the flood discharging tunnel, as shown in Fig. 2. B2 ~ B12 are the measurement sections for aeration concentration. The aeration concentration is measured using a resistive aeration concentration meter. Each section has three measurement points located at the bottom plate, the middle, and the water surface, respectively.

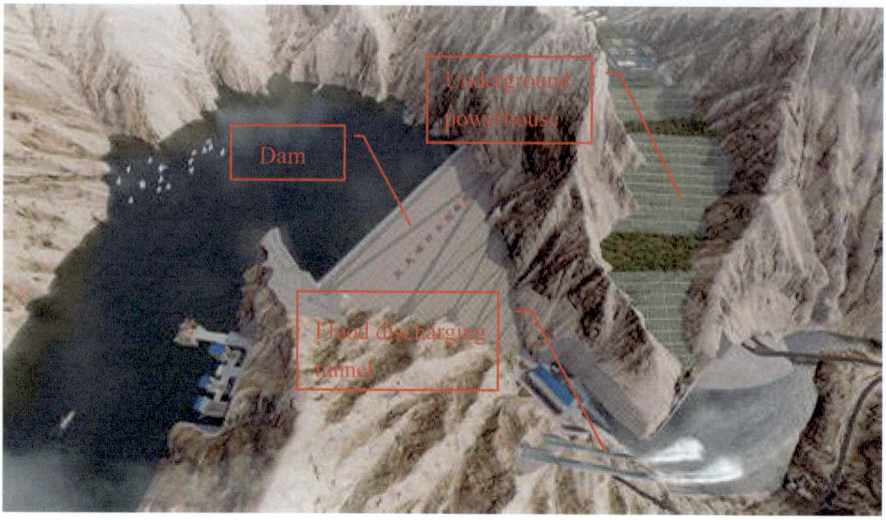

Fig. 1. The Yulongkashi Hydro Project

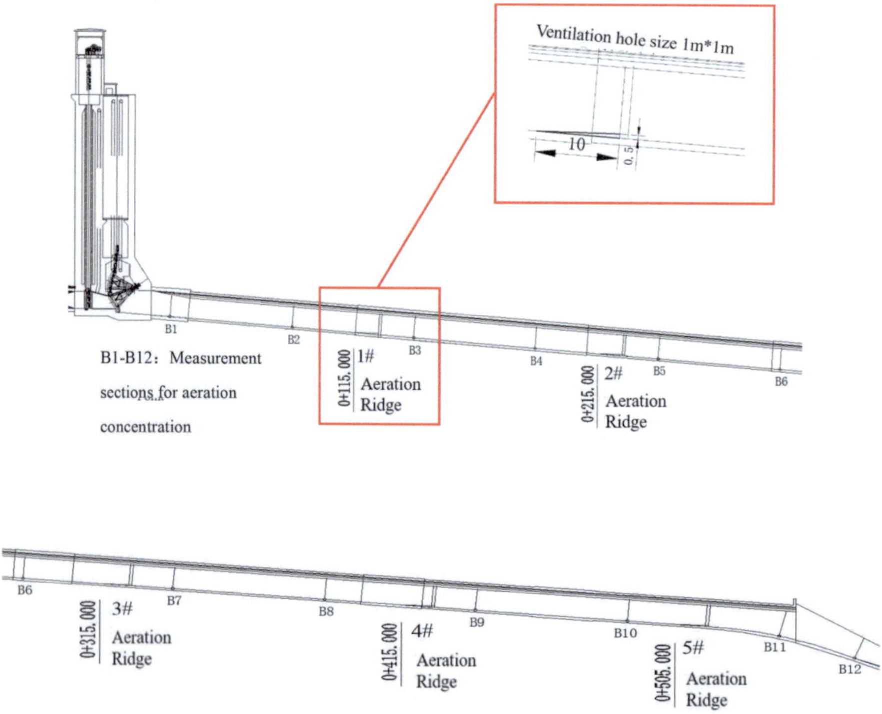

Fig. 2. Layout of flip bucket and measurement points along the route

Table 1. Working condition design

Working condition	Reservoir water level(m)	Outlet head(m)	Flow rate(m^3/s)	Fr
1	2170.00	75.00	1007.42	4.61
2	2160.00	65.00	933.23	4.29
3	2145.00	50.00	821.96	3.72
4	2130.00	35.00	710.68	3.24
5	2120.00	25.00	636.50	2.94

3 Analysis of Aeration Effects

3.1 Analysis of Water Flow Patterns

Figure 3 illustrates typical water flow patterns after the water passes through the aeration ridge under various working conditions. The red line in the figure represents the water surface line, and the area below the black line indicates the aerated cavity. When the discharge flows over the aeration ridge, the water tongue separates from the bottom of the discharge chute, forming an aeration cavity beneath it, which incorporates air through turbulent breaking. At a water head of 75m, there is a stable cavity without any backwater, indicating good aeration effects. At water heads ranging from 65m to 35m, backwater appears at the end of the cavity, and the lower the water head, the longer the range of the backwater. At water heads of 25m and below, the backwater has already reached the vent, causing intermittent or even complete submergence of the vent. The main reason for the backwater is that as the water head decreases and the jet velocity reduces, the jet impact angle exceeds the critical value for generating retrograde flow, becoming the primary source of water accumulation at the water tongue's landing point.

3.2 Aeration Cavity Length

The length of the aeration cavity is an important parameter to consider during the design of aeration facilities, as it affects both the ventilation volume and effective protection length of the aeration device. Generally speaking, the longer the cavity, the greater the ventilation volume, the more sufficient the water aeration, and the further the aeration protection distance [7]. Table 2 shows the length of the water tongue projection and the net cavity length under various working conditions. Analysis of cavity lengths under different conditions reveals that the higher the water head, the greater the outlet velocity, the further the landing point of the jet water tongue, and the greater the length of the aeration cavity formed. Additionally, the degree of water aeration increases. The length of the water tongue projection varies between 3.75m and 12.5m under different working conditions. When the water head is in the range of 25m to 70m, for every 10m decrease in water head, the net cavity length decreases by approximately 20.4% on average. When the water head is below 25m, a 10m decrease in water head results in an approximate 84.3% decrease in net cavity length, which is a significant reduction.

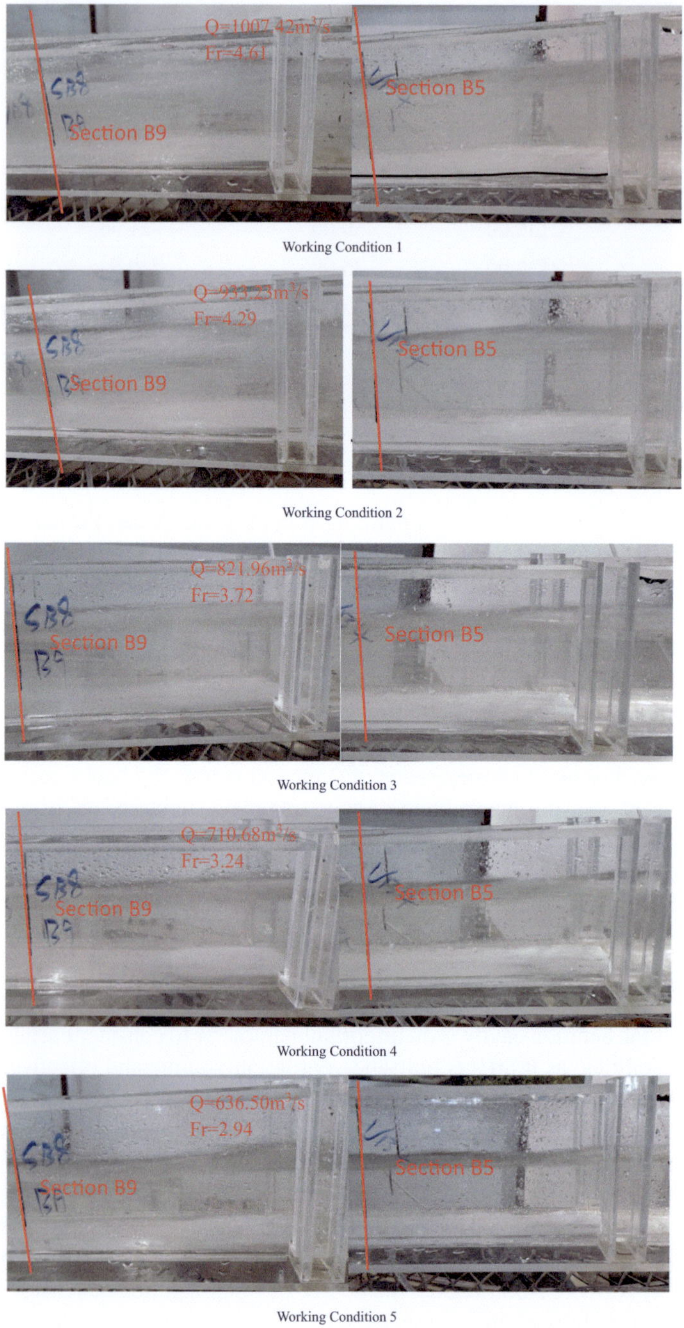

Fig. 3. Typical flow patterns after water passes through the aeration ridge under various working conditions

Table 2. Aeration cavity lengths

Aeration ridge number	Water tongue projection distance					Net cavity length				
	Condition 1	Condition 2	Condition 3	Condition 4	Condition 5	Condition 1	Condition 2	Condition 3	Condition 4	Condition 5
1#	9.00	6.75	5.50	4.50	4.00	9.50	6.50	4.00	3.25	—
2#	9.15	7.00	5.75	4.75	4.25	9.15	6.00	4.25	3.50	—
3#	9.35	7.50	6.25	5.00	4.50	9.35	6.50	4.50	3.75	0.25
4#	9.40	8.25	7.00	6.25	5.00	9.40	7.00	5.50	4.00	0.50
5#	9.50	8.50	7.50	6.50	5.50	9.50	7.50	5.75	4.50	1.25

3.3 Ventilation Hole Wind Speed and Aeration Concentration

The ventilation volume of the aeration facility and the aeration concentration in the downstream water body are important criteria for judging the effectiveness of aeration for erosion reduction. The ventilation volume of the aeration facility is the sum of the air supply from the two ventilation holes. The measured wind speeds at the ventilation holes for various working conditions are shown in Table 3. The table data indicates that higher water heads generally result in higher wind speeds at the ventilation holes, leading to better aeration effects. When the water head is low, the ventilation hole wind speed decreases due to the blockage of air intake by backwater, resulting in a weakened aeration effect. The ventilation volume decreases as the water head decreases. When the water head is in the range of 25m to 70m, the ventilation volume decreases by approximately 13.4% on average for every 10m decrease in water head. When the water head is below 25m, a 10m decrease in water head results in an approximate 46.1% decrease in ventilation volume, which is a significant reduction.

Table 3. Ventilation Hole Wind Speeds

Aeration ridge number	Condition 1	Condition 2	Condition 3	Condition 4	Condition 5
1#	4.03	3.59	2.88	2.03	—
2#	4.18	3.75	3.06	2.37	—
3#	4.38	3.85	3.17	2.50	1.36
4#	4.79	4.00	3.32	2.62	1.41
5#	5.10	4.38	3.75	2.86	1.53

The aeration concentration is defined as the ratio of gas volume to the total volume of the water-gas mixture in aerated water flow, and it can be used to measure the degree of aeration in the water flow. It is generally believed that when the aeration concentration of the water body is between 1.5% and 2.5%, the cavitation erosion damage to solid walls is greatly reduced. When the concentration reaches 3% to 5%, cavitation erosion damage can be avoided. In engineering design, it is considered that an aeration concentration of 2% to 3% can achieve the protective effect of aeration to reduce erosion [8]. The near-wall aeration concentration under various working conditions is shown in Fig. 4.

The distribution of aeration concentration along the downstream water body of each aeration ridge follows certain patterns: the highest aeration concentration is observed near the impact zone of the water tongue, where the water tongue collides with the bottom plate, causing intense water turbulence. The rolling water body sucks in air, which can be regarded as the aeration zone. The further away from the aeration ridge, the lower the aeration concentration becomes, as air continuously escapes from the water surface during the downstream movement with the water flow.

It can be seen that under a water head of 70 m or more, there is no backwater phenomenon in the cavity. The aeration concentration on the bottom plate behind the 1# aeration ridge exceeds 10%, then gradually decreases along the way, dropping to

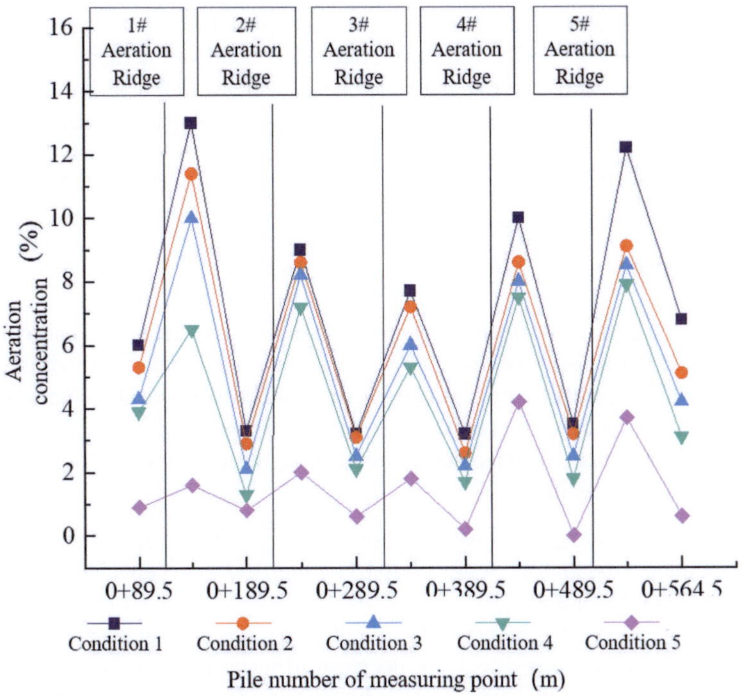

Fig. 4. Near-wall aeration concentration under various working conditions

about 3.3% before the 2# aeration ridge. After passing through the 2# aeration ridge, the aeration concentration rises to 9% and then gradually decreases again, remaining at 3.2% before the 3# aeration ridge. Under this condition, the aeration concentration before the outlet ridge is greater than 3%, and the lowest near-wall aeration concentration at each measurement point is 3.2%, showing a good aeration effect. The aeration effect is significant under high water head conditions because the high water head can provide sufficient motive force, and the water particles have enough turbulent energy to overcome the surface tension, allowing the water flow to carry a large amount of air and achieve the desired aeration effect.

When the water head is between 25 m and 70 m, there will be varying degrees of backwater phenomenon in the cavity. Despite this, the effect of aeration to reduce erosion still exists. Under such water head conditions, backwater hinders air from being mixed into the water. For example, the near-wall aeration concentrations downstream of the cavity behind the 1# and 2# aeration ridges in Working Condition 2 are 2.9% and 3.1%, respectively, but the erosion-reducing effect of aeration is not completely eliminated. The near-wall aeration concentration at each measurement point is above 2.6%. When the water head is less than 25 m, the flow velocity decreases significantly, and the backwater phenomenon in the cavity becomes severe, causing the entire cavity to be filled with backwater. The ventilation slot is blocked, making it difficult to mix air. The near-wall aeration concentration at each measurement point is low, and the near-wall aeration concentration downstream of the cavity of the 1# to 5# aeration ridges is less than

1%. Some measurement points even have undetectable aeration concentrations, failing to achieve the expected erosion-reducing effect. The near-wall aeration concentration shows a clear downward trend as the water head decreases. When the water head is between 25 m and 70 m, the near-wall aeration concentration decreases by about 12.1% for every 10-m drop in water head. When the water head is below 25 m, the near-wall aeration concentration decreases by about 69.0% for every 10-m drop in water head.

4 Conclusion

Small aeration ridges can effectively improve the flow pattern of discharge, but it is difficult to ensure the aeration effect when the water level varies greatly. Based on the Yulongkashi Hydropower Station, this paper studies the hydraulic characteristics of the flood discharging tunnel aeration facilities under various water heads through a 1:50 model test, and analyzes the adaptability of small aeration ridges under different water heads. The results show that:

1) The optimal water head for small aeration ridges is 70 m or more. There is no backwater in the cavity, and the water body has a high aeration concentration, resulting in a good aeration effect.
2) When the water head is between 25 m and 70 m, there is backwater in the cavity to varying degrees. For every 10-m decrease in water head, the average reduction in effective cavity length, ventilation volume, and aeration concentration is 20.4%, 13.4%, and 12.1%, respectively. The aeration effect generally decreases slightly with decreasing water head, indicating that small aeration ridges have a wide range of water head applicability.
3) When the water head is less than 25 m, due to the low flow velocity, the backwater intermittently floods the vents or even completely blocks them, significantly reducing the aeration effect and failing to achieve the expected erosion-reducing effect.

References

1. Liuyan, L., Liping, W.: Statistical analysis of large dams and reservoirs in China. Water Conserv. Constr. Manag. **36**(9), 12–16+32 (2016)
2. Weiwei, W., Jianhua, W., Shiping, R.: Research on the shape of aeration facilities in flat-bottomed spillways. J. Hydrodyn., Ser. A, (4), 397–402 (2007)
3. Shaobin, L., Cuiling, H., Juan, J.: Research on aeration and erosion reduction facilities on relatively gentle slopes. Water Conserv. Sci. Technol. Econ. **18**(10), 26–28 (2012)
4. Yongqin, L., Dongsheng, C., Bingxing, Z., et al.: Shape optimization and operational practice of aeration weir in high head and small slope spillway. Hydropower Gener. **48**(12), 98–102 (2022)
5. Shuangke, S., Jiawei, Y., Haitao, L.: Research on the shape of aeration and erosion reduction facilities under gentle slope conditions. Water Resour. Hydropower Eng. **11**, 26–29 (2004)
6. Chuan, W., Xudong, M., Lu, P.: Experimental study on the layout of aeration facilities in the spillway and emptying tunnel of Shuangjiangkou through hydraulic model tests. Water Resour. Power **40**(8), 136–138+207 (2022)

7. Chunfeng, Q.: Deep research on the aeration and ventilation characteristics of aeration facilities and supply systems in water release structures. Tianjin University (2019)
8. Nan Hailong, X., Yimin, Y.M., et al.: Experimental study on the distribution of water aeration concentration and aeration protection length behind the aeration weir. Water Resources and Power **33**(1), 173–176 (2015)

Insights into Renewable Energy Breakthroughs and Their Practical Applications

Main Circuit Parameter Design Research of Offshore Wind Farm DC Transmission Based on Grid-Forming Wind Turbines and Diode Rectifier Unit

Yingrui Liu[1], Jian Ning[2](✉), Taotao Qu[2], Xiaodong Qiu[2], and Kexin Wang[3]

[1] School of Automation, Central South University, Changsha 410083, China
Liu_yingrui@csu.edu.cn

[2] Zhongnan Engineering Corporation Limited, Changsha 410014, China
19397998118@163.com

[3] School of Energy, Power and Mechanical Engineering, North China Electric Power University, Beijing 102206, China

Abstract. The offshore wind power gradually develops towards the direction of far-reaching sea and large capacity, and the offshore wind power DC transmission system based on diode rectifier unit has a great development prospect, and its main circuit parameter design is an important part of the HVDC engineering design. Combined with the basic principle of wind power system and converter station, the working characteristics of offshore wind power DC transmission system based on diode rectifier unit are analyzed. For the important equipment, such as converter transformer, smoothing reactor and AC filter, the calculation formulas of main circuit parameters are derived, and the complete design idea and calculation process of main circuit parameters are given. Lastly, the parameters designed are verified by PSCAD/EMTDC simulation which can provide reference for the future design of offshore wind power DC transmission system based on diode rectifier unit.

Keywords: Offshore wind power · Diode rectifier unit · Direct current power transmission · Main circuit parameter design

1 Introduction

As clean energy improves by leaps and bounds, China has put forward a series of strategies and policies to promote the development of offshore wind power industry.

In order to solve the problem of poor economy of flexible DC transmission, offshore wind power transmission scheme based on diode rectifier unit (DRU) can be used, that is, diode rectifier replaces modular multilevel converter (MMC) in offshore converter station. Compared with MMC converters, diode rectifiers show great advantages in device, system design, transportation, installation and commissioning, operation and maintenance.But at present, the design and research of the main circuit parameters of the offshore wind power delivery system based on the grid-forming wind turbines and the diode rectifier unit is almost blank.

S. Zheng et al. (Eds.): IHDC 2024, LNCE 487, pp. 285–293, 2025.
https://doi.org/10.1007/978-981-97-9184-2_25

In view of the above problems, this paper first analyzes the topology structure of the offshore wind power DC transmission system based on diode rectifier unit, selects the appropriate topology structure for research, and then analyzes the basic principle and operation characteristics of the system, and designs the parameters of each electrical equipment in the main circuit on the basis of the analysis. Including converter transformer parameters, smoothing reactor parameters and AC filter parameters. Finally, the calculation principle of main circuit parameters is obtained through theoretical derivation and verified by PSCAD/EMTDC simulation which provides a design idea for the research of main circuit parameters of offshore wind power DC transmission system based on diode rectifier unit.

2 Topology of Offshore Wind Power Delivery System

The offshore wind power transmission system mainly includes four parts: offshore wind farm, offshore diode rectifier station, DC transmission line and onshore MMC inverter station. After the boost, the offshore wind turbine is connected to the collector line in groups, the main transformer is boosted again to the offshore diode rectifier station, and the power is transmitted to the MMC inverter station on land through the DC transmission line, and the MMC inverter station will transmit the power to the grid. The topology of the system is shown in Fig. 1.

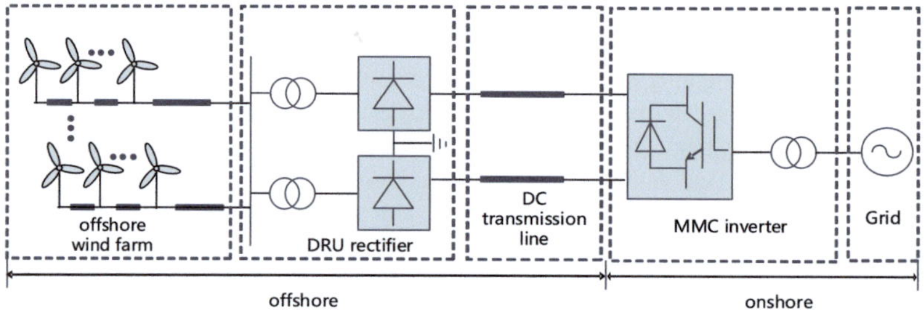

Fig. 1. Topology of the system

3 Design of Main Circuit Parameters of the Offshore Wind Power Delivery System

The important equipment in the main circuit of the system is DRU. Firstly, the operating characteristics of DRU are analyzed.

Equivalent circuit diagram of diode rectifier can be obtained by equivalent wind power system as AC power supply, and 6-pulse rectifier is taken as the analysis object, as shown in Fig. 2. e_a、e_b、e_c are the equivalent three-phase potential of wind power generation system respectively. L_γ is the equivalent inductance of each phase of the wind power system (from the power supply calculation to the AC end of the bridge); L_d is a DC smoothing reactor.

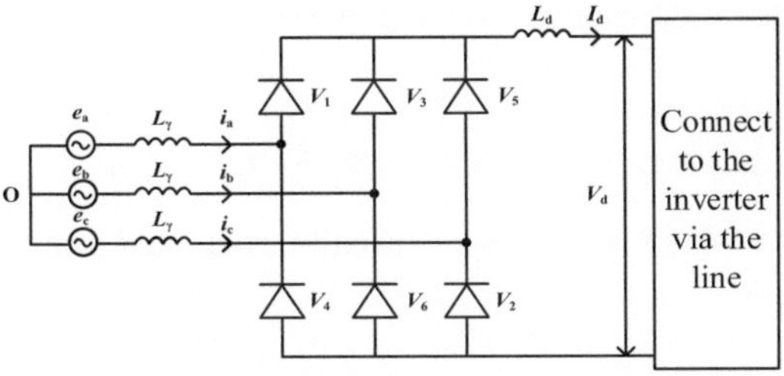

Fig. 2. Equivalent circuit diagram of diode rectifier

By analyzing the commutation waveform, the commutation current of the diode rectifier can be obtained:

$$i_\gamma = \frac{\sqrt{2}U_{v0}}{2X_r}(1 - \cos \omega t) \tag{1}$$

In the formula, U_{v0} is the effective value of the side line voltage of the no-load valve, and X_r is the equivalent reactance of each phase between the power supply and the bridge, also known as commutation reactance.

After a certain Angle γ, the commutation ends, γ is called the commutation Angle, γ expression is as follows:

$$\gamma = \arccos(1 - \frac{2X_r I_d}{\sqrt{2}U_{v0}}) \tag{2}$$

When the converter is unloaded, the ideal no-load DC voltage can be obtained by calculating the curve area:

$$U_{di0} = \frac{\sqrt{2}U_{v0}}{\frac{\pi}{3}} \approx 1.35U_{v0} \tag{3}$$

When the converter is not idle, the DC voltage obtained by calculating the curve area is:

$$V_d = U_{di0} - \frac{3}{\pi}X_\gamma I_d = \frac{3\sqrt{2}}{\pi}U_{v0} - \frac{3}{\pi}X_\gamma I_d \tag{4}$$

The DC voltage of the 12-pulse converter can then be obtained:

$$U_d = \frac{6\sqrt{2}}{\pi}U_{v0} - \frac{6}{\pi}X_\gamma I_d \tag{5}$$

12 Pulse converter DC power:

$$P_d = U_d I_d = \frac{\sqrt{2}}{X_\gamma}U_{vo}U_d - \frac{\pi}{6X_\gamma}U_d^2 \tag{6}$$

6-pulse converter rated relative inductive pressure drop d_{xN} in engineering calculation:

$$d_{\mathrm{xN}} \approx u_{\mathrm{k}}/2 \tag{7}$$

The rated ideal no-load DC voltage is usually calculated using the DC voltage, the formula is as follows:

$$U_{\mathrm{diON}} = \frac{\frac{U_{\mathrm{dN}}}{n} + U_{\mathrm{T}}}{1 - (d_{\mathrm{xN}} + d_{\mathrm{rN}})} \tag{8}$$

where, n is the series number of six-pulse converter Bridges in a single pole, which is taken as 1 in this paper; U_{T} is the forward pressure drop of the converter valve, usually 0.2kV; d_{xN} is $u_{\mathrm{k}}/2$; d_{rN} is 0.3%.

The no-load DC voltage in the non-rated state is:

$$U_{\mathrm{di0}} = U_{\mathrm{d}} + U_{\mathrm{T}} + (d_{\mathrm{x}} + d_{\mathrm{r}}) \cdot \frac{I_{\mathrm{d}}}{I_{\mathrm{dN}}} \cdot U_{\mathrm{diON}} \tag{9}$$

3.1 Converter Transformer Parameter Calculation

The main parameters of converter transformer are rated capacity and short-circuit impedance.

3.1.1 Rated Capacity

Three-phase capacity of a double-winding converter transformer connected to a 6-pulse converter:

$$S_{\mathrm{N}} = \sqrt{3} U_{\mathrm{vN}} \cdot I_{\mathrm{vN}} = \sqrt{3} \frac{\frac{\pi}{3}}{\sqrt{2}} U_{\mathrm{diON}} \cdot \sqrt{\frac{2}{3}} I_{\mathrm{dN}} = \frac{\pi}{3} U_{\mathrm{diON}} I_{\mathrm{dN}} \tag{10}$$

In the formula, U_{vN} is the rated voltage of the converter transformer valve side; U_{diON} is rated ideal no-load DC voltage; Subscript N indicates the rated status.

3.1.2 Short-Circuit Impedance

In engineering, the maximum DC short-circuit current value can be considered as follows:

$$I_{\mathrm{k\_max}} = \frac{2 I_{\mathrm{dN}}}{u_{\mathrm{k}} + \frac{S_{\mathrm{N}}}{S_{\mathrm{SC\_max}}}} \tag{11}$$

where, u_{k} is the short-circuit impedance of the converter transformer; $S_{\mathrm{SC\_max}}$ indicates the maximum short-circuit capacity of the AC system.

3.2 Parameter Calculation of Smoothing Reactor

The smoothing reactor is used to suppress DC current ripple to limit the harmonics of DC current. For 12-pulse rectifier, the smoothing reactor is mainly aimed at the lowest 12th characteristic harmonic, so that its per unit value relative to the rated DC current is no more than 0.01. The calculation formula is as follows:

$$L_1 = \frac{U_{\mathrm{d(n)}}}{n\omega I_{\mathrm{d}} \times \frac{I_{\mathrm{d(n)}}}{I_{\mathrm{d}}}} \tag{12}$$

where $U_{\mathrm{d}(n)}$ is the RMS value of the lowest subcharacteristic harmonic voltage on the DC side; I_{d}: rated DC current; $I_{\mathrm{d}(n)}/\, I_{\mathrm{d}}$ is the relative value of the lowest characteristic harmonic current on the DC side. n is the lowest characteristic harmonic, where 12 is chosen; ω is the fundamental angular frequency.

3.3 AC Filter Parameter Calculation

The reactive power of the diode rectifier station can be obtained from the following formula:

$$\begin{cases} Q_{\mathrm{d}} = P_{\mathrm{d}} \tan\varphi \\ \tan\varphi = \dfrac{2(\pi/180)\mu - \sin 2\mu}{(1 - \cos 2\mu)} \end{cases} \tag{13}$$

Common filters are single-tuned filters, double-tuned filters and triple-tuned filters. The 12-pulse rectifier station mainly has 11, 13, 23 and 25 characteristic harmonics. A double-tuned filter is selected and the tuning point is set at 12 and 24 times to simplify the number of filters (Fig. 3).

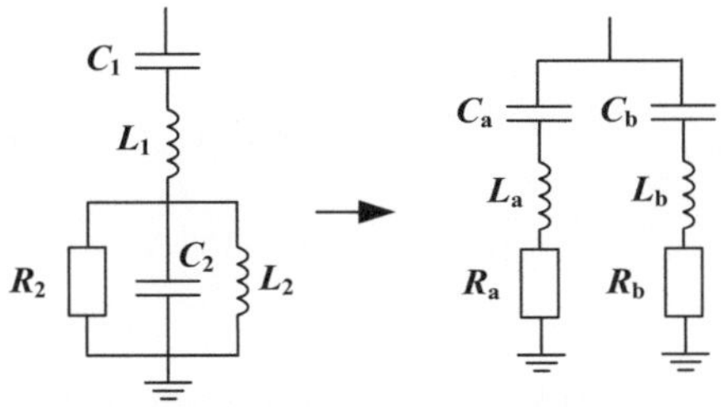

Fig. 3. Double tuned filter structure

In the calculation of component parameters, the double-tuned filter can be equivalent to two single-tuned filters, and the single-tuned filter calculation formula is as follows:

$$\begin{cases} C_x = \frac{Q_1(N_x^2-1)}{\omega N_x^2 U_1^2} \\ L_x = \frac{1}{\omega^2 N_x^2 C_x} \end{cases} \tag{14}$$

In the formula, x = a and b represent single-tuned filters that filter out the 12th harmonics and the 24th harmonics respectively; Q_1 is the reactive power output of the single-tuned filter; N_x is the number of harmonics suppressed by the filter, $N_a = 12$, $N_b = 24$; U_1 is the fundamental voltage of the bus.

Then the parameters of the double-tuned filter can be obtained:

$$\begin{cases} C_1 = C_a + C_b \\ L_1 = \frac{L_a L_b}{L_a + L_b} \\ C_2 = \frac{C_a C_a (C_a + C_b)(L_a + L_b)^2}{(C_a L_a - C_a L_b)^2} \\ L_2 = \frac{(C_a L_a - C_b L_b)^2}{(C_a + C_b)^2 (L_a + L_b)} \end{cases} \tag{15}$$

4 Simulation Results

A simulation model was built in PSCAD/EMTDC to verify the rationality and effectiveness of the parameter calculation method proposed in this paper. Combined with the default system parameters, and the system parameters were calculated according to Sect. 3, as shown in Table 1.

Table 1. Parameters of the system

Categories	Values
Rated capacity	2000 MVA
Fundamental frequency	50Hz
DC-side rated voltage	500 kV
Capacity of the transformer	1142.49 MVA
Short-circuit impedance of the transformer	16%
Smoothing reactor	0.386 H
AC filter	$Q_1 = 45$MVar、$C_1 = 32.741$ uF、$C_2 = 91.232$uF、$L_1 = 8.5828$e-4 H、$L_2 = 4.8278$e-4 H、$R = 1500\Omega$

The waveforms of no-load DC voltage U_{di0} and valve side voltage U_{v0} of the system in steady state are shown in Fig. 4. The DC voltage has harmonics and fluctuates up and down at 500kV, while the valve side voltage is 370kV, satisfying the Eq. (4).

To verify the parameter design of the smoothing reactor, two conditions with or without smoothing reactor are simulated. After the smoothing reactor is added, the DC current harmonics are significantly reduced and the waveform is smoother, as shown in Fig. 5.

Figure 6 shows the waveforms before and after adding the filter, which can reflect the reactive power compensation and filtering effects of the filter. The filter device is put into operation at 3s, and the reactive power curve shows that the reactive power

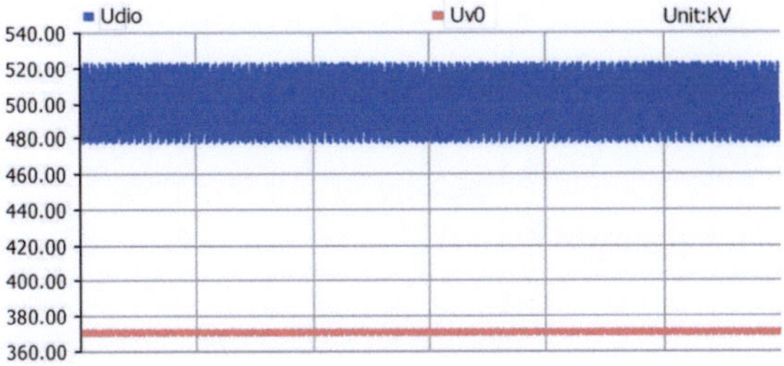

Fig. 4. No-load DC voltage U_{di0} and valve side voltage U_{v0} waveform

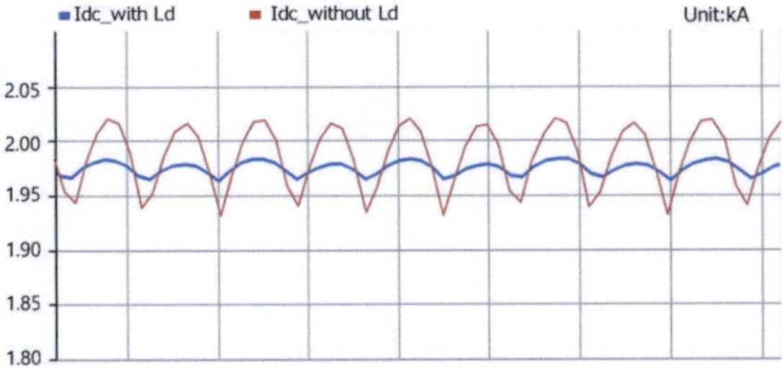

Fig. 5. DC current waveform

demand of DRU is compensated, and the compensation capacity is 45MVar, as shown in Fig. 6(a). The reactive power calculated by Equation (24) is 775MVar. Compared with the simulation result, there is a margin of error of 0.01%, which can reflect the actual system. The characteristic harmonic amplitudes of 11th, 13th, 23rd and 25th were obtained, as shown in Fig. 6(b), which verified the effectiveness of the AC filter parameter design.

5 Conclusions

1. For the offshore wind power DC transmission system based on grid-forming wind turbines and diode rectifier unit, the complete design idea of the main circuit parameters of the system is proposed, including the calculation process and method.
2. Based on the theoretical analysis of the principle of the offshore wind power DC transmission system of the diode rectifier unit, the feasibility of the system was verified, and the power characteristics of the diode rectifier circuit were analyzed.
3. A design method is proposed for the main equipment of the main circuit, such as converter transformer, smoothing reactor and AC filter. The analytical results are verified by PSCAD/EMT simulation.

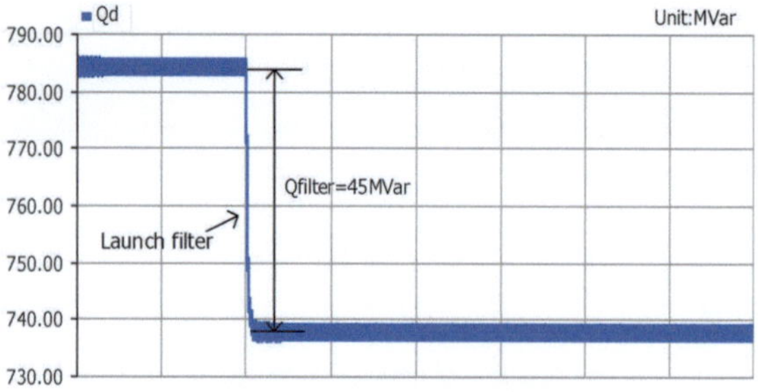

(a) DRU reactive power waveform

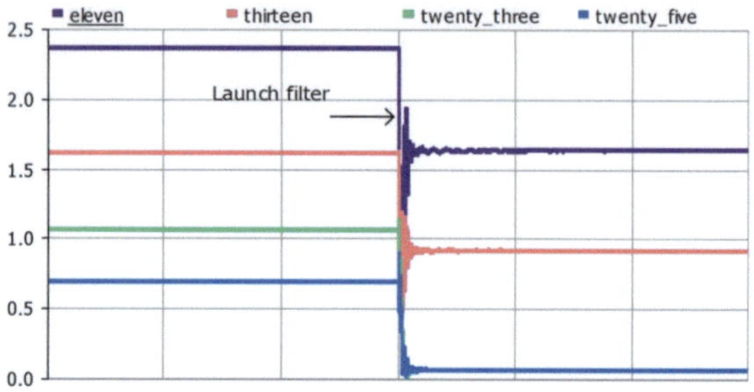

(b) 11th, 13th, 23rd and 25th characteristic harmonic amplitudes

Fig. 6. Waveforms before and after adding the AC filter

References

1. Bowles, J.P.: Multiterminal HVDC transmission systems incorporating diode rectifier stations. IEEE Trans. Power Appar. Syst. **100**(4), 1674–1678 (1981)
2. Chen, D., Yue, B., Mei, N., Wu, Y.K., Yu, S.F.: System design of Xiamen bipolar VSC-HVDC transmission project. Automation of Electric Power Systems. **42**(14), 180–185 (2018)
3. Chen, X.L., Zhou, H., Shen, Y., Ding, J., Qian, F., Zhou, Z.C., et al.: Study and Design of Main Circuit Parameters for 800 kV/7 500 MW DC Power Transmission Project From Xiluodu to Zhexi. Power System Technology. **35**(12), 7 (2011)
4. CPECC. HVDC transmission design manual. China electric power press, Beijing (2017)
5. Deng, X., Wang, D.J., Shen, Y., Zhou, Hao, Chen, X.L., Sun, K.: Main circuit parameter design of Zhundong-Sichuan ±1100kV UHVDC power transmission project. Electric Power Automation Equipment. **34**(04), 133–140 (2014)
6. Huang, Y., Huang, W.X., Xiao, H.Q., Liu, T.: Black start srategy of diode rectifier unit based flexible low-frequency transmission system for offshore wind power. High voltage engineering (2023)
7. Jin, Y.Q., Zhang, Z.R., Wu, H.Y., Pei, X.Y., Chen, J.F., Xu, Z.: Black start strategy of offshore wind farm based on grid-forming wind turbines and diode rectifier unit. High voltage eng. Ineering. **49**(09), 3730–3740 (2023)
8. Li, Z.C., Hu, P., Ma, J.X., Gao, M., Huang, H.L., Liu, X.Y., et al.: Analysis and prospect of offshore wind power development in China. China Offshore Oil and Gas **34**(05), 229–236 (2022)
9. Luo, H.W., Le, J., Mao, T., Li, M.K., Xu, X.Y., Cui, S.G.: Study and design of main circuit parameters for ±800kV UHVDC transmission project from Zhalute to Qingdao. High Voltage Apparatus. **54**(04), 128–134 (2018)

10. Machida, T., Ishikawa, I., Okada, E., Karasawa, E.: Control and protection of HVDC systems with diode valve converter. Electrical Engineering in Japan. **98**(1), 62–70 (2010)
11. Min, B., Wang, M.C., Fu, X.R., Zhao, C.: Offshore wind power as the development trend of wind industry developments of global offshore wind power. International Petroleum Economics. **26**(4), 29–36 (2016)
12. Xiao, H.Q., Huang, X.W., Huang, Y., Liu, Y.L.: Self-synchronizing control and frequency response of offshore wind farms connected to diode rectifier based HVDC system. IEEE Transactions on Sustainable Energy. **13**(3), 1681–1692 (2022)
13. Xin, B.A., Guo, M.Q., Wang, S.W., Li, X.: Friendly HVDC transmission technologies for large-scale renewable energy and their engineering practice. Automation of electric power system. **45**(22), 1–8 (2021)
14. Xu, J., Dong, D.P.: Rethinking of China's new energy development under the background of "double carbon." Energy **10**, 32–37 (2023)
15. Yang, K. Reasearch on calculation of main circuit parameters for multi-terminal hybrid HVDC transmission system. Shandong University (2022)
16. Yu, L.J., Fu, Z.Y., Zhu, J.B., Li, rui, Peng, G.P., Zhao, C.Y.: Review on grid-forming control and start-up method of diode-rectifier-unit based HVDC transmission system for remote offshore wind farm. Automation of electric power systems **47**(24), 63–79 (2023)
17. Zhang, Z.R., Tang, Y.J., Xu, Z.: Medium frequency diode rectifier unit based HVDC Transmission for offshore wind farm integration. Electric Power. **53**(07), 80–91 (2020)
18. Zhao, L.: Whether the grid type wind power technology can promote the high proportion of wind power development. Wind Energy **10**, 38–40 (2023)
19. Zhou, G.W., Liu, D., Gu, Y.D., Liu, J., Zhou, J.P.: Parameter design of main circuit of ±800kV UHVDC transmission project in Lingzhou-Shaoxing. Electrotechnical Application. **35**(01), 30–35 (2016)

Effects of Rotating Stall on Flow Patterns and Pressure Pulsation in Clearance Flow Channels of Pump-Turbines

X. X. Hou[1], S. F. Teng[1], C. X. Xiong[2], and Z. Y. Yang[3](✉)

[1] Department of Hydraulics, Changjiang River Scientific Research Institute, Wuhan 430010, China
xxhouwh@whu.edu.cn
[2] Chang Jiang Survey, Planning, Design and Research Co., Ltd, Wuhan 430010, China
[3] Shanghai Investigation, Design and Research Institute Co., Ltd, Shanghai 200434, China
mry@whu.edu.cn

Abstract. The clearance flow channel (CFC) of pump-turbine is a thin cavity composed of a runner and head cover or bottom ring, but excessive pressure pulsation in a flat CFC is easy to cause the head cover excitation. At present, there are insufficient studies on the flow patterns, and pressure pulsations in CFC. In this paper, the 3D CFD numerical simulation method was used to reveal the flow patterns, and pressure pulsations in the CFC of a low specific speed pump-turbine under rotational stall condition. The results showed that the stall vortexes rotated in the vaneless region of the main flow channel (MFC) when the rotating stall occurred in the pump mode; while that rotated in the runner flow channels, causing circumferentially imbalanced pressure at the runner inlet, as well as a low-frequency and high-amplitude pulsation in MFC. In CFC, uneven pressure distribution in the external cavity was formed when the clearance inlet pressure was unbalanced in the circumferential direction in turbine braking mode. The pulsation generated by the rotating stall in the MFC could be transmitted to the CFC, but the pulsation amplitude dropped sharply at the clearance inlet position, then increased first and ultimately decreased inward along the clearance flow channel. This study provides a reference for the study of hydraulic excitation of the head cover.

Keywords: Pump-turbine · Clearance flow channel · Rotating stall · Flow patterns · Pressure pulsatio · Propagation laws

1 Introduction

Against the backdrop of the "double carbon" target, the new power system with renewable energy as the mainstay requires more clean energy sources such as wind and photovoltaic power. However, wind and photovoltaic power generation are characterized by randomness, intermittency, and fluctuation. Pumped storage units, with their short start-up time and fast adjustment rate, can respond quickly to power demand, undertaking tasks such as peak shaving, valley filling, energy storage, and voltage regulation

S. Zheng et al. (Eds.): IHDC 2024, LNCE 487, pp. 294–307, 2025.
https://doi.org/10.1007/978-981-97-9184-2_26

for regional power grids. They play a crucial role as "regulators" and "stabilizers" in ensuring power security and enhancing power system performance, making them one of the most efficient and economical energy storage methods currently available. Due to the special tasks undertaken by pumped storage power stations, pump turbines often operate under off-design loads with a wide range of operating conditions [1, 2]. These operating conditions can easily induce hydraulic instability issues in the units. In recent years, numerous high-head units have experienced abnormal vibrations in their head covers when operating off-design loads [3]. Most of these issues stem from hydraulic excitation in the runner flow channel, which can induce hydraulic excitation in the unit's head cover, shafting, and powerhouse, affecting the safe, stable, efficient, and flexible operation of the power station [4].

The CFC of a pump-turbine is a thin cavity composed of the runner, head cover, and bottom ring. In addition, the lower the specific speed, the flatter the CFC. Although the CFC is small and complex in shape, the pressure pulsations within the CFC are in direct contact with the head cover, making it susceptible to structural vibrations. Therefore, this work reveals the sources of pressure pulsations in both the MFC and the CFC, as well as the propagation patterns of these pulsations, which is fundamental to addressing the issue of hydraulic excitation of units (Fig. 1).

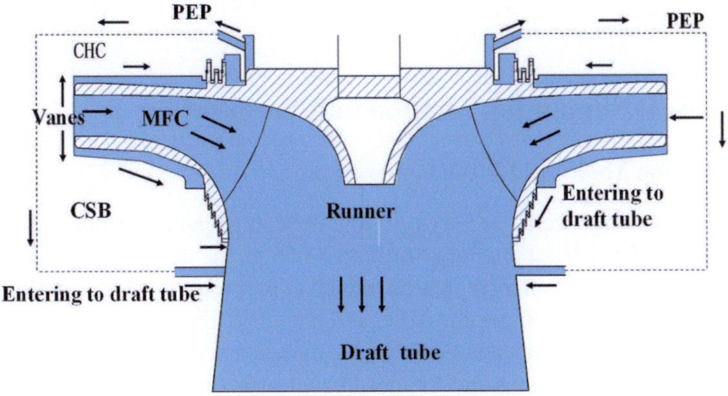

Fig. 1. Cross-sectional diagram of the pump-turbine

Rotating stall is a common source of pulsation in pump turbines, which mainly arises from the uneven pressure distribution in the flow channel when the unit operates at a condition that deviates from the design load. Due to the existence of pressure gradients, stall vortices rotate in the circumferential direction [5, 6]. The stall rotation is mostly the same as the rotation direction of the runner at a slower speed. Current research on rotating stalls mainly focuses on pump operating conditions. Braun observed that a rotating stall appeared in the guide vane domain under medium flow pump operating conditions (40%-80% rated flow) [5, 7]. The pulsation generated by rotating stalls is low-frequency and high-amplitude pulsation. Under turbine braking mode, the pulsation frequency is approximately 0.5–0.7 times the rotational frequency [8], while is approximately 0.1–0.3 times the rotational frequency under small flow pump operating conditions

[9]. The frequency of rotating stall pulsation varies with flow rate [10]. Under pump operating conditions, when the flow rate is below $0.3Q$, the pulsation frequency band becomes wider, and the pulsation spectrum is more complex [9]. Additionally, some studies suggest that rotating stall under pump conditions is the main cause of pump hump characteristics [11, 12]. Due to the small size and complex geometry of the CFC, model test scaling is limited, and prototype observation is difficult. Current research on pressure pulsations from rotating stalls mainly focuses on the runner in MFC. Hu et al. found low-frequency pressure pulsations in the CFC of high-specific-speed pump turbines, which originate from the uneven pressure distribution [3]. When the rotating stall occurs in the MFC, how the flow pattern and pressure pulsations evolve in the CFC, and how the pressure pulsations generated by the rotating stall propagate within the CFC, all remain to be further investigated.

This work utilizes 3D CFD numerical simulation methods to reveal the phenomenon of rotational stall in a low-specific-speed pump-turbine under both low-flow pump mode and turbine braking mode. The simulated result indicates that stall primarily occurs in the guide vane channel during pump operation, while it occurs in the runner channel during turbine braking. Subsequently, the study analyzes the flow patterns and pressure pulsations in CFC when rotational stall occurs in the MFC, as well as the propagation patterns of the pulsations generated by rotational stall throughout the entire flow channel. The study may provide a potential reference value on the excitation of the head cover caused by pressure pulsations in the CFC.

2 Numerical Simulation Method and Verification

2.1 Calculation Model and Mesh Generation

This work takes a low specific speed (n_s is 90.2) prototype pump-turbine as the research object. The rated head H_r of the pump-turbine is 655 m, the rated output P_r is 357 MW, and the rated flow Q_r is 62.43 m^3/s. The 3D model of the pump-turbine is shown in Fig. 2. The calculation domain includes the MFC and the CFC. The MFC calculation domain involves the spiral casing, stay and guide vanes, runner, and draft tube. The CFC is a thin cavity composed of the runner and the head cover or bottom ring, and can be divided into the upper clearance flow channel (CHC) and the lower clearance (CSB). The CHC is a thin layer flow channel between the upper surface of the runner (rotating surface) and the lower surface of the head cover (stationary surface). The CSB is a thin layer flow channel between the lower surface of the runner (rotating surface) and the upper surface of the bottom ring (stationary surface). The CHC includes the clearance inlet, external cavity, sealing ring, internal cavity, outlet, and pressure balance pipe (PEP), and the leaked water body flows to the draft tube through the PEP. The CSB includes the lower ring clearance external cavity and sealing ring, and the leaked water enters the draft tube directly through the sealing ring.

2.2 Grid Independence Verification

The model was divided into regional grids using the ANSYS ICEM software. All regions used hexahedral grids except the guide vane area which adopted wedge grids. The key

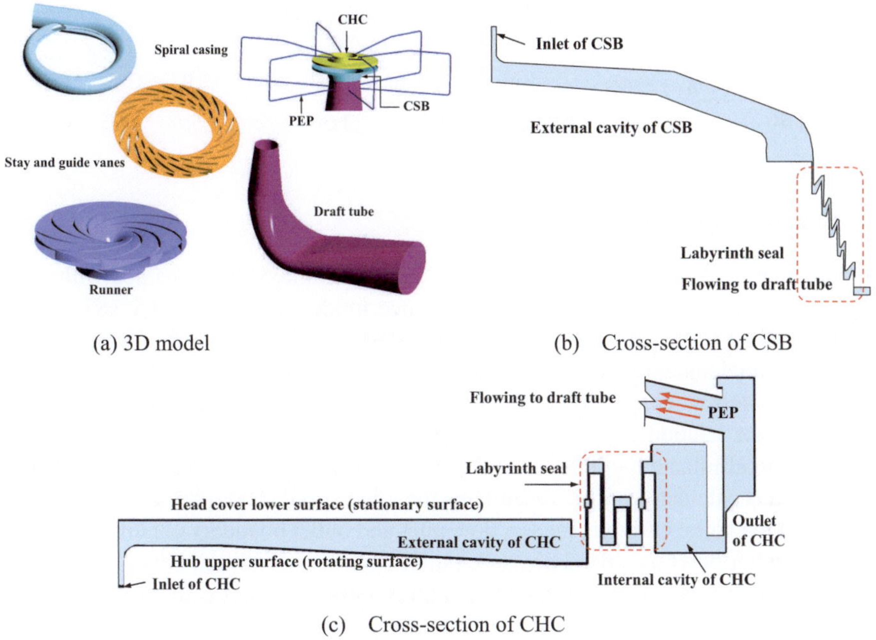

Fig. 2. Model of the pump-turbine

computational domain grid detail is shown in Fig. 3. Due to the complex flow conditions within the runner at the rotating stall condition, a finer grid was required. The calculation result showed that, at the turbine runaway condition, when the number of runner grids was greater than 2 million and the total grid number was greater than 10 million, the torque variation was within 0.2%. Therefore, a fine grid number of 12.52 million was adopted, and the distribution of grids in each computational region is shown in Table 1.

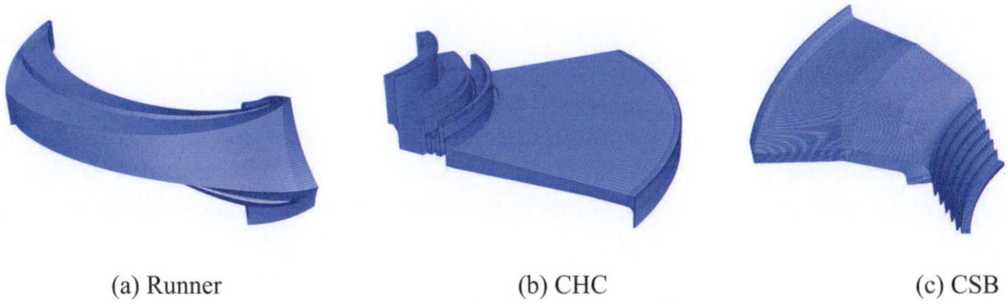

Fig. 3. Grid diagram of calculation domain

Table 1. Number of grid elements (million)

Grids	Spiral casing	Vanes	Transition	Runner	Draft-Tube	CHC	CSB	PEP
12.52	1.30	3.80	0.158	2.33	1.50	1.50	1.60	0.336

2.3 Numerical Simulation Method

The computational domain was calculated by using the finite volume method with commercial software ANSYS Fluent. The turbulence model adopts the SAS-SST model, and the SIMPLEC pressure-velocity coupling algorithm is used for the iterative solution. Both spatial and temporal discretization adopt the second-order difference format, and the time step is set to 0.001 s. Boundary condition settings are shown as follows, due to the close relationship between CFC patterns and pressure pulsation with runner speed [13], to eliminate the influence of speed, a constant speed calculation method is adopted in this calculation, which maintains the rated speed (500 rpm) unchanged [14]. The inlet of the spiral casing is set as a flow inlet and outlet boundary condition, and the draft tube outlet is set as a fixed pressure outlet boundary condition with a water head of 100 m. Much research found that the compressibility of water cannot be ignored in numerical calculations [15, 16], so the influence of water compressibility is included in this calculation. To universally describe pressure, the pressure pulsation amplitudes obtained in this paper are dimensionless processed using formula (1), which represents the ratio of the calculated pressure pulsation amplitude to the water pressure under the operating water head:

$$A^* = \frac{A}{\rho g H} \tag{1}$$

where A and A^* represent the pulsation amplitudes before and after dimensionless processing, respectively; and H is the water head under the corresponding operating conditions.

2.4 Pressure Monitoring Points

To monitor the pressure pulsation and its propagation characteristics in the whole flow channel, monitoring points are set up in the MFC, CFC, and CSB, as shown in Fig. 4. Additionally, pressure monitoring points in the runner are set to rotate with the runner.

2.5 Operating Conditions and Verification of Calculation

The operating condition involves the turbine braking mode (Q is 10 m^3/s) and the pump mode with small-medium flow (Q is 20 m^3/s). To verify the accuracy of the numerical simulation, a comparison was made between the calculated values and the model test for the micro parameters, as shown in Fig. 5 The results of the numerical simulation and model test are relatively well matched.

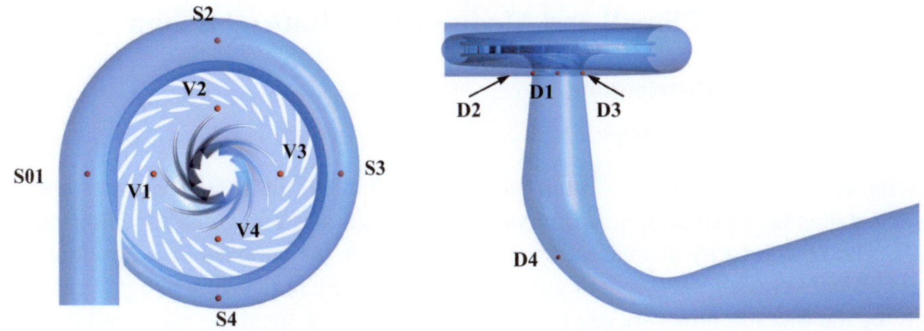

(a) Pressure monitoring points in MFC

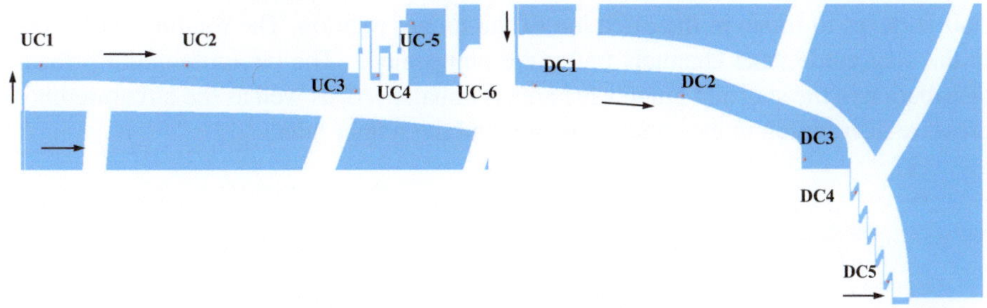

(b) Pressure monitoring points in CFC

Fig. 4. Layout of pressure monitoring points for whole flow channels

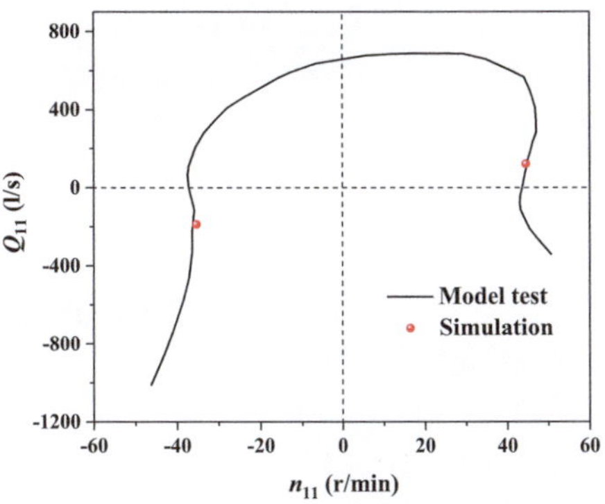

(a) n_{11} - Q_{11} curves

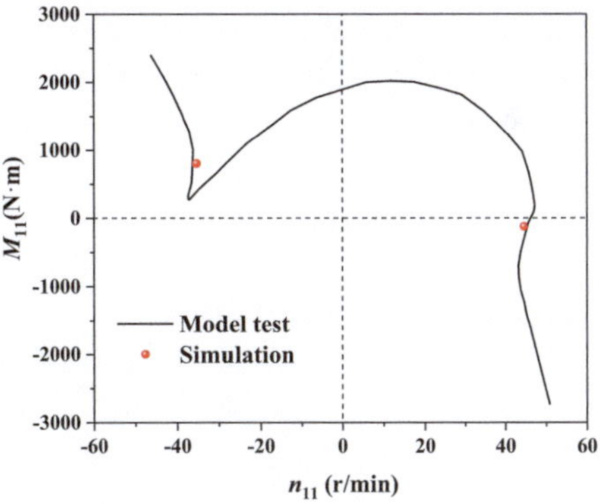

(b) n_{11} - M_{11} curves

Fig. 5. Macro parameters validation for numerical simulation and model test

3 Pump Mode of Small and Medium Flow Rate (Q is $20m^3/s$)

3.1 Flow Patterns Distribution

3.1.1 Flow Patterns in MFC

As shown in Fig. 6, under the pump mode (Q is 20 m3/s), three stall vortexes evenly distributed in the circumferential direction appear in the stay and guide vane flow channels. The flow velocity decreases at the position where the vortexes pass through, and the channel between the stay and guide vanes is blocked and generates high pressure, forcing the water to flow between the stay guide vanes and the guide vane until it flows out of the non-stall channel. The flow velocity of the non-stall channel increases, and the pressure decreases. The existence of the circumferential pressure gradient causes the stall vortexes to rotate in the direction of the runner rotation. The rotating stall occurs in the stay guide vane channels under the pump mode. The circumferential pressure distribution in the stay and guide vane area is unbalanced, as well as the circumferential pressure distribution in the vaneless region and the runner inlet.

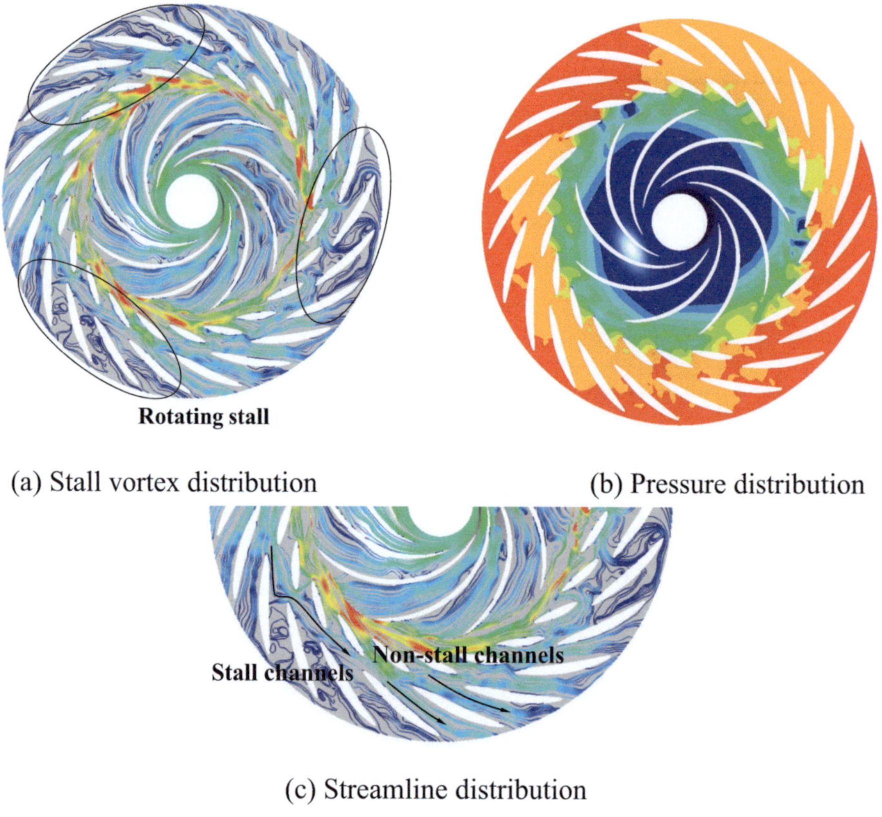

Fig. 6. Flow patterns of rotating stall in MFC under a pump mode of small and medium flow rate

3.1.2 Flow Pattern Distribution in CFC

Due to the CFC being directly connected to the runner inlet of the MFC, the pressure distribution inside the CFC is determined by the pressure at the runner inlet. In the case of pump mode, rotating stalls occur in the stay and guide vane channels, and the pressure in MFC at the runner inlet is evenly distributed in the circumferential direction. Therefore, after entering the CFC, the pressure distribution in the circumferential direction of the CFC is uniform, as shown in Fig. 7. A small range of fragmented vortex structures appear at the inlet of the CFC, while the vortices outside are evenly distributed in the circumferential direction. Vortex structures in CSB only appear at the clearance inlet.

(a) Pressure distribution in CHC (b) Vortex structure in CHC (c) Vortex structure in CSB

Fig. 7. Flow patterns in CFC under a pump mode of small and medium flow rate (stall in MFC)

3.2 Pressure Pulsations in Whole Flow Channels

3.2.1 Pressure Pulsation Laws in MFC

As shown in Fig. 8, under the condition of a small flow rate in the pump, there is a rotational stall pressure pulsation of 0.11 times f_n in the guide vanes flow channel, which belongs to a low-frequency and high-amplitude pulsation. Along the MFC towards the volute and draft tube, the pulsation amplitude decreases, so the rotational stall pulsation is the main frequency of pulsation in the guide vane region, but not the main frequency of pulsation in the spring casing and draft tube.

3.2.2 Pressure Pulsation Laws in CFC

As shown in Fig. 9, the MFC experiences a rotating stall and generates low-frequency high-amplitude pulsation in pump mode, but the rotating stall pulsation is not found in the CHC and CSB. The main reason is that the circumferentially uniform pressure distribution and no circumferential unbalanced vortex structure are formed in the CFC. However, a pulsation of 0.9 times f_n appears in both the CHC and CSB.

3.3 Propagation Laws of Rotating Stall Pulsation in CFC

Under the pump mode, the pressure pulsation of stall vortexes in the stay and guide vane flow channels can propagate upstream and downstream along the MFC, and the

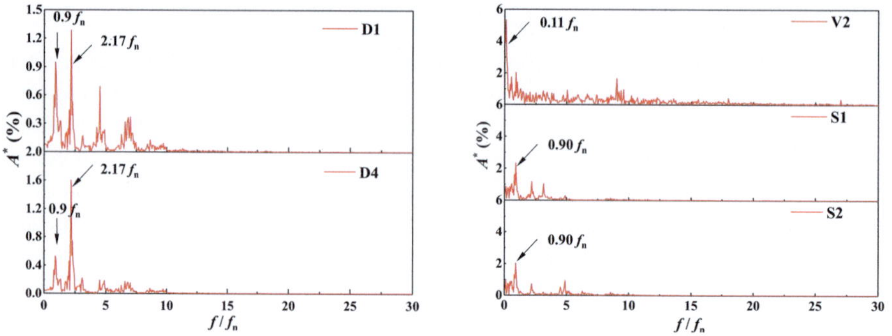

Fig. 8. Spectral characteristics of pressure pulsations in MFC under a pump mode

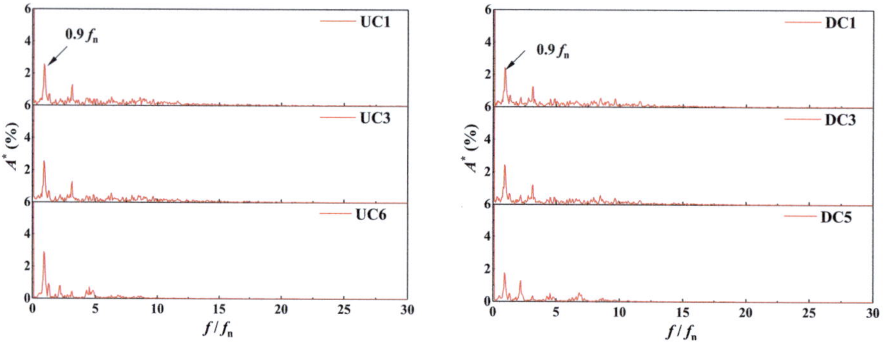

(a) In CHC (b) In CSB

Fig. 9. Spectral characteristics of pressure pulsations in CFC under a pump mode

pulsation amplitude gradually decreases. However, the rotating stall pulsation does not significantly propagate into the CFC.

4 Turbine Braking Mode (Q is 10m^3/s)

4.1 Flow Patterns Distribution

4.1.1 Flow Patterns in MFC

As shown in Fig. 10, under the turbine braking mode (Q is 10 m3/s), six blade channels in the runner suffer stall with various degrees, while three channels can enter the flow normally. For the stall runner channels, the water flows annularly between the stay and guide vanes until it flows into the non-stall channels. The stall flow channels rotate with the runner, with a rotational speed of 0.85 times the runner's rotational frequency. Due to the reduced flow velocity in the stall flow channels and the increased pressure, while the flow velocity in the non-stall channels increases and the pressure decreases, the pressure distribution from the inlet to the outlet of the runner is circumferentially uneven.

(a) Streamline distribution (b) Pressure distribution

Fig. 10. Rotational stall distribution in runner channels under a turbine braking mode

4.1.2 Flow Pattern Distribution in CFC

As shown in Fig. 11, due to the stall in the runner channels, the circumferential pressure distribution at the runner inlet is uneven. Since the CFC is directly connected to the runner inlet, the uneven circumferential pressure at the runner inlet leads to uneven inlet pressure in CHC and CSB, resulting in uneven pressure distribution in the CFC. The uneven circumferential pressure distribution leads to asymmetric vortexes along the circumferential direction in the external cavity in CHC and CSB, which rotates with the runner at a speed of 0.85 times f_n.

(a) Pressure distribution in CHC (b) Vortex structure in CHC (c) Vortex structure in CSB

Fig. 11. Flow patterns in CFC under a turbine braking mode

4.2 Pressure Pulsations in Whole Flow Channels

4.2.1 Pressure Pulsation Laws in MFC

The pressure pulsation spectrum characteristics of the measuring points in MFC are shown in Fig. 12. There is a high-amplitude pressure pulsation of 0.85 times f_n in both the bladeless region and the draft tube, which is caused by the rotating stall vortex in the runner blade channel. However, as the rotation stall pulse propagates upstream and downstream along the MFC, the amplitude gradually decreases.

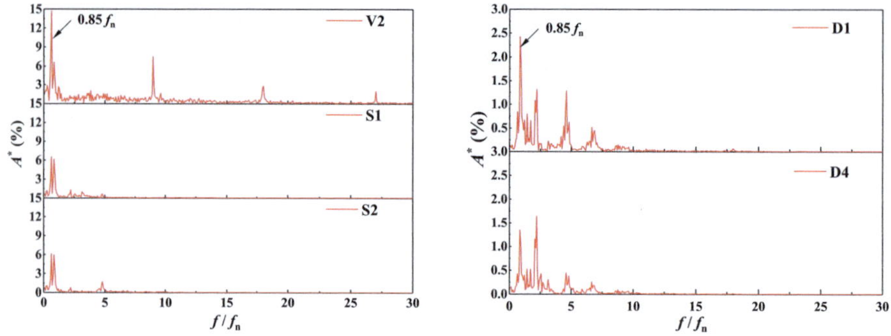

Fig. 12. Spectral characteristics of pressure pulsations in MFC under a turbine braking mode

4.2.2 Pressure Pulsation Laws in CFC

As shown in Fig. 13, under the turbine braking mode, low-frequency and high-amplitude pressure pulsations of 0.85 f_n appear in both CHC and CSB. The stall pulsation in the runner channel can be transmitted to the clearance flow channels. In addition, the unbalanced rotating vortex in the outer cavity of the clearance also generates pressure pulsations at the same frequency. The pulsation amplitude gradually decreases inward along the CFC.

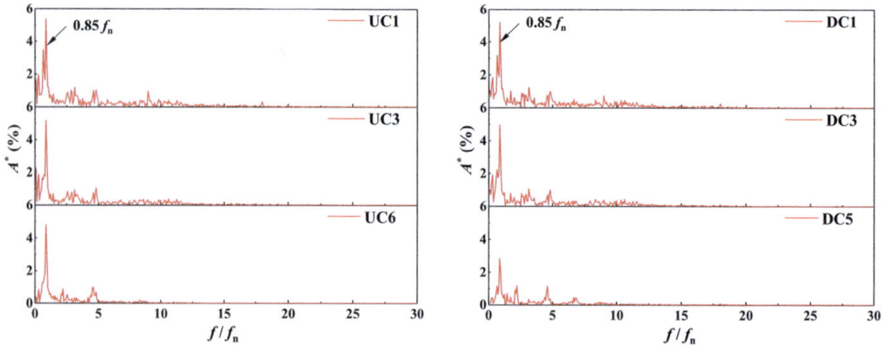

(a) In CHC (b) In CSB

Fig. 13. Spectral characteristics of pressure pulsations in CFC under a turbine braking mode

4.3 Propagation Law of Rotating Stall Pulsation in CFC

As shown in Fig. 14, during the turbine raking mode, partial blade channels stall, resulting in a maximum pulsation amplitude in the bladeless region. This pulsation propagates upstream and downstream along the MFC with a gradually decreasing amplitude. In the CFC, the amplitude of the rotational stall pulsation amplitude dropped sharply at the clearance inlet position, then increased first and ultimately decreased inward along the clearance flow channel because of the unbalanced rotating vortex structure in external cavity.

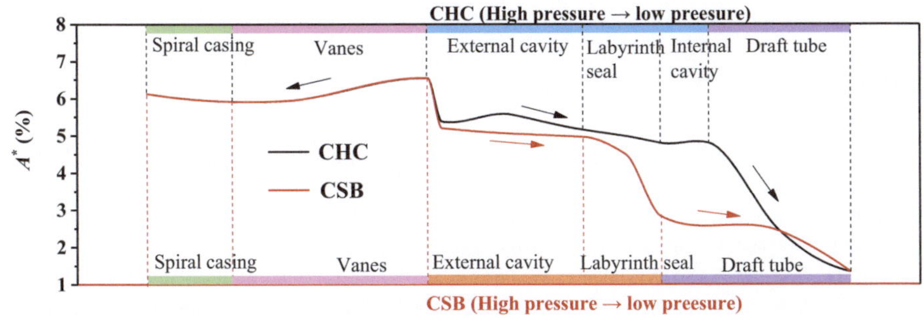

Fig. 14. Propagation law of rotating stall pressure pulsation in whole flow channels under a turbine braking mode

5 Conclusion

This work reveals the flow patterns and pressure pulsation characteristics in the MFC and CFC when a low specific-speed pump-turbine experiences a rotating stall under pump mode and turbine braking mode through 3D CFD numerical simulation method. It was found that stall mainly occurs in the stay and guide vane flow channel in pump mode, while that occurs in the runner blade channels under turbine braking mode. The specific conclusions are listed as follows:

(1) Under the small and medium flow conditions in pump mode, three stall vortexes appear in the stay and guide vane flow channel. Due to the uneven pressure distribution in the circumferential direction, the stall vortex rotates in the circumferential direction and generates low-frequency and high-amplitude pressure pulsations. The pressure imbalance caused by the rotating stall does not propagate significantly to the runner inlet and CFC, so there is no pulsation at the frequency of the rotating stall in the CFC.

(2) Under turbine braking mode, most of the runner blade channels stall, and the stall flow channel rotates with the runner, generating low-frequency and high-amplitude pulsations in the runner flow channels. Due to the runner stall, the pressure distribution at the runner inlet is uneven in the circumferential direction, resulting in uneven pressure distribution in the CFC. Asymmetric vortex structures appear in the clearance outer cavity of CFC in the circumferential direction, resulting in pulsations with the same frequency as the stall in MFC.

(3) The rotating stall pulsations generated in the MFC propagate upstream and downstream along the MFC, and the amplitude of the pulsations gradually decreases. Under turbine braking mode, the pulsations amplitude of the rotating stall drops sharply at the clearance inlet position, then increases first and ultimately decreases inward along the clearance flow channel. This study can provide a potential reference for the head cover excitation caused by pressure pulsations.

Acknowledgments. This work was sponsored by Shanghai Sailing Program [grant number 23YF1426300].

References

1. Landry, M., Gagnon, Y.: Energy storage: technology applications and policy options. Energy Procedia **79**, 315–320 (2015)
2. Kong, Y., Kong, Z., Liu, Z., Wei, C., Zhang, J., An, G.: Pumped storage power stations in China: The past, the present, and the future. Renew. Sustain. Energy Rev. **71**, 720–731 (2017)
3. Hu, H., Xia, M., Qiao, M., Wang, W., Wang, J., Wang, Z.: A simulation study of hydraulic vibration caused by clearance flow in a pump turbine. IOP Conf Ser: Earth Environ Sci **1079**, 012032 (2022)
4. Hou, X.: Study on pressure fluctuation and axial hydraulic thrust of pump-turbine considering clearance flow. Wuhan University (2023)
5. Braun, O., Kueny, J.L., Avellan, F.: Numerical analysis of flow phenomena related to the unstable energy-discharge characteristic of a pump-turbine in pump mode. American Society of Mechanical Engineers. pp 1075–1080 (2005)
6. Zhang, C., Xia, L., Diao, W., Zhou, J.: Pressure fluctuations characteristics and rotating stall propagation mechanism of a pump turbine in pump mode. J. Hydraul. Eng. **48**, 837–845 (2017)
7. Braun, O.: Part load flow in radial centrifugal pumps. Laboratory of Hydraulic Machinery, Institute of Technology (2009)
8. Liu, W., Qin, K., Xu, Y., Wang, L.: Pressure pulsation model test investigation of pump-turbine on the turbine-brake condition. Large Electric Machine and Hydraulic Turbine 41–45 (2016)
9. Xiao, Q., Zhang, C., Xia, L., Diao, W.: Pressure pulsation and runner mechanical property during pumping mode of pump-turbine. Water Resources and Hydropower Engineering **51**, 53–62 (2020)
10. Jese, U., Fortes-Patella, R.: Unsteady numerical analysis of the rotating stall in pump-turbine geometry. IOP Conference Series :Earth and Environmental Science. IOP Publishing, p 042005 (2016)
11. Sun, Y.: Instability characteristics and influencing factors of positive slope on pump performance curves of a low-specific-speed pump-turbine. Doctoral Thesis, Tsinghua University (2016)
12. Yin, J., Liu, J., WANG L,: Performance prediction and flow analysis in the vaned distributor of a pump turbine under low flow rate in pump mode. SCIENCE CHINA Technol. Sci. **53**, 3302–3309 (2010)
13. Hou, X., Cheng, Y., Yang, Z., Liu, K., Zhang, X., Liu, D.: Influence of clearance flow on dynamic hydraulic forces of pump-turbine during runaway transient process. Energies **14**, 2830 (2021)

14. Hu, J.: Research on the control of S-shaped characteristics and pressure pulsations of pump-turbines. Wuhan University (2021)
15. Wang, L., Liu, Y., Liu, W., Qin, D., Jiao, L.: Pressure fluctuation characteristics of pump-turbine at pump mode. Journal of Drainage and Irrigation Machinery Engineering **31**, 7–10 (2013)
16. Wu, G., Luo, X., Zhao, Q., Feng, J.: Research on characteristics of pressure fluctuation in a pump-turbine. Journal of Xi'an University of Technology **31**, 265–272 (2015)

Technical Challenges and Environmental Governance in the Construction of Pumped Storage Power Stations

Song Wang(✉) and Yongkang Yang

SINOHYDRO BUREAU 11CO.,LTD, Henan Province, China
643007481@qq.com

Abstract. With the continuous deepening of China's reform and opening-up, the coordinated development of environmental protection and economic development has become the focus of social attention. As a key new energy technology, pumped storage power stations have functions such as peak power regulation and energy storage, and play an important role in new energy construction. However, its application in China is still in its infancy and lags behind the international advanced level. This paper uses the methods of literature review and practical experience induction to conduct a detailed analysis of the technical issues in the construction of pumped storage power stations. Through an in-depth discussion of the development status of China's pumped storage power stations, as well as technical problems and governance measures that may arise during their construction, we will provide support for promoting ecological environmental protection, responding to the demand for new energy construction, and creating a green-friendly construction environment. This paper focuses on the technical difficulties encountered during the construction process and proposes corresponding management measures. At the same time, an in-depth analysis of the challenges faced by pumped hydro storage technology and construction was conducted. Through research, it is found that the development of pumped storage power stations in China has made some progress, but there are still some necessary technical challenges. In response to these problems, a series of effective governance measures are proposed, and future development prospects are forecast. Comprehensive research results show that pumped storage power stations occupy an important position and have great potential in China's new energy construction. Through scientific planning and effective management, problems that may arise during their construction and operation can be overcome, and their wider application in China can be promoted, making positive contributions to ecological environmental protection and sustainable development.

Keywords: Pumped storage power stations · Water conservancy and hydropower · Technological challenges · Environmental protection

S. Zheng et al. (Eds.): IHDC 2024, LNCE 487, pp. 308–316, 2025.
https://doi.org/10.1007/978-981-97-9184-2_27

1 Introduction

The development of pumped storage power technology in China has undergone five stages: the initial stage, stagnation stage, breakthrough stage, scale formation stage, and rapid development stage. Compared with developed countries, China started the development of pumped storage power stations later, with the construction of the small mixed pumped storage power stations Gangnan and Miyun in 1968 and 1973, respectively. In the late 1980s and early 1990s, the successive construction of the Shisanling, Guangzhou, and Tianhuangping pumped storage power stations marked significant breakthroughs in China's pumped storage technology. Research shows that pumped storage power stations currently have the highest energy storage conversion efficiency, with a storage cycle efficiency of 75% to 80% [1]. As a critical component of energy transition, the construction of pumped storage power stations is not only a technology-intensive project but also a profound consideration and significant challenge for ecological sustainability. During project implementation, the technical challenges encountered are complex and variable. However, it is even more crucial to carefully assess and effectively address the potential impacts on the surrounding environment, particularly the stability of ecosystems and environmental quality.

This paper focuses on the reservoir dam in a specific region. The upper reservoir is located on the Paofanggou section of the Matie River, a left-bank tributary of the Huangya River, while the lower reservoir is situated on the Luotuotou section of the Sangua Temple Village, part of the main upper reaches of the Huangya River, a major tributary of the Bai River, itself a tributary of the Han River (Yangtze River system). The upper reservoir has a concrete-faced rockfill dam with a crest elevation of 1068.40 m and a maximum dam height of 118.40 m. Its normal storage level is 1063.000 m, with a corresponding storage capacity of 14.05 million cubic meters. The lower reservoir also has a concrete-faced rockfill dam with a crest elevation of 540.60 m and a maximum dam height of 100.60 m. Its normal storage level is 537.500 m, with a corresponding storage capacity of 16.34 million cubic meters. This first-class (I) project includes the upper reservoir, water conveyance system, underground powerhouse system, lower reservoir, and surface switch station.

Based on extensive practical engineering experience and cutting-edge research results accumulated in the industry, this paper aims to analyze some key technical issues faced in the construction of pumped storage power stations and propose a series of practical solutions. These solutions are intended to provide scientific and practical guidance and recommendations for the sustainable development of the pumped storage power stations, thus promoting the healthy, stable, and green development of this sector.

2 Major Technical Issues in the Construction of Pumped Storage Power Stations

2.1 Multiple Construction Projects and Broad Professional Scope

Pumped storage power stations involve various disciplines, including civil engineering, hydraulic engineering, electrical engineering, mechanical engineering, and automation control. Each discipline has its unique technical standards and specifications. The construction process requires a large number of highly skilled technical personnel, and all participants on-site need to have a solid theoretical foundation and extensive practical experience to deal with the various technical issues that may arise during construction.

2.2 Numerous and Complex Temporary Construction Roads

Pumped storage power stations are typically built in suburban areas outside of urban centers, with a significant elevation difference between the upper and lower reservoirs. This requires the construction of numerous temporary roads. The complex terrain and variable climate increase the difficulty of designing and constructing these temporary roads. The design of temporary roads must consider factors such as road gradient, load-bearing capacity, and other elements to ensure that the roads can withstand the required excavation and filling strength. Concrete surfaces used for these roads should be at least 30 cm thick, and the concrete should have a flexural strength of at least R5.0 [2].

2.3 High Quality Requirements for Rockfill Dam Embankments

The quality of rockfill dam embankment determines the safety and stability of the project. First, the material used for embankment directly affects the compaction effect and stability of the dam body, so the raw materials for rockfill dam embankments must meet the design specifications, including particle size, gradation, and physical and mechanical properties. Second, the embankment must be layered according to the design requirements, with different materials requiring specific layer thicknesses, spreading thicknesses, and compaction degrees. Finally, professional machinery and equipment such as vibratory rollers and bulldozers are required for embankment construction and compaction, followed by compaction degree and layer thickness testing.

2.4 Long Peak Construction Time and Complex Construction Environment

The construction of pumped storage power stations involves multiple projects, often located in complex mountainous or aquatic environments. Geological conditions and climatic changes significantly impact construction, posing higher demands on material supply and storage. Construction schedules must consider rainy seasons, winter conditions, and flood control, leading to tight timelines. Peak construction periods require large numbers of workers, and prolonged high-intensity work can lead to worker fatigue, increasing the risk of accidents. Extended peak construction also places heavy demands on equipment, raising the risk of failure and damage.

2.5 Poor Air Circulation and Air Quality in the Construction Environment

Pumped storage power station construction often takes place in relatively closed environments, and construction workers are exposed to significant occupational health risks. Construction dust is a major contributor to various diseases. Poor air circulation can cause discomfort and lead to respiratory diseases such as pneumoconiosis and asthma. The mortality rate from dust-related diseases among construction workers is as high as 63%, with cerebrovascular and cardiovascular diseases accounting for 17% [3]. In addition, dust and airborne contaminants can enter machinery and affect its normal operation and life.

2.6 Large Reservoir Capacity and High Requirements for Seepage Prevention

Reservoir seepage is somewhat concealed, with various types, states, and complex causes. Due to the large storage capacity and complex geological conditions of reservoirs, the dam body and reservoir area are subjected to enormous water pressure. Reservoir seepage not only reduces storage capacity but may also affect the surrounding environment. Nationwide, about 20% to 30% of reservoir dams have varying degrees of seepage problems [4]. For example, the Shenkeng Reservoir in Daishan County, Zhejiang, experienced seepage before a dam failure on August 19, 2012, resulting in 10 deaths and 27 injuries. Similarly, the Lianteng Reservoir in Urumqi experienced a dam failure due to piping on February 2, 2013, causing 1 death and 18 injuries [5].

2.7 Severe Water Environment Pollution

During the construction of pumped storage power stations, wastewater mainly originates from tunnel wastewater, foundation pit wastewater, and domestic sewage [6, 7], as shown in Table 1. Tunnel wastewater primarily consists of construction process water, with suspended solids (SS) as the main pollutant. Foundation pit wastewater mostly includes groundwater seepage and concrete curing wastewater, with primary pollutants being suspended solids, biochemical oxygen demand (BOD), and chemical oxygen demand (COD). The main pollutants in domestic sewage are total phosphorus and ammonia nitrogen.

Pollutants in water can disrupt aquatic ecosystems, resulting in the death of fish and other aquatic organisms. When polluted water is used for irrigation, farmland can be adversely affected to varying degrees. When reservoirs or surrounding water bodies serve as local drinking water sources, water pollution directly affects the safety of drinking water for local residents. Pollutants entering the human body through drinking water can cause various health problems, such as digestive system diseases and skin conditions.

2.8 Flood Control Plan for Pumped Storage Power Stations

The construction period of the power station is long and spans multiple flood seasons. During these periods, heavy rainfall, floods, and extreme weather conditions may occur, posing threats to the power station dam and reservoir area. If the flood control plan is not comprehensive, flooding could result in equipment malfunctions that, at best, disrupt the construction process, or, at worst, have serious consequences such as dam damage and reservoir overtopping. In addition, torrential rains can trigger mudslides and landslides, threatening the safety of the power station and construction personnel on site.

3 Countermeasures for Major Technical Issues in the Construction of Pumped Storage Power Stations

3.1 Countermeasures for Numerous and Diverse Construction Projects

Introducing advanced project management tools: To address the numerous and diverse construction projects involved in the construction of pumped storage power stations, advanced project management tools can be introduced. The use of building

Table 1. Characteristics of wastewater from pumped storage power station construction [7]

No	Power plant name	Source of wastewater	Maximum wastewater quantity/(m^3/h)	Main pollutants
1	Xiangjiaba Power Station	Sand and gravel processing and mixing wastewater	450.0	SS
2	Jurong Pumped Storage Power Station	Sand and gravel processing and mixing wastewater	334.0	SS/BOD/COD
3	Yimeng Pumped Storage Power Station	Sand and gravel processing and mixing wastewater	480.0	SS
4	Liyang Pumped Storage Power Station	Domestic sewage	20.8	BOD/COD
		sand and gravel processing wastewater	200.0	SS
5	Fujian Xianyou Pumped Storage Power Station	Domestic sewage	13.2	SS/BOD/COD
		Sand and gravel processing and mixing wastewater	300.0	SS
6	Shisanling Pumped Storage Power Station	Tunnel wastewater	83.3	Petroleum/TNT/SS
7	Meizhou Pumped Storage Power Station	Sand and gravel processing and mixing wastewater	200.0	SS

information modeling technology and project management software for comprehensive management enables real-time tracking of project progress and coordination, ensuring an orderly construction process.

Build a professional team and conduct regular training: The formation of a comprehensive technical team consisting of civil, hydraulic, electrical and mechanical engineers is crucial. Prior to construction, a detailed construction organization design should be conducted, clearly defining the construction processes and procedures for each specialty and project. Scientific construction plans and schedules should be developed. Regular technical training and briefings should be conducted to promptly address technical and coordination issues that arise during construction to ensure construction quality. Dedicated coordination management positions should be established to facilitate communication and coordination among different specialties to ensure the orderly progress of all work on the site.

3.2 Countermeasures for Complex and Numerous Temporary Construction Roads

Optimize temporary road design scheme: Based on the terrain conditions and construction requirements, the design plan for temporary roads should be optimized. Detailed road planning should be carried out before construction using modern surveying technology and geographic information systems to improve the scientific and rational aspects of planning. An efficient traffic organization and management mechanism should be established, with reasonable arrangements for the travel routes and times of transportation vehicles to avoid traffic congestion and confusion. Traffic signs and safety warning signs should be set up, and maintenance plans and contingency plans for temporary construction roads should be formulated. Regular inspection and maintenance of roads should be conducted to repair damaged sections promptly to ensure smooth and safe roads.

3.3 Measures for Quality Control of Rockfill Dam Embankments

Prevention and in-process control: There are two main methods for dam quality control: pre-emptive prevention and in-process control. Pre-emptive control involves analyzing and judging potential quality issues in advance and taking preventive measures against possible quality problems. During the embankment construction of the Tianchi Pumped Storage Power Station, GPS satellite positioning systems were used for in-process quality control [8]. All quality indicators during the embankment process are monitored and detected in real time, so that quality problems can be identified and solved promptly, thus ensuring the quality of the embankment.

3.4 Countermeasures for Long Peak Construction Period and Complex Construction Environment

Shift System and Safety Awareness: Reasonable work schedules should be formulated to avoid prolonged continuous work by construction personnel, thereby reducing fatigue and ensuring their health. A shift system should be established to ensure that construction workers have sufficient rest periods. Regular safety training should be conducted to improve the safety awareness and skills of construction personnel. On-site safety management should be strengthened, and safety protection measures should be implemented to reduce the occurrence of safety accidents.

3.5 Countermeasures for Poor Ventilation and Air Quality in Construction Environment

Set different control measures based on construction air quality: An air quality monitoring system should be established to regularly check the air quality in the construction area. Based on the monitoring results, the air quality should be classified accordingly. Worker rest areas should be set up around trees, as studies have shown that high levels of greenery in rest areas significantly control dust levels [9]. In areas of poor air quality around the reservoir, a fresh air system should be installed. High-pressure, long-range spray nozzles should be used for dust control in areas where dust may be generated during construction [10].

Equip personal protective equipment upon entering the construction site: Construction workers should be provided with necessary personal protective equipment and must wear protective gear when entering areas with poor air quality. Protective awareness training should be enhanced to raise awareness of the importance of air quality and personal protection among workers. Regular health checks should be conducted for construction workers, focusing on their health status. Health records should be established to identify and address health problems promptly. For workers exposed to high levels of dust and harmful gases for long periods, job rotation and rest periods should be arranged to avoid long-term exposure to harmful environments.

3.6 Seepage Control Measures for Large Reservoirs

Establish a monitoring system that combines prevention and control: A comprehensive monitoring system should be established to monitor key parameters such as dam deformation, seepage volume, and water pressure in real time. In the reservoir seepage control design, a multi-layered seepage control system should be adopted to improve seepage prevention effectiveness. This typically includes a primary seepage control layer, a secondary seepage control layer, and a drainage layer. A multi-layered seepage control design can minimize the risk of seepage. Environmental protection measures should be taken to mitigate the impact of seepage control on the surrounding environment, such as the provision of drainage systems downstream of the dam to reduce the impact of seepage water on downstream soils and groundwater.

3.7 Measures for Water Environment Pollution Treatment

Establish wastewater treatment measures: Conduct wastewater inspections using drones, unmanned boats, intelligent robots, and other means to identify all water-related discharge outlets and wastewater discharge characteristics during construction. Wastewater treatment facilities should be established to treat construction wastewater before discharge to ensure that the wastewater meets discharge standards. Construction-generated slurry should be recovered and treated to prevent direct discharge into water bodies. Oil pollution collection and treatment devices should be installed for waste oil generated by mechanical equipment during construction and operation, and any leaked oil should be handled promptly to prevent it from entering water bodies.

Real-time water quality monitoring and regular testing and evaluation: Water quality monitoring points should be set up in the reservoir and surrounding water bodies, and a real-time water quality monitoring system should be established to monitor key pollutant indicators. Any abnormal water quality should be promptly addressed with appropriate measures. Regular water quality testing should be conducted in the reservoir and surrounding water bodies to assess water environment quality. Pollution control measures should be adjusted based on test results to ensure continuous improvement of water quality.

3.8 Flood Control Measures

Establish a comprehensive meteorological and hydrological monitoring system: Real-time monitoring of key parameters such as rainfall, flow rate, and water level, establishment of information exchange mechanisms with meteorological and hydrological departments, and development of scientifically sound reservoir scheduling plans. During the flood season, reservoir inflow and outflow should be dynamically adjusted based on real-time monitoring data to ensure sufficient reservoir capacity to reduce the pressure on the dam during high-flow floods.

Timely detection and reinforcement before training: Potential vulnerabilities should be repaired to ensure the safety and stability of the dam and embankments. Necessary flood control facilities, such as spillways and drainage ditches, should be built around the power station to ensure rapid drainage during floods and to reduce pressure on power station facilities. Detailed contingency plans for flood control should be developed, specifying the responsibilities and response actions of various departments. Contingency plans should be promptly activated and effective measures taken in the event of sudden floods or extreme weather conditions. Regular flood control emergency drills should be conducted to enhance the emergency response capabilities of power station staff. These drills help test the effectiveness and operability of contingency plans and identify and improve any issues. Effective communication and cooperation mechanisms should be established with local governments, flood control departments, and communities to promptly share flood control information, cooperate in flood control efforts, and jointly address potential flood risks during the flood season.

4 Conclusion

China is rich in wind and photovoltaic resources, with great potential for new energy development. Through continuous efforts and accumulation, China has achieved global leadership in many new energy technologies and equipment manufacturing levels, and establish the world's largest clean power supply system. With technological progress, renewable energy construction will gradually increase in the coming decades. However, due to the volatility of renewable energy, it cannot maintain or regulate continuous power supply, which requires a significant increase in energy storage to balance the variability of solar and wind power.

The construction and operation of pumped storage power stations face many technical challenges, including those mentioned above, as well as issues related to construction diversion, directional blasting, equipment assembly, structural seismic resistance, and earthworks. The key to effectively solving these problems lies in scientific planning, the application of advanced technologies, strict quality control, the enhancement of environmental protection measures, and the formulation of contingency plans. This will ensure that the project proceeds with high quality, safety, and environmental protection. In the future, with the development of intelligent and green construction technologies, pumped storage power station construction will become more efficient and sustainable, making important contributions to the optimization of the energy structure and the protection of the ecological environment.

References

1. Vennemann, P., et al.: Pumped storage plants – status and perspectives. VGB PowerTech **91**(4), 32–38 (2011)
2. Yan, B., Wang, Y.: Design and research on cement concrete pavement structure of roads in pumped storage power station. Hydropower Pumped Storage **5**(27), 93–97+101 (2019)
3. Li, X.D., Su, S., Huang, T.J.: Health damage assessment model for construction dust. J. Tsinghua Univ. (Sci. Technol.) **55**(1), 50–55 (2015)
4. Huang, S.Q.: Overview of methods for detecting dam seepage in reservoirs. Dam Saf. **2**, 42–50 (2021)
5. Guoping, D., Jiansheng, C., Liang, C.: Detection of reservoir seepage paths using isotope tracing technology. Jiangsu J. Agric. Sci. **1**, 57–60 (1998)
6. Liqing, Z.: Environmental protection design for Fujian Xianyou pumped storage power station. Hydropower Stn. Des. **32**(1), 57–62 (2016)
7. Xiaoxiao, M., et al.: Environmental and ecological issues and treatment measures during the construction period of pumped storage power stations. Water Conserv. Hydropower Express, 1–6 (2024)
8. Hongqi, M., Chuan, Z.: Research on the basic theory and key technologies of gravel-clay core wall rockfill dam of Nuozhadu Hydropower Station. J. Hydroelectr. Eng. **32**(2), 208–212 (2013)
9. Tianjian, H., Xiaodong, L., Shu, S., et al.: Monitoring and analysis of dust pollution during the earthwork construction phase of building engineering. J. Saf. Environ. **14**(3), 317–320 (2014)
10. Gang, T., et al.: Study on the spatial diffusion law of construction dust. Environ. Sci **1**, 259–262 (2008)

Research on Simulation and Prediction of Photovoltaic Power Generation Based on Radiation Models and Machine Learning Method

Jie Gao[1(✉)], Xu Wang[1], Jianwei Gu[1], Siwei Tang[2], Fangliang Zhu[1], Jingyi Li[3], and Yiming Zhu[4]

[1] China Renewable Energy Engineering Institute, Beijing, China
690676797@qq.com

[2] Power China Guiyang Engineering Corporation Limited, Guizhou, China

[3] Power China Northwest Engineering Corporation Limited, Xi'an, China

[4] Power China Beijing Engineering Corporation Limited, Beijing, China

Abstract. Focus on the Carbon Peaking and Carbon Neutrality Goals, new energy such as solar and wind power generation developed rapidly. In 2023, the installation of solar energy in China exceeded 0.6 Terawatt, accounting for over 20% of the total installed electricity capacity, surpassing hydropower for the first time, becoming the second largest power supply in China. Annual photovoltaic (PV) power generation achieved nearly 583.3 TWh, gradually towards the main power supply. An accurate simulation and prediction of PV power generation is of great significance for the safe and economical operation of the new power systems. In this paper, on 15-min measured irradiance and power generation data of PV plants within one year and the reanalysis meteorological hourly data of ERA5 derived from ECMWF (European Centre for Medium-Range Weather Forecasts), Firstly, we discover the characteristics of PV power generation by analyzing the daily insolation hours and hourly mean power output. Then physical mechanism method is used through radiation model, inclined plane radiation correction model and photoelectric conversion model. PV power output is simulated based on grid-type reanalysis meteorological data. Finally, according to the deviation sequence of simulated and measured power output, a machine learning method extreme gradient boosting (XGBoost) is introduced. After dividing the deviation time series into training set and test set, the training set is applied to learn the patterns to correct the test set. And the test set is fed back to modify the prediction. The results show that by using machine learning method, the determination coefficient (R-squared) of hourly PV power output of a certain station for medium and long-term could reach 0.9, which contribute to improve the accuracy of PV power output generation prediction effectively.

Keywords: photovoltaic power generation · machine learning · XGBoost

This work was supported by the National Natural Science Foundation of China (Grant No. U2243232) and Power Construction Corporation of China, Ltd Technology Project (DJ-HXGG-2022-01, DJ-ZDXM-2022-10, DJ-ZDXM-2021-26).

S. Zheng et al. (Eds.): IHDC 2024, LNCE 487, pp. 317–326, 2025.
https://doi.org/10.1007/978-981-97-9184-2_28

1 Introduction

China is committed to achieving the dual carbon goal (Wang et al. 2023), to realize carbon neutrality by 2030 and carbon peak by 2060. As the dominant clean and low-carbon energy, solar and wind resources are important elements to build the new power system and achieve carbon reduction. By the end of 2023, the installed capacity of new energy has exceeded 1TW (Terawatt) in China, accounting for 35% of the total installed power capacity. New energy generation is 1470TWh (Tera watt hour), accounting for more than 15% of the total power generation in China, achieving a high penetration rate of new energy in the power system. Among them, the installed capacity of photovoltaic (PV) power generation exceeds 0.6TW, and the annual power generation is 583TWh, which is vital to the development of new energy. PV power generation has the characteristics of randomness, volatility and intermittences, which pose a serious challenge on the safe and stable operation of the electric power system, and causes the "duck curve", which directly affects the price of PV power generation in electricity market. Therefore, the research on PV power simulation and prediction is of great practical significance.

PV power simulation and prediction in the time scale includes ultra-short-term prediction (0~4 h), short-term prediction (4 h~72 h), medium and long-term prediction (week-month-year), from the spatial scale involves point prediction and regional prediction. It can be calculated by physical mechanism method and statistical method which involves time method, machine learning method and etc. Moreover, the accuracy is evaluated by means of average absolute error, average absolute error percentage, root-mean-square error, root-mean-square error, correlation coefficient, determination coefficient (R-squared) and other indicators (Lai et al. 2019).

In this paper, from the physical mechanism of PV power generation, irradiance is converted to power through radiation model, inclined plane radiation correction model, and photoelectric conversion model, using grid-type reanalysis meteorological data input. In order to improve prediction performance, deviation sequence between the simulated and measured power output is divided into training set and test set. The machine learning method extreme gradient boosting (XGBoost) is used in the training set to correct the deviation for the test set which helps to improve the prediction accuracy. The hourly predicted PV power generation of a certain station for a medium-long term can reach R-squared of 0.9. The effect performs well.

2 Models

In our study, annual hourly PV power output is simulated by physical mechanism models, and then the hourly PV power output in the test set is corrected by machine learning model.

2.1 Physical Mechanism Method

2.1.1 Radiation Model

The energy of PV power generation originally comes from solar radiation outside the atmosphere, which reflects the solar constant vary with hours, days and seasons. PV power generation is closely related to radiation directly from the sun to the earth, involving the calculation of solar declination angle and solar hour angle (Eq. (1)). The solar hour angle also needs to be adjusted considering time difference (Li et al. 2019).

$$I_0 = \gamma E_{sc}(\sin\varphi\sin\delta + \cos\delta\cos\varphi\cos\omega) \quad (1)$$

where,

I_0	solar radiation at the top of the atmosphere [W/m^2]
E_{sc}	the solar constant, equals 1367 W/m^2
γ	correction coefficient of Earth-Sun distance, $\gamma = 1 + 0.033\cos[(360 \times (284 + n)/365]$
n	number of the day in the year
t	hour of the day
φ	latitude [rad]
λ	longitude [°]
ω	solar hour angle [rad]

where, $\omega = [(\lambda\text{-}120)/15 + (0.258\cos\theta - 7.416\sin\theta - 3.648\cos(2\theta) - 9.228\sin(2\theta)) + (t - 12)] \times (\pi/180)$

θ	day angle [rad]

where, $\theta = 2\pi(n\text{-}1)/365.242$

δ	solar declination angle [rad]

where, $\delta = 23.45\sin[(360 \times (284 + n)/365 \times (\pi/180)] \times (\pi/180)$

2.1.2 Inclined Plane Radiation Correction Model

The key factor of PV power generation is radiation, which is not only depended on the total radiation on the horizontal plane, but also directly related to the radiation on PV panels which is sloping (Li et al. 2022). The total radiation of inclined plane includes direct radiation received by the inclined plane, scattered radiation and reflected radiation (Eq. (2)~Eq. (5)).

$$I_T = I_{T,b} + I_{T,d} + I_{T,g} \quad (2)$$

$$I_{T,b} = I_b R_b \quad (3)$$

$$I_{T,d} = I_d R_d \quad (4)$$

$$I_{T,g} = I\rho[(1 - \cos\beta)/2)] \quad (5)$$

where,

I_T	total radiation of the inclined plane [W/m^2]
$I_{T,b}$	direct radiation [W/m^2]
I_b	direct radiation received by horizontal plane [W/m^2]
R_b	correction factor of direct radiation from horizontal plane to inclined plane

where, $R_b = (\sin\delta \sin\varphi \cos\beta - \sin\delta \cos\varphi \sin\beta \cos\gamma + \cos\delta \cos\varphi \cos\beta \cos\omega + \cos\delta \sin\varphi \sin\beta \cos\gamma \cos\omega + \cos\delta \sin\beta \sin\gamma \sin\omega)/(\sin\varphi \sin\delta + \cos\varphi \cos\delta \cos\omega)$

γ	PV panel azimuth, south is 0
β	inclined angle of PV panel, can using best installation angle of PV panel [rad]
$I_{T,d}$	scattered radiation [W/m^2]
I_d	scattered radiation received by horizontal plane [W/m^2]
R_d	correction factor of scattered radiation from horizontal plane to inclined plane
I	total radiation received by horizontal plane [W/m^2]
A	sky anisotropy index

where, $A = I_b/I_0$

$I_{T,g}$	reflected radiation from the ground [W/m^2]

where, $I_{T,g} = I\rho[(1\text{-}\cos\beta)/2)]$

ρ	surface albedo, preferably 0.2~0.25

2.1.3 Photoelectric Conversion Model

Photo-electric conversion calculates the power generation (Eq. (6)~Eq. (7)) from total irradiance received on the inclined PV panel by considering the influence of inverter efficiency, photoelectric conversion efficiency and other factors (Sarah et al. 2020).

$$P = I_T \times \eta_s \times [1 - \alpha(T_c - 25)] \times S \times K_1 \times K_2 \times K_3 \times K_4/1000 \quad (6)$$

$$T_c = T_a + [(NOCT - 20)/800] \times I_T \quad (7)$$

where,

P	power output [kW]
η_s	photoelectric conversion efficiency under standard test conditions, crystal silicon cells take 12%~18%
α	temperature coefficient, crystalline silicon material takes 0.003–0.005°C^{-1}
T_c	PV array panel temperature [°C]

(continued)

(*continued*)

T_a	air temperature [°C]
NOCT	rated solar cell operating temperature, crystalline silicon cell [47°C]
S	effective area of PV panels [m^2]
K_1	aging loss coefficient of PV array

where, $K_1 = 1 - k \times y_a$

k	number of years that grid-connected PV station went into operation
y_a	annual decay rate of solar module, crystalline silicon cells take 1%
K_2	loss coefficient of PV array mismatch, take 0.95~0.98
K_3	loss coefficient of dust hidden, 0.9~0.95
K_4	line loss coefficient of DC circuit, 0.95~0.98

2.2 Machine Learning Method

Extreme gradient boosting (XGBoost), an integrated learning algorithm based on decision tree, is widely used in supervised machine learning methods (Zhang et al. 2023). By iterating the CART decision tree several times (Eq. (8)), XGBoost fits and trains the deviation between the simulated and the actual values, in order to build a classifier with better classification performance for prediction in the test set.

$$y_i' = \sum_{i=1}^{K} f_k(x_i), f_k \in F \tag{8}$$

where,

f_k the kth decision tree

K total number of decision trees

x_i eigenvector of the sample i

y_i' prediction result of sample i

F set of all decision trees

XGBoost iterates decision tree by minimizing the objective function built from the residual loss function and model complexity (Eq. (9)).

$$Obj = \sum_{i=1}^{m} l(y_i, y_i') + \sum_{k=1}^{K} \Omega(f_k) \tag{9}$$

where,

$l(y_i, y_{i'})$	loss function of residual between predicted result $y_{i'}$ and the actual value y_i
m	number of samples
Ω	regular term, which indicates the complexity of the model

3 Data and Schedule

3.1 Data Introduction

- A PV power station in Hebei Province, China is selected for study, of which the installed capacity is 14 MW, and the measured PV power generation per 15 min (unit: W/m^2) are collected for the whole year of 2021.
- On reanalysis meteorological data, hour-step PV power output simulation is carried out through the radiation model, inclined plane radiation correction model and photoelectric conversion model. Meteorological input mainly uses ERA5 reanalysis data. ERA5 is a multi-factor meteorological data product published by ECMWF (European Centre for Medium-Range Weather Forecasts). The data is an update version of ERA-Interim, with a horizontal resolution of 31 km and time steps accurate to hour, covering historical data from 1950 to the present. Currently, ERA5 data products are available for sharing and open access.

3.2 Data Spatiotemporal Matching and Processing

- Space matching: hourly skin temperature, surface solar radiation downwards and total sky direct solar radiation at surface (unit: J/m^2) of reanalysis dataset are downloaded in the grid corresponding to the position of PV power plant during the year of 2021.
- Temporal matching: according to the reanalysis data with an interval of hour, the measured PV power output per 15-minute is calculated hourly of the whole year.
- Delete total 24 h data of the missing day.

3.3 PV Power Output Simulation

Using the total surface solar radiation (ssrd: I) and direct surface solar radiation (fdir: I_b) of the reanalysis dataset as input, hourly PV power output are computed by radiation model, inclined plane radiation correction model and photoelectric conversion model.

3.4 PV Power Output Prediction

Taking the deviation time series between the simulated and measured PV power output as input, considering the seasonal features of PV power generation, the training set and test set are divided following the proportion of 1:1 approximately, abstracting from summer, spring/autumn, and winter respectively. Conducting XGBoost in the training set and predicting for the test set, the feasibility of the method above for medium even long term prediction is analyzed.

4 Results

- There are 21 days with miss data, and all the data in those 21 days are deleted.

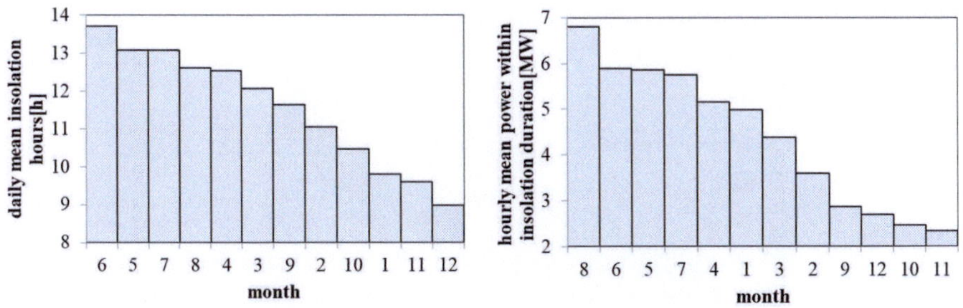

(a) mean daily insolation hours (b) mean hourly power output

Fig. 1. The statistics of insolation hours and hourly mean power output for 12 months

- For the remaining 8256 data throughout the year, the mean daily insolation hours and hourly power output of the 12 months are calculated and sorted (Fig. 1). It reveals that the insolation hours and hourly power output in summer (May to August) are at a higher level throughout the year, followed by spring (March to April), and in winter (November-December) they are at a lower level. Later, we shall divide the training set and test set on this basis.

- The remaining 8256 data are divided into the training set and the test set according to a proportion of about 1:1. Concerned with the seasonal characteristics, nearly 50% of the data in summer (May-August), spring and autumn (March-April, September-October), and winter (January-February, November-December) are located in the training set and the test set, respectively.
- The annual reanalysis meteorological data, radiation and correction, photoelectric conversion models are used to calculate the PV power generation hour by hour (Fig. 2) for the training set of 4296 data. Some parameters should be paid attention to keep within a certain range. For example, R_b should be controlled between 0 and 2.
- In order to improve the prediction effect of the test set, the deviation sequence between simulation and measurement of the training set is calculated to train the XGBoost model (Fig. 3).

- For the test set of 3960 data, after using radiation correction and photoelectric conversion models, the simulation is as follows with the R-squared equals 0.8338 (Fig. 4).

- Based on machine learning on the deviation of the training set, the results of the test set are analyzed as follows.

– After machine learning the deviation of the training set with 4296 hours the deviation of the test set of 3960 hours is predicted with season, date and hour as the main variables, then the original simulation based on physical models is corrected. The fitting effect is significantly improved, with the R-squared increasing from 0.8338 to 0.8823 (Fig. 5(a)).

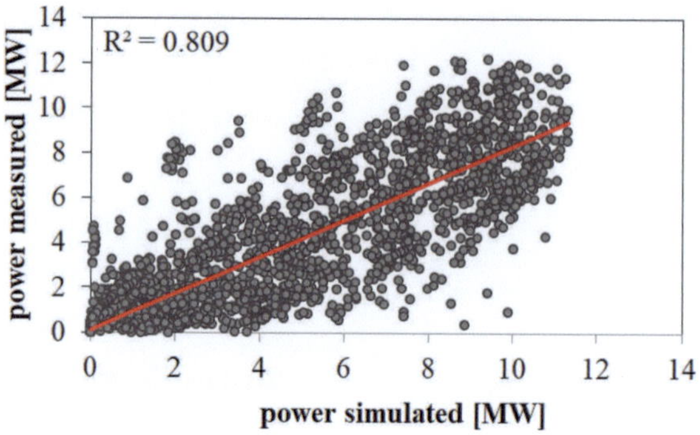

Fig. 2. Comparisons between PV power output simulated and measured

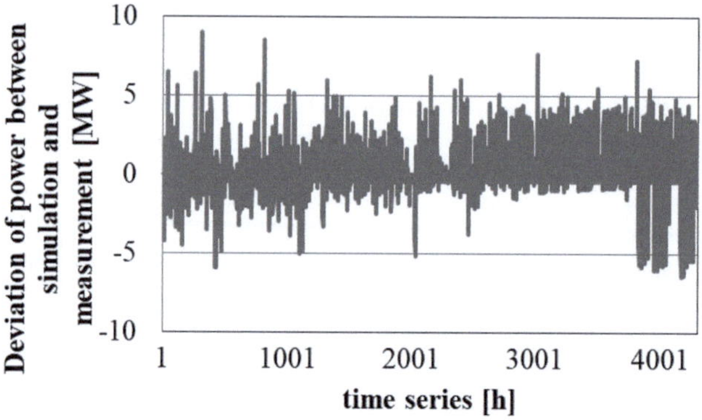

Fig. 3. Deviation of PV power output between simulation and measurement

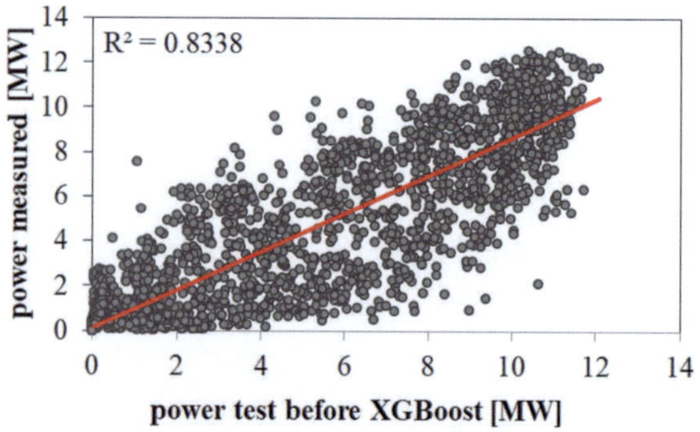

Fig. 4. Power simulation before XGBoost of the test set

– Insight into the results in different seasons, the correspondence between sunshine duration and power generation in summer and winter is obvious, and the regularity for machine learning is also relatively significant, and the R-squared is close to 0.9 (Fig. 5(b), 5(d)), with a good effect. However, due to the influence of more factors in spring and autumn (Fig. 5(c)), further more is needed to improve the forecasting result.

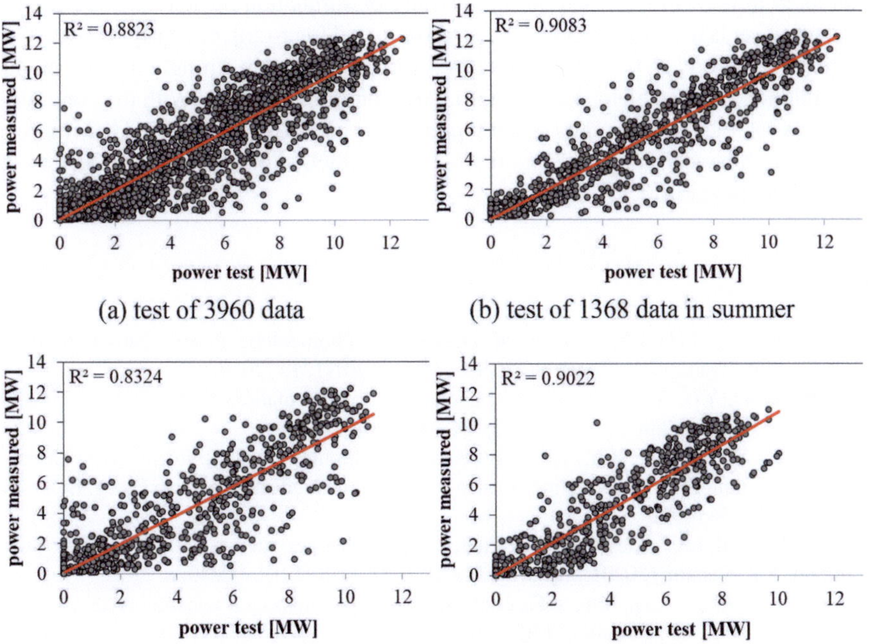

(a) test of 3960 data

(b) test of 1368 data in summer

(c) test of 1320 data in spring and autumn

(d) test of 1272 data in winter

Fig. 5. Power prediction by XGBoost of the test set

5 Conclusions

- Through physical mechanism method, a PV power simulation method with solar radiation, inclined plane radiation correction, and photo-electric conversion models based on reanalysis meteorological data is established.

The results show that, with the input of radiation data derived from ERA5 grid, modified by inclination correction of PV panel, converted considering temperature of reanalysis data and power generation loss, a long-term simulation of PV power generation can be basically realized hour by hour.

A PV power station (10~20 MW) perhaps covering an area of 0.1~0.2 km^2, especially for that with complex terrain and variable local climate, relying on the radiation and temperature data of a reanalysis grid with an area of 900 km^2, is very difficult to guarantee the accuracy. Therefore, the R-squared of annual hour-step statistics is only about 0.7~0.8.

- Based on the machine learning of deviation sequence between simulation and measurement, a long-term PV power generation prediction model is constructed.

In order to improve the results, the XGBoost method of supervised machine learning is adopted in this study. Based on the learning of simulation deviation in the training set, the test set is used to predict. For the medium and long-term of nearly 3 months, the R-squared could reach 0.85 level. However, machine learning strongly depend on data accumulation, is affected by series length, data conditions and etc. With the continuous accumulation of reanalysis data, contemporaneous local observation, and the improvement of machine learning model simulation and prediction of PV power generation can achieve a higher accuracy.

References

1. Lai, C.W., Li, J.H., Chen, B., et al.: Review of Photovoltaic Power Output Prediction Technology. Trans. China Electrotech. Soc. **34**(6), 1201-1217 (2019)
2. Li, W., Zhao, Y.C.: The improvement in solar position calculations in the ellipsoid model of the earth. J. Univ. Chin. Aca. Sci. **36**(3), 363-375 (2019)
3. Li, Y., Li, Z.R., Wang, X.Y., et al.: Research on optimum tilt angle for pv array based on combination models of calculating solar radiation on inclined surface. Acta Energ. Solar. Sin. **43**(5), 127–136 (2022)
4. Sarah, F., Raúl, R.C., Alessandro, D., et al.: Climate change extremes and photovoltaic power output. Nat. Sustain. (2020). https://doi.org/10.1038/s41893-020-00643-w
5. Wang, J.X., Chen, L.D., Tan, Z.F., et al.: Inherent spatiotemporal uncertainty of renewable power in China. Nat. Commun. **14**(5379), 1–11 (2023)
6. Zhang, X.Y., Su, Q.M., Zhao, C.S., et al.: A landslide susceptibility evaluation method using Bayesian algorithm to optimize XGBoost. Sci. Survey. Map. 48(6):140–150 (2023)

Strategies for the Integration of Energy Systems, Enhancing Efficiency and Sustainability

Role of the Hydro-Solar Hybrid Operation Mode in the Novel Power System

Haibo Du(✉), Hongyong Li, Kai Liu, Ruixian Chen, and Ying Cao

Manwan Hydropower Plant of Huaneng Lancang River Hydropower INC., Lincang 675800, People's Republic of China
654235711@qq.com

Abstract. Currently, photovoltaic technology is rapidly advancing, and competition in the industry is becoming increasingly intense. The decreasing prices of silicon wafers and modules have led to a growing demand for photovoltaic installations. The installed capacity and installed share of photovoltaics in power systems are increasing, promoting the transformation of traditional power systems to novel power systems; this maximizes the consumption of new energy as the main task and is supported by source-network-load-storage interactions and multi-energy complementation. Using the Manwan hydro-solar hybrid base as a model, the role of hydro-solar hybrids in source-network-load-storage interactions and multi-energy complementation in novel power systems are discussed.

Keywords: Hydro-solar hybrid power · Reservoir energy storage · Manwan hydropower plant · Photovoltaic

1 Photovoltaic Industry Ushered in the Third Major Technological Change

In 1955, Siemens AG successfully developed the technology of using H_2 to reduce high-purity $SiHCl_3$ and applied it to industrial production in 1957, that is, the original Siemens method. The hydrogenation method used was the thermochemical hydrogen production method, which has a low conversion rate, high comprehensive power consumption, high production costs, and serious byproduct emission pollution.

In 2004, the patent protection restrictions of the cold hydrogenation method expired, and the development of the cold hydrogenation process was rapid, significantly reducing the electrical energy loss in the hydrogenation process and saving more than 70% of energy consumption. At the same time, European countries such as Germany and Spain took the lead in implementing electricity price subsidies to stimulate demand for photovoltaic installations, ushering in the first technological revolution in the photovoltaic industry.

In 2015, the promotion of the PERC process significantly improved the conversion efficiency of monocrystalline silicon, driving a significant decrease in silicon wafer costs. Coupled with the release of domestic photovoltaic installation demand through

S. Zheng et al. (Eds.): IHDC 2024, LNCE 487, pp. 329–342, 2025.
https://doi.org/10.1007/978-981-97-9184-2_29

the introduction of electricity price subsidies in China in 2013, the photovoltaic industry ushered in a second technological transformation.

After 2019, with the end of policy intervention, the photovoltaic industry entered an era of parity and gradually evolved into a growth industry, with demand growth triggered by decreasing costs and increasing benefits. The efficiency of the current PERC process is close to the limit, while the efficiency of TOPCon, HJT, IBC, calcium titanium ore and other photovoltaic cell technologies is increasing, thus causing a third technological change in the photovoltaic industry.

2 Technological Advances Spurred Growth in Demand for Photovoltaic Installation

In the absence of national policies and electricity price subsidies, the development of the photovoltaic industry is pursuing a continuous reduction in the levelized cost of energy, which is influenced by two key factors: the construction cost and power generation of photovoltaic power plants. Technological progress can, on the one hand, drive the prices of photovoltaic modules by reducing production costs and, on the other hand, increase power generation by improving the conversion efficiency of photovoltaic modules.

The price of photovoltaic modules continues to fall due to decreasing costs and increasing benefits caused by the third technological change in the photovoltaic industry. According to the recent execution of the photovoltaic module framework agreement by the China Huaneng Group Co., Ltd., the price of the P-type module was reduced to 0.83 yuan/W, and the price of the N-type double-sided module was reduced to 0.85 yuan/W. In addition, according to industry estimates, by the end of 2024, silicon production capacity will exceed 4.2 million tons, which will trigger a decrease in polysilicon prices; this will create a further decrease in the price of photovoltaic modules. According to the proportion of 3.5 GW modules corresponding to 10,000 tons of silicon material, the photovoltaic module production capacity will exceed 1400 GW by the end of 2024. Globally, the number of new photovoltaic installations is expected to reach 460 GW by 2024. According to the ratio of installed capacity to rated capacity of 1.25, the demand for photovoltaic modules is approximately 575 GW; thus, there will be severe overcapacity of the modules. With respect to the overcapacity and falling prices in the industrial chain, the price war of photovoltaic enterprises will enter a white-hot state, and the price of photovoltaic modules will decrease further, promoting the growth of downstream photovoltaic installed demand.

3 Changes in China's Electric Energy Structure

According to the national power industry statistics published by China's National Energy Administration, China's cumulative installed power generation capacity at the end of 2023 was approximately 29.1965 million kilowatts, representing a year-on-year growth rate of 13.9%. The installed capacity, share and year-on-year growth rate of each type of power source are provided in Table 1.

From the statistics in 2023, for the first time, the proportion of clean energy installed capacity, such as hydropower, wind power and photovoltaic power, exceeded the thermal

Table 1 China's installed capacity, share, and growth rate in 2023 by energy source type (as of 2023)

Type of energy source	Capacity/Million kilowatts	Share	Growth rate in 2023
Thermal Power	139032	47.62%	4.1%
Hydro Power	42154	14.44%	1.8%
Wind Power	44134	14.12%	20.7%
Photovoltaic Power	60949	20.87%	55.2%
Nuclear Power	5691	1.95%	2.4%

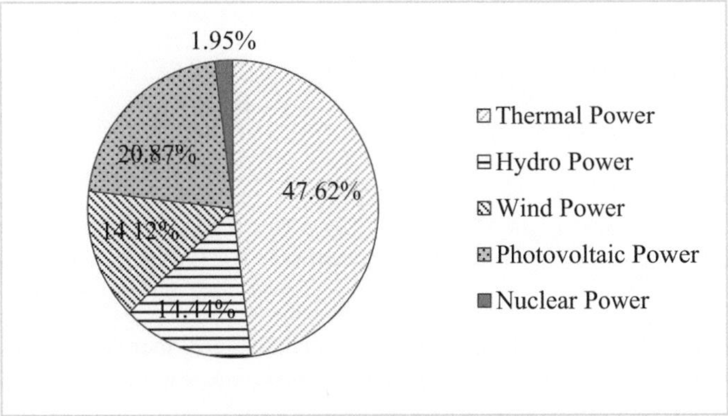

Fig. 1. Share of China's installed capacity in 2023 by energy source type

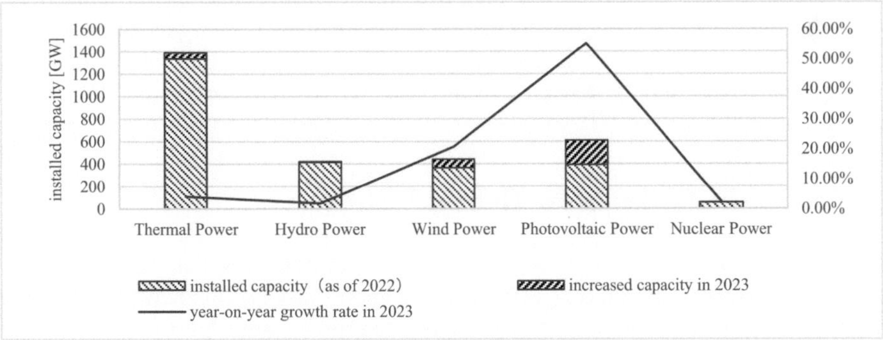

Fig. 2. Installed capacity and growth rate of China in 2023 by energy source type

power installed capacity, and the installed capacity of the new energy showed a rapid growth trend; in particular, the growth rate of photovoltaic installed capacity reached 55.2%. With the rapid growth of new energy installations, the characteristics of large installed capacity, small power generation, and power generation is greatly affected by external factors, and this aspect has become increasingly apparent. As shown in Figs. 1 and 3, the installed photovoltaic capacity in 2023 is 20.87%, and its power generation accounts for only 8.4%. Moreover, the thermal power capacity accounts for 47.62%, and its share of power generation is 63.2%, which shows that thermal power still carries a large amount of the base load (Fig. 2).

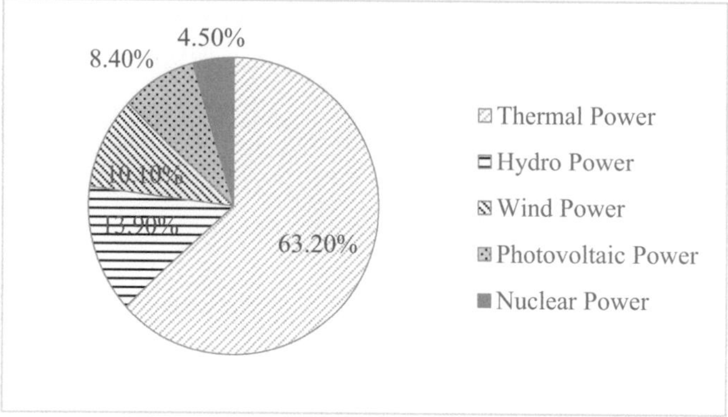

Fig. 3. Share of China's electricity generation in 2023 by energy source type

The installed capacity of the power system is constantly increasing, and the proportion of new energy installed is also increasing, which causes higher requirements for the transmission and regulation capacity of the power grid. In terms of improving the transmission capacity, the utilization rate of the transmission channel should be improved while increasing the transmission channel of the grid. In terms of improving the regulation capacity, the role of energy storage will be further reflected, and its demand will further increase. Due to the large fluctuation of photovoltaic power generation with the intensity of sunlight and to turn the fluctuating photovoltaic output into a stable and high-quality power supply delivered to the grid, the demand for energy storage will increase further.

4 Role of the Hydro-Solar Hybrid System

In the absence of significant breakthroughs in lithium iron phosphate battery technology, which is mainly reliant on electrochemical energy storage, the new power system requires other forms of energy storage. Some enterprises have built pumped storage power stations to convert new energy power into reservoir storage. Some enterprises, such as the Longyangxia and Laxiwa hydropower plants in Qinghai, the Dongqing and Guangzhao hydropower plants in Guizhou, and the Manwan hydropower plant in Yunnan, have adopted the hydro-solar hybrid operating mode; in this hybrid mode, the new energy plant is connected to the hydropower plant, and then this system is connected to the power grid through the transmission channel of the hydropower plant. With the help of the storage capacity of hydropower station reservoirs and the rapid adjustment capacity of hydropower units, this mode combines photovoltaic and hydroelectric power generation by bundling them together to improve the utilization of transmission channels. This section uses the Manwan hydro-solar hybrid base as a model to discuss the role of hydro-solar hybrids in novel power systems.

4.1 Basic Information of the Manwan Hydro-Solar Hybrid Base

The Manwan Hydropower Plant has seven hydropower units (No. #1, #2, #3, #4, #5, #6, and #8) with a total installed capacity of 1,670 MW and has two 500 kV and three 220 kV power transmission channels; these are responsible for the base-load, frequency modulation, peak load regulation and accidental backup tasks in the system. Since 2023, the Paling photovoltaic power station has been connected to the 35 kV side of the Manwan #7 contact transformer (750 MW), and the Azhutian photovoltaic booster station has been connected to the 220 kV side of the Manwan #7 contact transformer. The power they generate is boosted and fed into the main 500 kV Yunnan power grid through the #7 contact transformer. As of December 31, 2023, a total of 503 MW of photovoltaic installed capacity has been connected, and the wiring diagram is shown in Fig. 4.

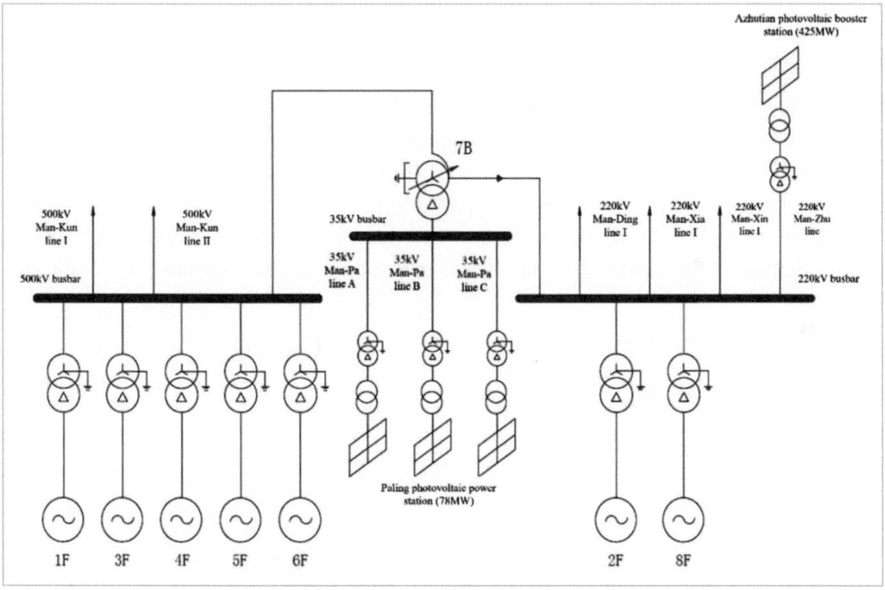

Fig. 4. Wiring diagram of the Manwan hydro-solar hybrid base

4.2 Model of the Manwan Hydro-Solar Hybrid Base

4.2.1 Symbol Specification in the Model

The symbol specifications in the model are shown in Table 2.

Table 2. Symbol specifications

Symbol	Specification	Unit
$P_{\sum}$	Real-time power transmitted from Manwan hydro-solar hybrid base (500 kV side)	MW
P_a	Real-time power of Azhutian photovoltaic booster station	MW
P_P	Real-time power of Paling photovoltaic power station	MW
P_f	Real-time power of 1F and 3F-6F	MW
P_2	Real-time power of 2F	MW
P_8	Real-time power of 8F	MW
P_7	Power penetration of 7B	MW
P_m	Sum of real-time power of three 220 kV transmission channels	MW
a	Power generation water consumption rate of 2F	m^3/MW
b	Power generation water consumption rate of 8F	m^3/MW
H	Total water consumption of 2F and 8F	m^3

4.2.2 Basic Mathematical Model

The total installed capacity of the Manwan Hydropower Plant is 1670 MW; the transmission capacities of 500 kV Man-Kun lines I and II are both 1670 MW; the rated power of 2F is 250 MW; the rated power of 8F is 120 MW; the rated power of 7B is 750 MW; the rated power of the Paling photovoltaic power station is 78 MW; and the rated output power of the Azhutian Booster Station is 425 MW. Based on the above data and the wiring situation in Fig. 4, the active power calculation formula can be obtained as follows:

$$\begin{cases} P_{\sum} = P_f + P_2 + P_8 + P_m + P_a + P_d \\ 0 \le P_{\sum} \le 1670 \\ 0 \le P_a \le 425 \\ 0 \le P_p \le 78 \\ 0 \le P_2 \le 250 \\ 0 \le P_8 \le 120 \end{cases} \quad (1)$$

According to the actual situation of the Manwan Hydropower Plant, due to the system's breakpoint operation mode needs, the 220 kV Man-Ding line I, the Man-Xia line I and the opposite side form a mutually hot standby state. Coupled with the fact that the 220 kV Man-Xin line I is in a state of tidal inflow during the peak period of new energy generation, 7B becomes the key juncture for the Manwan hydro-solar hybrid base, as follows:

$$\begin{aligned} & P_7 = P_2 + P_8 + P_m + P_a + P_b \\ & 0 \le P_7 \le 750 \end{aligned} \quad (2)$$

With the objective of increasing the reservoir storage efficiency and reducing the water consumption rate for power generation, the following objective function can be obtained:

$$\min H = aP_2 + bP_8 \tag{3}$$

The above mathematical model is based on the relationship between the active power as a constraint, with the goal of reducing water consumption for power generation; in reality, the constraints are more complex, and a variety of goals need to be achieved. For example, P is the real-time load issued by the grid dispatch, and the hydropower unit also has a minimum forced output, a vibration zone, no-load operation regulation, etc. Mathematical modeling can also be carried out with other constraints and objectives. The actual operating conditions of the hydro-solar hybrid base are the result of balancing after the superposition of a variety of models; thus the role of the hydro-solar hybrid system also needs to be summarized from the actual operating conditions of the accumulated operational data.

4.3 Role of the Hydro-Solar Hybrid Base in Actual Operation

4.3.1 Staggered Generation Boosts Peak Power Capacity

At the hydro-solar hybrid base, during the daytime, the photovoltaic output increases, the hydropower output decreases (under normal conditions), and the reservoir stores water; in the morning and evening peak hours, the hydropower output increases, creating a reasonable interleaving of photovoltaic output and hydropower output in the system demand.

The daily load curves are compared for April 4, 2022 (before photovoltaic access) and December 21, 2023 (after photovoltaic access), which have similar daily generation capacities, and the results are shown in Fig. 5.

In 2023, the total annual power generation of photovoltaics was 276 million kWh; thus, photovoltaic access will increase the average daily energy storage for hydropower by 3,637,200 m^3 (calculated by the average annual water consumption rate of the Manwan hydropower unit in 2023 of 4.81 m^3/kWh). The hydro-solar hybrid system improves the peak capacity of the hydropower units in the morning and evening peak hours through staggered power generation and accordingly increases the peak power of hydropower; this process improves the gain of peak regulation and achieves reasonable transformation of the two kinds of energy in the form of reservoir storage.

Under the hydro-solar hybrid operation mode, 2F runs no-load for a long time to regulate the 220 kV system, and 8F is shut down to stand by for the peak; this process transfers the power generation from 2F and 8F to the 500 kV system of the hydropower units such as 1F, 3-6F, and at the same time gives way to transmission channels for the photovoltaic power generation to ensure that it is able to send close to 750 MW of new energy power through 7B.

After the operation of the hydro-solar hybrid system, the annual power generation and power generation shares of 2F and 8F are reduced. The annual generation capacity of 2F decreases from 1.494 billion kWh (2022) to 1.277 billion kWh (2023), and the annual generation capacity of 8F decreases from 266 million kWh (2022) to 151 million kWh (2023). In the case of the limited margin of the original hydropower transmission channel, the hydro-solar hybrid system utilizes the property that water can be stored in actual operation to attain staggered power generation of the two energy sources.

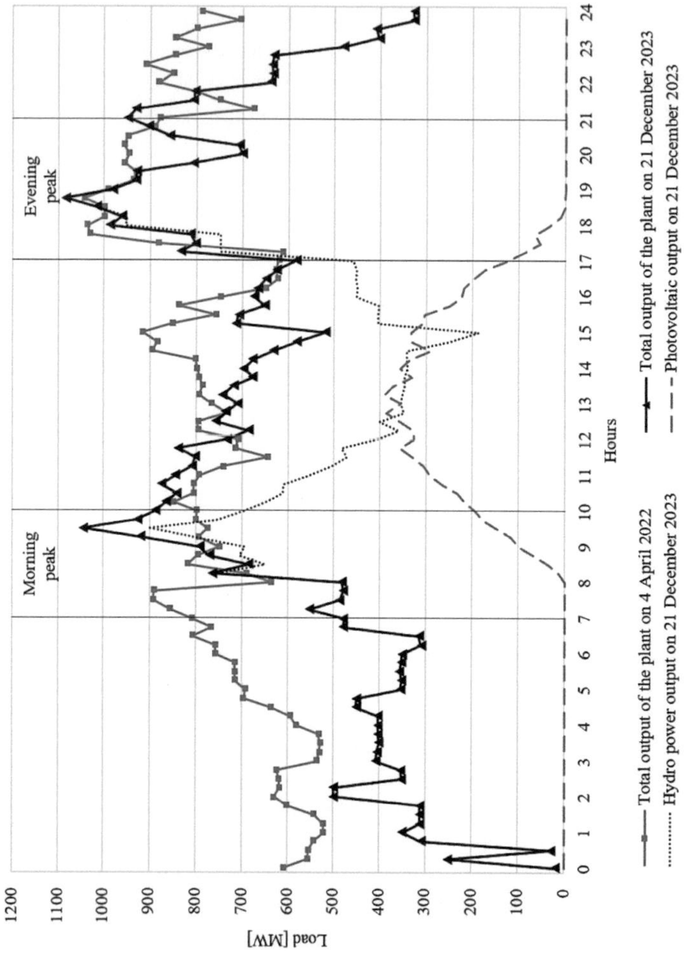

Fig. 5. Comparison between the original daily load curve and the current daily load curve

4.3.2 Effectively Optimizing the Energy Structure of the Power Grid by Utilizing the Changes in Water Inflow Throughout the Year

Hydroelectric power generation is affected by the flood season and dry season. During the flood season, a large amount of water inflow comes from upstream areas, and to avoid water abandonment, a certain amount of hydropower output needs to be ensured. During the dry season, the upstream water inflow is low. While ensuring the minimum forced output of hydropower, the transmission channel can be transferred to photovoltaic power generation to the greatest extent, and the reservoir can be used for storage. Using the maximum photovoltaic output day during the flood season, the maximum annual photovoltaic power generation day, and the maximum photovoltaic output day during the dry season of the Manwan hydro-solar hybrid base as examples, the daily output processes are analysed (Figs. 6, 7).

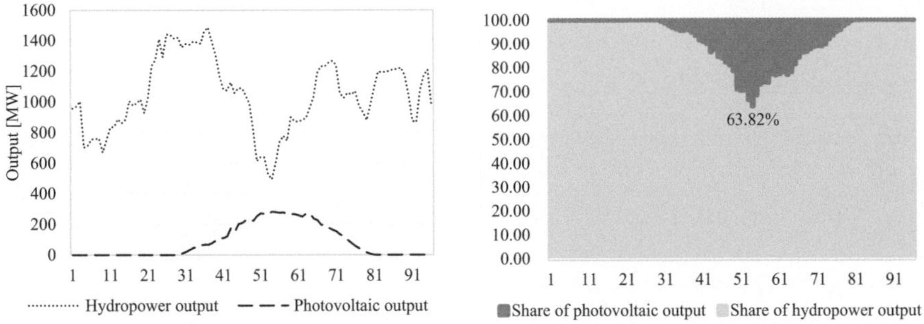

(a) Output curve of the maximum photovoltaic output day in the flood season (August 10, 2023)

(b) Proportion of maximum photovoltaic output day in the flood season (August 10, 2023)

Fig. 6 **(a)** Output curve of the maximum photovoltaic output day in the flood season (August 10, 2023). **(b)** Proportion of maximum photovoltaic output day in the flood season (August 10, 2023)

During the flood season, when the sunlight intensity is sufficient and the capacity of the channel is limited, the hydropower unit reduces the output and provides short-term water storage, prioritizing photovoltaic power generation. The hydropower unit adjusts the output according to the system needs, and the hydropower output is shown to track the photovoltaic output during the day. When the sunlight intensity is insufficient, the hydropower output increases according to the water inflow and system demand, and the storage capacity is fully absorbed (Fig. 8).

During the dry season, the transmission channel is transferred to photovoltaic output, which is not limited. The hydroelectric units have the characteristics of flexible starts and stops, providing peak shaving and frequency regulation capacity for the system and new energy. Hydropower units utilize their flexible start-up and shutdown characteristics to provide peak shaving and frequency regulation capacity for the system and new energy.

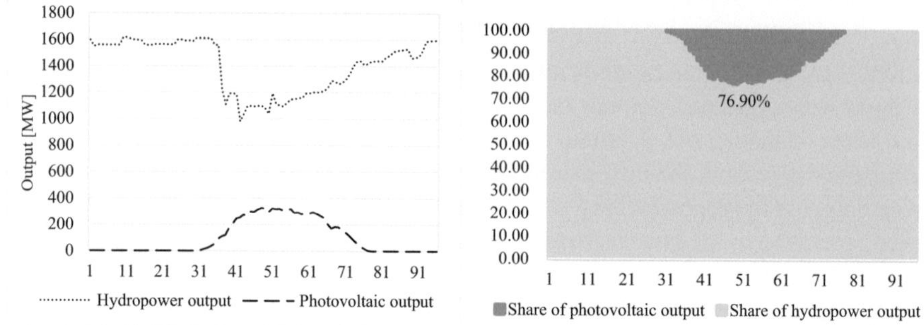

(a) Output curve of the maximum photovoltaic output day throughout the year (September 22, 2023)

(b) Output proportion of the maximum photovoltaic output day throughout the year (September 22, 2023)

Fig. 7 (**a**) Output curve of the maximum photovoltaic output day throughout the year (September 22, 2023) (**b**) Output proportion of the maximum photovoltaic output day throughout the year (September 22, 2023)

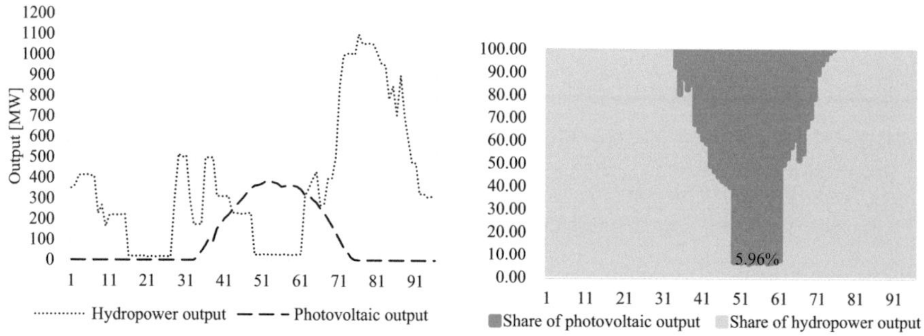

(a) **(b)**

Fig. 8 (**a**) Output curve of the maximum number of photovoltaic output days in the dry season (December 20, 2023). (**b**) Proportion of maximum photovoltaic output days in the dry season (December 20, 2023)

4.3.3 Active Power Regulation Capability of Hydroelectric Units Can Stabilize New Energy Load Fluctuations During the Day

As shown in Figs. 1 and 3, photovoltaic power generation is characterized by a large installed capacity and low power output, and it is easily affected by weather, temperature, solar radiation intensity and other factors. The hydropower regulation capacity in hydro-solar hybrid systems can be fully utilized to stabilize intraday photovoltaic output fluctuations.

As shown in Fig. 9, the output curves of the Azhutian and Paling photovoltaic plants are smooth and have relatively constant peak occurrence times on sunny days, while they have significant fluctuations and irregular peak occurrence times on cloudy and rainy days.

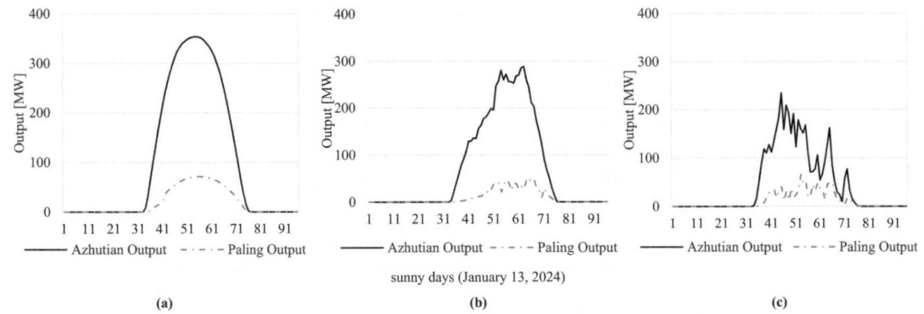

Fig. 9 (**a**) Photovoltaic output curves of Azhutian and Paling on sunny days (January 13, 2024) (**b**) Photovoltaic output curves of Azhutian and Paling on cloudy days (January 17, 2024) (**c**) Photovoltaic output curves of Azhutian and Paling on rainy days (January 16, 2024)

As shown in Fig. 10, the active regulation capability of the Manwan hydropower units can stabilize the intraday output fluctuation of the accessing photovoltaic system. According to the AGC performance test data of the Manwan hydropower units, the active regulation rate is 12.5 MW/s under all AGC inputs; thus, these can fully manage extreme situations, such as an instantaneous decrease in photovoltaic output caused by extreme weather and a portion of the photovoltaic matrix going off-grid due to collector line failure caused by the fast and flexible response capability of the hydropower unit.

4.3.4 Regulating Characteristics of Hydropower Units Strongly Support New Energy Consumption

Different from the traditional power source-to-grid mode, the hydro-solar hybrid operation mode of Manwan is from a photovoltaic power source to a hydroelectric power source. In this mode, hydropower units, as large power sources, rapidly adjust and can provide effective frequency modulation technology support for the operation of the surrounding photovoltaic power generation. At the same time, by rotating the hydroelectric unit as a backup, the response of the synchronous generator increases, providing a regional inertia response capability. Accelerating the construction of a multi-energy hybrid integrated energy base can fully exploit the reservoir storage effect of the existing hydropower stations, increase the efficiency of the stock of regulating resources, and provide a "base effect demonstration" of the use of traditional energy to regulate the consumption of new energy.

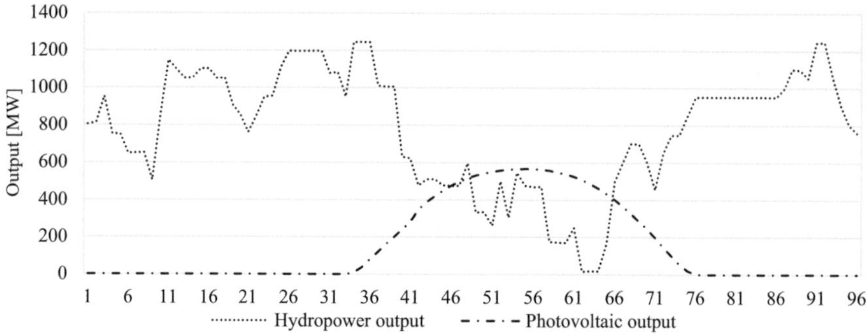

(a) Output curves of the hydropower and photovoltaics on sunny days (January 13, 2024)

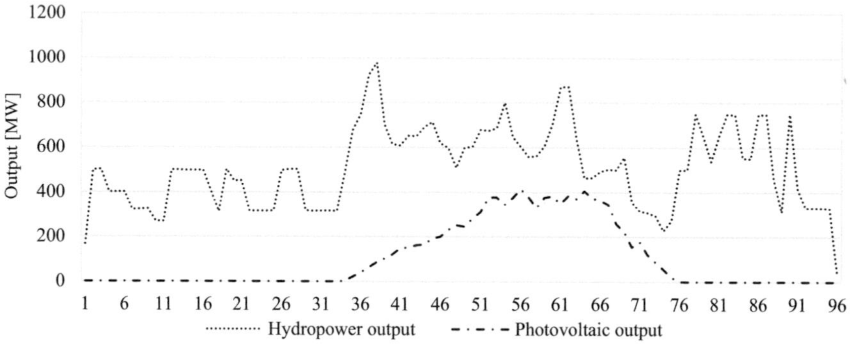

(b) Output curves of the hydropower and photovoltaics on cloudy days (January 17, 2024)

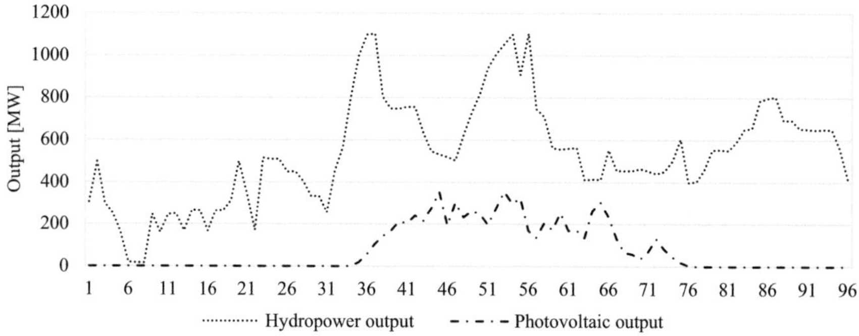

(c) Output curves of the hydropower and photovoltaics on rainy days (January 16, 2024)

Fig. 10 (**a**) Output curves of the hydropower and photovoltaics on sunny days (January 13, 2024) (**b**) Output curves of the hydropower and photovoltaics on cloudy days (January 17, 2024). (**c**) Output curves of the hydropower and photovoltaics on rainy days (January 16, 2024)

5 Conclusion

With respect to the advances in the photovoltaic technology and intensified industry competition, the prices of photovoltaic modules continue to decline, leading to a sustained increase in demand for photovoltaic installation. The power supply structure in our country is also undergoing significant changes. Photovoltaic power plants have the characteristics of "large installed capacity, small power generation," and their access to the power grid also causes greater requirements on the transmission capacity and regulation capacity of the power grid. The hydro-solar hybrid system combines photovoltaic and hydroelectric power generation by bundling them together to improve the utilization of transmission channels. It also makes full use of the seasonal characteristics of the reservoir water inflow to store water to increase the reservoir capacity during the dry season and reasonably consumes the reservoir capacity during the flood season. This system also utilizes the rapidity and flexibility of the hydropower units to stabilize new energy load fluctuations within the day, achieving stable control of active power and on-site balance of reactive power, effectively sharing the regulation pressure of the power grid. At the same time, based on the regulation performance of the hydropower units and the power-to-power features of the hydro-solar hybrid system, the system technically meets the requirements of the power grid for photovoltaic "inertia response" and "primary frequency regulation," thereby exerting the effect of reservoir energy storage and effectively supporting the consumption of new energy around the hydropower system. However, the application of hydro-solar hybrid systems is still constrained by the geographic conditions because hydroelectric power stations are mostly located in deep mountain valleys, and the photovoltaic plots are dispersed in the surrounding mountainous areas, with the long route construction paths and high investment.

References

1. Guo, X.-Y., Cui, Q.-R., Wang, W.-B., et al.: Key technologies and research prospects for hydro-photovoltaic hybrid system. Hydropower Pumped Storage **5**(45), 48–53 (2022)
2. Jia, Y.-F., Lin, M.-R., Dong, Z.-C.: Research on optimal operation of hydro photovoltaic complementarity in Longyangxia hydropower station. Water Resour. Power **10**(38), 207–209 (2020)
3. Ma, X., Hu, X.-L., Wang, Q.-Y., et al.: Analysis of daily changes in operation mode of hydropower with large-scale photovoltaic connection. Northwest Hydropower **3**, 1–6 (2019)
4. Ming, B., Guo, X.-R., Cheng, L., et al.: Grid integration priority of large-scale photovoltaic power and hydropower within a hybrid generation system. J. Hydraul. Eng. **11**(54), 1287–1296 (2023)
5. Pang, X.-L., Zhang, W.: Research and application of large-scale hydro-solar hybrid coordinated operation technology. Qinghai Sci. Technol. **2**, 24–28 (2017)

Optimization of Ultra-High Voltage Direct Current Power Transmission Curve for High Proportion New Energy Bases

Bo Yi[1,2], Yunhe Liu[1], and Xiao Wang[2](✉)

[1] State Key Laboratory of Eco-Hydraulics in Northwest Arid Region of China, Xi'an University of Technology, Xi'an 710048, China
01828@nwh.com, Liuyhe@xaut.edu.cn

[2] Power China Northwest Engineering Corporation Limited, Xi'an 710065, China
wx1213904@163.com

Abstract. With the increase in demand for the construction of high proportion new energy base, the power transmission scale of Ultra-High Voltage Direct Current(UHVDC) is growing rapidly, and the characteristics of inter-provincial and inter-regional power transmission are affected by multiple factors, such as the randomness and volatility of new energy in the sending province, the peaks and valleys of the loads, the process of power shortages, and the new time-of use electricity pricing policy in the receiving province. Uncertainty issues are more prominent, and it has changed significantly compared with the traditional UHVDC power transmission methods. It is urgent to study the UHVDC transmission curve planning method in new period. Taking a new energy base in Northwest China as a typical example, a UHVDC power transmission system optimization model is constructed. A UHVDC power transmission curve optimization method that takes into account the power shortages process and the time-of use electricity pricing is proposed, and the Cplex12.6 Solver is used to carry out the 8760 h time series power production simulation. The results show that the obtained 12 curves by the new method can cope with the price-based response demand. Compared with the traditional method, the total cost present value is reduced by RMB 6.10 billion, the power shortage rate is reduced to 0.56%, and the proportion of new energy power is increased to 88.58%. Especially, the power output of pumped storage can track the transmission curve well, and alleviate the requirements for peak regulation flexibility of thermal power units. The power configuration scheme is technically feasible and economical, which verifies the effectiveness of the proposed method. It is suitable for the planning of large-scale UHVDC power transmission schemes. In the meantime, it plays an important role in promoting the level of new energy consumption and helping the low-carbon energy transformation.

Keywords: Power shortages process · Time-of use price · UHVDC power transmission · curve 8760 h time series production simulation · Pumped storage

S. Zheng et al. (Eds.): IHDC 2024, LNCE 487, pp. 343–353, 2025.
https://doi.org/10.1007/978-981-97-9184-2_30

1 Introduction

Accelerating the construction of new energy supply and consumption system is an important measure to promote the high-quality development of new energy, build a new power system and achieve the double carbon target (Luo M, 2023; Li H, 2022). The way is to focus on Gobi and desert areas, based on large-scale wind and solar power bases, supported by clean, efficient, advanced and energy-saving thermal power, and based on stable, safe and reliable UHVDC transmission and transformation lines. New energy resources are abundant in Gobi and desert areas in China. A large-scale new energy base is built, and the clean power of the sending provinces is sent to the receiving provinces with power shortage through the UHVDC project. It plays an important role in optimizing the allocation of regional resources, promoting the level of new energy consumption, reducing the proportion of non-fossil energy consumption.

In recent years, the scale of power transmission in new energy bases has grown rapidly (Li M, 2024; Liu Y, 2024; Xiao L, 2023; Liu Z, 2023; Cavazzini G, 2021; Gao R, 2022). The characteristics of inter-provincial and inter-regional power transmission are affected by multiple factors, the great changes have been undergone compared with the traditional UHVDC method. The formulation of transmission curve considering multiple factors is the basis of transmission channel planning of new energy base, and it is also of great significance to improve the transmission capacity, economy and reliability of the channel (Li H, 2024).

At present, for the optimization of the power transmission curve, Chen et al. (2013) analysed the characteristics of the incoming water in Sichuan power grid, optimized the power transmission curve, and evaluated the benefits of reducing the amount of abandoned hydropower in the wet season. Based on the thermal power capacity, Sun et al. (2023) analysed the output curves of different supporting thermal power capacity considering the new energy output and energy storage at the sending province. About the load at the receiving end and the peak-shaving pressure, Gao et al. (2017) proposed four transmission curves, and used the interconnected consumption capacity analysis model to compare the utilization hours and new energy curtailment corresponding to different curves.

In 2023, the new time-of-use electricity pricing policies in Shandong, Fujian, Yunnan, Guizhou, northern Hebei, Guangxi and other places will be implemented one after another, and will be fully implemented on January 1, 2024. The traditional transmission curve formulation is generally based on the output characteristics of new energy and regulating power sources, while partially taking into account the load characteristics of the provinces that may be affected, ignoring the time series fluctuation of new energy output and the actual power shortage process of the receiving province, and lacking the comprehensive consideration of price-based demand response under the new policies (Huang J, 2024). It is difficult to apply to the planning of new energy base delivery schemes that mainly serve wind, photovoltaic and other energy transmission needs that affect the intraday peak-valley structure.

In this paper, the new energy base A and the receiving province B are taken as typical cases, the optimization model of UHVDC transmission system is constructed, and the method of UHVDC transmission curve considering time-of-use electricity price is proposed. The 8760 h sequential production simulation is carried out to obtain the optimal configuration scheme of power capacity. It can promote the coordination of the transmission curve with the receiving province as much as possible, provide technical support for the planning of the new energy base.

2 Optimization Model of UHVDC Transmission System

2.1 Drafting Principle of UHVDC Power Transmission Curve

The following principles are considered in the formulation of UHVDC transmission curve:

- Referring to the load characteristics of the receiving province, the DC power adjustment is not more than 6 times a day according to the requirements of the dispatching operation department.
- Referring to the new energy output in the sending province, it is ensured that the sum of the various power sources output can meet the requirements of the UHVDC transmission curve, the annual utilization hours and utilization rate of new energy is reasonable.
- Referring to the power shortage and time-of use electricity price in the receiving province, it is ensured that the transmission curve matches the changes of price and the of typical daily load trend.
- It is ensured that the curve optimization strategy is in line with reality and has significant benefits.
- It is ensured that the curve optimization strategy is simple and easy to implement.

2.2 Objective Function

The objective function of the model: (1) minimize the new energy curtailment and thermal power fuel cost; (2) Maximize the amount of electricity transmitted to meet the power shortage in the receiving province; (3) Maximize the matching degree between the UHVDC power transmission curve and the time-of use electricity pricing and load change of the power grid in the receiving province. The above objectives are given a certain weight respectively, so that the multi-objective problem is transformed into a single-objective problem (Jiang M, 2024).

$$\min \sum_t \left[\sum_i k^f P_{it}^f + k^{re} \sum_k (\tilde{P}_{kt}^w - P_{kt}^w) + k^{re} \sum_m \left(\tilde{P}_{mt}^{pv} - P_{mt}^{pv}\right) \right. \\ \left. + \alpha P_t^c + \beta \left| P_t^{DC} - P_t^d \right| + \gamma \left(P_t^{dc} - P_{t-1}^{dc} \right) \right] \tag{1}$$

P_{it}^f is the output of thermal power station i in t period; k^f is the unit fuel cost and carbon emission reduction cost of thermal power; k^{re} is the cost of unit new energy curtailment loss; P_{kt}^w is the grid-connected output of wind farm k in t time period; P_{mt}^{pv} is the grid-connected output of photovoltaic power station m in t period; $\tilde{P}_{kt}^w$ is the power generation output of wind farm k in t period; $\tilde{P}_{mt}^{pv}$ is the power output of photovoltaic power station m in t period; P_t^c is the power shortage that the system fails to meet the transmission curve in t period, α is weight parameters; P_t^{dc} is the transmission power of t period, P_t^d is the lack of power in the receiving power grid; β is the weight to measure the matching degree between the transmission curve and the power shortage of the receiving end power grid; γ is the weight to measure the matching degree of the transmission curve with the price of the receiving province and the typical daily load change trend. In this paper, under the premise of unified dimension, the weight parameters are determined by equal weight.

2.3 Constraint Condition

2.3.1 Power Output Model

1. System constraint

$$\sum_i P_{it}^f + \sum_k P_{kt}^w + \sum_m P_{mt}^{pv} + \sum_n (P_{nt}^{sh} - P_{nt}^{sp}) + P_t^c = P_t^{dc} \tag{2}$$

   There, $\forall t \in T$, P_{nt}^{sh}, P_{nt}^{sp} are the power generation and pumping output of pumped storage power station n in t period; P_t^c is the power shortage in the system t period.
2. Thermal power station

$$0 \le P_{i,t}^f \le N_{i,t}^f v_i^f \tag{3}$$

$$-d_i \Delta t \le P_{i,t+1}^f - P_{i,t}^f \le u_i \Delta t \tag{4}$$

$$0 \le N_i^f \le \overline{N}_i^f \tag{5}$$

   u_i, d_i are the output ramp rate and load shedding rate of thermal power station i; $N_{i,t}^f$ is the units number; v_i^f is the unit capacity; $\overline{N}_i^f$ is the number of installed units.
3. Wind and photovoltaic power plants

$$0 \le P_{kt}^w \le \tilde{P}_{kt}^w \tag{6}$$

$$0 \le P_{mt}^{pv} \le \tilde{P}_{mt}^{pv} \tag{7}$$

4. Pumped storage station

$$V_{l,t+1}^u = V_{lt}^u + (Q_{lt}^{sp} - Q_{lt}^{sh})\Delta t, \forall l, t \tag{8}$$

$$V_{l,t+1}^d = V_{lt}^d + (Q_{lt}^{sh} - Q_{lt}^{sp})\Delta t, \forall l, t \tag{9}$$

$$Q_{lt}^{sp} = \eta_{lp} P_{lt}^{sp} \Delta t / H_{lt}^p, \forall l \in L, t \in T \tag{10}$$

$$Q_{lt}^{sh} = P_{lt}^{sh} \Delta t / \eta_{lh} H_{lt}^h, \forall l \in L, t \in T \tag{11}$$

$$P_{lt}^{sp} - MX_{lt}^{sp} \leq 0 \quad (12)$$

$$P_{lt}^{sh} - MX_{lt}^{sh} \leq 0 \quad (13)$$

$$X_{lt}^{sp} + X_{lt}^{sh} \leq 1 \quad (14)$$

There, X_{lt}^{sp} and X_{lt}^{sh} are 0,1 variables. When $X_{lt}^{sp} = 1$, which represents that the pumping output of pumped storage station l in t period, When $X_{lt}^{sh} = 1$, which represents that the power generation output of pumped storage station l in t period. η_{lp} and η_{lh} are pumping and power generation efficiency respectively. Q_{lt}^{sp}, Q_{lt}^{sh}, P_{lt}^{sp}, P_{lt}^{sh} are the pumping flow, power generation flow, pumping output and power generation output of pumped storage station l in t period respectively. V_{lt}^{u}, V_{lt}^{d} are the upper and lower reservoir storage capacity of pumped storage power station l in t period respectively; H_{lt}^{p}, H_{lt}^{h} are the pumping head and power generation head of pumped storage power station l in t period.

2.3.2 UHVDC Transmission Operation Model

$$\sum\nolimits_{r=t}^{t+\sigma} z_r \leq 1, \forall t < T - \sigma \quad (15)$$

$$-z_t v^{dc} \leq P_t^{dc} - P_{t-1}^{dc} \leq z_t u^{dc}, \forall t \in T \quad (16)$$

$$P_t^{dc} \leq P^{dc,\max}, \forall t \in T \quad (17)$$

$$\sum\nolimits_t P_t^{dc} \geq h^{\min} P^{dc,\max} \quad (18)$$

$$\sum\nolimits_t (P_{kt}^{w} + P_{kt}^{pv}) \geq \gamma^{pen} \sum\nolimits_t (P_t^{dc} - P_t^{c}) \quad (19)$$

$$\sum\nolimits_t P_t^{c} \leq \theta \sum\nolimits_t P_t^{dc} \quad (20)$$

There, z_r is 0,1 variables. When $z_t = 1$, the UHVDC transmission power is adjusted in t period. σ is the minimum constant operation time of UHVDC transmission power, u^{dc}, v^{dc} are the maximum upward and downward ramping rates of UHVDC transmission power, respectively. $P_t^{dc,\max}$ is the channel capacity. $h^{\min}$ is the minimum utilization hours of the channel, γ^{pen} is the proportion of new energy, θ is the maximum allowable power shortage rate.

3 Case Study

Taking a large-scale new energy base A in Northwest China as an example, the UHVDC channel transmission capacity is 8000 MW. Power types include wind power, photovoltaic, thermal power, pumped storage. A multi-energy complementary power generation system is constructed to transmit power to the receiving area B. It can meet the power demand of the receiving province, and support the safe and stable operation of the transmission channel.

According to the local power construction conditions and cost level (Geng X, 2023), the basic parameters, investment cost, construction and operation cycle are given in Table 1. The social discount rate is 8%. Since the construction and operation period of wind and thermal power is 20 years, the reconstruction is included in the construction cost in the 21st year. On the whole, the system is apportioned proportionally over a 40-year cycle. The standard coal consumption is 300 g/kWh, and the coal price is 1000 RMB/ton. The carbon dioxide emission coefficient of thermal power coal is 0.741 kg/kWh, and the carbon emission cost is calculated according to the above. The system power shortage is solved by purchasing electricity, and the cost of purchasing electricity is converted according to the equivalent thermal power on-grid price. The maximum allowable power abandonment rate of the system is 12%.

Table 1. Analysis of each power supply capacity parameter

Power	Maximum capacity (10 MW)	Investment per kilowatt (RMB)	Construction running time (year)	Cost of operation (%)
Wind	800	4100	20	2.5
Photovoltaic	1600	3500	20	2.0
Thermal power	264	3800	20	4.5
Pumped storage	360	6100	40	2.5

Combined with the historical electricity consumption of the receiving province B and the judgment of the future economic development trend, it is estimated that the electricity consumption of the whole society is 1130 billion kilowatt hours, the maximum load is 225 million kilowatts in 2030.

The maximum load of the power grid in summer occurs in July and August, and the maximum load in winter occurs in November. The maximum load occurs alternately in summer and winter. The daily load change rule usually has two kinds of winter and summer. The typical daily load in summer has three peaks, generally at 10:00 ~ 11:00, 14:00 ~ 17:00 and 20:00 ~ 21:00. The maximum daily load appeared in the noon peak, that is, 14:00 ~ 17:00. The typical daily load in winter also presents three peaks, generally at 0:00 ~ 11:00, 15:00 ~ 18:00 and 19:00 ~ 21:00. The daily maximum load appears at about 10:00 ~ 11:00 (Fig. 1).

The time-of-use electricity price map of the receiving province B is shown in Fig. 2. The peak of July-August appears at 11:00–12:00 and 15:00–17:00, and there is no peak and deep valley in other months.

The daily adjustment times of UHVDC transmission curve are not more than 6 times, and is drawn up according to five steps. The utilization hours are not less than 4500 h, the proportion of new energy power is not less than 75%, and the power shortage rate is not more than 5%. The model is solved by using the scheduling optimization method with monthly cycle in the year and hourly cycle in the month. The 8760 h hourly production simulation is realized, and the transmission curve is drawn up according to the quarter. The optimization model is solved by Cplex12.6 Solver.

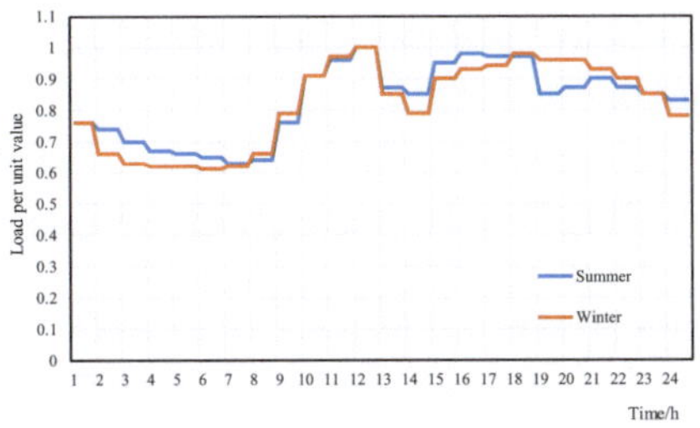

Fig. 1. Typical daily load curves

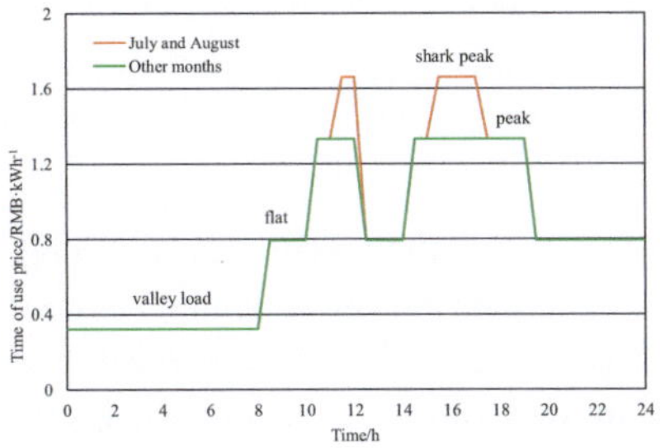

Fig. 2. Time-of-use price map of B power grid

4 Results and Discussion

The optimization results of 12 power transmission curves are shown in Fig. 3 after solving the model.

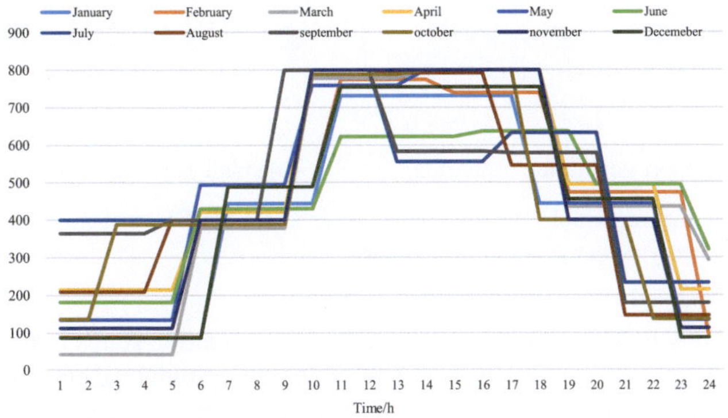

Fig. 3. Power transmission curve results

At the same time, the power generation system of a large new energy base A needs to support the construction of wind power 6000 MW, photovoltaic 1400 MW, thermal power 1320 MW and pumped storage 3600 MW, channel capacity 8000 MW. The installed capacity, annual power generation and annual utilization hours of different types of power supply are shown in Table 2.

Table 2. The results of installed capacity and annual power generation

Power	Capacity /10 MW	Proportion /%	Annual power generation /(10^8 kWh)	Proportion /%
Wind	600	24.1	117.5	28.65
Photovoltaic	1400	56.2	245.8	59.93
Thermal power	132	5.3	64.12	15.63
Pumped storage	360	14.4	51.84	12.64

The total installed capacity is 24920 MW. The installed capacity of new energy accounts for 80.3%, and the annual power generation accounts for 89%. The annual utilization hours of thermal power are 4857 h. The permeability of new energy is 89.92%, and the total cost is 16.42 million yuan, which is the best result.

The typical days of continuous large and small generation periods of new energy are selected respectively, and the adaptability of transmission curve and various power supply operation are analyzed, as shown in Fig. 4.

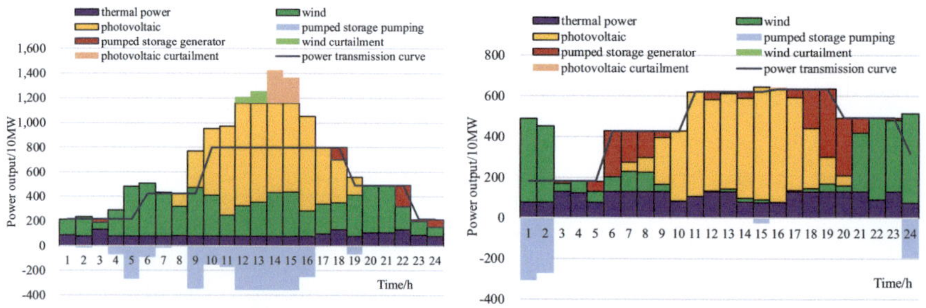

(a) large power generation scenarios (b) small power generation scenarios

Fig.4. Output of the renewable energy power base and UHVDC transmission curve under large/small new energy power generation scenarios

The typical day in March is selected to reflect the operation status in the continuous large generation period of new energy. At 1:00–5:00, the system load curve is gentle, and the wind power basically meets the system load demand. Once insufficient, it is satisfied by thermal power peak shaving and pumped storage power generation, and the excess wind is stored by pumped storage. At 20:00–24:00, the photovoltaic output is basically 0. The combined output of wind power, pumped storage and thermal power units meets the system load demand. In this scenario, there are more new energy power, and more energy can be stored in pumped storage power stations, and there may be a situation that pumped storage is full and can no longer store power.

The typical day in May is selected to reflect the operation status in the continuous small generation period of new energy. On this day, wind power fluctuates greatly, and there is no wind at 10:00 and 16:00. In addition, at 6:00, 14:00 and 19:00, the output of pumped storage power station can better track the load curve, which can effectively alleviate the frequent start-stop and fluctuation of thermal power units. At 1–2:00, 15:00 and 24:00, the pumped storage power station can store the excess wind and solar energy, release energy in the period of wind and solar anti-peak regulation characteristics, and alleviate the requirements for the peak regulation flexibility of thermal power units. In this scenario, there is less energy available for pumped storage. Compared with the large output of new energy, the matching power supply has a relatively poor degree of satisfaction with load demand. More extreme, there may be a situation where the pumped storage power station has no storage capacity and cannot support the channel.

The results of the traditional method and the proposed method in this paper are compared, as shown in Table 3. According to the traditional method, the total cost is 170.327 billion RMB, the utilization hours are 5155 h, the annual power generation is 41.241 billion kWh, the proportion of new energy power is 86.51%, and the power shortage is 307 million kWh. Compared with the method in this paper, the annual power generation and channel utilization hours are higher.

Table 3. Comparison of results of different methods

Indicators	Proposed method	Traditional method
Cost/10^8 RMB	1642.24	1703.27
Annual power produced /(10^8 kWh)	410.14	412.41
New energy abandon ratio /%	10.08	10.14
The proportion of new energy power /%	88.58	86.51
Power shortage rate /%	0.56	0.73
Channel utilization hours /h	5127	5155

The traditional method ignores the 8760 h time series fluctuation of new energy output, and does not analyze the process of power shortage in the receiving provinces, so the peaking pressure of power system is low. Therefore, the higher hours of UHVDC channel is obtained and the result is relatively optimistic.

The proposed method in this paper can better simulate the delivery of new energy base and the access of receiving power grid, and also take into account the time-of-use electricity price, which further verifies the effectiveness of the method.

5 Conclusions

Taking the new energy base as a typical case, the optimization of UHVDC transmission curve is studied. Considering the factors such as power output, load characteristics, and the new policy of time-of-use electricity price, the optimization model of UHVDC transmission system is constructed. The 8760 h time series production simulation is carried out, and the UHVDC transmission curve and the optimal configuration scheme of power capacity are obtained. The proposed method is suitable for the UHVDC channel delivery planning of new energy bases of different scales. It can be used as the basis for the optimal configuration of power supply capacity in new energy base, and provide an effective analysis tool for the formulation of power transmission curve and economic evaluation of power supply configuration in UHVDC project.

The model constructed in this paper considers factors such as power output, load characteristics, and new time-of-use electricity price policies. In fact, the influence weight of each factor is inconsistent. It is necessary to further study the factors and improve the feasibility of the UHVDC delivery planning scheme for new energy bases.

References

1. Luo, M.: Analysis of the development policy situation of large-scale wind and photovoltaic power bases and relevant suggestions. Water Power **49**(05), 1–3 (2023)
2. Li, H., Zhang, R., Mahmud, M.A., et al.: A novel coordinated optimization strategy for high utilization of renewable energy sources and reduction of coal costs and emissions in hybrid hydro-thermal-wind power systems. Appl. Energy **320**, 119019 (2022)
3. Li, M., Fan, Y., Wang, Y., et al: Coordinated optimal configuration of wind-photovoltaic-energy storage capacity for large-scale renewable energy bases. Electr. Power Autom. Equip., 1–13 (2024)
4. Liu, Y., Chen, D., Zhang, M., et al: Evaluation of hydropower, coal and electricity coordinated development to promote low-carbon transformation of power system. Yangtze River, 1–11 (2024)
5. Xiao, L., Jiang, H., Liu, C.: Economic analysis of the new energy export base in the Gobi and desert areas of China. Coal Qual. Technol. **38**(01), 21–26 (2023)
6. Liu, Z., Yuan, B., Jin, C.: Optimization strategy study on installation mix of renewable energy power base for supporting outbound delivery. J. Glob. Energy Interconnect. **6**(02), 101–112 (2023)
7. Cavazzini, G., Benato, A., Pavesi, G., et al.: Techno-economic benefits deriving from optimal scheduling of a virtual power plant: pumped hydro combined with wind farms. J. Energy Storage **37**, 102461 (2021)
8. Gao, R., Wu, F., Zou, Q., et al: Optimal dispatching of wind-PV-mine pumped storage power station: A case study in Lingxin Coal Mine in Ningxia Province, China. Energy **243** (2022)
9. Li, H., Liu, D., Qin, J., et al: Stochastic planning method for UHVDC transmission of renewable energy power base considering wind and photovoltaic output uncertainties. Power Syst. Technol., 1–9 (2024)
10. Chen, H.: Transmission curves optimization of Sichuan grid's seasonal electric energy. Electr. Power **46**(12), 144–150 (2013)
11. Sun, Y., Li, J., Ji, R., et al.: Research on power transmission and consumption scheme of multi-energy complementary base containing wind, solar, thermal and power storage. Inner Mongolia Electr. Power **41**(05), 62–68 (2023)
12. Gao, C., Niu, D., Ma, M., et al.: Accommodating capability analysis and comprehensive assessment method of large-scale new energy areas interconnected. Electr. Power **50**(07), 56–63 (2017)
13. Huang, J., Yu, X., Zhao. D., et al: Bi-layer optimization model of peak-valley prices and peak-valley periods considering benefits of source-grid-load. Electr. Power Autom. Equip., 1–13 (2024)

14. Jiang, M., Wang, X., Dong, C., et al.: Optimal capacity configuration for hydroelectric-thermal-wind-photovoltaic-storage multi-energy complementary system based on sequential power generation simulations. J. Hydroelectr. Eng. **43**(3), 71–83 (2024)
15. Hu, X., Wu, L., Yuan, Y., et al.: Study on capacity configuration of pumped storage, wind power and photovoltaic combined power generation system. Yangtze River **55**(4), 244–251 (2024)
16. Geng, X., Xu, B.: Capacity configuration method of wind-photovoltaic-thermal-pumped storage multi-energy system considering reliability and economy. Hydropower and Pumped Storage **9**(05), 53–60 (2023)

Optimal Scheduling of Wind-Thermal-Hydro-Storage Multi-Energy Complementary System with Pumped Hydro and Battery Storage

Zehua Zou(✉), Quan Zhao, Miao Deng, Chong Gao, and Liangsong Zhou

Three Gorges Cascade Dispatch and Communication Center, Chengdu 610095, China
zou_zehua@ctg.com.cn

Abstract. With increasing scale of renewable energy integrated into the power system, the power system needs more flexible regulating resources. At present, besides traditional thermal and hydro power plants, pumped hydro storage and battery storage are the most commonly used resources, and they form a wind-thermal-hydro-storage multi-energy complementary system. This paper proposes an optimal scheduling strategy to dispatch the resources in the multi-energy complementary system. First, models of diverse types of resources. i.e., hydro power, pumped hydro storage, and battery storage, are established. Then, a day-ahead optimization scheduling model is proposed for the multi-energy complementary system. Finally, case study is conducted on a revised IEEE 30 node system. Simulation results demonstrate that the proposed method can fully utilize the characteristics of different kinds of power resources to consume renewable energy and enhance the safety and economy of the multi-energy complementary system.

Keywords: Hydro power · Renewable energy · Energy storage · Multi-energy complementary system · Day-ahead schedule

1 Introduction

On the way of pursuing the goal of "achieving carbon dioxide emissions peak by 2030, carbon neutrality by 2060", the power system is experiencing a profound change [1]. The transformation pace towards low-carbon, cleaning, and green of the power system is accelerating to build a New Power System [2]. In the New Power System, the capacity of renewable energy, such as wind and photovoltaic power, will be in a dominating position on the power supply side. However, the power of renewable energy has great randomness, intermittency, and volatility, which threatens the safety of the power system [3]. To consume renewable energy power and ensure reliable power supply, the demand for flexible regulation resources is increasingly crucial [4].

At present, thermal power, hydro power, and energy storage are the most used regulating resources in the power system. Thermal and hydro power are traditional resources and have a great proportion in the current power system. Energy storage has fast response

S. Zheng et al. (Eds.): IHDC 2024, LNCE 487, pp. 354–363, 2025.
https://doi.org/10.1007/978-981-97-9184-2_31

speed and flexible energy shaving ability, and its capacity in the power system is increasing at a high speed in recent years. Pumped hydro storage and electric battery storage are the most used energy storage. The renewable energy resources and different types of regulating resources form a multi-energy complementary system, which helps to consume renewable energy by a coordinated control to fully utilize the individual characteristics of different types of power resources and meet the demand of the power system [5, 6].

Some research has covered the topic of the operation of the multi-energy complementary system. Peng et al. [7] establishes a RO-AUB dispatch model for reliability and economy on a large-scale wind-photovoltaic-hydro-thermal power system. Yanmeng et al. [8] proposes a bi-level optimal scheduling of wind-PV-hydro-thermal-storage multi-energy complementary systems, which optimizes hydro power in the upper level, and optimizes thermal power in the bottom level. Zhengshuo et al. [9] builds a comprehensive energy generation model for wind-solar-hydrothermal power system dispatching and uses an extended crisscross optimization algorithm to solve the model. Wenting and Hua [10] focuses on the optimal operation of power systems with hydro, thermal, wind, photovoltaic, and nuclear power, and proposes a data-driven robust day-ahead unit commitment framework based on RKDE. Yuge et al. [11] proposes a two-stage stochastic optimization scheduling model for virtual power plants combining wind power, photovoltaic, small hydropower, battery, and flexible load. Uncertainties of renewable energy in a wind-PV-battery storage-pumped storage combining system are considered in [12], which uses GAN and DPC algorithm to generate day-ahead scenarios and proposes a multi-time scale joint optimal scheduling method. The effect of multi-energy complementary system on reducing carbon emission is considered in [4, 13], which uses carbon emission as an optimization objective. Other problems of multi-energy complementary system are also considered, such as network planning [1, 14], power plant allocating [11, 15–17], and evaluation index design [18]. Among all the research above, there's no research considering the scheduling problem of a multi-energy complementary system containing hydro power, pumped hydro storage, and battery storage, which is the topic of this paper.

This paper proposes an optimization scheduling model of a wind-thermal-hydro-storage multi-energy complementary system. Two types of storage, i.e., pumped hydro storage and electric battery storage, are considered in the model with their detailed cost and operation model. First, models of the power resources in the system are introduced. Then, the scheduling optimization model is proposed, which considers all the operating constraints of power resources and minimizes the total cost. Finally, the method is tested on an improved IEEE 30 node system.

2 Power Resource Models

2.1 Hydro Power

The operating cost of hydro power units is comparatively small and usually can be neglected [19], so only operating constraints are considered.

(1) Power constraints

$$P_{H,min} \le P_{H,t} \le P_{H,max} \tag{1}$$

where $P_{H,max}$, $P_{H,min}$ denote the maximum and minimum power of the hydro power unit, respectively; $P_{H,t}$ is the output power at time t.

(2) Water constraints

$$Q_{H,min} \le \sum_t Q_{H,t} \le Q_{H,max} \tag{2}$$

where $Q_{H,max}$, $Q_{H,min}$ denote the maximum and minimum volume of consumed water within the optimization period; $Q_{H,t}$ denotes the consumed water at time t.

2.2 Pumped Hydro Storage

(1) Cost Model

The operational maintenance cost of a hydro pump storage C_{PS} is expressed as:

$$C_{PS} = f_{PS,f}\left(P_{PSP,max} + P_{PSG,max}\right) + f_{PS,v} \cdot \sum_t \left(P_{PSP,t} + P_{PSG,t}\right) \tag{3}$$

where $f_{PS,f}$, $f_{PS,v}$ denote the unit fixed and variable operational maintenance cost; $P_{PSP,t}$, $P_{PSG,t}$, $P_{PSP,max}$, $P_{PSG,max}$ denote the pump and generator power at time t and their maximum value, respectively.

(2) Operation Constraints
1) Power constraints

$$P_{PSP,min}u_{PSP,t} \le P_{PSP,t} \le P_{PSP,max}u_{PSP,t} \tag{4}$$

$$P_{PSG,min}u_{PSG,t} \le P_{PSG,t} \le P_{PSG,max}u_{PSG,t} \tag{5}$$

where $P_{PSP,min}$, $P_{PSG,min}$ denote the minimum power of the pump and generator, respectively; $u_{PSP,t}$, $u_{PSG,t}$ are the operation status denoting the pumped storage is in pump or generator mode.

2) Reservoir constraints

$$\begin{cases} W_{PSU,min} \le W_{PSU,t} \le W_{PSU,max} \\ W_{PSD,min} \le W_{PSD,t} \le W_{PSD,max} \end{cases} \tag{6}$$

$$\left| W_{PSU,t_0} - W_{PSU,t_e} \right| \le \Delta W_{PSU,max} \tag{7}$$

where t_0, t_e are the beginning and end time instant of the optimization period; $W_{PSU,max}$, $W_{PSD,max}$, $W_{PSU,min}$, $W_{PSD,min}$ denote the maximum and minimum capacity of the upper and lower reservoir, respectively; $\Delta W_{PSU,max}$ denotes the maximum limit of capacity variation within the optimization period; $W_{PSU,t}$, $W_{PSD,t}$ denote the capacity of the upper and lower reservoir at time t, respectively.

3) Operation status constraints

The pumped hydro storage can only operate in up to one mode:

$$u_{PSP,t} + u_{PSG,t} \leq 1 \tag{8}$$

2.3 Battery Storage

(1) Cost Model

The operational maintenance cost of battery storage C_{BS} is expressed as:

$$C_{BS} = f_{BS,f} P_{BSG,max} + f_{BS,v} \cdot \sum_{t} \left(P_{BSC,t} + P_{BSG,t}\right) \tag{9}$$

where $f_{BS,f}$, $f_{BS,v}$ denote the unit fixed and variable operational maintenance cost; $P_{BSC,t}$, $P_{BSG,t}$ denote the charging and discharging power at time t, respectively.

(2) Operation Constraints
1) Power constraints

$$P_{BSC,min} u_{BSC,t} \leq P_{BSC,t} \leq P_{BSC,max} u_{BSC,t} \tag{10}$$

$$P_{BSG,min} u_{BSG,t} \leq P_{BSG,t} \leq P_{BSG,max} u_{BSG,t} \tag{11}$$

where $P_{BSC,max}$, $P_{BSC,min}$, $P_{BSG,max}$, $P_{BSG,min}$ denote the maximum and minimum power in charging and discharging mode, respectively; $u_{BSC,t}$, $u_{BSG,t}$ are the operation status denoting the battery mode.

2) State of charge (SOC) constraints

$$SOC_{min} \leq SOC_t \leq SOC_{max} \tag{12}$$

$$\left|SOC_{t_0} - SOC_{t_e}\right| \leq \Delta SOC_{max} \tag{13}$$

where SOC_{max}, SOC_{min} denote the maximum and minimum SOC limit, respectively; ΔSOC_{max} denotes the maximum limit of SOC variation within the optimization period; SOC_t denote the SOC at time t.

3) Operation status constraints

The battery cannot be working in both charging and discharging mode:

$$u_{BSC,t} + u_{BSG,t} \leq 1 \tag{14}$$

The thermal power model can be found in [20] and is omitted in this paper due to space limitations.

3 Optimal Scheduling Model of the Wind-Thermal-Hydro-Storage Multi-Energy Complementary System

3.1 Optimization Objective

The objective the proposed model is to minimize the overall system operating cost:

$$\min C_T + C_{BS} + C_{PS} + C_{RE} \tag{15}$$

where C_T denote the cost of thermal power, C_{RE} denote the punishment for wind power curtailment, which can be expressed as:

$$C_{RE} = \sum_t f_{RE}\left(P_{REE,t} - P_{REO,t}\right) \tag{16}$$

where f_{RE} denote the unit punishment for wind power curtailment; $P_{REE,t}$, $P_{REO,t}$ are the estimated and output wind power at time t, respectively.

3.2 Optimization Constraints

Besides power resource constraints introduced in Sect. 1.2, the optimization constrains include:

1) Wind power constraints

$$0 \le P_{REO,t} \le P_{REE,t} \tag{17}$$

2) System power balance constraints

$$\begin{aligned} &\sum P_{T,t} + \sum P_{H,t} + \sum P_{PSG,t} + \\ &\sum P_{BSG,t} + P_{REO,t} = \sum P_{PSP,t} + \sum P_{BSC,t} + P_{L,t} \end{aligned} \tag{18}$$

where $P_{T,t}$, $P_{L,t}$ denote the thermal power and the estimated system load at time t, respectively. In this paper, the network constraints and losses are not considered.

4 Case Study

4.1 Case Parameters and Scenario Design

The case study is conducted on an improved IEEE-30 bus system. Wind, thermal, hydro, and storage (pumped hydro & battery) power units all integrate into the system. The maximum pump power of the pumped hydro storage is 330MW/320MW, and the generating capacity is 300MW/180MW; The capability of the battery storage is 100MW/200MWh. Parameters of the 4 thermal power units (G1-G4) and a hydro power unit (H1) are shown in Table 1. The volume limit of consumed water is 172000m^3/20000m^3. The unit punishment for wind power curtailment is 2000 \$/MW. Other parameters can be found in [20].

In the simulation, 4 typical days are used to represent 4 seasons, respectively. The typical days are constructed based on the actual data of one city located in Southwest China. Figures 1 and 2 show the wind power and load of the typical days.

In the economy analysis, 5 scenarios are constructed based on different resource compositions of the multi-energy complementary system, as shown in Table 2.

Table 1 Parameters of the thermal and hydro power units

	G1	G2	G3	G4	H1
P_{max}^{G}/MW	1000	1000	600	300	800
P_{min}^{G}/MW	400	400	300	180	0

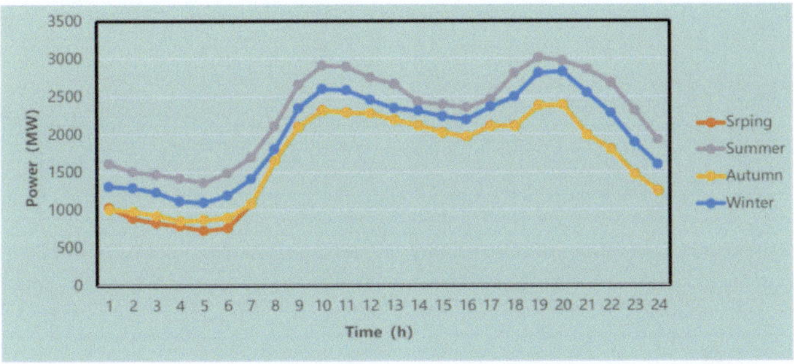

Fig. 1. Wind power of the 4 seasons

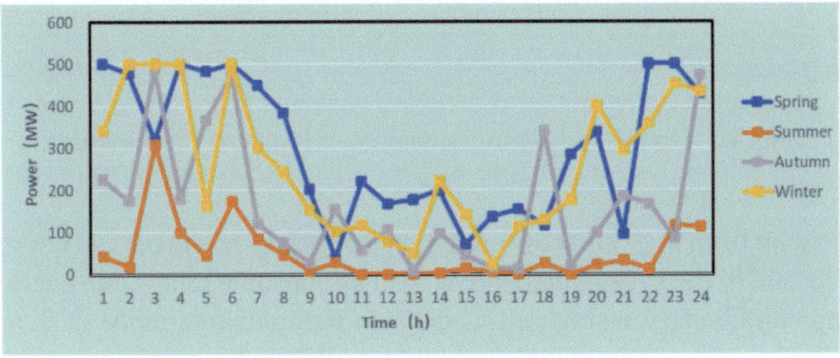

Fig. 2 Load of the 4 seasons

Table 2 Resource composition of the 5 scenarios

	Wind	Thermal	Hydro	Pumped Hydro	Battery
Scenario 1	√	√	√	√	√
Scenario 2	√	√		√	√
Scenario 3	√	√	√		√
Scenario 4	√	√	√		
Scenario 5	√	√			

4.2 Simulation Results

The optimal schedule strategies for the Scenario 1 in 4 seasons are shown in Fig. 3. The optimization period is 24h, and the time step is 1h.

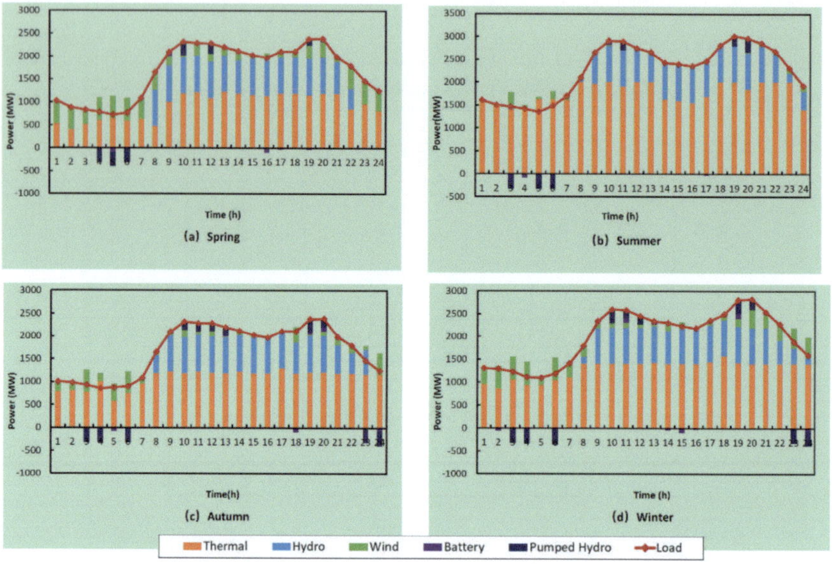

Fig. 1.3 Optimal scheduling strategies of Scenario 1 in 4 seasons

It can be seen from Fig. 3 that different types of power resources can coordinate with each other to satisfy the system demand as well as consume renewable energy. Pumped hydro storage and battery storage are usually in pump/charge mode at 3–6, when the load is low, and the wind power is relatively high. When the load becomes higher, such as at 10–11 and 19–20, the storages are usually in discharge/generator mode to supply the load. Hydro power is usually allocated at 8–21 for similar reasons. Since the thermal power serves as the base load, its output power is higher in summer, and lower in other seasons. The usage of storage is also less frequency in summer due to a less severe renewable energy consumption problem.

To make further analysis on the scheduling strategy, the scheduling strategy for autumn (Fig. 3(c)) is detailed in Fig. 4. It can be seen that in the early morning, only 2 thermal units (G1, G4) are operating. The storages are operating in pump/charge mode within this period. As the load increases at 8, the hydro unit (H1) and the storages begin to generate power to supply the load. At 18, the wind experiences a sudden increase, and the output power of H1 and G4 decreases for renewable energy consumption. The battery also charges at that time.

Figure 5 shows the overall cost of the scenarios with different resource compositions. It can be seen that Scenario 1, i.e., with the most kinds of regulating resources, has the lowest cost, while Scenario 5, i.e., with the least kinds of regulating resources, has the highest cost. The cost of other scenarios is in the middle. The results demonstrate that

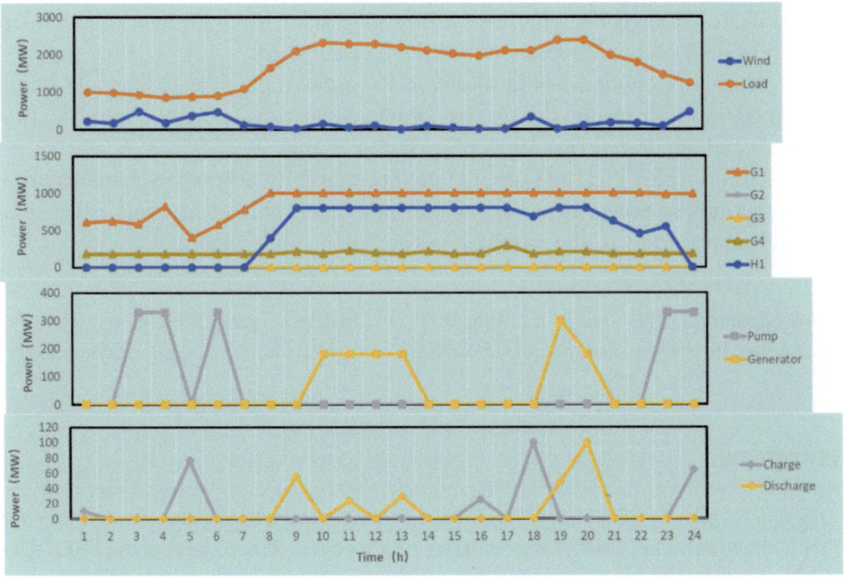

Fig. 4 Detailed optimal scheduling strategies for Scenario 1 in autumn

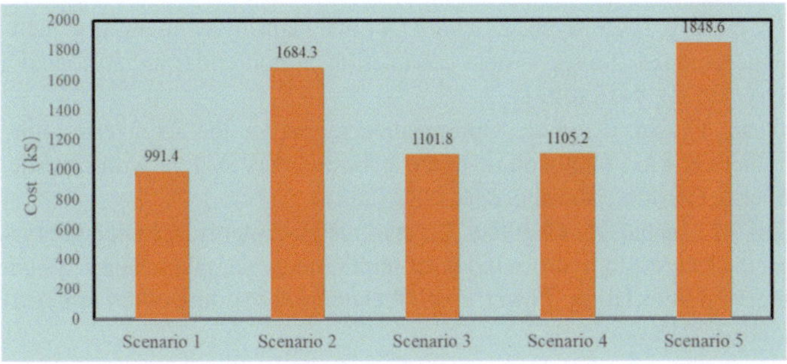

Fig. 5 Overall cost of scenarios with different power resource composition

the proposed schedule strategy can make good arrangement of diverse kinds of power resources to achieve an overall economy optimization. With more regulating resources, the scheduling strategy can use them to decrease the system cost for better economy efficiency.

5 Conclusion

Facing the increasing integration capacities of renewable energy, this paper proposes an optimal scheduling method for wind-thermal-hydro-storage multi-energy complementary system.

(1) Two kinds of storage, i.e., pumped hydro storage and battery storage, are considered. Their detailed cost and operation model are established.
(2) An optimization model is built to obtain the optimal day-ahead schedule strategy. The model minimizes the cost of the complementary system while ensuring power balance and operation constraints of each power unit.
(3) Case study verifies that the proposed method can fully utilize the characteristics of various kinds of power resources to consume renewable energy and enhances the safety and economy of the multi-energy complementary system.

Acknowledgments. This work is supported by Science and Technology Project of China Yangtze Power Co., Ltd. (NO.4323020009).

References

1. Jianjian, S., Yue, W., Chuntian, C., et al.: Research status and prospect of generation scheduling for hydropower-wind-solar energy complementary system. Proceedings of the CSEE **42**(11), 3871–3884 (2022)
2. Xianbo, K., Chen, W., Pan, L. et al.: Impact of wind and solar energy consumption on the operating water level of cascade reservoirs. China Rural Water and Hydropower (2024) in press
3. Zhong, L., Xingyu, C., Shuyun, Z., et al.: Optimal capacity configuration method for wind-photovoltaic-pumped-storage system considering carbon emission. Automation of Electric Power Syst. **45**(22), 9–18 (2021)
4. Lu, L., Yuan, W., Su, C., et al.: Optimization model for the short-term joint operation of a grid-connected wind-photovoltaic-hydro hybrid energy system with cascade hydropower plants. Energy Convers. Manage. **236**, 1–12 (2021)
5. Changxing, M., Juntao, Z., Chuntian, C., et al.: High-dimensional uncertainty scenario generation method for basin hydro-wind-solar multi-energy complementary system considering spatio-temporal correlation. Power Syst. Technol. (2024). https://doi.org/10.13335/j.1000-3673.pst.2023.2213
6. Wei, G., Xiaomeng, A., Jiakun, F., et al.: Coordinated optimal operation of the wind, coal, hydro, gas units with energy storage. Trans China Electrotec. Soc. **32**, 11–20 (2017)
7. Peng, C., Xie, P., Pan, L., et al.: Flexible robust optimization dispatch for hybrid wind/photovoltaic/hydro/thermal power system. IEEE Trans. on Smart Grid **7**(2), 751–762 (2016)
8. Yanmeng, S., Jun, R., Yu, Z.: Research on bi-level optimal scheduling of wind-PV-hydro-thermal-storage multi-energy complementary systems. China Rural Water and Hydrop. **3**, 34–40 (2024)
9. Zhengshuo, L., Hao, D., Yue, L. et al.: Application of an extended crisscross algorithm in the wind-solar-hydrothermal power system dispatching. In: Panda Forum on Power and Energy: 1886–1895 (2023)
10. Wenting, H., Hua, W.: Data-driven robust day-ahead unit commitment model for hydro/thermal/wind/photovoltaic/nuclear power systems. Int. J. Electr. Power Energy Syst. **125**, 1–11 (2021)
11. Yuge, C., Qianyun, D., Mengkai, W., et al.: Two-stage optimal scheduling of virtual power plant with wind-photovoltaic-hydro-storage considering flexible load reserve. Energy Rep. **8**, 848–856 (2022)

12. Weidong, H., Yang, L., Jingyan, L., et al.: Multi-time scale joint optimal scheduling for wind-photovoltaic-electrochemical energy storage-pumped storage considering renewable energy uncertainty. Electric Power Automation Equipment **43**(4), 91–98 (2023)
13. Xiaolin Z, Chang L, Yi M et al (2023) Optimal operation of wind-solar-fire synergy considering carbon trading and energy storage system. Journal of North China Electric Power University (Natural Science Edition). In press
14. Yandong, S., Jie, L., rigetu J, et al.: Research on power transmission and consumption scheme of multi-energy complementary base containing wind, solar, thermal, and power storage. Inner Mongolia Electric Power **41**(5), 62–68 (2023)
15. Anfeng, Q., Yiping, W., Ping, L., et al.: Study on hydro-wind-photovoltaic-pumped storage hybrid electrical power generating system in Qinghai at the upper reaches of Yellow River. Northwest Hydropower **1**, 85–92 (2024)
16. Mengyan, J., Xiao, W., Chuang, D., et al.: Optimal capacity configuration for hydroelectric-thermal-wind-photovoltaic-storage multi-energy complementary system based on sequential power generation simulations. J. Hydroelectric Eng. **43**(3), 71–83 (2024)
17. Koholé, Y., Ngouleu, C., Fohagui, F., et al.: Quantitative techno-economic comparison of a photovoltaic/wind hybrid power system with different energy storage technologies for electrification of three remote areas in Cameroon using Cuckoo search algorithm. J. Energy Storage **68**, 1–29 (2023)
18. Danni, S., Jie, L., Bo, C. et al.: Modeling of evaluation index system for coordinated operation of wind-solar-hydro multi energy complementary systems at multiple time scales. In: 13th International Conference on Power and Energy Systems, pp. 558–564 (2023)
19. Kaiyan, W., Xianjue, L., Ling, W., et al.: Optimal dispatch of wind-hydro-thermal power system with priority given to clean energy. Proceedings of the CSEE **33**(13), 27–35 (2013)
20. Songyan, Z., Shihong, M., Binxin, Y., et al.: Economic analysis of multi-type energy storages considering the deep peak-regulation of thermal power units. Electric Power Constr. **43**(1), 132–142 (2022)

Detailed Analyses of the Ecological Impacts Stemming From Hydropower Projects

Exploring the Impacts of Large Hydroelectric Projects on Downstream Wetland Ecosystems: A Case Study of the Impact Zone of the Jingwei Wetland Reserve

Weifeng Wan[1,2], Feng Zeng[1,2], Liqun Sun[1,2(✉)], and Weidong Zhou[1,2]

[1] Yellow River Engineering Consulting Co., Ltd. (YREC), Zhengzhou 450003, China
lqsun7@163.com

[2] Key Laboratory of Water Management and Water Security for Yellow River Basin, Ministry of Water Resources (Under Construction), Zhengzhou 450003, China

Abstract. The construction of a hydroelectric project will impact the interaction between surface water and groundwater downstream, potentially disrupting the ecological balance of the downstream wetland water bodies. This study focuses on the effects of the Dongzhuang Reservoir construction on the downstream Jingwei Wetland and conducts numerical modeling calculations of unstable three-dimensional groundwater flow under various operating conditions. It simulates and predicts the variations in water level burial depth at different sections of the wetland protection area and at different distances from the riverbank. The study determines the area of water level changes with and without hydraulic engineering under various conditions, and analyzes the impact of groundwater level changes on vegetation area under different adverse scenarios. The research reveals that the construction of the Dongzhuang Reservoir will have a certain impact on groundwater levels in the Jingwei Wetland. Under the most unfavorable conditions of five consecutive years of drought, the maximum water level drop can reach up to 0.54 m. The range of groundwater burial depth variations under different conditions is between 0.249 km^2 to 0.432 km^2, with the impact on the vegetation area of the protection area relatively small (all less than 2%). The most extreme adverse effects only occur in individual months, and overall, the operation of the project has a minimal impact on the protection area. This study provides important theoretical support for the harmonious and healthy development of the Dongzhuang hydropower project and the ecological balance of the Jingwei Wetland.

Keywords: Hydroelectric engineering · Ecological environment · Vegetation impact · Groundwater

1 Introduction

The ecological impacts of constructing large-scale hydraulic projects in the context of extensive hydropower development cannot be overlooked (Wohl, 2012; Zarfl et al., 2015). Construction of hydraulic projects on natural river channels directly alters river

S. Zheng et al. (Eds.): IHDC 2024, LNCE 487, pp. 367–379, 2025.
https://doi.org/10.1007/978-981-97-9184-2_32

flow, affecting the ecological environment evolved over long periods in rivers. The utilization of water flow for power generation changes the local morphology and discontinuity of river segments, leading to homogenization and discontinuity, consequently altering the diversity of riverine ecosystems and further impacting the interaction between downstream surface water and groundwater (Elcin and Emre, 2013; Wood, 2006), disrupting the balance of wetland aquatic ecosystems (Uehlinger et al., 2011; Yi et al., 2012; Matt et al., 2012; Liu et al., 2013; Deng, 2016).

The construction of hydroelectric projects significantly impacts groundwater levels in downstream wetlands. These projects alter the annual flow variations and pulsed hydrological cycles of natural rivers, reducing groundwater recharge and indirectly affecting the groundwater flow system's replenishment and discharge relationships (Dong et al., 2009; Pan et al., 2003; Zhang et al., 2011). The level of threat to different categories of ecological vegetation in wetlands varies due to this impact. Therefore, it is essential to study fluctuations in groundwater levels and analyze vegetation categories in wetlands as a prerequisite for investigating the extent of hydroelectric projects' impact on wetlands (Mallik et al., 2009; May et al., 2014; Vesipa et al., 2017).

Based on the flow data from five sections and water level monitoring data from 28 sections of the Wei River, Ba River, and Jing River within the Jingwei Wetland Reserve, this study statistically analyzed and predicted the water level variations during the sediment trapping period, normal operation period, and low-flow period before and after the construction of the Dongzhuang Reservoir. Groundwater level numerical simulations were conducted focusing on three adverse water level conditions in wetlands. By estimating the capillary rise heights of different categories of plants (aquatic plants, mesophytes, xerophytes) based on statistics, the study explored the impact range on wetlands of different categories within the protection zone under varying water level conditions. Understanding the response mechanism of groundwater and wetland systems to future changes is a crucial prerequisite for achieving sustainable utilization of downstream water resources in the harmonious ecological coordination of large-scale hydroelectric project construction and operation with wetland ecosystems.

2 Study Area

2.1 Location and Wetland Formation Conditions

The Dongzhuang Hydraulic Hub Project is located on the main stream of the Jing River in Dongzhuang Township, Liquan County, Shanxi Province, China. The project is designed with a dam height of 230 m and a reservoir capacity of 3 billion cubic meters, making it the largest hydraulic hub project in the Guanzhong region of China. The project serves multiple functions including flood control, sediment retention, irrigation, and power generation. The Xi'an Jingwei Wetland Nature Reserve is situated approximately 84 km downstream from the dam site, serving as a riverine wetland-type natural reserve primarily focused on conserving the wetland ecosystem. The location is shown in Fig. 1.

The total area of the nature reserve is 30.30 km^2, with the Jing River subarea covering 2.77 km^2, accounting for 9.14% of the reserve area. The remaining portions (Wei River and Ba River subareas) cover 27.53 km^2, representing 90.86% of the reserve area, as illustrated in Fig. 2. Formed over geological epochs through erosion and deposition

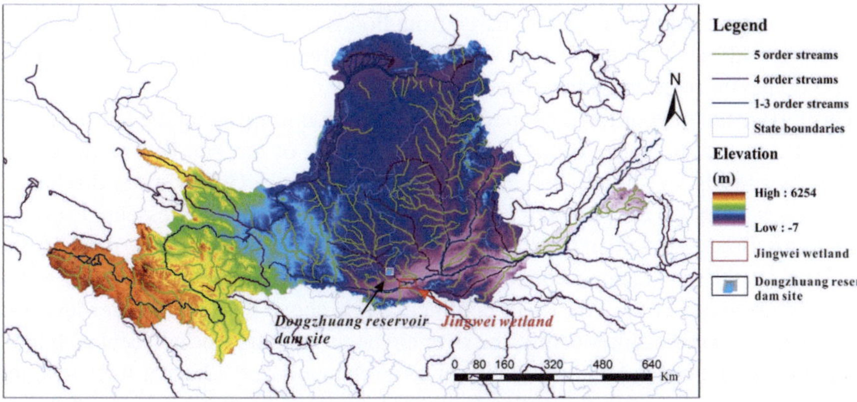

Fig. 1. Position map of the Dongzhuang hydraulic hub project relative to the Yellow River Basin

processes of the Wei River, Jing River, and Ba River, the reserve primarily consists of the floodplains and river mouth areas of the Jing River and Wei River, characterized by subtle pitted microtopography. Periodic or random flood pulses hold significant importance for rivers and floodplain wetlands. They not only establish lateral connections between rivers and floodplains but also govern the spatial patterns and layouts of riparian floodplain wetlands.

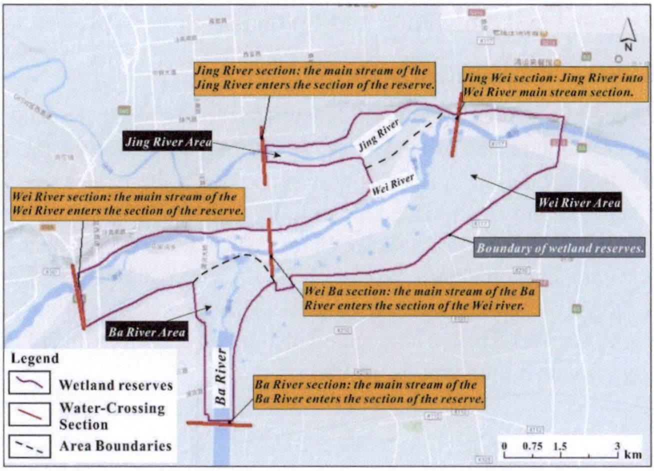

Fig. 2. Map of the Jingwei Wetland Nature Reserve Area and the location of flow monitoring sections

2.2 Hydrogeological Conditions

The wetland groundwater and river water exhibit complex interactions. Three sources contribute to groundwater recharge in the Jingwei Wetland Nature Reserve: atmospheric precipitation, pulsed groundwater recharge from floods during flood periods, and lateral groundwater inflow. At different time periods and locations, these three sources exhibit varying dominant roles and interactive effects.

During normal flow or low-flow periods (without flooding), along the north bank of the Wei River and the banks of the Jing River upstream of the Wei River confluence, groundwater levels are slightly higher than river water levels. Groundwater discharges into the river, with the river serving as a drainage channel, providing essential support for and sustaining the wetlands. When flooding occurs during high-flow periods, the water levels in the river channels on the north bank of the Wei River and the banks of the Jing River upstream of the Wei River confluence rise above the groundwater levels in the adjacent wetlands. Consequently, the river channels replenish the adjacent wetlands. The rate of groundwater exchange between river channels and adjacent wetlands, the fluctuation in water levels in the wetlands, and the extent of flood impact are influenced by factors such as the topography, geology, vegetation, and river flood processes within the wetland area. Along the south bank of the Wei River, at the confluence of the Ba River with the Wei River, at the junction of the Jing River and Wei River, and on the downstream banks of the Wei River, groundwater levels are generally lower than river water levels during normal flow, low-flow, and high-flow periods. The Wei River and Ba River contribute to groundwater recharge.

3 Materials and Methods

3.1 Groundwater Level Simulation and Prediction

3.1.1 Prediction and Evaluation Scope

At the confluence of the Jing River with the Wei River and along the banks of the Wei River, river water replenishes groundwater. After the completion of the Dongzhuang Reservoir, due to a reduction in river flow in most years, the river's role in replenishing groundwater is slightly diminished, potentially affecting ecological vegetation. The most significantly affected areas are those where surface water replenishes groundwater, specifically the core and some buffer zones of the Jingwei Wetland upstream and downstream of the Jing River confluence with the Wei River, as well as at the confluence of the Ba River with the Wei River. The wetland evaluation area determined for analysis is depicted in Fig. 3, with a total area of 13.96 km^2.

3.1.2 Simulation Condition

By collecting relevant actual engineering data, the stratigraphic structure and various hydrological parameters of the study area were obtained. A three-dimensional hydrogeological model of the study area was established to predict the changes in river water levels before and after the construction of Dongzhuang under different operating conditions. In order to more accurately simulate the range of groundwater level changes caused by surface water variations, this study employed a non-steady flow model for calculation and prediction. Based on the comparison of water quantities before and after the construction project, during the sediment interception period, normal operation period, and

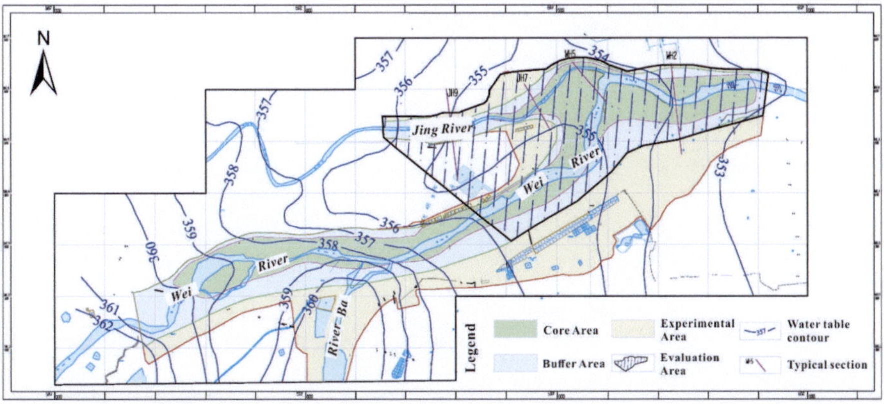

Fig. 3. Schematic diagram of prediction and evaluation scope of the Jingwei Wetland Nature Reserve

five consecutive years of dry season segment, distinct reduction quantities in specific years and the continuous dry season segment were selected as the most unfavorable scenarios for calculation and evaluation. These scenarios include the abundant water year during the sediment interception period, the average water year during normal operation, and the historical period of five consecutive years of dry season (utilizing river flow data from July 1993 to June 1997).

3.1.3 Simulation Methods and Parameters

Simulation Methods This study utilized the Visual Modflow numerical simulation software to create a computer model that visually represents the groundwater characteristics of the Jingwei Wetland, enabling the prediction of water levels. The model was stratified into three layers vertically based on the actual geological and hydrogeological features of the research area. The top of the model receives atmospheric precipitation as recharge, while the bottom is composed of silt clay and considered an impermeable aquiclude. The western boundary of the model is generalized as a lateral inflow boundary, while the northern and southern boundaries are treated as streamline boundaries. The eastern boundary is generalized as a lateral outflow boundary, with initial water levels at each boundary determined by the initial flow field of the aquifers.

For each scenario in typical years and during a 5-year consecutive drought period, input the surface water levels of typical cross-sections for each month of the year into the model, obtain the differences in surface water levels between the sections, and calculate the changes in groundwater levels under conditions with and without the Dongzhuang Reservoir. One month serves as a stress period for non-steady-state groundwater flow simulation. In the model, for the selected typical profiles, a groundwater observation well point is placed along the riverbank to output dynamic curves of groundwater level changes. Based on the groundwater level change curves, identify the most adverse month, output the groundwater flow field map for that month, and create depth-to-groundwater maps under conditions with and without the Dongzhuang Reservoir (Fig. 4).

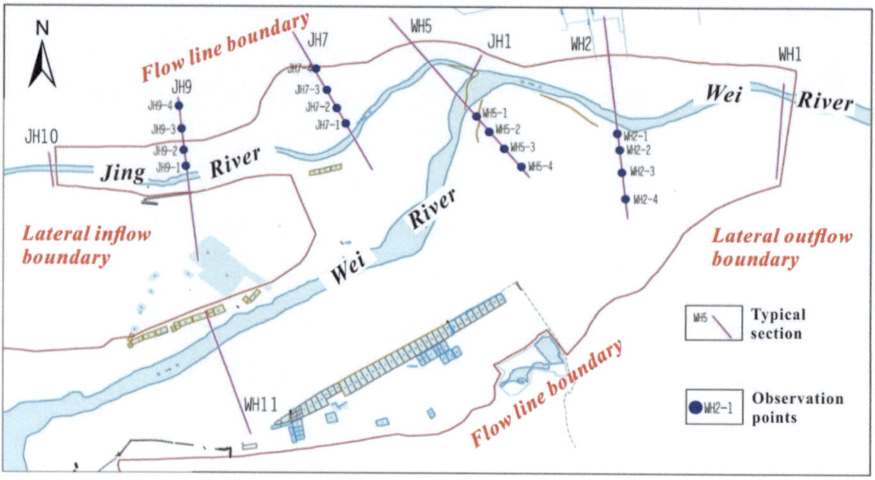

Fig. 4. Boundary conditions of the model, locations of typical profiles, and distribution map of observation wells

Hydrogeological Parameters (1) River Seepage Recharge Coefficient

The inter-annual variations of river seepage coefficients in the evaluation area are significant. Based on data on river seepage recharge in the northern plain area, a comprehensive analysis suggests that the annual average river seepage recharge coefficients under three conditions (sand trapping period during high-flow years, normal operation period during average flow years, and a consecutive 5-year drought period) can be approximately 0.16, 0.22, and 0.30, respectively.

In the simulation process, the surface water levels of various typical cross-sections predicted by hydrological conditions are input into the model to calculate the groundwater level changes between sections under conditions with and without the Dongzhuang Reservoir.

(2) Permeability Coefficient and Specific Yield

The initial permeability coefficient used in this calculation fully considers factors such as topography, lithology of formations, and the attenuation effects of fill materials and aquifers on permeability coefficients. It is a comprehensive assessment based on actual experimental values and empirical data. The values for regional parameters are shown in Fig. 5 and Table 1.

3.2 Investigation of Vegetation Types and Hydrotropic Characteristics

3.2.1 Survey of Vegetation Types in the Reserve

The ecosystems in the Jingwei Wetland Nature Reserve include terrestrial and aquatic environments. Based on plant adaptation to water availability, terrestrial plants are further categorized into hydrophilic plants, mesophytic plants, xerophytes, and other types.Hydrophilic plants are mainly found in riverbanks, ditches, embankments, marshes with water depths below 0.5 m or shallow groundwater levels, such as reeds, false reed, and knotweed. Their root system depth is generally less than 0.3 m.Mesophytic plants and

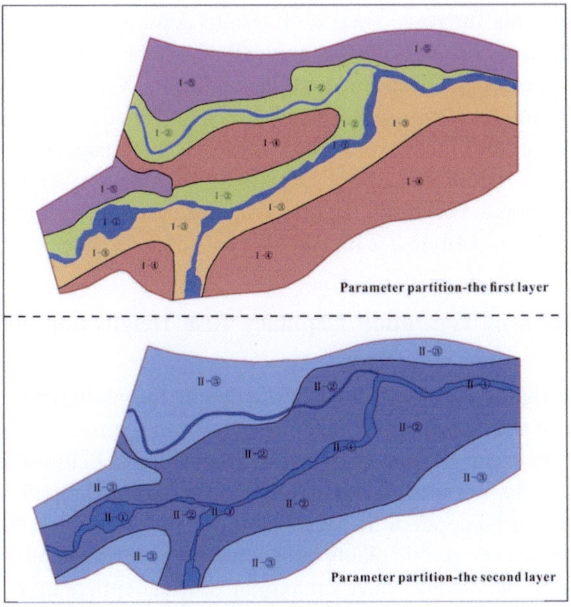

Fig. 5. Zoning map of parameters for the first and second aquifer layers

Table 1. Overview table of main hydrogeological parameters zoning for aquifer layers

Layers	Number	Main Aquifer Lithology	Permeability Coefficient (m/d)	Specific Yield
1	I-①	Fine Sand Mixed with Gravel	30	0.2
	I-②	Medium Fine Sand	25	0.15
	I-③	Fine Sand Mixed with Medium Sand	15	0.13
	I-④	Silty Sand	1.5	0.1
	I-⑤	Loess	0.3	0.08
2	II-①	Sand and Gravel	75	0.3
	II-②	Pebbles	90	0.32
	II-③	Fine Sand	13	0.12
3	III	Silty Clay	0.05	0.06

meso-xerophytes exhibit morphological structures and adaptability between hydrophilic plants and xerophytes. They cannot endure severe droughts or prolonged waterlogging and usually thrive in moderately moist environments. Common mesophytic plants in the reserve include dogtail grass, foxtail grass, wild taro, datura, mandrake, and desert willow, while typical meso-xerophytic species comprise Artemisia sacrorum, Tribulus terrestris, and Adlumia fungosa. Their root system depth typically ranges from 0.3 to 1.0 m. Xerophytes and xero-mesophytes are well adapted to grow in arid environments and can withstand prolonged or severe drought conditions. Various xerophytes and xero-mesophytes are found on steep slopes, sandy loess, and sandy wastelands on the north bank of the Wei River and Jing River in the reserve. Common species include Artemisia lactiflora, Cucurbita foetidissima,, and Lagotis brachystachya. Along the Jing River, there are extensive poplar forests on the first-level terraces with a significant variation in root system depth from 0.4m to 2.2 m, and the roots of some trees extend beyond 4 m.

3.2.2 Investigation of Vegetation Capillary Rise Height and Maximum Burial Depth

Based on the investigation of capillary water rise height, the geological formations within the wetland area of the conservation zone consist of sand, gravel, and loam. There is a significant variation in capillary rise heights, and according to field surveys, the capillary water rise height in the wetland area generally ranges from 1.0 to 1.5 m.

The maximum burial depth of the water table for hydrophilic plants is 1.3 to 1.8 m; mesophytic and meso-xerophytic plants have a maximum burial depth of 1.3m to 2.5 m; xerophytes and xero-mesophytes have a maximum burial depth of up to 3.7 m. For some woody vegetation in the conservation area, such as poplar and ginkgo trees, the root system can extend to depths of 4 to 6m, with a maximum burial depth typically ranging from 6 to 7 m.

4 Results and Discussion

4.1 Results of Unsteady Flow Calculations Under Various Adverse Operating Conditions

For the three aforementioned adverse operating conditions, the groundwater level variations at model observation wells located at distances of 75 m, 225 m, 475 m, and 725m from the riverbank were statistically analyzed for the selected four typical cross-sections (refer to Fig. 4 for the locations of the cross-sections and observation wells). Based on the variations in the observation wells, the most unfavorable month was identified, and the groundwater level difference at the Dongzhuang observation well in the most unfavorable month was output for evaluation.

4.1.1 Sand Trapping Period in a Flood Year

The groundwater level variations at model observation wells located at distances of 75 m, 225 m, 475 m, and 725m from the riverbank under the conditions of presence and absence of Dongzhuang during the sand trapping period in a flood year (adverse typical year) are shown in Fig. 6. It can be observed that the time periods with substantial groundwater level fluctuations before and after the presence of Dongzhuang are May and September to November. A comparison between the presence and absence of Dongzhuang reveals the largest difference in November at the JH9 cross-section of the Jing River (with a groundwater level change value of -0.41m). However, overall, the water level fluctuations in September are generally significant. In September, the groundwater level variations at different distance observation wells for the JH9 cross-section of the Jing River range from -0.39 m to 0.0 m, while for the JH7 cross-section, it ranges from -0.31 m to -0.13 m. The differences in the groundwater levels at various distance observation wells for the downstream Wei River's WH5 and WH2 cross-sections in September are -0.26 m to 0.0 m and -0.28 m to 0.0 m, respectively. The groundwater level variations at each cross-section in September are depicted in Fig. 7.

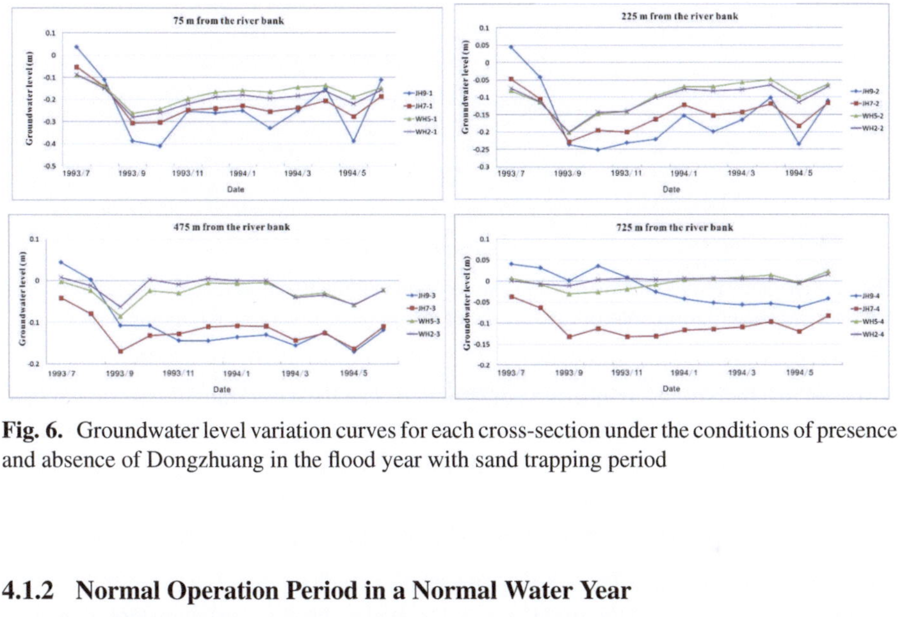

Fig. 6. Groundwater level variation curves for each cross-section under the conditions of presence and absence of Dongzhuang in the flood year with sand trapping period

4.1.2 Normal Operation Period in a Normal Water Year

The groundwater level variations under the conditions of presence and absence of Dongzhuang during the normal operation period in a normal water year were analyzed using the same research method. The largest difference still occurred in September, with groundwater level variations at different distance observation wells for the JH9 cross-section of the Jing River ranging from -0.27 m to 0.0 m. For the JH7 cross-section, the variations ranged from -0.14 m to 0.0 m. The differences in groundwater levels at various distance observation wells for the downstream Wei River's WH5 and WH2 cross-sections were -0.17 m to 0.0 m and -0.13 m to 0.0 m, respectively.

4.1.3 Consecutive 5-Year Dry Period (July 1993 to June 1996)

Over a Consecutive 5-Year Dry Period, comparing the presence and absence of Dongzhuang, the most unfavorable month was September 1995. The corresponding groundwater level variations at different distance observation wells for the JH9 cross-section of the Jing River ranged from -0.54m to -0.08 m. For the JH7 cross-section, the variations ranged from -0.47 m to -0.20 m. The differences in groundwater levels at various distance observation wells for the downstream Wei River's WH5 and WH2 cross-sections were -0.23 m to -0.12 m and -0.29 m to -0.11 m, respectively. It can be

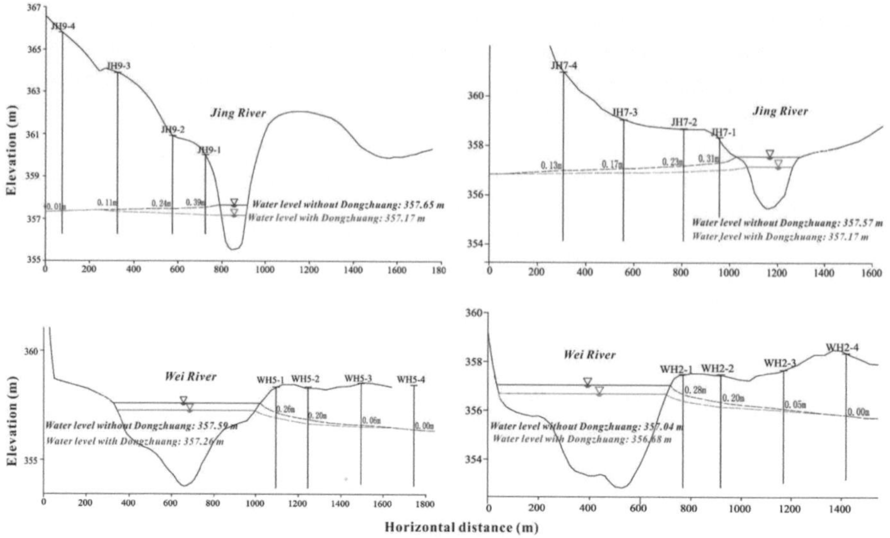

Fig. 7. Schematic representation of groundwater level variations at different distance observation wells for various cross-sections of the Jing River in a flood year with sand trapping period

observed that hydraulic engineering constructions can have a certain impact on wetland water levels, especially during a consecutive 5-year dry period, where the maximum impact at the JH9 cross-section can reach up to 0.54 m.

4.2 Statistical Analysis of Water Level Changes for Various Operating Conditions

Based on the groundwater level change prediction results, under the most unfavorable conditions, the decrease in the Jing River water level results in a certain extent of decline in the groundwater level on both banks of the Jing and Wei Rivers. The ranges of groundwater burial depth variations under three different scenarios - sand trapping period in a high water year in September, normal operation period in a normal water year in September, and the consecutive dry period in September 1995 - are 0.318 km^2, 0.249 km^2, and 0.432 km^2, respectively. The impact areas are mainly concentrated in the core and buffer zones, as shown in Figs. 8a to 8c.

4.3 Analysis of the Impact of Groundwater Changes on Vegetation

Based on the maximum burial depth of vegetation in the protection area, overlay the simulated groundwater results with the distribution of vegetation in the protection area using GIS to calculate the area of impact of groundwater burial depth changes on different vegetation types before and after dam construction.

4.3.1 Sand Trapping Period in a High Water year in September

After the operation of the project, the total area of vegetation affected by the decrease in groundwater burial depth in September of the sand trapping period in a high water year amounts to 0.318 km^2, representing 1.05% of the protection area. Among them, wetland vegetation (reeds, bulrush) is affected over an area of 0.09 km^2, accounting for 0.30% of the protection area; mesophytic vegetation (sedge, dogtail grass) is affected over an area of 0.072 km^2, representing 0.24%; xeromesophytic vegetation (goldenrod) is affected over an area of 0.041 km^2, accounting for 0.109%; and tree forest (artificial poplar) is affected over an area of 0.047 km^2, accounting for 0.15% (Fig. 8d-8f).

4.3.2 Normal Period in a Moderate Water Year in September

The total area of vegetation affected by the decrease in groundwater burial depth in September of the normal operation period in a moderate water year amount to 0.249 km^2, representing 0.28% of the protection area. Among them, wetland vegetation (reeds, bulrush) is affected over an area of 0.085 km^2, accounting for 0.48% of the protection area; mesophytic vegetation (sedge, dogtail grass) is affected over an area of 0.057 km^2, representing 0.19%; xeromesophytic vegetation (goldenrod) is affected over an area of 0.073 km^2, accounting for 0.24%; and tree forest (artificial poplar) is affected over an area of 0.034 km^2, accounting for 0.11%.

4.3.3 Third year in September of the Dry Season Segment Over Five Consecutive Years

After the operation of the project, the total area of vegetation affected by the decrease in groundwater burial depth in September of the third year of the dry season segment over five consecutive years amounts to 0.432 km^2, representing 1.43% of the protection area. Among them, wetland vegetation (reeds, bulrush) is affected over an area of 0.111 km^2, accounting for 0.37% of the protection area; mesophytic vegetation (sedge, dogtail grass) is affected over an area of 0.118 km^2, representing 0.39%; xeromesophytic vegetation (goldenrod) is affected over an area of 0.151 km^2, accounting for 0.50%; and tree forest (artificial poplar) is affected over an area of 0.052 km^2, accounting for 0.17%.

Overall, after the operation of the project, the most extreme adverse effects on the protection area only occurred in individual months, and the impact on the vegetation area of the protection area is relatively small (all less than 2%), therefore the overall impact of the project operation on the protection area is minimal.

5 Conclusions

In conclusion, the Dongzhuang hydraulic engineering projects have a discernible influence on groundwater levels in wetlands, particularly during consecutive 5-year low-flow periods, with the most significant impact observed up to 0.54 m at the JH9 section. The changes in groundwater burial depth under adverse conditions varied during the sediment trapping period in the wet year of September, normal operation period in the average year of September, and consecutive low-flow period in September 1995, amounting to 0.318 km^2, 0.249 km^2, and 0.432 km^2, respectively. The impacts were predominantly concentrated in the core and buffer zones. Post-engineering operation, the most extreme adverse effects on the conservation area only occur in isolated months, with the affected vegetation area relatively limited (less than 2% in all cases), indicating minimal overall

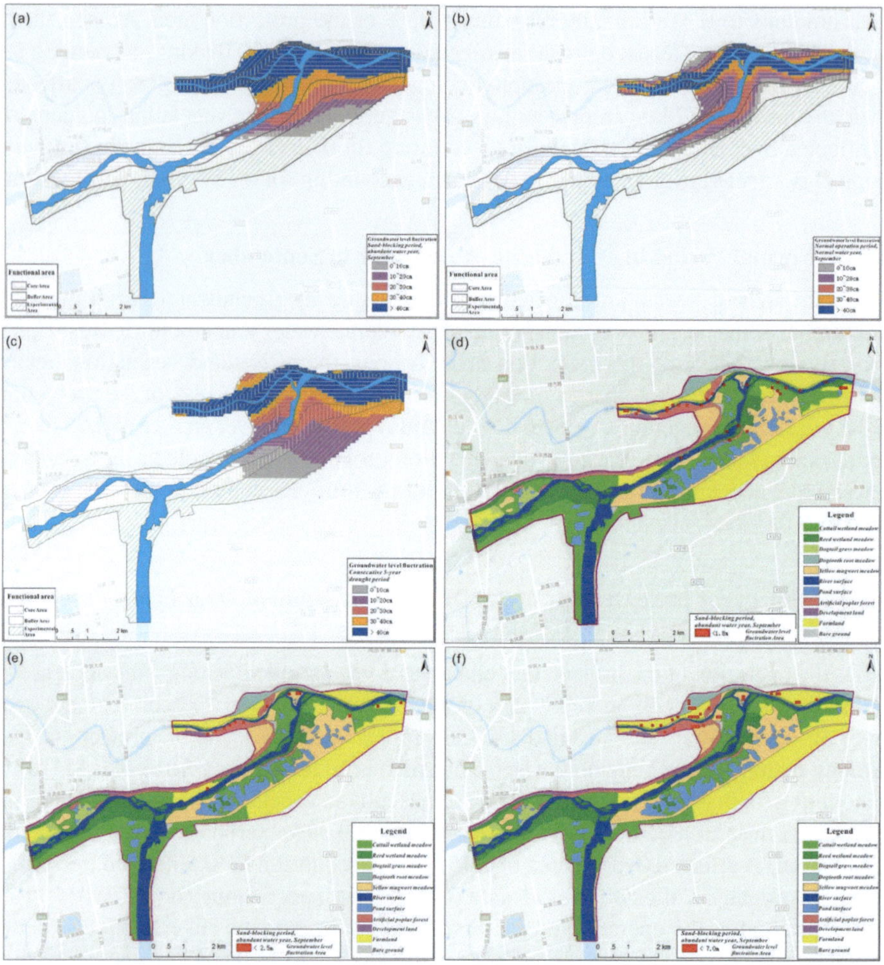

Fig. 8. (a), (b), (c): Groundwater level changes in various functional zones within the protection area under different working conditions. (d), (e), (f): Map of the area where the groundwater burial depth is less than 1.8m, 2.5m, and 7m in September of the sand trapping period in a high water year

impact on the conservation area. The operation of the reservoir has brought adverse effects on the Jing River region of the Jingwei Wetland Nature Reserve, while its impact on the overall protection area is limited. After implementing measures such as ecological flow guarantee and wetland inundation flow, the adverse effects can be mitigated to a certain extent.

Acknowledgements. This study was supported by Major Science and Technology Projects of the Ministry of Water Resources of China (No. SKS-2022062) and Postdoctoral Research Funding Project in Henan Province.

References

Deng, X.Y.: The effects and control measures of the river wetland ecology form the hydropower construction in Hilly area-based on Angu hydropower station, Thesis Doctor in sciences, Sichan Agricultural University Dissertation, China (2016)

Dong, Z.R., Zhang, J.: Ecological effect of flood pulses. J. Hydraul. Eng. **40**(3), 281–288 (2009). (In Chinese with English abstract)

Elcin, K., Emre, A.: Hydropower in Turkey: Economical social and environmental aspects and legal challenges. Environ Sci Policy **31**, 34–43 (2013)

Liu, J.G., Zang, C.F., Tian, S.Y., et al.: Water conservancy projects in China: Achievements, challenges and way forward. Glob. Environ. Chang. **23**, 633–643 (2013)

Mallik, A.U., Richardson, J.S.: Riparian vegetation change in upstream and downstream reaches of three temperate rivers dammed for hydroelectric generation in British Columbia, Canada. Ecol. Eng. **35**(5), 810–819 (2009)

Matt, F., Clinton, N.J.: Proliferation of hydroelectric dams in the Andean Amazon and implications for Andes-Amazon connectivity. PLoS ONE **7**(4), 1–9 (2012)

May, R., Mazlan, N.S.B.: Numerical simulation of the effect of heavy groundwater abstraction on groundwater-surface water interaction in Langat Basin, Selangor, Malaysia. Environ. Earth Sci. **71**(3), 1239–1248 (2014)

Pan, H.L., Wang, Q.: Negative impacts of reservoirs on water and sands problem of Yellow River. Bull. Soil Water Conserv. **23**(2), 73–76 (2003). (In Chinese with English abstract)

Uehlinger, U.: Effects of experimental floods on periphyton and stream metabolism below a high dam in the Swiss Alps (River Spol). Aquat. Sci. **65**, 199–209 (2011)

Vesipa, R., Camporeale, C., Ridolfi, L.: Effect of river flow fluctuations on riparian vegetation dynamics: processes and models. Adv. Water Resour. **110**, 29–50 (2017)

Wohl, E.: Identifying and mitigating dam-induced declines in river health: three case studies from the western United States. Int. J. Sed. Res. **27**(3), 271–287 (2012)

Wood, P.J.: Biological effects of fine sediment in the lotic environment. Environ. Manage. **21**, 203–217 (2006)

YI, Y.J., Yang, Z.F., Zhang, S.H.: Ecological influence of dam construction and river-lake connectivity on migration fish habitat in the Yangtze River basin, China. Procedia Environ. Sci. **2012**(2), 1942–1954 (2012)

Zarfl, C., Lumsdon, A.E., Berlekamp, J., et al.: A global boom in hydropower dam construction. Aquat. Sci. **77**(1), 161–170 (2015)

Zhang, H.B., Huang, Q., Zhang, S.H.: Cumulative effects of operation of cascade reservoirs on hydrological conditions in upper reaches of Yellow River. J. Hohai Univ. (Nat. Sci.) **39**(2), 137–142 (2011). (In Chinese with English abstract)

Assessing the Impact of Dongzhuang Water Conservancy Hub on Vegetation Ecological Distribution Based on Numerical Simulation and Machine Learning

Mengyan Ge(✉)

Yellow River Engineering Consulting Co., Ltd., Zhengzhou 450003, Henan, China
gostn@outlook.com

Abstract. The current assessment of the ecological benefit of reservoirs commonly lacks quantitative calculation of actual ecological distribution. The spatial distribution prediction of vegetation growth generally requires the concurrent application of multiple numerical models, which are complicated and involve numerous parameters. This research comprehensively utilized a surface hydrodynamics numerical model and machine learning method to construct a flow-based vegetation growth prediction model. The ecological impact during the storage period and regulation period after the completion of the Dongzhuang Water Conservancy Hub was calculated and analyzed. The results indicate that the regulation period after the construction of the reservoir has a minor impact on vegetation growth in comparison to pre-construction flow conditions in the normal flow years. The limited extent and magnitude of regional NDVI decline induced by the reservoir storage period will gradually recover after the beginning of reservoir regulation. The formulated vegetation growth prediction model can reflect the relevant influence processes on NDVI to a certain extent and alleviate the complexity of prediction.

Keywords: Remote sensing · Machine learning · NDVI · Large scale water conservancy · Yellow River Basin

1 Introduction

Reservoir construction is an important method to achieve optimal allocation of water resources. Reservoir construction generally aims at one or more of the engineering construction objectives, including flood control, water supply, ecology, power generation, sediment reduction, and navigation [1]. Upon completion, the reservoir will play a significant role in economic, social, and ecological benefits.

Ecological impact is one of the important aspects that is required to be considered in reservoir construction, and contemporary research on reservoir construction and operation also regards ecological goals as important objectives. The current ecological impact

S. Zheng et al. (Eds.): IHDC 2024, LNCE 487, pp. 380–388, 2025.
https://doi.org/10.1007/978-981-97-9184-2_33

assessment and scheduling research of reservoirs principally focuses on the satisfaction of ecological flow and the calculation of water level effects, lacking quantitative prediction and analysis of actual ecological distribution.

Generally, numerical modeling methods such as surface water models, groundwater flow models, and ecological hydrological models are required to be comprehensively utilized to attain precise spatial distribution prediction of vegetation ecology [2, 3]. The construction and operation of numerical models are complicated, involving numerous parameters [4, 5], and the simulation process is lengthy and cumbersome.

Machine learning methods can substitute complex processes to a certain extent, and feature selection is crucial. Identifying key processes and influencing factors can enhance the reliability and fitting degree of machine learning methods. This research focuses on the Dongzhuang Water Conservancy Hub Project. Based on the identification of key influencing processes and factors, machine learning methods were utilized to construct a prediction model for the spatial distribution of vegetation growth based on river flow, providing a reference for relevant research.

2 Field Site

The Jing River is the largest tributary of the Wei River, a tributary of the Yellow River in China, with a total length of about 455km. The Jing River originates from Ningxia Autonomous Region and flows into the Wei River in Xi'an City, Shaanxi Province in China. The Dongzhuang Water Conservancy Hub is located in the lower reaches of the Jing River in Liquan County, Shaanxi Province, China, and is one of the major water conservancy projects of the Yellow River basin. The Dongzhuang Water Conservancy Hub has a dam height of 230 m and a storage capacity of 3 billion cubic meters. Dongzhuang Reservoir undertakes important tasks, including flood control, reduction of sediment deposition, improvement of water environment and ecology, as well as power generation. The Dongzhuang Reservoir is still under construction and will enter the storage period and subsequent normal regulation period after completion.

The section of the Jing River extending from the head of Jinghui Canal to its intersection with the Wei River was selected as the research area as shown in Fig. 1. Jinghui Canal is a large-scale irrigation project on the Jing River. The canal head of Jinghui Canal is located 20km downstream of Dongzhuang Water Conservancy Hub. Important wetlands are distributed along the Jing River, commencing from the head of the Jinghui Canal, serving as significant ecological protection objectives.

3 Methods

3.1 Remote Sensing Interpretation

The Normalized Vegetation Index (NDVI) has been widely used to reflect regional vegetation growth. Based on Landsat 8 remote sensing images with red and near-infrared bands, the NDVI distribution in the study area from 2021 to 2023 was interpreted, The NDVI calculation formula is as below.

$$\mathrm{NDVI} = \frac{\rho_{NIR} - \rho_R}{\rho_{NIR} + \rho_R} \tag{1}$$

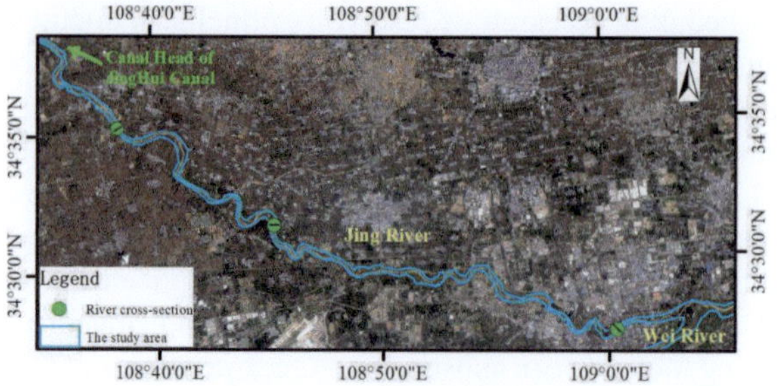

Fig. 1. The locations of the study area and river cross-sections

where ρ_{NIR} is NIR surface reflectance, and ρ_R is the R surface reflectance.

3.2 Surface Hydrodynamic Model Simulation

A surface hydrodynamic model was used to simulate the distribution of river water level and flow velocity in the study area under different flow rates at the Jinghui Canal section. For the x and y directions, there are two-dimensional shallow water flow equations as below.

$$\frac{\partial h}{\partial t}+\frac{\partial hu}{\partial x}+\frac{\partial hv}{\partial y}=hS \tag{2}$$

$$\frac{\partial hu}{\partial t}+\frac{\partial hu^2}{\partial x}+\frac{\partial huv}{\partial y}=fvh-gh\frac{\partial \eta}{\partial x}-\frac{gh^2}{2\rho_0}\frac{\partial \rho}{\partial x}+\frac{\tau_{ax}}{\rho_0}-\frac{\tau_{bx}}{\rho_0}+\frac{\partial}{\partial x}(hT_{xx})+\frac{\partial}{\partial y}\left(hT_{xy}\right)+hu_sS \tag{3}$$

$$\frac{\partial hv}{\partial t}+\frac{\partial hv^2}{\partial y}+\frac{\partial huv}{\partial x}=-fuh-gh\frac{\partial \eta}{\partial y}-\frac{gh^2}{2\rho_0}\frac{\partial \rho}{\partial y}+\frac{\tau_{ay}}{\rho_0}-\frac{\tau_{by}}{\rho_0}+\frac{\partial}{\partial y}\left(hT_{yy}\right)+\frac{\partial}{\partial x}\left(hT_{xy}\right)+\mathrm{hv}_s\mathrm{S} \tag{4}$$

where h is the still water depth, t is time, u and v are the velocity components along the x and y directions, s is the source and sink term, g is the gravitational acceleration, f is the Coriolis force parameter, η is the water level, ρ is the fluid density, ρ_0 is the reference water density, u_s and v_s are the flow velocities of the source and sink terms, T_{ij} is the stress term, τ_{ij} is the component of the shear stress of the water flow at the boundary between the water surface and the riverbed in the x and y directions.

3.3 The Random Forest Regression

The Random Forest Regression algorithm is a parallel ensemble learning algorithm developed by Breiman based on regression trees [6]. It uses the Bootstrap method and constructs a strong model composed of numerous independent and weak regression trees.

Compared to other machine learning regression methods, the random forest algorithm is less prone to overfitting and has the advantages of fast speed and a good tolerance for outliers and noise.

In this study, the random forest regression method and surface hydrodynamic model were utilized to construct a vegetation growth prediction model. RMSE and Bias were used to evaluate the error between predicted and measured results.

$$Bias = \sum_{i=1}^{n} \left(S_{p,i} - S_{a,i}\right)/n \tag{5}$$

$$RMSE = \sqrt{\sum_{i=1}^{n} \left(S_{p,i} - S_{a,i}\right)^2/(n-1)} \tag{6}$$

where $S_{p,i}$ is the predicted result, $S_{a,i}$ is the actual result, and n is the sample size.

4 Results and Discussion

4.1 Current Status of Regional Vegetation Distribution

According to Fig. 2, the overall NDVI ranges from 0 to 0.8. The NDVI distribution does not reveal a significant correlation with the distance from the river channel. The NDVI of the area adjacent to the river channel is lower than the NDVI in some areas far away from the river channel. Apart from spatial differences, the overall NDVI in the region is the smallest in 2022 and the highest in the summer of 2021.

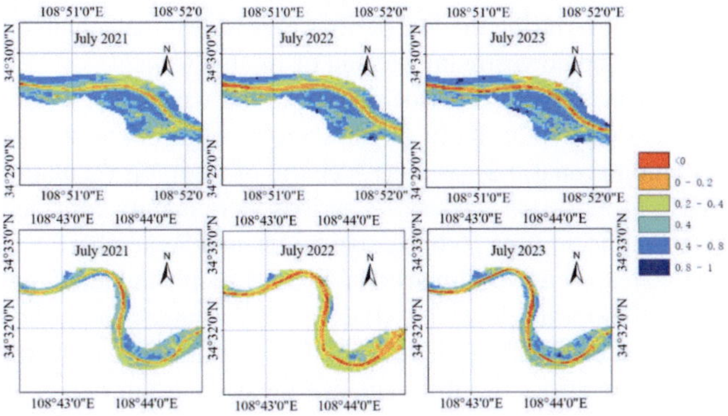

Fig. 2. The spatial distribution of NDVI in July of each year from 2021 to 2023

According to Fig. 3, the overall NDVI of the cross-section exhibits an increasing trend from the river center to both sides. For areas close to river channels, the groundwater depth and the NDVI were both small. Herbal plants with shallow groundwater depth can achieve an NDVI of 0.8 or above under suitable conditions, indicating that the growth of plants in this area has been suppressed, and it is speculated that it was affected by the suppression effect of river water immersion during the summer flood season.

The surface elevation of the high NDVI area exceeds the river water level by more than 5m, which signifies that the appropriate groundwater depth for the vegetation is generally above 5m, and the corresponding main vegetation type should be woody plants. The NDVI in this area is influenced by various factors, including river water level, surface elevation, vegetation type, previous vegetation growth status, etc. Consequently, the study area presents a complex spatiotemporal distribution of NDVI.

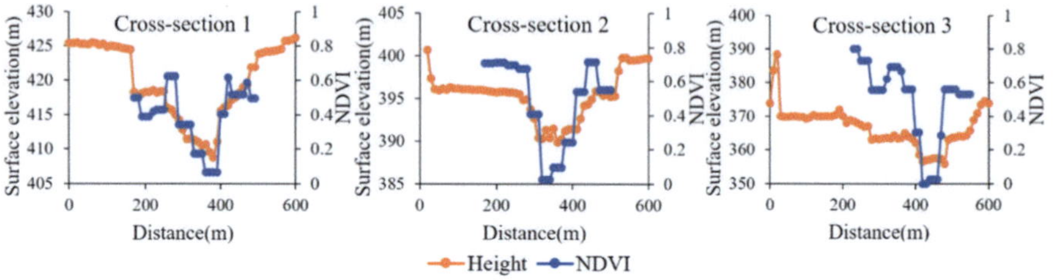

Fig. 3. The distribution of surface elevation and NDVI of the river cross-sections

4.2 The Impact of Reservoir Construction on Water Level Distribution

After the completion of the construction of Dongzhuang Reservoir, the reservoir will enter the water storage period and discharge according to the ecological base flow rate. After the reservoir storage period, Dongzhuang Reservoir will enter the normal regulation period and the discharge will be regulated based on the reservoir scheduling plan. The changes in river flow during the two periods may have an undeniable impact on the vegetation ecology in the downstream wetland of Jinghui Canal.

This study constructs three scenarios, including the scenario of a normal flow year before reservoir construction(scenario 1), the scenario of reservoir storage period after reservoir construction(scenario 2), and the scenario of reservoir regulation in a normal flow year(scenario 3). For each scenario, 3 periods were mainly considered, including early April 1st to June 15th, June 16th to August 31st, and September 1st to March 31st.

For the three scenarios, a two-dimensional hydrodynamic model of the river channel was constructed and the water level and flow velocity changes along the river channel were simulated. According to Fig. 4, compared to scenario 1, the overall decrease in river water level of scenario 2 is within 1m, and the decrease in river water level from mid-June to the end of August is relatively greater. Compared to scenario 1, the river water level during the summer period of scenario 3 can rise by nearly 1 m. In other periods, the river water level displays a downward trend compared to scenario 1. The aforementioned difference in average water level of each period caused by changes in the discharge flow of the reservoir is much smaller than the surface elevation difference between high and low NDVI areas.

The variation values in the upstream and downstream sections of the studied river section are relatively higher than those in the middle sections, which indicates that the differences in river section morphology along the river are also one of the unignoring influencing factors on NDVI distribution in the study area.

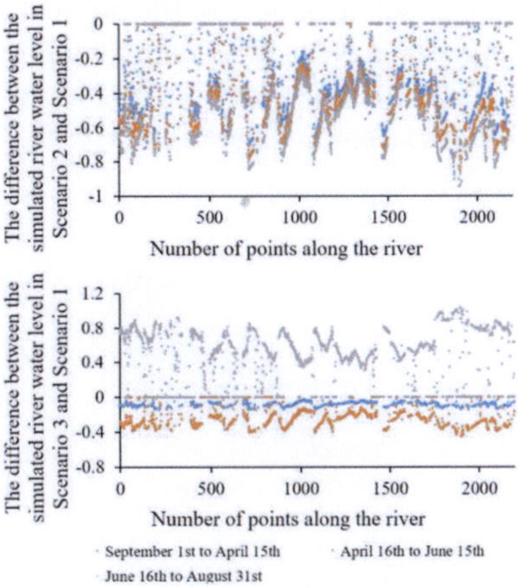

Fig. 4. The difference between the simulated river water levels from 3 scenarios along the river channel

4.3 Prediction of the Impact of Reservoir Construction on Regional Vegetation Growth

4.3.1 Model Feature Selection

The discharge flow of reservoirs varies over time, controlling the water level and flow distribution of the river. The interaction between surface water and groundwater affects the spatiotemporal distribution of groundwater depth. Surface water and groundwater, alongside other factors, collectively influence vegetation growth. The related process involves numerous parameters.

Based on the influencing factors of NDVI distribution in the area, in conjunction with the parameters of surface water flow models, groundwater flow models, and vegetation prediction models [3, 7], the characteristic feature of this prediction model needs to encompass information including river flow, river section morphology, groundwater depth, and past vegetation growth status.

In this study, the river cross-section morphology was derived by simulating the water level of the river with different discharge flows utilizing the constructed surface water model. The characteristic parameters are determined as the distance from the river, the elevation difference with the riverbed, the distribution of water level along the river under different discharge flows of Jinghui Canal (based on surface water numerical simulation), the maximum NDVI of the previous year, the mean discharge flow from April to June, and the mean discharge flow from June to August. These parameters can reflect the relevant mechanism processes to a certain extent.

4.3.2 Prediction Model Construction

Collected data including remote sensing interpretation data and cross-sectional flow monitoring data from 2021 to 2023 were used as samples to construct the NDVI prediction model for the study area. Randomly, 60% of the original dataset was set as training samples, while the remaining 40% was set as validation samples. The grid search method was employed to identify the optimal parameter configuration. The optimal number of trees was set as 200, and the max depth was set as 5 (Fig. 5).

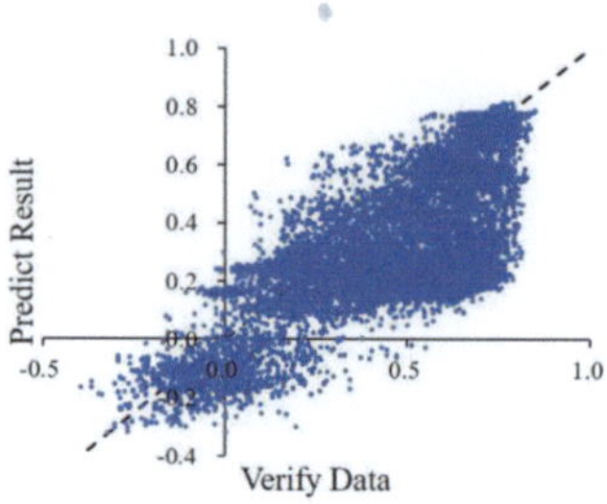

Fig. 5. Comparison between the calculation results of the model and the remote sensing interpretation results

RMSE and Bias were used to evaluate the model results. Specifically, RMSE is 0.15, and Bias is 0.24. The model calculation results have a good fit with the validation samples, indicating that the constructed model can reflect the comprehensive effects of various regional factors on NDVI. To a certain extent, this model can serve as a substitute for the intermediary process between the surface flow model, the groundwater flow model and the vegetation model.

4.3.3 Scenario Prediction

Based on the constructed NDVI spatial distribution prediction model, utilizing the NDVI spatial distribution in 2023 as the initial value, corresponding parameters for 3 scenarios were employed for continuous calculation for 5 years, and the spatiotemporal distribution of NDVI was obtained.

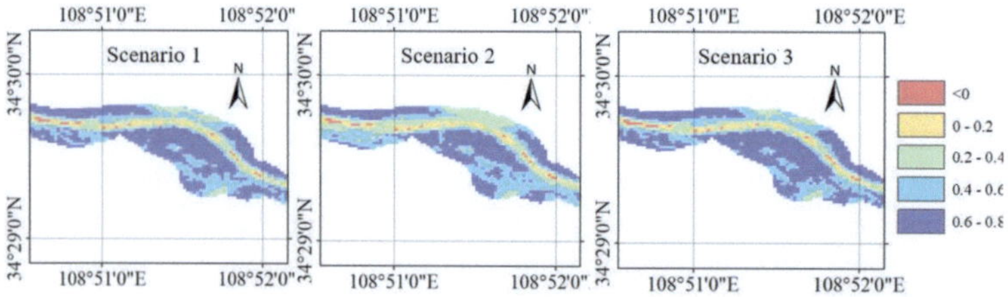

Fig. 6. The distribution of the calculated NDVI in the 3 scenarios

The Figs. 6 and 7 separately illustrate the distribution and histogram of NDVI prediction results for each scenario. In each scenario, the distribution characteristics of regional NDVI are generally similar. Scenario 2 exhibits the smallest NDVI overall, with some areas undergoing a decrease in NDVI. However, there is no significant difference in the overall distribution and the degree of reduction, indicating that the ecological base flow during the reservoir storage period can sustain regional vegetation growth to a considerable extent and prevent significant degradation in vegetation.

Compared to scenario 1, the overall difference in NDVI distribution of scenario 3 is relatively small, indicating that the impact of regulation of Dongzhuang Reservoir in normal flow year on vegetation growth is similar to the water flow process before the construction of the reservoir. Moreover, the adverse effects of the water storage period on vegetation growth will gradually recover after the end of the water storage period and the start of reservoir regulation.

After the completion of Dongzhuang Reservoir's construction, in addition to the benefits of flood control and power generation, the impact on vegetation growth is relatively slight and the construction of Dongzhuang reservoir has comprehensive positive significance.

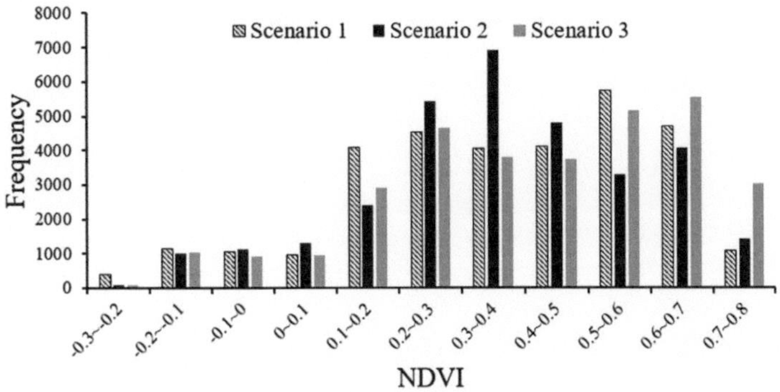

Fig. 7 Histogram of calculated NDVI of the entire study area in the 3 scenarios

5 Conclusion

This study comprehensively considered the influencing process and factors of regional vegetation growth and constructed the prediction model of spatial vegetation distribution. The constructed model can reflect the relevant process to a certain extent, mitigate the intricacies of forecasting endeavors, and ensure a certain degree of fitting. It has reference significance for other related studies. The influence of Dongzhuang Reservoir has a limited impact on vegetation growth compared to the scenario before the construction of the reservoir. The construction of Dongzhuang Reservoir has comprehensive beneficial implications.

References

1. Hongrui, L.: Cascade Reservoirs Optimization in the Middle Yellow River Based on Multi-Objective Evolutionary Algorithm. Master, North China University of Water Resources and Electric Power (2022)
2. Mengyan, G.E.: The Coupling Mechanism Between Groundwater and Vegetation Ecosystem of Terminal Lake Area of Inland River in Arid Area. Doctor, China University of Geosciences (Wuhan) (2022)
3. Han, M., Zhao, C., Feng, G., et al.: An eco-hydrological approach to predicting regional vegetation and groundwater response to ecological water conveyance in dryland riparian ecosystems. Quatern. Int. **380–381**, 224–236 (2015)
4. van der Tol, C., Dolman, A.J., Waterloo, M.J., et al.: Optimum vegetation characteristics, assimilation, and transpiration during a dry season: 2. Model evaluation. Water Resour. Res. **44**(3) (2008)
5. Sivapalan, M., Schymanski, S.J., Roderick, M.L.: Transpiration as the leak in a carbon factory: a model of self-optimising vegetation. In: AGU Fall Meeting Abstracts (2001)
6. Breiman, L.: Random forests. Mach. Learn. **45**(1), 5–32 (2001)
7. Muneepeerakul, C.P., Miralles Wilhelm, F., Tamea, S. et al.: Coupled hydrologic and vegetation dynamics in wetland ecosystems. Water Resour. Res. **44**(7) (2008)

Examination of the Environmental Footprint Associated with Renewable Energy Sources

Analysis of Meteorological Situation in Different Regions and Its Impact on Power Generation of Different Types of Solar Modules

Fengqin He[1,2,3](✉), Qi Yang[1,2,3], Xuelin Ding[1,2,3], and Weniun Lei[1,2,3]

[1] Qinghai Hydroelectric Engineering Society, Xining, China
fq_he@126.com
[2] Huanghe Hydropower Development Company Ltd, Xining Shi, China
[3] SPIC PV Industrial Innovation Center, Xi'an, China

Abstract. This paper starts from the key factors affecting photovoltaic power generation and first studies the lighting characteristics of different regions. It then conducts an analysis of the characteristics of the massive spectral data collected, determines the extraction method of spectral data, and calculates the difference in short-circuit current of photovoltaic modules due to spectral differences for specific regions. This has guiding significance for further research on the differences in the operation of novel and efficient photovoltaic modules in different regions.

Keywords: meteorological situation · spectrum · latitude division · temperature change · humidity change

1 Introduction

With the progress of photovoltaic technology, photovoltaic cell and module technology has also continued to develop, new photovoltaic cells and modules occupy more and more important market share, at the same time due to the differences in battery structure and working mechanism, the laboratory efficiency of each high-efficiency battery module and the actual operation of the power station there is a large difference. To solve this problem, starting from the key factors affecting photovoltaic power generation, this paper first studies the lighting characteristics of 13 different regions, such as Xi'an, Daqing, Garze, Gong He, Dali, Hangzhou, Ordos, Yinchuan and Emin, and analyzes the characteristics of the collected massive spectral data. The solar spectrum changes in different periods of a day, different days in January and different months in a year were studied to establish the solar spectrum analysis and value method for photovoltaic power generation. Meanwhile, the characteristics of temperature and humidity changes in different latitudes in China were analyzed and summarized, and the short-circuit current differences of photovoltaic modules due to spectral differences were calculated.

S. Zheng et al. (Eds.): IHDC 2024, LNCE 487, pp. 391–398, 2025.
https://doi.org/10.1007/978-981-97-9184-2_34

2 Experiment and Result

The short-circuit current of a photovoltaic cell module is calculated using formulas 1 and 2, where the quantum efficiency is defined as the number of particles undergoing photoelectric reaction within a fixed wavelength range divided by the number of incident particles, and the short-circuit current is the sum of the currents generated by the particles undergoing quantum effect within the same fixed wavelength range.

$$QE_\lambda = \tilde{I}_{QE,\lambda} / \tilde{I}_{Light,\lambda} \tag{1}$$

$$J_{SC} = \int QE_\lambda \cdot G_\lambda d\lambda \tag{2}$$

2.1 Analysis of Meteorological Data in Different Regions

Figure 1 presents variation of ambient temperature, average daily global solar radiation and annual average air humidity in the 13 different locations, including Xi'an, Shaanxi, Daqing, Heilongjiang, Ganzi, Sichuan, Gonghe, Qinghai, Shannan, Tibet, Hotan, Xinjiang, Dali, Yunnan, Bijie, Guizhou, Hangzhou, Zhejiang, Ordos, Inner Mongolia, Yinchuan, Ningxia, Emin, Xinjiang, and Shenzhen, Guangzhou. From Fig. 1, it can be seen that the lowest ambient temperature is ~ – 30 °C, which exists in Daqing, Heilongjiang and Emin, Xinjiang, while the highest ambient temperature is ~ – 45 °C, which exists in Xi'an, Shaanxi, Ordos, Inner Mongolia, Yinchuan, Ningxia, and Emin, Xinjiang. Therefore, variation of ambient temperature of Emin, Xinjiang is the most severe. For annual average ambient temperature, it is the highest for Shenzhen, Guangzhou, Hangzhou, Zhejiang, Dali, Yunnan, Xi'an, Shaanxi, and Bijie, Guizhou, followed by Ordos, Inner Mongolia, Emin, Xinjiang, and Yinchuan, Ningxia, it is the lowest for Hotan, Xinjiang, Daqing, Heilongjiang, Shannan, Tibet, Ganzi, Sichuan, and Gonghe, Qinghai. For average daily global solar radiation, the highest is ~ 5.0 kWh/m^2, which exists in Ganzi, Sichuan, Shannan, Tibet, Gonghe, Qinghai, Yinchuan, Ningxia, and Ordos, Inner Mongolia. The lowest daily radiation is ~ 3.5 kWh/m^2, existing in Bijie, Guizhou, and Hangzhou, Zhejiang. For annual average air humidity, the highest is ~ 80%, existing in Shenzhen, Guangzhou, Bijie, Guizhou, Hangzhou, Zhejiang, and Ganzi, Sichuan, while the lowest is ~ 39%, existing in Hotan, Xinjiang.

Figure 2 shows the monthly average global solar radiation in the 13 different locations, including Xi'an, Shaanxi, Daqing, Heilongjiang, Ganzi, Sichuan, Gonghe, Qinghai, Shannan, Tibet, Hotan, Xinjiang, Dali, Yunnan, Bijie, Guizhou, Hangzhou, Zhejiang, Ordos, Inner Mongolia, Yinchuan, Ningxia, Emin, Xinjiang, and Shenzhen, Guangzhou. From Fig. 2, it can be seen that the highest monthly global radiation exists in May to July for these 13 locations. The lowest monthly global radiation exists in January and December. Additionally, the variation of solar radiation in the whole year is smaller for Ganzi, Sichuan, Dali, Yunnan, Shannan, Tibet, compared with the other 10 locations, which can be attributed to the difference in latitude.

Figure 3 presents the annual average global solar radiation in the 13 different locations, including Xi'an, Shaanxi, Daqing, Heilongjiang, Ganzi, Sichuan, Gonghe, Qinghai, Shannan, Tibet, Hotan, Xinjiang, Dali, Yunnan, Bijie, Guizhou, Hangzhou, Zhejiang,

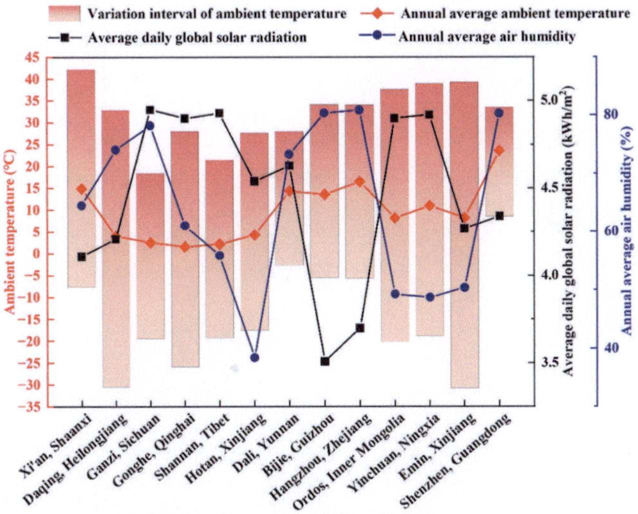

Fig. 1. Ambient temperature variation, average daily global solar radiation and annual average air humidity in 13 different locations, including Xi'an, Shaanxi, Daqing, Heilongjiang, Ganzi, Sichuan, Gonghe, Qinghai, Shannan, Tibet, Hotan, Xinjiang, Dali, Yunnan, Bijie, Guizhou, Hangzhou, Zhejiang, Ordos, Inner Mongolia, Yinchuan, Ningxia, Emin, Xinjiang, and Shenzhen, Guangzhou.

Ordos, Inner Mongolia, Yinchuan, Ningxia, Emin, Xinjiang, and Shenzhen, Guangzhou. From Fig. 3, it can be found that the variation of the annual average global solar radiation is generally stable for each location among different years. Furthermore, the radiation is the highest for Shannan, Tibet, Ganzi, Sichuan, Gonghe, Qinghai, Yinchuan, Ningxia, Ordos, Inner Mongolia, Dali, Yunnan, and Hotan, Xinjiang, followed by Emin, Xinjiang, Daqing, Heilongjiang, and Shenzhen, Guangzhou, it is the lowest for Xi'an, Shaanxi, Hangzhou, Zhejiang, and Bijie, Guizhou.

Figure 4 shows solar altitude angle variation as a function of time in the 13 different locations, including Xi'an, Shaanxi, Daqing, Heilongjiang, Ganzi, Sichuan, Gonghe, Qinghai, Shannan, Tibet, Hotan, Xinjiang, Dali, Yunnan, Bijie, Guizhou, Hangzhou, Zhejiang, Ordos, Inner Mongolia, Yinchuan, Ningxia, Emin, Xinjiang, and Shenzhen, Guangzhou. From Fig. 4, it can be found that the solar altitude angle increases from morning to noon, then decreases from noon to afternoon. Furthermore, the solar altitude angle is the largest for Shenzhen, Guangzhou, is the smallest for Emin, Xinjiang, and Daqing, Heilongjiang. This is directly caused by the difference in latitude (Shenzhen, Guangzhou (22.5° N), Emin, Xinjiang (46.5° N), Daqing, Heilongjiang (46.32° N)). In addition, it can be found that, for the same location, the solar altitude angle is the largest in summer solstice, followed by spring equinox and autumn equinox, it is the smallest in winter solstice. The variation among the different seasons is attributed to the periodic movement of the sun in the whole year.

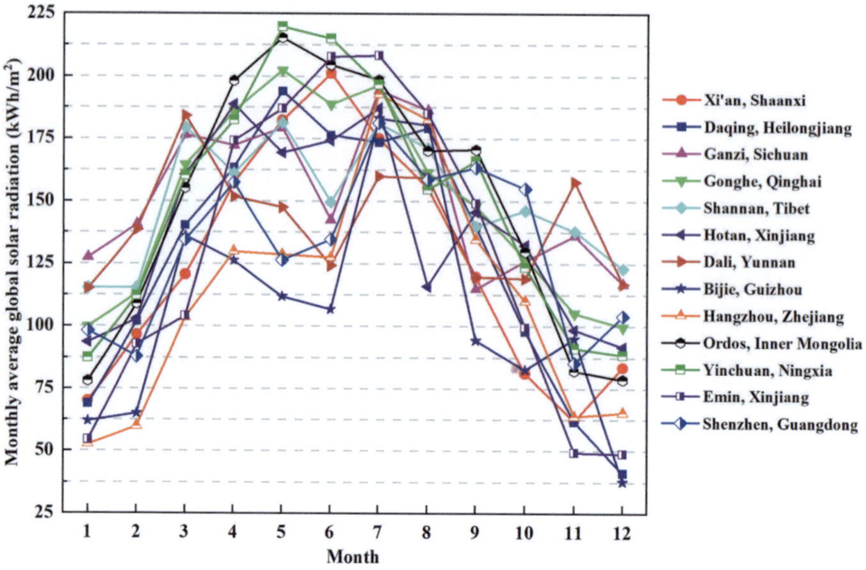

Fig. 2. Monthly average global solar radiation in 13 different locations, including Xi'an, Shaanxi, Daqing, Heilongjiang, Ganzi, Sichuan, Gonghe, Qinghai, Shannan, Tibet, Hotan, Xinjiang, Dali, Yunnan, Bijie, Guizhou, Hangzhou, Zhejiang, Ordos, Inner Mongolia, Yinchuan, Ningxia, Emin, Xinjiang, and Shenzhen, Guangzhou.

2.2 Analysis of Meteorological Data in Different Regions

Next, we selected approximately 52560 spectra from the Daqing area in China to study methods for extracting spectral data. Figure 5 Analyzing the spectral differences of Daqing every day at different hours, one set of data was taken every 1 h, a total of 12 sets of data were taken every day, and 3 days were selected every month, resulting in a total of 396 sets of spectral data analysis for the whole year. The analysis results show that the spectrum is unstable in the early morning and evening, while it maintains good convergence at other times. The representative data of the spectrum can be selected from any time other than the early morning and evening. The daily data shows the same change pattern.

Figure 6 Analyzing the spectral differences between different days of each month in Daqing, one day of data is selected every 10 days, with a total of three sets of data per month and 33 sets of spectral data analyzed annually. The spectra of each day in a month show good convergence, and a typical weather spectrum can be selected as the representative data for that month.

Figure 7 Analyzing the spectral differences of Daqing every month, one day of data is selected for each month and 12 sets of data are selected for the entire year for analysis. In general, in the summer in Daqing, blue light dominates and red light proportion decreases, while the opposite is true in the winter.

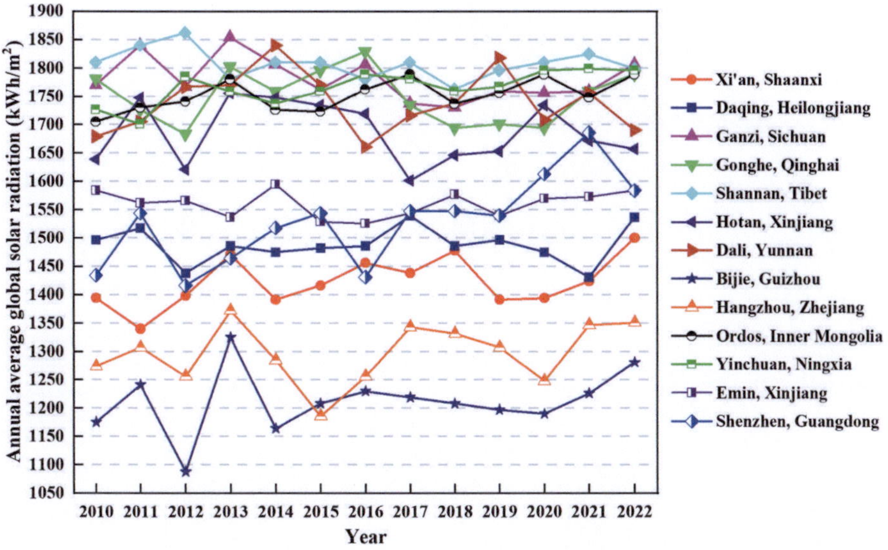

Fig. 3. Annual average global solar radiation in 13 different locations, including Xi'an, Shaanxi, Daqing, Heilongjiang, Ganzi, Sichuan, Gonghe, Qinghai, Shannan, Tibet, Hotan, Xinjiang, Dali, Yunnan, Bijie, Guizhou, Hangzhou, Zhejiang, Ordos, Inner Mongolia, Yinchuan, Ningxia, Emin, Xinjiang, and Shenzhen, Guangzhou.

2.3 Analysis of the Impact of Different Spectral Distributions on Short-Circuit Current of Different Components

Figure 8 Quantum efficiency curves for P-type crystalline silicon PERC cells and N-type crystalline silicon heterojunction cells. The short-circuit current differences between PERC cells and HJT cells under laboratory test conditions (AM1.5) and Daqing actual spectrum were calculated using formula 2-1 and 2-1. The calculation results show that the short-circuit current of the HJT cell under the AM1.5 spectrum is 40.36 mA/cm2, and the short-circuit current under the Daqing actual spectrum is 36.11 mA/cm2, with a current difference of 11.7%. The short-circuit current of the PERC cell under the AM1.5 spectrum is 39.01 mA/cm2, and the short-circuit current under the Daqing actual spectrum is 34.79 mA/cm2, with a current difference of 12.3%.

2.4 Conlusion

A comprehensive analysis of the climate differences in different regions was conducted, and a spectral data extraction method was established based on the analysis. Taking into account the characteristic that the spectral distribution of different regions has differences, the differences in short-circuit current brought by the PERC and HJT cells in the Daqing area and the laboratory test spectrum were calculated. This study provides guidance for further research on the differences in the performance of new and efficient photovoltaic components in different regions.

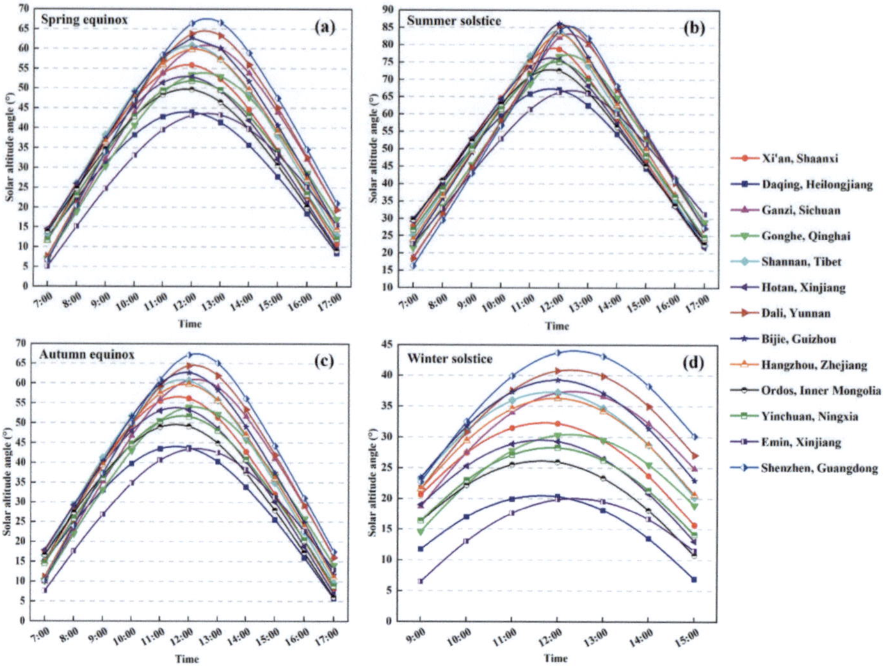

Fig. 4. Solar altitude angle variation as a function of time in 13 different locations, including Xi'an, Shaanxi, Daqing, Heilongjiang, Ganzi, Sichuan, Gonghe, Qinghai, Shannan, Tibet, Hotan, Xinjiang, Dali, Yunnan, Bijie, Guizhou, Hangzhou, Zhejiang, Ordos, Inner Mongolia, Yinchuan, Ningxia, Emin, Xinjiang, and Shenzhen, Guangzhou. (a) Spring equinox, (b) summer solstice, (c) autumn equinox, (d) winter solstice.

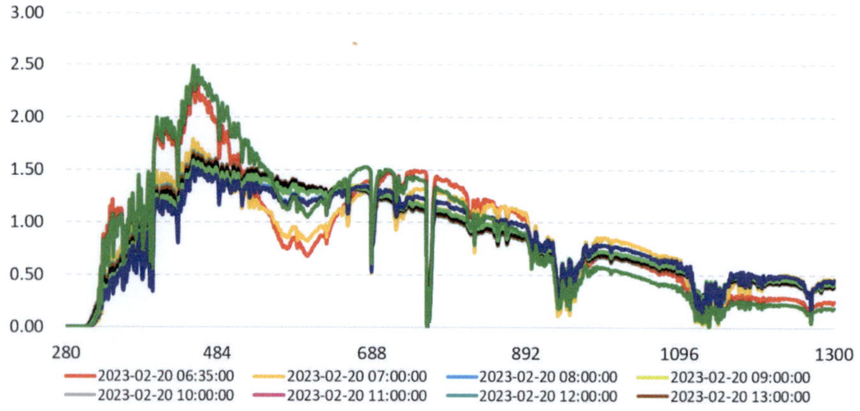

Fig. 5. Spectral data at different times of the day in Daqing

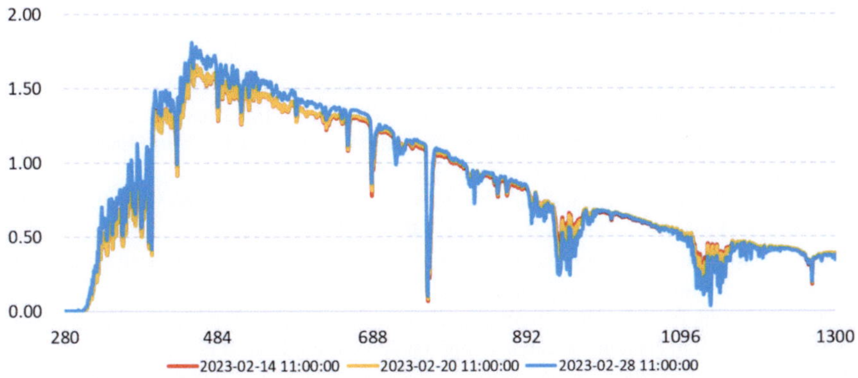

Fig. 6. The spectral signatures of different days in a certain month in Daqing

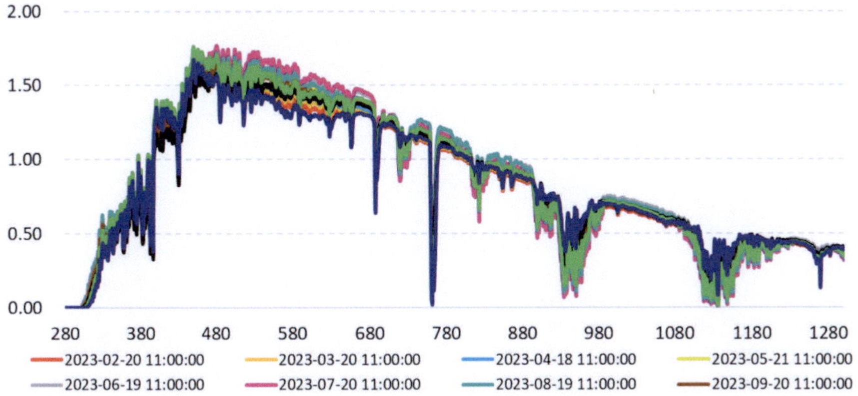

Fig. 7. The spectral signatures of different months in Daqing

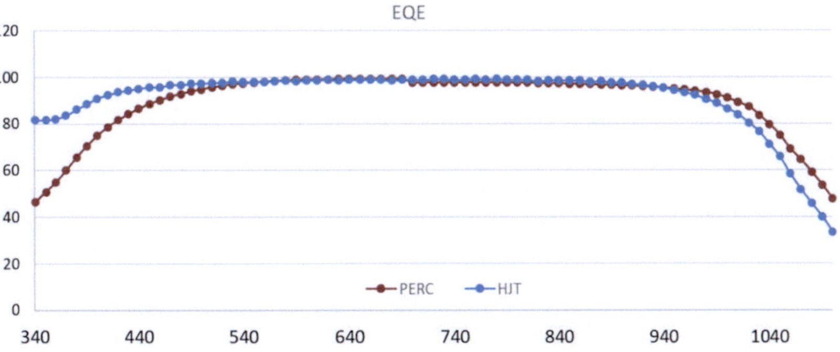

Fig. 8. Quantum efficiency of Perc cell and HJT cell

References

1. Becquerel, A.E.: Photoelctrochemical effect. CR Acad. Sci. Paris **9**, 14 (1839)
2. Theologitis, I.T., Jaeckel, B.: SolarPower Europe: O&M Best Practices Guidelines (2016)
3. Greenpeace, Europe S.P.: GWEC. Energy [r]evolution - A Sustainable World Energy Outlook: 100% Renewable Energy for All (2015)
4. Hegedus, S.: Thin film solar modules: the low cost, high throughput and versatile alternative to Si wafers. Prog. Photovoltaics Res. Appl. **14**(5), 393–411 (2010)
5. Chapin, D.M., Fuller, C.S., Pearson, G.L.: A new silicon p-n junction photocell for converting solar radiation into electrical power. J. Appl. Phys. **25**, 676–677 (1954)
6. Bothe, K., Hezel, R., Schmidt, J.: Recombination-enhanced formation of the metastable boron–oxygen complex in crystalline silicon. Appl. Phys. Lett. **83**, 1125–1127 (2003)
7. Schutz-Kuchly, T., Dubois, S., Veirman, J., et al.: Light-induced degradation in compensated n-type Czochralski silicon solar cells. Physica Status Solidi A, **208**(3), 572–575 (2011)
8. Singha, B., Solanki, C.S.: N-type solar cells: advantages, issues, and current scenarios. Sol. Energy Mater. Sol. Cell. **4**, 072001 (2017)

Comprehensive Studies on the Combined Environmental Effects of Integrated Energy Projects

Construction of a Full Process Evaluation for the SocialIntegration of Migrants in Water Conservancy and Hydropower Projects

Sun Zhonggen[1], Shao Ziting[1(✉)], Zhang Huazhong[2], Chen Jihua[1], Yu Qingnian[1], and Wang Yifei[1]

[1] National Research Center for Resettlement, Hohai University, Nanjing 21009, Jiangsu, China
ziting_shao@163.com

[2] Changjiang Water Resources and Hydropower Development Group, Wuhan 430014, Hubei, China

Abstract. China's water conservancy and hydropower enterprises continue to face the challenge of lagging hydropower standards despite being a major player in the construction of water conservancy and hydropower projects. This paper presents a theory for evaluating the social integration process of immigrants in water conservancy and hydropower projects. The theory aims to promote China's 'Belt and Road' strategy, enhance China's modernization, and establish China's right to speak. The paper explains the overall framework, process evaluation, and result evaluation of the theory. The evaluation process can be divided into three periods: planning, placement, and later support. The result evaluation is divided into three aspects: environmental adaptation, population development, and cultural integration. Using the Three Gorges Project as an example, this text preliminarily tests the science, rationality, and feasibility of the entire assessment theory. In the future, research on engineering immigration should expand the application scope of the whole process assessment. Sufficient attention should be paid to the problem of social restructuring, and efforts should be made to build an assessment system with Chinese characteristics.

Keywords: Engineering immigration · Social integration · Full process evaluation

1 Introduction

The construction of water conservancy and hydropower projects can result in the disintegration and fragmentation of migrant communities, with the original social structure of migrants being broken up and the platform of social organisation and interpersonal relationships being destroyed. Attention to the resettlement and subsequent development of reservoir migrants is related to the overall situation of social stability and development. With the development of economy and society, the resettlement and integration of migrants have received particular attention from the academic community [1]. Generally speaking, the construction of water conservancy and hydropower projects takes a long

S. Zheng et al. (Eds.): IHDC 2024, LNCE 487, pp. 401–422, 2025.
https://doi.org/10.1007/978-981-97-9184-2_35

time, and in the face of such a huge project, there is an urgent need to review the whole process of assessment of the three stages of planning, resettlement and post-assistance, and overcome the limitations of stage-by-stage assessment, in order to promote the social integration of immigrants, and then complete the restoration and reconstruction of the immigrant social system, to realise the harmonious coexistence of immigrants and the hydraulic project, to bring the benefits of the project into full play, and ultimately to achieve the goal of common prosperity in the new era.

Internationally, under General Secretary Xi Jinping's "One Belt, One Road" initiative, China's water conservancy and hydropower engineering-related enterprises have begun to go out and explore overseas business. Social integration is a core component of the work of water conservancy and hydropower project migrants, and is also an important element to consider in evaluating the success or failure of water conservancy and hydropower project migrants, and even the entire project. In the process of social integration of water conservancy and hydropower project immigrants, the assessment of the integration status is particularly critical. Internationally, regarding the assessment of the social integration status of water conservancy and hydropower project immigrants, the International Hydropower Association (IHA), the International Association for Impact Assessment (IAIA), the International Association of Dams (WCD) and other Western-led international organisations have put forward the Sustainability Assessment Protocol (SAP) and other technical systems, which include social integration [2]. However, the corresponding technical system in China is still in a blank stage. As General Secretary Xi Jinping pointed out, "standards boost innovation and development, and standards lead the progress of the times. In the international market, who mastered the standard, who means the first to get the ticket to enter the international market, seize the high ground of the international market [3]. Based on the strategic overall situation of the great rejuvenation of the Chinese nation and the world's great changes that have not occurred in a hundred years, to help China's hydropower "go out", there is an urgent need for China's technical standards of condensation, construction and promotion of the use of China's technical standards, and to enhance China's right to speak in the international arena.

Accordingly, this paper aims to construct a social integration system for engineering migrants with Chinese characteristics, analysing the Three Gorges Project, a mega project with great Chinese characteristics, as an example, and introducing the Full Process Evaluation for Social Integration, referred to as FPESI, which conducts a whole-process evaluation of the planning period, the resettlement and implementation period and the post-support period), and combines an all-around evaluation with the process evaluation by the experts and the final effect by the migrant households. Overall, this study aims to promote the social integration of engineering migrants and ultimately achieve common prosperity, innovate the social integration system of engineering migrants with Chinese characteristics through the introduction of the whole-process assessment method, and prove the feasibility and applicability of the theory with the help of the Three Gorges Project as a case study, so as to help China's hydropower engineering standards "go out" and establish the right to speak in Chinese hydropower project development. The project is expected to help China's hydropower project standards "go out" and establish China's discourse power in hydropower project development.

2 Literature Review

Shi and Chen took the lead in pointing out that the destruction of primary social networks over blood and geography would cause adaptation difficulties for immigrants, and called for the strengthening of sociological research in engineering immigration [4]. Around the twenty-first century, the Three Gorges Project attracted world attention, and academics have paid extra attention to the measurement of immigrants' social integration brought about by this project. Feng Xiaotian's study divided the social adaptation of immigrants into three dimensions: economic adaptation, psychological adaptation and cultural adaptation [5]. According to Li and Jiang, the main factors affecting the social integration of immigrants in the Three Gorges Project are political, economic and cultural, and the main signs of immigrants' social integration are political equality, economic synchronisation and cultural integration [6]. Wang also adopted the same classification for the influencing factors and integration signs of social integration [7]. Gao and Xu explored the integration and adaptation problems of "landless resettlement" migrants in Beiyuan resettlement community from four dimensions: residential space, livelihood space, social space and cultural and psychological space [8].

From the experience of social integration of engineering migrants abroad, the World Bank, in the engineering migrant projects it finances, places special emphasis on the fact that development projects should minimise the number of migrants, attaches importance to the use of sociological research results in resettlement, and focuses on the planning of migrant resettlement and the social development of its reservoir areas [9]. The World Bank's view is that the mark of success of any resettlement is whether the standard of living of the migrants and the resettlement area is maintained or improved. The process is that migrants through contact, interaction, communication, penetration and mutual acceptance with the society in the place of relocation, and finally achieve economic integration, cultural adaptation, social integration and identity [10]. Cernea [11] believes that rural migrants who leave their land and homes due to engineering construction often enter the city because of the disintegration of the original social structure and the destruction of social order caused by the impact of the urban and rural living environment, culture, customs, differences in production and living styles, breaking the balance of humanities and ecology, which makes the migrants face the dilemmas and maladjustment of production and life.

Domestic assessment of engineering migrants can be divided into two categories, one is for the assessment of planning, resettlement or back-up [12–14] and other stages of work; the other is for different research themes such as the risk of social stability of migrants, protection of migrants' rights and interests [15, 16] and so on. In addition, research methods such as fuzzy comprehensive evaluation, hierarchical analysis method, participatory assessment [17–20] and so on are also widely used in the field of social assessment of engineering migrants. Guideline OD4.30 "Involuntary Migration" of the World Development Bank and Chapter 50 "Involuntary Migration" of the Operations Manual of the Asian Development Bank are common references for scholars at home and abroad. The International Hydropower Institute (IHE) has proposed the Sustainability Assessment Programme (SAP), which assesses the different stages of the project cycle in terms of criteria in four parts: strategic assessment of projects providing energy and water services; hydropower project preparation (i.e., the various studies and plans carried out prior to the awarding of the construction contract); hydropower project implementation; and hydropower project operation. However, the linkage of this system with other assessment frameworks, the scope and adequacy of the themes, and the importance of strategic planning are subject to further debate [2].

The success of a project is a natural result after the whole process of pre, middle and post sequentially takes effect [21]. The existing whole-process evaluation follows the classification of the three major blocks of the project before, during and after, and is refined in the division of the specific procedures according to the disciplines of investment science, environmental science and administrative science to which it belongs [22–25]. From abroad, Rossi et al. [26] put forward an implementation assessment model for policy assessment, pointing out that in the process of governance, policy assessment must pay attention to the development of the situation in the process of governance. The implementation assessment model consists of two types: process assessment and outcome assessment, with process assessment referring to the assessment of the governance process, focusing on the way in which the policy provides services and on some internal factors affecting the implementation of the policy to fulfil the set objectives; outcome assessment refers to the assessment of the results of the governance, with assessment indicators usually set beforehand, and the use of policy assessment standards to measure the results of the governance or the operation of the project.

In view of the involuntary, long-cycle and whole-household relocation characteristics of engineering migrants, this paper's assessment of their social integration status is summarised in three dimensions. First, environmental adaptation is the foundation of social integration. Social adaptation of the living environment and basic public service facilities can guarantee that migrants can "move out". Secondly, demographic development is a further requirement for the social integration of migrants. Following the law that "human beings are the most active elements in a social system", the political and economic factors that have been studied should be grouped into the dimension of demographic development, and specific indicators should be used to measure the livelihoods of migrants' families, their social interactions and social equity, so as to ensure that migrants can "hold on to their homes". Lastly, cultural integration reflects spiritual integration; only when customs and habits and other cultural aspects are adapted can migrants truly achieve social integration.

3 Theoretical Analysis of Social Integration of Migrants in Water Conservancy and Hydropower Projects

3.1 Concept of Social Integration of Migrants in Water Conservancy and Hydropower Projects

Social integration is the gradual acceptance and adaptation to the social culture of the place of relocation by the migrant population, and the construction of benign interaction, and ultimately the formation of mutual recognition, mutual "penetration, intermingling, reciprocity and complementarity" [27]. The social integration of water conservancy and hydropower project immigrants refers to a social state in which immigrants are relocated and, on the basis of guaranteed living environment and public service facilities, gradually adapting to the environment, pursuing demographic development and cultural integration, and finally integrating into the resettlement place. Environmental adaptation, demographic development and cultural integration constitute a unified whole for the social integration of migrants.

3.2 Content of Social Integration of Migrants in Water Resources and Hydropower Projects

Based on the whole-process assessment framework, the whole-process assessment of social integration of migrants in water conservancy and hydropower projects will be carried out in the three major aspects of environmental adaptation, demographic development, and cultural integration in the three different phases of planning, implementation, and post-support, respectively. The content of the assessment is shown in Table 1.

Table 1 Content of social integration of migrants in water resources and hydropower projects

Primary index	Secondary index	Implication
Environmental adaptation	Living environment	Safe living environment, housing area, housing quality
	Basic public service	Residential water supply, power supply, road traffic, commercial retail, medical and health care, leisure and entertainment, basic education service supply
Population development	Family livelihood	Allocation of land quantity and quality, allocation of land ownership, non-agricultural employment opportunities, employment discrimination, existing livelihood skills, productive skills training, immigrant income, etc.
	Social interaction	Close contact with relatives and friends, and social integration with indigenous people
	Social equity	Gender equality and vulnerable groups
Cultural fusion	Religion	Freedom of religious belief, places for religious activities, etc.
	Customs and habits	Differences in customs and habits between the original residence and the resettlement place, original customs and habits of immigrants, etc.
	Minority nationality	Ethnic minority festival activities have been effectively carried out

The evaluation of social integration is mainly carried out from three subsystems: environmental adaptation, population development and cultural integration. Specifically, environmental adaptation refers to whether the living environment is safe and adaptive, and whether the social service function is complete after the relocation. Population development refers to population quality, family livelihood and social interaction. Cultural integration refers to the reasonable and orderly integration of basic religions, customs and national systems contained in social activities. Environmental adaptation, population development and cultural integration constitute a unified whole of immigrant social integration. The three subsystems constantly exchange material, energy and information, interact and restrict each other, and move towards the common goal.

3.3 Basic Process of Social Integration of Water Conservancy and Hydropower Engineering

In the process of social integration of water conservancy and hydropower immigrants, the main work of the planning stage is centered on the "relocation" of immigrants, and the final results of the planning stage are reflected in the adequate preparation for the implementation of resettlement, and the social integration of immigrants is a crucial part of it. Social integration in the planning stage is reflected in the environmental adaptation of immigrants, which is reflected in the adaptation of the environmental capacity of the planning area to the actual situation of immigrants [28], and the protection of the living environment and basic public services. In terms of population development, based on the policy of development migration and the resettlement method combining early compensation and subsidies with late support, the planning stage takes the family as a unit to comprehensively consider the livelihood, social communication and social equity of immigrants. In terms of cultural integration, due to the people-oriented engineering concept, cultural factors such as religion, customs and ethnic minorities are considered.

"Stability" is the focus of work in the implementation stage of resettlement, and the environmental adaptation, population development and cultural integration of immigrants are preliminarily tested in the implementation stage. The technical work flow of this stage includes the preparation of the implementation plan of migrant resettlement, the preparation of the periodic report of the implementation plan of migrant resettlement, the implementation organization, the design change, the acceptance of migrant resettlement, and the adjustment of the budget estimate. The social integration work of immigrants in the implementation stage is based on the regulations and norms formulated in the planning stage. If there are no special circumstances, it will be implemented. For those that do not meet the requirements of social integration, corresponding adjustments will be made according to the process.

After the implementation is completed, the follow-up work of immigrants enters the later stage of support, which focuses on "gradually getting rich". In a period of time after the completion of resettlement, the production and living of the immigrants shall be continuously supported, so that the production and living standards of the immigrants can reach or exceed the original level, and create conditions for their sustainable development and synchronous development with the regional economy and society. In the later stage of support, the social integration of immigrants pays more attention to development [29]. In terms of environmental adaptation, the focus on basic public services increases the indicators of the provision of health services. In terms of population development, indicators such as income and consumption of immigrants and their contacts with relatives and friends have been taken into account. In terms of cultural integration, more emphasis is placed on the integration of immigrants and regional cultural backgrounds.

4 Evaluation Framework for the Full Process of Social Integration of Migrants in Water Conservancy and Hydropower Projects

4.1 Overall Framework of the Full Process Evaluation

4.1.1 General Ideas

MichaelM.Cernea once pointed out that immigration research needs the participation of two aspects of social science, one is the academic research on the basic process of immigration, and the other is the evaluation research on the actual operation results of immigration. Only the cross-application of these two aspects of social science can construct the framework of social action and guide the smooth progress of work. In July 2006, The State Council promulgated the Regulations on Land Compensation and Resettlement of Migrants for the construction of Large and medium-sized Water Conservancy and hydropower Projects (Decree No. 471 of The State Council), which put forward the general requirements for the supervision and evaluation of the whole process of resettlement, which clarified the purpose, nature and objective content of the supervision and evaluation of resettlement, and ensured that the resettlement activities were carried out as planned. So as to achieve the goal of resettlement and ensure the smooth construction of water conservancy and hydropower projects. Shao Kan's evaluation of engineering resettlement projects is also divided into post-evaluation of early work, post-implementation, post-implementation effect, post-impact and post-sustainability. Therefore, the whole process assessment of the social integration status of water conservancy and hydropower migrants includes before, during and after the project, conducting academic research on the basic work in the planning stage, implementation stage and later support stage, and constructing an "environmental adaption-population development-cultural integration" system in order to evaluate the results of the immigration work. At the same time, it also defines the evaluation subject, evaluation index, evaluation standard and evaluation method. The overall frame diagram is as follows (Fig. 1).

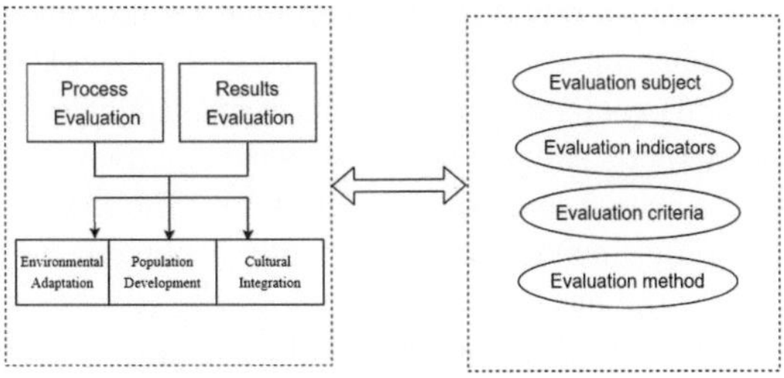

Fig. 1 General framework diagram

4.1.2 Process Evaluation Ideas

The process evaluation is carried out from three aspects: planning period, placement period and later support period, and the evaluation method is to invite relevant personnel to score. The same index system was adopted in the assessment of environmental adaptation, population development and cultural integration in the planning period and the resettlement period, and the index of the later support period increased slightly with the economic and social development.

4.1.3 Results Evaluation Ideas

The result evaluation is the evaluation of the overall completion of the resettlement after implementation. Using the information obtained from literature collection, data review, colloquiums and questionnaires conducted by households, the "goal-result" comparison method was adopted to evaluate the specific situation of the three dimensions of environmental adaptation, population development and cultural integration of immigrants.

4.2 Process Evaluation

4.2.1 Assessment Subject

1. Implementation subject – independent third-party evaluation agency

In the evaluation of water conservancy and hydropower projects, the project owner is often the main body of responsibility. In order to avoid "self-evaluation", the project owner entrusted social intermediary organizations as independent third parties to carry out evaluation work in accordance with relevant regulations. In the evaluation of the social integration status of engineering immigrants, an independent third party, as an evaluation agency, designs a special questionnaire according to the actual situation of the evaluated project in the planning period, resettlement period and later support period, and invites relevant social organizations, professional institutions, experts and scholars, as well as representatives of the masses involved in the project to evaluate.

2. Participants - all kinds of experts

The participants in the three stages of planning, placement and post-support are generally similar, but they are slightly different according to the actual work progress. In the whole process of evaluation, according to the needs of the work, the project will also consult academicians and experts with attainments in relevant fields, including appropriate reference to young experts and old experts who have participated in the early work, to participate in the evaluation and consultation work. The major involves immigration management, sociology, economics, engineering management, water conservancy engineering and other disciplines.

The evaluation subjects in the planning period are experts in planning preparation, experts in immigration argumentation, experts in immigration discussion and government officials. The main body of the evaluation during the implementation period is experts in planning, immigration argumentation, immigration discussion, local government officials, implementation supervision and evaluation experts. The main body of integration evaluation in the late support period is planning experts, immigration argumentation experts, immigration thematic discussion experts, local government officials, supervision and evaluation experts in the later support and follow-up work planning.

4.2.2 Evaluation Indicators

According to the theoretical framework of the whole process evaluation, the whole process evaluation of the social integration of immigrants in water conservancy and hydropower projects will be carried out from three aspects: environmental adaptation, population development and cultural integration in three different stages: planning, implementation and later support. With the development of economy and society, some indicators of the later stage of support should be updated accordingly. According to the evaluation content, the corresponding index system is constructed in the following Table 2.

Table 2 Whole process evaluation index system of social integration

Evaluation content	Index system		
	Environmental adaptation	Population development	Cultural fusion<
Integration process			
Planning stage	(1) Living envionment 1. Safe living environment 2. Housing area 3. Housing quality (2) Basic public services 4. Residential water service 5. Residential power supply service 6. Residential road traffic services 7. Residential commercial retail services 8. Health services in the place of residence 9. Residence leisure and entertainment services 10. Supply of basic education services	(1) Family livelihood 1. Allocation of land quantity and quality 2. Allocation of land ownership 3. Non-farm employment 4. Employment discrimination 5. Existing livelihood skills 6. Training of production skills (2) Social exchanges 7. How close friends and relatives are to each other 8. Social integration with indigenous peoples (3) Social equity 9. Gender equality 10. Vulnerable groups	(1) Religion 1. Freedom of religious belief 2. Places for religious activities (2) Customs and habits 3. Differences in customs and habits between the original residence and the place of resettlement 4. Original customs and habits of immigrants (3) Ethnic minorities 5.The festival activities of ethnic minority immigrants were effectively carried out
Implementation phase	Same as planning stage indicators	Same as planning stage indicators	Same as planning stage indicators

(*continued*)

Table 2 (*continued*)

Evaluation content	Index system		
	Environmental adaptation	Population development	Cultural fusion<
Integration process			
Late support stage	Increase the supply of health services on the basis of indicators in the planning and implementation stages	Changes on the basis of indicators in the planning and implementation stages (1) family livelihood 1. Education level of immigrants 2. Life expectancy of migrants 3. The spiritual life of immigrants 4. Allocation of quantity and quality of land 5. Stable tenure of alloted land 6. Non-farm employment 7. Employment discrimination 8. Existing livelihood skills 9. Production skills Training 1 (Training or not) 10. Production Skills Training 2 (Training effect) 11. Income of migrants 12. Immigrant consumption (2) Social exchanges 13. How close friends and relatives are to each other 14. Support from friends and family 15. Relations between migrants and indigenous peoples of resettlement 16. Intermarriage between migrants and indigenous peoples in settlements (3). Social equity 17. Gender equality 18. Vulnerable groups	Changes on the basis of indicators in the planning and implementation stages (1) Religion 1. Conflicts between immigrants and indigenous peoples of different faiths 2. Religious activities (2) Customs and habits 3. Language acceptance 4. Familiarity with customs (3) Ethnic minorities 5. Ethnic minority immigrants carry out festivals 6. Customs of minority immigrants 7. Conflicts between immigrant and indigenous peoples

4.2.3 Evaluation Criteria

According to the concept of whole-process assessment, we set different evaluation criteria for the three indicators of environmental adaptation, population development and cultural integration in the three different stages of planning, implementation and later support, so as to evaluate the migration work of hydropower projects at different stages. For example, in terms of the living environment safety index of environmental adaptation in the planning stage, the planning fully considered the living environment safety of immigrants in the resettlement place was rated 5 points; In the planning, more requirements for the safety of the living environment of immigrants in the resettlement place were rated 4 points; The planning put forward certain requirements for the safety of the living environment of the immigrants in the resettlement place and was rated 3 points; The requirements for the safety of the living environment in the resettlement place in the planning were less rated 2 points; The planning did not consider the safety of the living environment of the migrants in the resettlement place at all, and was rated 1. Overall, the criteria are divided into five levels, with 5 representing optimal performance, 4 representing sub-optimal performance, 3 representing basically satisfactory performance, 2 representing slightly inferior performance, and 1 representing the worst performance.

4.2.4 Evaluation Method

With reference to Shao Kan's research on immigration policy evaluation [30], the Delphi Method is used in the process evaluation. An independent third party of the evaluation implementer designs a three-stage expert questionnaire to consult the opinions of provincial experts in related fields and government leaders at all levels in the form of a questionnaire, improve the index system and evaluation content, construct the final social integration index system and explain the scoring criteria. Finally, the participants in the evaluation, namely various experts, are invited to score the planning, resettlement and post-support stages.

4.3 Result Evaluation

4.3.1 Evaluation Subject

The implementation subject of the result evaluation is similar to that of the process evaluation, both of which are commissioned by an independent third party or conducted by a qualified and competent social consultation and evaluation intermediary unit through bidding. The difference is that in terms of participants, immigrant groups are included.

4.3.2 Evaluation Indicators

For the consistency of evaluation, process evaluation and outcome evaluation adopt the same index system, which are the evaluation of three subsystems: environmental adaptation, population development and cultural integration.

4.3.3 Evaluation Process

Through literature research, seminars, field observations, household interviews and questionnaire surveys, an independent third party collects information on the sub-systems of environmental adaptation, population development and cultural integration, and uses the evaluation idea of "goal-result" comparison to conduct targeted evaluation based on goals and results.

The assessment of environmental adaptation is generally divided into two aspects: living environment and basic public services. The goal of living environment is not only to ensure the geological safety and the quality of houses, but also to ensure that the per capita area of resettlement houses is not lower than that of surrounding residents. In terms of basic public services, the supply of water, electricity, roads, commercial retail, education, medical care, leisure and health services should meet the basic needs of immigrants. In the analysis of the implementation results, the living environment should be evaluated by integrating the monitoring and evaluation reports of the project over the years. For example, the migration planning and design of the Three Gorges Project of the

Yangtze River pointed out that considering the moderate development of immigrants, the urban planning and construction land area should be calculated according to 70m^2/ person, which is higher than the current 30–50m^2 / person [31]. This indicator will be verified during the outcome evaluation process. The evaluation of basic public services is measured by the primary data obtained from questionnaires.

The goals of population development include three aspects: family livelihood, social interaction and social equity. Family livelihood is related to the education level of migrants, the life expectancy of the population, the state of spiritual life, the quantity and quality of land, the stability of land tenure, non-agricultural employment opportunities, and employment discrimination. Social interaction is related to the degree of close contact with relatives and friends and support, the relationship between immigrants and the indigenous people in the resettlement place, and intermarriage. Social equity is related to gender equality and vulnerable groups. In the analysis of implementation results, the official data and questionnaire survey data are also combined to evaluate.

The goal of cultural integration is from three aspects: religion, customs and minority nationalities. In the religious context, the goal is that there is no conflict between immigrant and indigenous peoples of different faiths and that religious activities are carried out normally. The goal of custom is a certain degree of familiarity with language and custom. For ethnic minorities, the festival activities of ethnic minorities immigrants are carried out normally, customs and habits are respected, and there is no conflict between immigrants and indigenous people. In the analysis of implementation results, the evaluation method of combining official data and questionnaire survey is adopted.

4.3.4 Source of Data

Evaluation data sources are divided into two parts, one is to carry out discussions and collect statistical yearbooks and statistical communiques, monitoring and evaluation reports, acceptance reports and other materials as units involved in the project; the other is to carry out questionnaires on immigrant households as units of villages.

5 Case Analysis of the Full Process Evaluation of Social Integration of Migrants in Water Conservancy and Hydropower Projects

5.1 Three Gorges Project and Three Gorges Migration

According to the "Yangtze River Three Gorges Project reservoir inundation treatment and resettlement planning Report", the Three Gorges Project reservoir area planned to relocate a total population of about 1.245,500 people, in the subsequent development and implementation, the total number of migrants has expanded to 1.27 million. There are many challenges and difficulties in the reorganization of reservoir productivity, the adjustment of economic structure, the social integration of immigrants and the management of immigration work. Migration is the key to the success of the Three Gorges Project. Based on this, this paper takes the rural migrants in the Three Gorges reservoir area of Hubei as a case study, and uses the whole-process assessment method to evaluate their social integration status.

5.2 Evaluation Process

The whole process evaluation is divided into process evaluation and result evaluation. The process evaluation sets the target of each index as 3 grades, and the index system, standard, subject and method refer to the evaluation system of the whole process of social integration of water conservancy and hydropower project immigrants. Based on the questionnaire designed for the evaluation of the whole process of social integration of migrants in water conservancy and hydropower projects, 30 migration experts, officials and scholars who are familiar with the planning, resettlement and post-support processes of the Three Gorges Project are selected for the evaluation. They are from Yangtze River Survey, Planning and Design Research Institute of Yangtze River Water Resources Commission, Yangtze River Three Gorges Group Co., LTD., Yangtze River Project Supervision and Consulting Company; Yichang City, Yiling District, Zigui County, Xingshan County, Badong County; Hohai University, Nanjing University, Three Gorges University and other units, and finally recovered 28 effective expert questionnaires. Results The overall objective of the evaluation is to balance the benefits of the project and the benefits of the immigrants. Considering that the data comes from both primary and secondary data, the structural evaluation method combining quantitative and qualitative is adopted. In the evaluation of specific indicators, the primary data is the main, the evaluation goal is to set the satisfaction of immigrants at 85% or above, and some indicators are based on the requirements of official documents such as the "Report on reservoir inundated treatment and resettlement planning of the Three Gorges Project on the Yangtze River" (1998) and the "Regulations on the construction of the Three Gorges Project on the Yangtze River" (2001).

5.3 Evaluation Results and Conclusions

5.3.1 Evaluation Results

(1) Process evaluation

Comparing the statistics of the experts' scores with the goals set in the planning period, the Three Gorges resettlement planning is relatively comprehensive, most of the indicators in environmental adaptation and population development can be achieved, and more than half of the goals in the five indicators in cultural integration can be achieved. The goal of the difference of customs between the place of origin and the place of settlement, and the original customs of the immigrants has not been achieved. In the implementation stage and the later support stage of the Three Gorges resettlement, the environmental adaptation, population development and cultural integration have been improved compared with the planning stage, and all indicators can be achieved, and some indicators have been achieved to a high degree (Figs. 2, 3, 4, 5, 6, 7, 8, 9, and 10).

(2) Result evaluation

First of all, in terms of environmental adaptation, for the geological safety of the living environment of immigrants, Article 24 of the "Regulations on the Construction of the Three Gorges Project on the Yangtze River" (2001) requires that "the site selection

Fig. 2 Degree of environmental adaptation in the planning stage

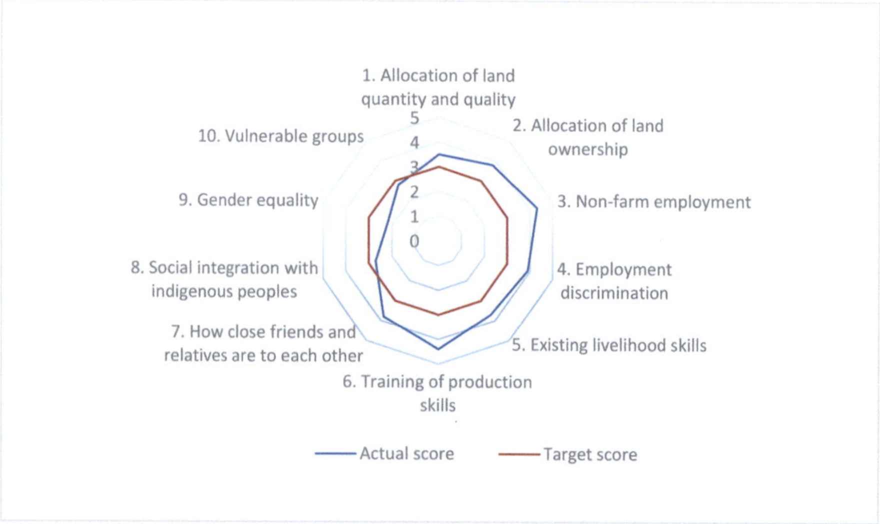

Fig. 3 Degree of realization of population development goals in the planning stage

and relocation of towns, rural settlements, industrial and mining enterprises, and infrastructure should be hydrogeological, engineering site survey and geological disaster risk assessment." In terms of living environment, as of 2014, slope control and landslide

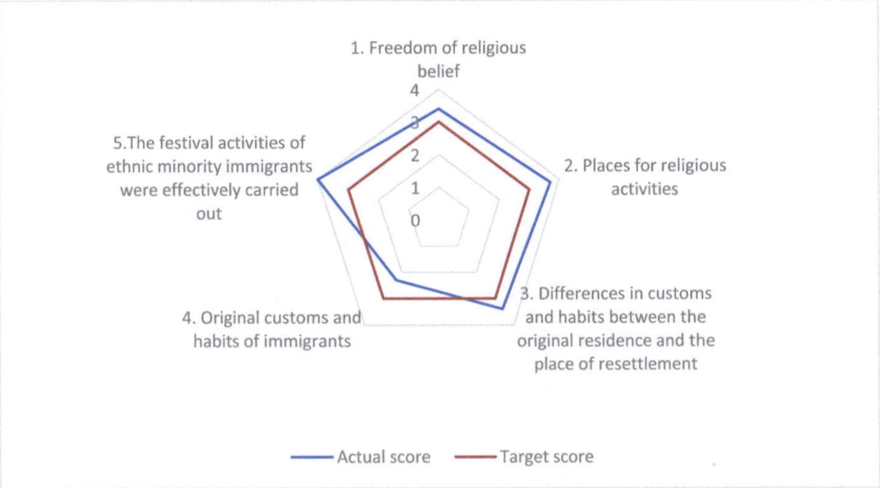

Fig. 4 Degree of cultural integration goal realization in the planning stage

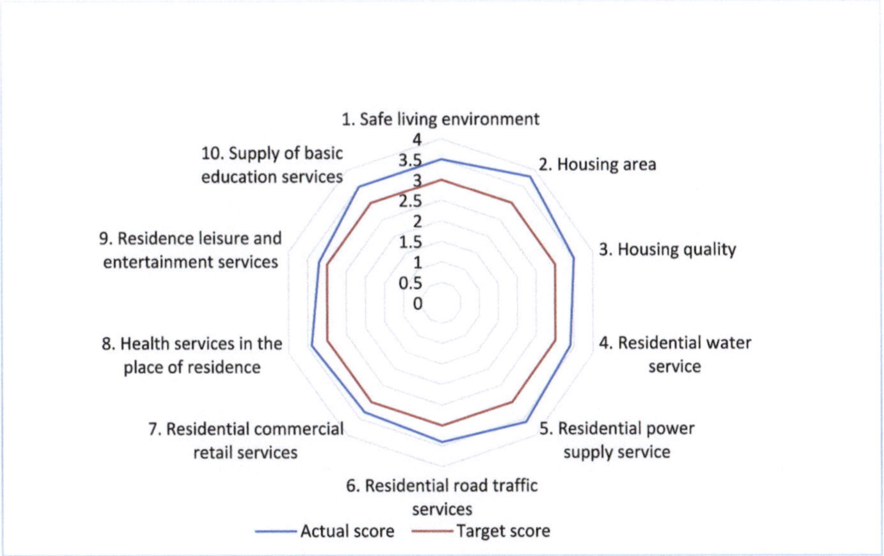

Fig. 5 Degree of realization of environmental adaptation goals in the implementation phase

control projects in the Three Gorges Reservoir area of Hubei have been completed and accepted. In terms of the evaluation of the living environment, 92.1 percent of the immigrants think that the geology of the place is "relatively safe" or "very safe", and all the immigrants think that the security is "very good" or relatively good. For the resettlement of rural migrants, Article 18 of the Regulations on the Construction of migrants for the Three Gorges Project on the Yangtze River (2001) requires that "the construction of

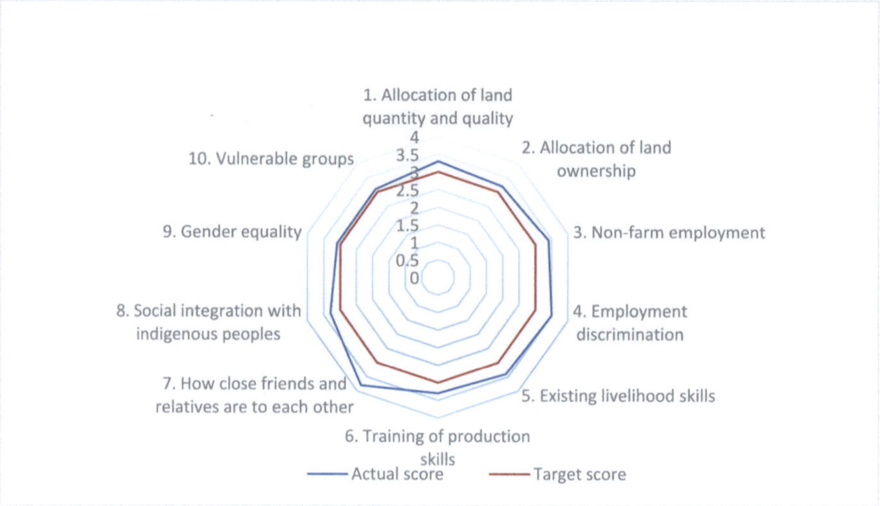

Fig. 6 Implementation stage population development goals achieved

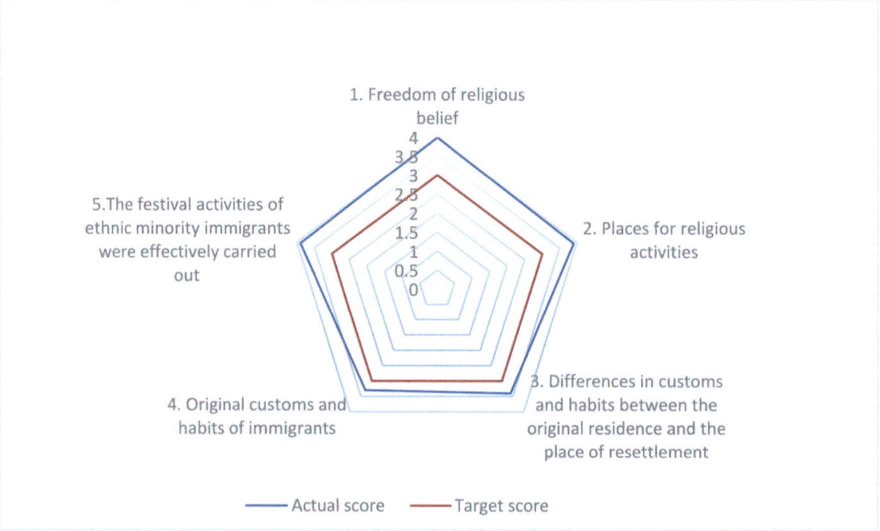

Fig. 7 Degree of realization of cultural integration goals in implementation stage

housing by migrants can be built by separate families, or it can be built in accordance with the principle of voluntary unification." The relevant local people's governments and villagers' committees shall not impose standards for building houses." 85.46% of the respondents were "very satisfied" or "satisfied" with their living conditions. 87.46% of the respondents were "very satisfied" or "satisfied" with the quality of the housing structure. In terms of basic public services, 100% of migrants have access to water and electricity, and 94.77% consider transport "very convenient" or "relatively convenient."

Fig. 8 Achievement of environmental adaptation goals in the late support stage

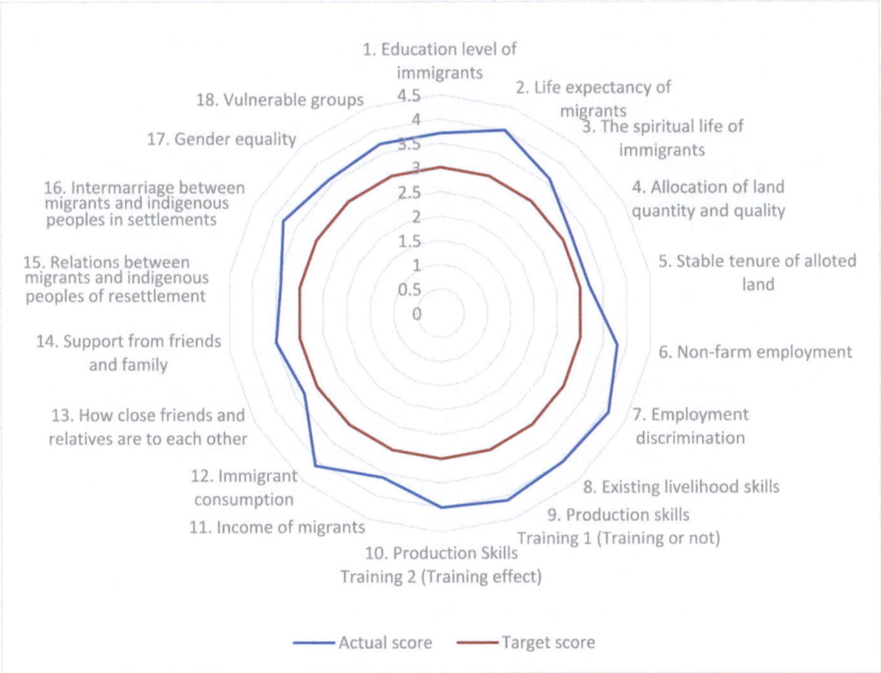

Fig. 9 Population development goals achieved in the late support stage

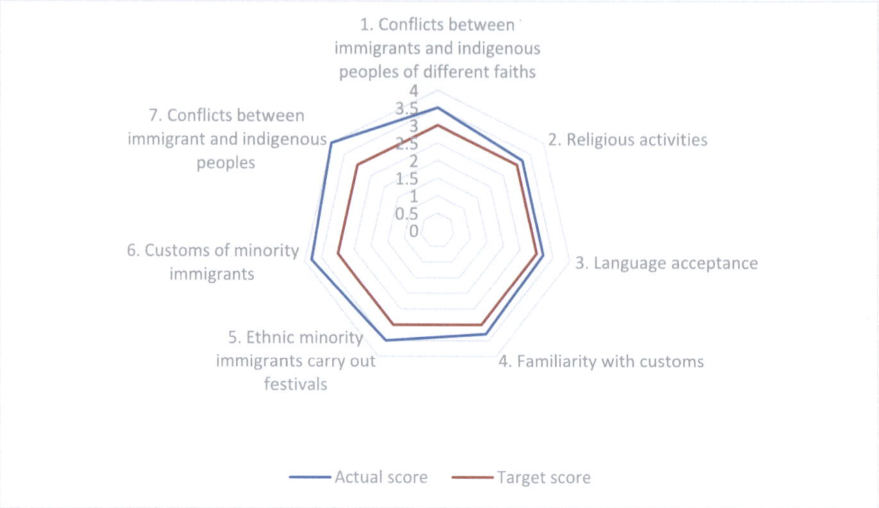

Fig. 10 Achievement degree of cultural integration goal in the late support stage

84.74% of villages have a market within 3 km. 85.79% of immigrants think that the public health service supply in their place of residence is "very good" or "relatively good". 91.91% of immigrant households have activity squares around them for leisure and entertainment. 85.03% of immigrant villages have primary schools within 3 km, and more than 60% of immigrants think the facilities are "relatively good" or "very good".

Secondly, in terms of population development. As for the amount of land, the Report on the Planning of the Flood Treatment and Resettlement of the Reservoir of the Three Gorges Project of the Yangtze River (1998) put forward the principle that "under the premise of conserving soil and water and improving the ecological environment, rational development of resources should be carried out by adjusting cultivated land, transforming medium and low yield fields, and reclaiming land suitable for agriculture." So that immigrants have a stable and high-yield farming land, optimize and adjust the planting structure, and actively develop the aquaculture industry." Article 11 of the Regulations on the Construction of the Three Gorges Project on the Yangtze River (2001) requires that "the construction land of the Three Gorges Project shall be approved at one time and allocated in stages in accordance with the approved plan, and the registration formalities for the change of land ownership shall be handled according to law." The scale of the resettlement land for the Three Gorges project construction shall be strictly controlled, and according to the general plan for land use and the annual plan for land use, it shall be reported to the people's governments at or above the provincial level step by step for the conversion of agricultural land and land acquisition procedures according to law." In terms of family livelihood, the satisfaction of land allocation is 60.44%, and the per capita housing area of immigrants is 32.66 square meters, reaching or exceeding the average level of resettlement. On average, migrant workers are mainly migrant workers, followed by local odd jobs, and there is no unemployment phenomenon. Zigui County promotes farmers' income through the e-commerce value chain [32], and the proportion of local casual labor is 41.60%, which exceeds the proportion of migrant workers by 16%. Basically, there is no employment discrimination, 55.17% of immigrant households have received training in production skills, and more than half of immigrant families have been transferred to employment because of training. The income of families actively participating in agricultural technology training increased as high as 62.44 million yuan. In terms of social communication, the communication with relatives and friends is relatively close, but the scope of interpersonal communication is relatively general, the acceptance degree of local residents to immigrants "easy to accept" and "relatively easy to accept" is 76.47%. In terms of social justice, 94 percent said men and women were "completely equal" or "fairly equal." In terms of how easy it is for vulnerable groups, such as the poor, to get help from the government, migrants see little change before and after relocation.

Finally, in terms of cultural integration. Article 14 of the Regulations on Migration for the construction of the Three Gorges Project on the Yangtze River points out that immigrants are first settled in their counties and districts. This is a forward-looking consideration of the cultural integration of immigrants. According to the satisfaction survey results, 97.44% of immigrants believe that religious activities can be carried out normally after immigration. 90.1% of immigrants think that customs are acceptable. Of the 246 immigrants surveyed, 15 were from the Tujia family, and the customs and habits of ethnic minorities were also maintained.

5.3.2 Evaluation Conclusion

From the perspective of process assessment, although the early planning time of Hubei Three Gorges migrant resettlement is not comprehensive enough to consider the aspects of immigrants such as gender equality, attention to vulnerable groups, and suitable integration of emigrants, certain problems and difficulties are faced in the implementation, but after feedback, reflection and measures are taken to solve the problems in the process of resettlement practice. At the end of the implementation period, the planned tasks can basically be completed and social integration can basically be achieved. In the post-immigrant support stage, through the further development of post-support and follow-up work, continue to improve the living environment of immigrants, improve the livelihood of immigrants' families, pay more attention to social equity and respect for religious beliefs and customs. According to the result evaluation, immigrants have a high degree of satisfaction in environmental adaptation, population development and cultural integration, which is consistent with the conclusion of some scholars, that is, the overall social adaptability of immigrants is good and normal [33].

6 Conclusion and Prospect

6.1 Conclusion

In line with the strategic background of the Belt and Road Initiative, this paper proposes to construct evaluation criteria with Chinese characteristics, makes a preliminary exploration of the evaluation of the whole process of social integration of immigrants in water conservancy and hydropower projects, and discusses the evaluation of the process and result of social integration of immigrants in water conservancy and hydropower projects. In addition, on the basis of comprehensively collecting relevant experts' scores on the normalization degree of the whole process of the Three Gorges migration planning stage, implementation stage and later support stage, as well as the immigrants' satisfaction with the results, and combining with the summary of existing literature and critical thinking, the author innovatively constructs an analytical framework for the evaluation of the whole process of social integration of engineering immigrants. This paper uses the research paradigm of theory and demonstration to clarify the theoretical basis and direction of the research on social integration of engineering immigrants. Through the process assessment, we can establish the work flow of problem occurrence - problem feedback - taking measures to solve it, and avoid the problems in the early stage of immigration work. The results of the assessment also promoted the rights and interests of migrants in the aspects of environmental adaptation, population development and cultural integration.

6.2 Prospect

Future research on engineering immigration can also be considered from the following three aspects.

First, expand the application scope of the whole process assessment. Other water conservancy and hydropower projects should also pay attention to the social integration of immigrants, and evaluate the social integration of engineering immigrants according to this standard. In the future, the relevant research framework should draw on the whole process analysis, and carry out targeted evaluation on the planning period, resettlement period and later support period of engineering immigrants. Timely correct possible problems, strive to be scientific, reasonable and standardized, and help build the "three major" evaluation system with Chinese characteristics [34].

The second is to pay attention to the social reconstruction of immigrants. Moving to a new region means being cut off from the social networks of the place of departure. It is difficult for the original social network to give strong support to immigrants in the place of relocation. To do so, migrants must rebuild social networks [35]. While paying enough attention to social integration, the social reconstruction of immigrants is also the direction that needs to be studied in the next step.

The third is to establish an international voice and promote Chinese standards to "go out". China's international influence is increasing [36]. The establishment of the evaluation system for the whole process of social integration of migrants in water conservancy and hydropower projects is a beneficial attempt for the hydropower industry to build Chinese standards. In the next step, it is necessary to play a leading role in the transformation of strategic awareness and global awareness, establish an international voice, and promote Chinese standards to "go out".

References

1. Shi, G., Yan, D., Sun, Z.: A study of impact and reconstruction of resettlement social system pertainging to water conservancy and hydropower project construction. J. Hohai Univ. **17**(1), 36–41 (2015)

2. Foran, T.: Making hydropower more sustainable? A sustainability measurement approach led by International Hydropower Association. cpwfbfp pbworks com (2021)
3. Yue, L.: Increasing international research on hydropower technical standards helps China Hydropower "go global." Stand. Eng. Const. **6**, 2 (2017)
4. Shi, G., Chen, A.: Discussion on sociological issues in engineering immigration. J. Hohai Univ. **01**, 23–28 (1999)
5. Feng, X.: "Put down roots"? – Social adaptation of rural immigrants in Three Gorges. Sociol. Res. **5**, 19–27 (2004)
6. Li, H., Jiang, H.: On social integration and social stability of immigrants in the Three Gorges Project. J. Chongqing Uni. Soc. Sci. Ed. **9**(2), 4 (2003)
7. Wang, S.: A study on social integration of immigrants from the Three Gorges Project: Xi'an Technological University (2007)
8. Gao, X., Xu, J.: Spatial reconstruction and immigrant community integration: Sociological thinking based on the "soilless resettlement" project. Soc. Dev. Res. **1**, 73–93 (2017)
9. He, X.: Research on the sustainable development of the Three Gorges migrants. Southwest University (2005)
10. Park: Introduction to the science of sociology. University of Chicago Press, Chicago (1969)
11. MichaelM.Cernea: Migration and development: compiled by Reservoir Migration Economics Research Center. Hohai University Press (1996)
12. Pang, Z., Tan, Z., Yue, C.: Danjiangkou reservoir migration planning and practice. Water Resour. Hydropower Express **44**(5), 112–6 (2023)
13. Jiao, H., Li, Q., Zhang, Q., (eds.): Discussion on ways of getting rid of poverty and getting rich by reservoir immigrants from the perspective of precise poverty alleviation. Proceedings of the 2016 Annual Conference of the Chinese Hydraulic Society (Part II) (2016)
14. Shen, R.: Research on the implementation of supporting policies in the later period of migration of Fengtou Reservoir in Zhangzhou City (2022)
15. Chen, S., Xiong, F., Zhao, S.: Research on implementation mechanism of social stability risk assessment for reservoir migrants. People's Yangtze River **48**(B06), 291–293 (2017)
16. Li, X.: Research on the rights and interests system and guarantee mechanism of rural migrants in hydropower projects. Hubei Soc. Sci. **06**, 56–59 (2010)
17. Chen, W., Li, H., Li, S.: Application of fuzzy comprehensive evaluation in independent evaluation of hydropower project resettlement. Hydroelectric Power Generation (2020)
18. Li, J., Bu, M.: Application of AHP method in resettlement assessment of the project of diversion of Yellow River to Jibudian Lake. Water Conservancy Hydropower Technol. **48**(10), 133–5+41 (2017)
19. Wu, Z., Wang, Z.: The practical application of participatory assessment in the study of ethnic minority reservoir migration. Ethnic Stud. Guizhou **42**(03), 80–86 (2021)
20. Liu, X., Xin, Y.: The application of MATLAB in the independent evaluation and quantitative analysis of hydropower project resettlement. Northwest Hydropower **5**, 4 (2019)
21. Shang, H., Liu, J.: Scientific evaluation of the whole process of public policy: logical system, technical pedigree and application strategy. Acad. Res. **3**, 47–57 (2023)
22. Zou, Y.: The whole process evaluation is the key to improve the economic benefit of investment projects. J. Nanjing Univ. Philos. Humanit. Soc. Sci. **3**, 8 (1995)
23. Li, F., Bi, J., Qu, C., Huang, L., Yang, J., Wan, W.: Research and application of environmental risk assessment and management model. Chinese Environ. Sci. **6**, 7 (2010)
24. Zhou, S., Zhai, G., Wu, T., Shi, Y., 鲁钰雯.: Comprehensive assessment of the whole process of urban disaster risk response: a case study of Xiamen, Fujian Province. Urban Plann. Shanghai **006**(006), 99–105 (2021)
25. Jiang, A., Yang, Q.: Research on performance evaluation of "whole process" of government purchase of public services. J. Central Univ. Finance Econ. **3**, 8 (2020)

26. Rossi, P.H., Lipsey, M.W., Henry, G.T.: Evaluation: a systematic approach. Sage Publications (2018)
27. Zhou, H.: Measurement and theoretical consideration of social integration of floating population. Popul. Study **36**(3), 11 (2012)
28. Jiang, J.: Discussion on reservoir resettlement planning system based on high-quality development. People's Yangtze River **53**(07), 232–236 (2022)
29. Zhang, G., Sun, Y.: Promote the high quality development of reservoir migration from the perspective of late support. Water Conservancy Dev. Res. **23**(05), 34–37 (2023)
30. Shao, K.: Aggregation assessment model in immigration policy assessment. Nanjing University (2016)
31. Yin, Z., Yuan, Y.: Yangtze River Three Gorges project migration planning and design. People's Yangtze River (08), 49–52+66 (2003)
32. Zeng, Y., Zhao, X., Duan, Y.: Study on the impact of e-commerce value chain renewal on the income increase of rural migrants in the reservoir: based on the analysis of Zigui, the first county in the Three Gorges Reservoir Area. Agric. Econ. Prob. **01**, 131–144 (2023)
33. Tan, P., Peng, H., Zhang, G.: Investigation and reflection on social adaptability of immigrants in Three Gorges of Hubei Province. J. China Three Gorges Univ. (Humanit. Soc. Sci. Ed.) **31**(03), 15–18 (2009)
34. Group NDaRCEaSDR, Wang, Q., Li, D.: Some thoughts on constructing the "Three major" evaluation system with Chinese characteristics. Manage World **38**(12), 76–83+91+84 (2022)
35. Zhong, Z., Du, Y.: A review of immigration studies. World Peoples **01**, 68–72 (2009)
36. Yu, Q., Ren, Y., Song, P., Xia, T.: A preliminary study on the evaluation system of energy power. Hydroelectric Power Gener. **49**(03), 1-4–108 (2023)

Research on Urbanization Resettlement of Reservoir Projects Under High-Quality Development in China

Jing Wu[1,2], Shaojun Chen[1,2](✉), Langxing Xu[3], and Jinjin Sun[1,2]

[1] School of Public Administration, Hohai University, Nanjing 210098, China
825464712@qq.com, shaojun_chen@126.com
[2] National Research Center for Resettlement, Hohai University, Nanjing 210098, China
[3] Xi'an Jiaotong-Liverpool University, Suzhou 215123, China

Abstract. The era of socialism with Chinese characteristics and the century-long unprecedented changes in the world are intertwined and mutually stimulating. With the coordinated promotion of the strategic deployment of achieving carbon peaks and carbon neutrality and the overall layout of ecological civilization construction, the national policy dividend is favorable to hydroelectric power development, and hydroelectric energy development presents huge advantages and development prospects. However, at present, hydropower development is constrained by non-engineering technical issues, especially the resettlement and relocation activities caused by the submergence of reservoirs have become one of the most concerned issues. In the context of China's vigorous promotion of new urbanization, it is worth conducting in-depth research on how to seize this historical opportunity and integrate the resettlement of rural migrants from reservoir projects into the development process of new urbanization. This article takes the basic elements of reservoir resettlement as a starting point, conducts an in-depth analysis of three key elements, and uses this as the basis to construct an urbanization resettlement analysis framework alled "production resettlement-living resettlement-institutional arrangement." Based on this, it focuses on the core issue of production resettlement, and endeavors to propose an enclave economic model in areas with more developed secondary and tertiary industries. Simultaneously, this model is applied to the "NA" reservoir in Zhejiang, calculating the value of land resources in the reservoir area, proposing specific purchase plans for the resettlement area by cross-township, and analyzing the effects of immigrant resettlement. This study found through empirical research that: 1) By trading submerged resources (land requisition comprehensive area price) in the reservoir area plus non-submerged resources (land transfer and custody price) for industrial land, the preferred purchase solution for industrial land is 1298.83 acres of industrial land plus 0 acres of standard factories, while the preferred purchase solution for standard factory is 0 acres of industrial land plus 116.80 acres of standard factories. By trading submerged resource in the reservoir area plus non-submerged resources (land requisition comprehensive area price) for industrial land, the preferred purchase solution for industrial land is 3,827.03 acres of industrial land plus 0 acres of standard factories, while the preferred purchase solution for standard factory is 0 acres of industrial land plus 344.15 acres of standard factories. 2) The submerged resources (land requisition comprehensive

S. Zheng et al. (Eds.): IHDC 2024, LNCE 487, pp. 423–445, 2025.
https://doi.org/10.1007/978-981-97-9184-2_36

area price) and non-submerged resources (land transfer and custody price) in the reservoir area can be purchased with 115.03 to 122.02 mu of standard factories. The per capita annual rental income in the base year is 5879.02 to 6236.27 yuan, and the per capita annual rental income in the planning year is 5991.81 to 6355.91 yuan. The submerged resources and non-submerged resources (land requisition comprehensive area price) in the reservoir area can be purchased with 338.93 to 359.52 mu of standard factories. The per capita annual rental income in the base year is 173,222.3 to 183,744.5 yuan, and the per capita annual rental income in the planning year is 176,545.7 to 187,270.8 yuan. Accordingly, whether using the calculation method of submerged resources (land requisition comprehensive area price) and non-submerged resources (land transfer and custody price) or submerged resources and non-submerged resources (land requisition comprehensive area price), the per capita rental income of immigrants exceeds the per capita agricultural net income of immigrants from the base year to the planning year. Therefore, the enclave economic model has a great promoting effect on the future production recovery and development of immigrants and can fully ensure the improvement of their production income level and sustainable development after resettlement.

Keywords: Reservoir project · Rural immigrant · Urbanization Resettlement · Enclave economic model · Resettlement effects analysis

1 Introduction

The era of socialism with Chinese characteristics and the century-long unprecedented changes in the world are intertwined and mutually stimulating. With the coordinated promotion of the strategic deployment of achieving carbon peaks and carbon neutrality and the overall layout of ecological civilization construction, the national policy dividend is favorable to hydroelectric power development, and hydroelectric energy development presents huge advantages and development prospects. A reservoir is a key supporting facility and storage medium for hydropower development projects, with significant ecological and environmental characteristics. So far, China has built a total of more than 98,000 reservoirs, with over 80% of them being medium-sized and small reservoirs constructed before the reform and opening-up policy was implemented (Yao, 2020). The "13th Five-Year Plan" proposed that large hydropower projects should be developed in an orderly manner, and that the construction of pumped-storage hydroelectric reservoirs should be vigorously promoted, while the development of small and medium-sized hydropower projects should be controlled (Wang & Hu, 2011). The "14th Five-Year Plan" and the "2035 Vision Plan" approved by the 19th CPC Central Committee put forward clear requirements for enhancing the functions of major water conservancy projects. The value of multipurpose reservoirs has been validated in practice in many countries around the world, as the World Commission on Dams has estimated that one-third of the world's large dams have multipurpose functions (Wilmsen, 2016). Therefore, driven by the demand for hydroelectric energy and the trend of hydroelectric construction, China's water conservancy and hydropower construction has entered the "second half" and entered the "fast lane" of development. However, at present, hydropower development is constrained by non-engineering technical issues, especially the resettlement and relocation activities caused by the submergence of reservoirs have become one of the most concerned issues (E, 2021; Fan, Lu, Zhang, & Li, 2020). As academician Wu Liangyong remarked on the construction of the Three Gorges Project, the resettlement of reservoir residents entails not only population relocation but also the intricate process of "constructing living environments" and facilitating "urbanization development" post-relocation, making it a complex "social and cultural project"(Wu & Zhao, 1997). Based on this, it is evident that the development of various parts in the reservoir area is contingent upon regional development, and the resettlement of the reservoir serves as not only a mechanism for immigrants but also a comprehensive framework encompassing both immigrants and the intricate social structure elements within their respective regions.

There are various types of reservoir resettlement, and various resettlement methods are intertwined. Since the main body of reservoir relocatees in China are farmers, more than 90% of the total relocatees are rural migrants (Jia & Shi, 2012). Therefore, for a long time, the majority of reservoir migrants in China have been settled through a combination of agricultural resettlement and other resettlement methods. In this resettlement model, land plays an important role, which maximizes the continuation of the production and lifestyle of rural migrants and provides them with the maximum survival and psychological assurance (Zheng, Zhang, & Shi, 2011). However, agricultural resettlement is becoming increasingly difficult due to social economic and natural environmental conditions, as evidenced by the following two points: First, the cultivated land resource is becoming increasingly scarce, with the actual area continuously shrinking and the overall quality of the land being low and showing a downward trend. Second, the countryside is already facing a lot of pressure from labor migration, and further agricultural resettlement will not help alleviate the pressure, but will only exacerbate the hidden employment problems in the primary sector (Rozelle, Guo, & Shen, 1999; Banister & Taylor, 1989; Cai, 2002; Solinger, 1999; North, 1990; Vendryes, 2011; Cai, 1995). The dramatic changes in social and economic environments have led to the fact that agricultural resettlement can no longer meet the diverse development needs of reservoir rural migrants (Yang, 2004; Du & Li, 2016), and is also not conducive to the development of the national urbanization strategy (Zheng, Zhang, & Shi, 2011). In this context, with the vigorous promotion of new-type urbanization in China, it is worth exploring how to seize this historical opportunity and integrate the resettlement of rural migrants from reservoir projects into the process of new-type urbanization, which not only helps break the constraints of the current resettlement method mainly focused on agricultural settlement for reservoir migrants, but also promotes the development of new-type urbanization.

2 Conceptual Definition and Analysis Framework

2.1 Conceptual Definition

China has a history of resettling reservoir migrants through urbanization for decades, dating back to the Three Gorges Dam project in the 1960s. Under the planned economy system, the government resettled migrants by converting a large number of farmers into non-farmers and established the city of Sanmenxia. Since then, some reservoir projects have begun to adopt urbanization resettlement methods, and some of them have had unsatisfactory results, such as the Qingjiang Geheyan Reservoir immigrants in Hubei Province (Sun, 2014). While others have achieved better results, such as the Qingshan Zui Reservoir immigrants in Yunnan Province. Overall, despite some efforts, the resettlement of rural migrants from reservoir areas in terms of both theory and practice is still in the exploration stage at present. On the theoretical level, urbanization resettlement is more complex than agricultural resettlement, which requires not only consideration of hard constraints such as land availability and infrastructure, but also attention to soft constraints such as labor absorption capacity and industrial development. Meanwhile, in the context of new urbanization, the two processes of urbanization resettlement cannot be separated from each other. The first process is the process of moving to the city, and the second process is the process of integrating into urban life. Only by completing these two processes can urbanization resettlement be considered successful. Agricultural resettlement does not need to consider the second process. These factors have increased the difficulty of theoretical research. On the operational level, the current policy and technical documents guiding water conservancy and hydropower engineering resettlement primarily focus on agricultural resettlement, with no detailed requirements for urban resettlement. This has resulted in a lack of policy basis and technical support for urban resettlement in various places when implemented. Additionally, in the "Acceptance Criteria for Resettlement of Water Conservancy and Hydropower Projects (SL682–2014)" issued in 2015, only the first stage of urban and rural settlement was required, without considering the subsequent integration into urban life, which may lead to residual problems in urbanization resettlement and affect social stability.

From the perspective of living and production resettlement, resettlement of reservoir migrants is divided into two categories: agricultural resettlement and non-agricultural resettlement. The latter can also be called urbanization resettlement, which is a type of resettlement associated with urbanization. Urbanization, itself is a multidimensional and complex evolutionary process (Jedwab & Vollrath, 2015; Friedmann, 2006), but at its core, as pointed out by Wirth, a representative figure of the Chicago School (Wirth, 1938), it is a transformation of production and lifestyle, as well as the migration of rural populations to urban areas. The close link between urbanization and population and industrial concentration indicates that urbanization relocation is not feasible in societies with restricted population mobility and agricultural-based economies. Therefore, before the reform and opening-up of China, due to the low level of urbanization rate, very little urbanization resettlement was carried out for reservoir migrants. After the reform and opening-up, with the construction of the Three Gorges Project as an opportunity, urbanization resettlement began to appear in the resettlement planning for migrants. A group of literature (Zhu, 1996; Han, 1997; Yang, 1995; Gu & Zhang, 1992) that studied the urban resettlement of reservoir migrants in China earlier believed that urban resettlement refers to the transfer of agricultural employment to secondary and tertiary industries by resettling rural migrants into existing cities. The limitation of this concept lies in equating rural resettlement with urban employment for rural migrants, while ignoring the institutional factors that influence the transformation of agricultural population in the dual urban-rural structure, and failing to clarify the relationship between rural resettlement and urbanization development. Based on this, this paper believes that the urbanization resettlement of reservoir migrants refers to the transfer of rural migrants to towns for production and living settlement through a combination of social, economic, legal, and administrative measures, within the premise of not exceeding the population carrying capacity of the resettlement area's towns, with the focus on solving the re-employment of agricultural population in the second and third industries, and with the guarantee of a reasonable institutional arrangement.

2.2 Analysis Framework

The resettlement of reservoir migrants in urban areas is an inevitable result of the increasing scarcity of land resources and the trend of urbanization in China. Urbanization is not equivalent to a simple increase in the proportion of urban population. On the surface, it manifests as the migration of rural population to urban areas, while in essence, it reflects a living space centered on non-agricultural employment, and its essence is the geographical expansion of urban lifestyles and urban civilization. The urban resettlement of reservoir migrants is a breakthrough and innovation of the long-standing agricultural resettlement method. Its goal is consistent with urbanization, which aims to complete the transformation of rural migrants in terms of production and lifestyle. From agricultural resettlement to urbanization resettlement, the basic elements are the same, but because of the transformation of production and lifestyle and the different needs for resource allocation, these basic elements will change. The paper provides a comprehensive overview of the key factors influencing urbanization resettlement and establishes an analytical framework for rural resettlement in reservoir projects. Within this framework, living resettlement is considered foundational, production resettlement is deemed central, and institutional arrangement is regarded as essential.

Reservoir rural migrants are a group of people who lose their land, houses and other property due to the construction of infrastructure projects, and then rebuild their production and living systems with the help of external forces. Before urbanization relocation, they were rural residents. When land expropriation and demolition occur, the land, houses, social relationships, and collective assets that they used to rely on for their production and livelihood gradually disappear. After relocating into the city, with the support of national compensation and subsidy policies, the migrants' production and living conditions gradually returned to normal. In this process of loss and gain, the basic elements affecting the resettlement of migrants, whether through agricultural resettlement or urbanization resettlement, are consistent. Of course, the specific composition of each element will change due to the different needs of migrants. Broadly speaking, from the perspective of meeting the needs of migrants for settlement and production, the basic elements of reservoir migration urbanization settlement include the following six aspects of infrastructure and public facilities, Residential, Means of production and livelihood, basic public services, social security, and community management and service (Zhang, 2013; Zhou, 2017).

Given the characteristics of urbanization resettlement for reservoir migrants, the basic elements of urbanization resettlement are actually influenced by key factors such as production resettlement, living resettlement, and institutional arrangement. The primary issue in urbanization resettlement is living resettlement. When migrants move from rural areas to urban areas, it is crucial to provide them with a stable living space in the resettlement area so that they can adapt to their new life in the resettlement area and achieve the goal of settling down and becoming wealthy. And after the resettlement, how to reasonably plan the production resettlement to restore and surpass the immigrants' income sources before the relocation has always been the core issue of resettlement. The failure of livelihood recovery and reconstruction can lead to impoverished immigrant lives, triggering various conflicts and inharmonious phenomena and thus producing a series of social problems (Shi, Yan, & Sun, 2015; Zhao, Xiao, & Duan, 2018). From the perspective of implementation, whether the production and living resettlement for immigrants can be implemented smoothly, whether there are corresponding policies to help them better adapt to the city after resettlement, and a series of other related issues ultimately require a suitable institutional arrangement to ensure their implementation. Institutional arrangements can establish a stable structure for people's interactions and reduce uncertainty by providing rules for daily life, both formal and informal. Therefore, appropriate institutional arrangements mean that the risk of immigrant settlement can be reduced.

Based on the above analysis, this paper constructs an analytical framework for rural resettlement of reservoir project migrants, which consists of "production resettlement-living resettlement-institutional arrangement", as shown in Fig. 1. Among them, living resettlement is the prerequisite for urbanization resettlement, production resettlement is the core issue of urbanization resettlement, and institutional arrangement is the guarantee for urbanization resettlement. Therefore, the focus should be on the core issue of production resettlement, which refers to a state of employment that is fully transitioned from agricultural to non-agricultural modes, and the livelihood model of immigrants will complete the transition from the first to the second and third industries. In this model, the following features are primarily included: **(1)** The dependence of production activities on natural capital is decreasing, while the impact of social and economic factors on employment is increasing. **(2)** The demand for human capital is increasing, and the social division of labor within the family is becoming more distinct. **(3)** The sources of income are becoming increasingly narrow. **(4)** The relative non-overlapping of production activities and living spaces in terms of geographical location.

However, there is currently a lack of official national statistics on the urbanization resettlement rural migrants in the reservoir area. This article can depict the current situation and existing problems of production resettlement of reservoir migrants in the context of urbanization through a survey and analysis of representative reservoirs such as the Three Gorges Project, Xiangjiaba Hydropower Station, and Yunnan Qingshanzui Reservoir. Firstly, there is a lack of relevant policies to support immigrants in securing non-agricultural employment. There is a lack of guiding documents at the national level for reservoir urbanization resettlement and the post-support policy also pays little attention to the employment support for reservoir urbanization resettlement migrants who have settled in urban areas. Preferential employment policies for general rural migrants do not benefit those who have been resettled as non-agricultural residents, and the current resettlement policy is seriously outdated and cannot meet the needs of urban resettlement. Secondly, the reservoir migrants themselves are not well-equipped to find non-agricultural employment opportunities. The overall quality of immigrant labor is generally lower, especially for middle-aged and older immigrants who are not competitive in the job market. Immigrants lack the ability to collect and analyze employment information, and tend to blindly seek job opportunities in big cities. Finally, the lack of effective employment support programs limits the re-employment of immigrants in urban areas. The resettlement planning for reservoir relocation lacks specific measures in terms of industry and employment, especially in urban relocation, where the related content is even more vague and empty. There are also no corresponding policy requirements and design specifications at the central and local levels. Detailed industrial and employment plans are usually drawn up after relocation, but the problem is that migrants' demand for employment is immediate upon entering the town, rather than delayed. Once they have been settled, they need to immediately find employment to ensure the sustainability of their livelihood, while the current industrial and employment plans are clearly lagging behind. Furthermore, the most important aspect of traditional urbanization resettlement is that it is landless relocation. Residents of reservoir areas give up their land use rights, receive equal-value financial compensation in a lump sum, and are responsible for solving their subsequent livelihood problems on their own. Generally speaking, this resettlement method carries risks and is not sustainable. Therefore, in order to effectively solve the resettlement problem, safeguard the interests and long-term development of migrants, and promote regional urbanization and economic development, this paper attempts to propose an enclave economic model in areas where the second and third industries are relatively developed. This model refers to the purchase of industrial park standard factories in economically strong towns across township enclaves by calculating the total value of land resources in the immigrant reservoir area, and set up or rely on the relevant rural stock economic cooperatives to operate and manage them, through renting to achieve sustainable income increase of immigrant groups.

3 Materials and Methods

3.1 Case Introduction and Data Source

The "NA" Reservoir is the leading project proposed for the management of the Nanxi River in Zhejiang Province in several river basin plans. It holds a position of utmost importance in the governance of the river basin system and was included in the national "172" major project list in May 2014. Currently, it is still undergoing deep research and preliminary work processes. The "NA" reservoir is planned primarily for flood control and water supply, with the integration of tourism development to improve the overall utilization of the Nanxi River water environment. The reservoir controls a watershed area of $311.9\mathrm{m}^2$, representing 27.3% of the total catchment area above Shizhu. The total storage capacity of the reservoir is 36,331 million cubic meters, with a normal storage capacity of 29,220 million cubic meters and a flood control capacity of 9,634 million cubic meters. After the completion of the "NA" reservoir, implementing the scheduled staggered flood discharge can enhance the flood control capacity of Xikou and Lixi from less than a 10-year occurrence to a 20-year occurrence. It can also alleviate the flood pressure on Yantou and Shatou, reducing the 20-year peak flow to $1080\mathrm{m}^3/\mathrm{s}$. In the event of typical historical major floods, such as those in 1960, 1962, 1965, and 1982, the reduction of flood peaks at the Shatou control section upstream of the county town is

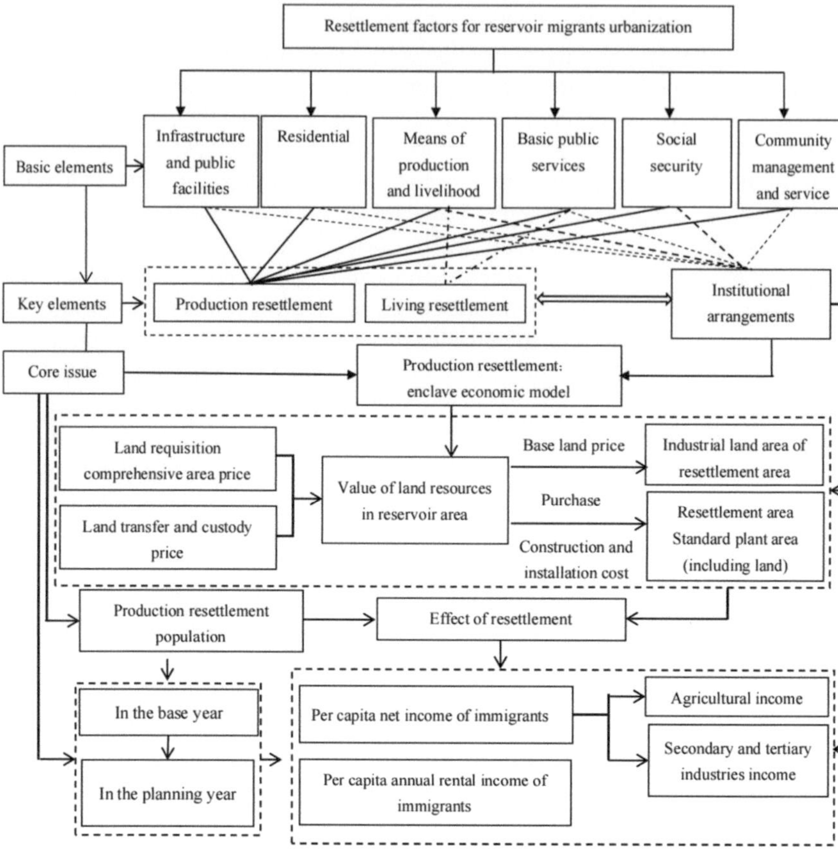

Fig. 1. Analysis framework for rural migrants' urbanization in reservoir projects

significant (Fang, Qiu, & Dai, 2022). The construction project involves land acquisition in two townships and 17 administrative villages in T County, with a total land area of 878.83 hectares (Zhou & Gu, 2015). The data presented in this article are sourced from the research report on the resettlement policy for urbanization resulting from the "NA" Reservoir immigration in T County.

3.2 Research Methods

3.2.1 Calculation of Production Resettlement Population

When calculating the population to be accommodated by production, this paper employs a standard method that involves identifying the types of land resources, including cultivated land, forest land, and garden land. During the field investigation, it was found that the cultivated land and garden land in County T were different, but their land use was essentially similar, with almost identical crops being planted. Therefore, for the purpose of categorizing land resources, cultivated and garden lands were combined for calculation. However, there is a clear disparity in the per-acre yield between forested and cultivated land. Therefore, separate calculations are required when determining the population to be resettled for production purposes (Zhang, Liu, & Zhu, 2017).

First, the calculation of production resettlement population in pure agricultural or forestry zones. For pure agricultural or forestry areas, the calculation of the production resettlement population is based on the remaining cultivated land or forest land after reservoir inundation.

The calculation formula is as follows:

$$P = R - \frac{Q}{r} \tag{1}$$

In the formula: P represents the production resettlement population; R represents the original resident population of the calculation unit; Q represents the remaining cultivated or forest land area; r represents the average cultivated or forest land area per person.

Secondly, the calculation of production resettlement population in integrated agricultural and forestry zones. The income from agriculture and forestry differs, thus it is necessary to calculate the ratio between agricultural and forestry income in order to determine a consistent dimension for calculation using the same standard. Based on the ratio of income categories, forest land is converted into a certain amount of cultivated land and then calculated based on the quantity of cultivated land.

The calculation formula is as follows:

$$D = \frac{P_m S_g}{P_z S_m} \tag{2}$$

In the formula: D represents the equivalent area of forest land converted into cultivated land;P_m represents the income from forest land;S_g represents the cultivated land area;P_z represents the income from cultivated land;S_m represents the area of forest land.

$$S_d = S_g + S_m D \tag{3}$$

In the formula: S_d represents the actual total equivalent cultivated land area in the submerged region; S_g represents the cultivated land area; S_m represents the forest land area;

D represents the equivalent cultivated land area of the forest.

$$S_y = S_{gy} + S_{my} \tag{4}$$

In the formula: S_y represents the total equivalent cultivated land submerged by the reservoir; S_{gy} represents the area of submerged cultivated land; S_{my} represents the area of submerged forest land.

$$S_r = \frac{S_d}{R} \tag{5}$$

In the formula: S_r represents the per capita equivalent cultivated land in the submerged region; S_d represents the actual total equivalent cultivated land area in the submerged region; R represents the original resident population of the calculation unit.

$$R_a = \frac{S_y}{S_r} \tag{6}$$

In the formula: R_a represents the calculation of production resettlement population based on the equivalent cultivated land submerged affected by the reservoir; S_y represents the total equivalent cultivated land submerged by the reservoir; S_r represents the per capita equivalent cultivated land in the submerged region.

However, formula 6 only considers the loss of resources and does not consider the impact of remaining resources on the production and living standards of non-productive production resettlement population. Therefore, when calculating the production resettlement population, it is necessary to consider the environmental capacity of the remaining resources.

The calculation formula is as follows:

$$S_s = S_g - S_{gy} + (S_m + S_{my})\,D \tag{7}$$

In the formula: S_s represents the total equivalent cultivated land area remaining after the reservoir is submerged; S_g represents the cultivated land area; S_{gy} represents the area of submerged cultivated land; S_m represents the area of forest land; S_{my} represents the area of submerged forest land; D represents the equivalent cultivated land area of the forest.

$$R_s = \frac{S_s}{S_r} \tag{8}$$

In the formula: R_s represents the remaining environmental capacity; S_s represents the total equivalent cultivated land area remaining after the reservoir is submerged; S_r represents the per capita equivalent cultivated land in the submerged region.

$$R_a = R - R_S \tag{9}$$

In addition, formula 9 only takes into account the production resettlement population in the base year, without considering the lengthy resettlement period for reservoir migrants and their natural population growth. Therefore, the formula for calculating the actual production resettlement population in the planning year is as follows:

$$P' = R_a(1 + j)^l \tag{10}$$

In the formula: $P\prime$ represents the production resettlement population in the planning year; j represents the natural population growth rate; l represents the number of years spanned from the base year to the planning year.

3.2.2 Assessment of Land Resources in the Reservoir Area

Based on the inundation line involved in the reservoir project, this paper categorizes the land resources in the reservoir area into two types: submerged resources and non-submerged resources. The submerged resources in the reservoir area are calculated based on the land requisition comprehensive area price, while the non-submerged resources in the reservoir area are calculated using either the land transfer and custody price or land requisition comprehensive area price. The land transfer and custody price ncludes the following two methods: land trust transfer and public welfare forest custody. The land trust transfer system generally refers to the establishment of a land trust service center at the township level by the government, where farmers' wishes are fully respected. This involves entrusting the non-submerged resources to be managed and operated by the government-established land trust service center. The farmers entrust their right to operate and contract the land to this center. The government manages the land through land trust contracts with farmers, while agricultural companies or large-scale farmers engage in leasing land from the Land Trust Service Center to achieve large-scale agricultural cultivation. The public welfare forest custody involves the collective ownership and use rights (including contractual rights) of the forested non-submerged resources thorough negotiations with the municipal or county government. This process delineates the scope of public welfare forests. After obtaining approval for the construction of national and provincial public welfare forests, both parties sign an agreement on the protection of public welfare forests. The government is responsible for managing the public welfare forests and provides regular subsidies to the owners and users of these forests.

Based on the land requisition comprehensive area price, the calculation formula for assessing the value of land resources in the reservoir area is as follows:

$$P = \sum_{i}^{n} S_i \times T_i \tag{11}$$

In the formula: P represents the total value of the affected village's land resources; S_i represents the area of different types of land resources in the affected village; T_i represents the land requisition comprehensive area price of different types of land resources in the affected village.

Based on the land transfer and custody price, the calculation formula for assessing the value of land resources in the reservoir area is as follows:

$$P = \sum_{i}^{n} S_i \times M_i(1 + r)^N \tag{12}$$

In the formula: P represents the total value of the affected village's land resources; S_i represents the area of different types of land resources in the affected village; M_i represents the land transfer and custody price of different types of land resources in the affected village; r represents the annual growth rate; N represents the remaining land contract management period when different types of land resources in the affected village are requisitioned.

3.2.3 Purchase of Factory in Enclave of Resettlement Area

Based on the benchmark land price for industrial land of different grades in the resettlement area, the formula for calculating the area of purchase industrial land is as follows:

$$S = {}^{P}/_{V_i} \tag{13}$$

In the formula: S represents the industrial land area in the resettlement area; P represents the total value of the land resources affected in the village; V_i represents the benchmark land price of industrial land of different grades in the resettlement area.

Based on the benchmark land price for industrial land of different grades in the resettlement area and the construction cost of standard factories, the formula for calculating the area (including land) of purchase standard factory is as follows:

$$T = \frac{P}{(V_i + M)} \tag{14}$$

In the formula: T represents the standard factory building area (including land) in the resettlement are; P represents the total value of land resources in affected villages; V_i represents the benchmark land price of different levels of industrial land in the resettlement area; M represents the construction and installation cost of the standard factory building in the resettlement area.

3.2.4 Calculation of Resettlement Effectiveness

For the convenience of analysis, this article mainly determines the annual rental income of standard factory buildings per unit area in the resettlement area by analyzing the statistical yearbooks of each township in the resettlement area for many years and conducting on-site investigations. Using this rental income as the benchmark, the annual rental income of standard factory buildings after the benchmark year is determined by considering the annual growth rate. The calculation method is shown in the formula:

$$F_h = \frac{T \times V(1 + r)^h}{R_h} \tag{15}$$

In the formula: F_h represents the per capita rental income of immigrants in the hth year after the base year; T represents the the standard factory building area (including land) in the resettlement area; V represents the annual rental income per unit area of standard factory buildings; r represents the annual growth rate; R_h represents the production resettlement population in the hth year after the base year.

For the convenience of analysis, this article mainly determines the annual per capita agricultural net income in the reservoir area by analyzing the statistical yearbooks of each township in the resettlement area for many years and conducting on-site investigations. Using this agricultural net income as the benchmark, the annual per capita agricultural net income after the benchmark year is determined by considering the annual growth rate.

The calculation method is shown in the formula:

$$b_h = b_0 \times (1 + a)^h \tag{16}$$

In the formula: b_h represents the per capita agricultural net income in the hth year after the base year; b_0 represents the per capita agricultural net income of the base year; a represents the growth rate of per capita net income.

4 Results

4.1 Calculation of Production Resettlement Population

According to the survey, the resettlement of "NA" reservoir immigrants involves a total of 8 administrative villages, divided into two types: whole village relocation and partial relocation. Among them, the affected villages of whole village relocation involve 6 administrative villages. When calculating the resettlement population for the "NA" reservoir project, two situations were considered: the first is land directly submerged by the reservoir, and the second is land indirectly affected by the reservoir. This project takes 2023 as the base year and 2028 as the planning level year, with a natural population growth rate of 4.6%. According to the physical survey indicators, the land resources in the "NA" reservoir area mainly involve two categories: cultivated land and forest land. The specific submerged and non-submerged areas of cultivated and forest land are shown in Table 1. After calculating from formula 1 to 10, the base year production resettlement population of Zhejiang "NA" reservoir is 4432 people, and the planning y year production resettlement population is 5550 people.

Table 1. Summary of land resources area in the reservoir area (Unit: mu)

Town	Administrative village	Submerged resources		Non-submerged resources	
		Cultivated land	Forest land	Cultivated land	Forest land
Town 1	Village 1	39.41	799.46	174.64	5241.64
	Village 2	106.53	754.35	457.32	1249.50
Town 2	Village 3	187.94	1348.89	246.01	4916.16
	Village 4	217.95	1045.96	162.90	1908.29
	Village 5	153.85	632.27	167.00	2447.83
	Village 6	110.192	158.37	440.61	1991.83
	Village 7	20.79	622.08	1245.21	9224.07
	Village 8	218.16	777.67	740.19	9187.13
	Total	1054.82	6139.05	3633.88	36166.05

4.2 Accounting of Land Resources in the Reservoir Area

According to the investigation, the non-submerged resources of the affected villages of the "NA" reservoir are handed over to the village's joint-stock economic cooperative for operation, while residents who have relocated to the resettlement area can still enjoy the same rights and interests as those who have not relocated. Therefore, this paper primarily calculates the land resources of villages impacted by whole village relocation and the submerged land resources of villages impacted by partial village relocation. Among them, submerged resources in the reservoir area are calculated based on the land requisition comprehensive area price, while non-submerged resource are calculated based on the land transfer and custody price or land requisition comprehensive area price.

From Table 1, the submerged cultivated land area of the “NA” reservoir area is 1054.82 mu, and the forest area is 6139.05 mu; the non-submerged cultivated land area of the “NA” reservoir area is 1931.35 mu, and the forest area is 25692.48 mu. According to the comprehensive land price of T County’s land acquisition area in 2023, cultivated land is priced at 5.8 ten thousand yuan/mu, and forest land is priced at 3.5 ten thousand yuan/mu. 1) The total value of submerged resources in the reservoir area (land requisition comprehensive area price) is calculated by formula 11: the value of submerged cultivated land is 6117.96 ten thousand yuan, the value of submerged forest land is 21486.68 ten thousand yuan and the total value of offline resources is 27604.64 ten thousand yuan. 2) The total value of non-submerged resources in the reservoir area (land transfer and custody price/land requisition comprehensive area price). Firstly, calculate based on the and transfer and custody price. According to the survey, the local cultivated land transfer price in T County in 2023 is 1200 yuan /mu· year. And the second round of land contracting for the administrative village involved in the project took place in 1999, with about 8 years remaining for the remaining contracting period. However, according to the relevant provisions of the Rural Land Contracting Law, rural land belongs to the collective ownership of farmers in accordance with the law. Unless expropriated by the state, its ownership permanently belongs to the rural collective economic organization. Therefore, for the calculation of cultivated land transfer prices, the annual limit is 50 years. In addition, the local land transfer prices have increased by a certain amount every year. In order to simplify the calculation, this article uses the discount rate to offset the land transfer prices. The discounted total value of 50 years of non-submerged cultivated land transfer is 11588.10 ten thousand yuan by formula 12. And the national subsidy for local ecological public welfare forests in T County in 2023 is 35 yuan/mu · year, for the sake of simplicity in calculation, this article does not consider the impact of subsidy price growth and discount rate. The total amount of non-submerged forest land subsidy for 50 years is 4496.18 ten thousand yuan. Therefore, calculated based on the transfer and custody price, the total value of non-submerged resources is 16084.28 ten thousand yuan. Secondly, calculated based on the land requisition comprehensive area price. According to formula 11, the non-submerged cultivated land value is 11201.83 ten thousand yuan, the non-submerged forest land value is 89923.68 ten thousand yuan, and the total non-submerged resource value is 101125.51ten thousand yuan. In summary, the total value of submerged resources (land requisition comprehensive area price) and non-submerged resources (land transfer and custody price) in the reservoir area land resources is 43688.92 ten thousand yuan. The total value of submerged resources and non-submerged resources (land requisition comprehensive area price) in the reservoir area land resources is 128730.15 ten thousand yuan.

4.3 Factory in Enclave of Resettlement Area

According to the 2023 benchmark land price list in T County, the average benchmark land price for industrial land of different grades in three townships, Town 1, Town 2, and Town 3, is calculated by formula 13. Using the submerged resource (land requisition comprehensive area price) and non-submerged resource (land transfer and custody price) method, Class I, Class II and Class III industrial land that can be purchased are 1112.13 mu, 1298.83 mu and 2460.49 mu respectively. Using the submerged resource and non-submerged resource (land requisition comprehensive area price) method, Class I, Class II and Class III industrial land that can be purchased are 3276.90, 3827.03, and 7249.87 mu, as shown in Table 2.

According to the investigation, the construction and installation cost of local standard factories in T County in 2023 is around 2200 yuan/m^2. The volume ratio of industrial land in the local industrial development zone is generally around 2.32. Therefore, the construction cost per mu of standard factory building on 1 mu of industrial land is approximately 340.27 ten thousand yuan. Due to the fact that the per-mu investment of standard factory buildings is the sum of construction costs and land transfer fees, considering the benchmark land prices for industrial land of different grades in the resettlement area, calculated by formula 14, the average value is adopted. The method of submerged resources (land requisition comprehensive area price) and non-submerged resources (land transfer and custody price): the land resources in the reservoir area can be purchased with a minimum of 115.03 mu and a maximum of 122.02mu of standard factory buildings (including land) in the resettlement area; the method of submerged resources and snon-ubmerged resources (land requisition comprehensive area price): the land resources in the reservoir area can be purchased with a minimum of 338.93 mu and a maximum of 359.52 mu of standard factories (including land) in the resettlement area (Table 3).

Table 2. Area of purchase industrial land of different grades in resettlement areas

Purchase plan	Immigrant assets		Region	Purchase area		
	Accounting type	Amount (in ten thousand yuan)		Minimum (mu)	Average (mu)	Maximum (mu)
Purchase plan 1	Submerged resource (land requisition comprehensive area price) and non-submerged resource (land transfer and custody price)	43688.92	Town 1	1129.88	1191.51	2520.50
			Town 2	1213.58	1394.32	2520.50
			Town 3	992.93	1310.66	2340.47
			Average	1112.13	1298.83	2460.49
Purchase plan 2	Submerged resource and non-submerged resource (land requisition comprehensive area price)	128730.15	Town 1	3329.21	3510.80	7426.70
			Town 2	3575.82	4108.39	7426.70
			Town 3	2925.67	3861.89	6896.22
			Average	3276.90	3827.03	7249.87

Table 3. Area (including land) of purchase standard factory buildings in resettlement areas

Purchase plan	Immigrant assets		Region	Purchase area		
	Accounting type	Amount (in ten thousand yuan)		Minimum (mu)	Average (mu)	Maximum (mu)
Purchase plan 1	Submerged resource (land requisition comprehensive area price) and non-submerged resource (land transfer and custody price)	43688.92	Town 1	115.29	115.90	122.17
			Town 2	116.11	117.56	122.17
			Town 3	113.69	116.93	121.71
			Average	115.03	116.80	122.02
Purchase plan 2	Submerged resource and non-submerged resource (land requisition comprehensive area price)	128730.15	Town 1	339.70	341.50	359.97
			Town 2	342.11	346.40	359.97
			Town 3	334.98	344.55	358.63
			Average	338.93	344.15	359.52

4.4 Calculation of Resettlement Effectiveness

According to the 2022 T County National Economic and Social Development Statistical Bulletin, the industrial structure of the resettlement area is 3.7:44.3:52.0, with the focus on the secondary and tertiary industries, and residents mainly earning income from these industries. Furthermore, due to the geographical situation of "eight mountains, one river, and one field" in T County, land resources are scarce and land prices are high, resulting in an overall shortage of factory buildings. According to the survey, the rental price of local standard factories in County T in 2023 is 28.3 yuan/m^2·month, with a rental growth rate of 5%.

According to formula 15, the calculation of per capita annual rental income for immigrants using the method of submerged resources (land requisition comprehensive area price) and non-submerged resources (land transfer and custody price) is as follows: for the base year (2023), the range of per capita annual rental income for immigrants is 5879.02 to 6236.27 yuan; for the planning year (2028), the range of per capita annual rental income for immigrants is 5991.81 to 6355.91 yuan. The calculation of per capita annual rental income for immigrants using the method of submerged resources (land requisition comprehensive area price) and non-submerged resourcess (land requisition comprehensive area price) is as follows: for the base year (2023), the range of per capita annual rental income for immigrants is 17322.23 to 18374.55 yuan; for the planning year (2028), the range of per capita annual rental income for immigrants is 17654.57 to 18727.08 yuan, as shown in Table 4 below.

According to the outline of the 14th Five-Year Plan for National Economic and Social Development of T County, and combined with field investigations, the average annual growth rate of per capita net income for local immigrants is 8%, while the average annual growth rate of per capita net income from agriculture is 4.74%. Based on formula 16, in the base year, the per capita net income for immigrants was 12,284.25 yuan, with agriculture accounting for 36.34% at 4,463.78 yuan; at the planning year, the per capita net income for immigrants is projected to be 18,049.38 yuan, with agriculture accounting for 29.42% at 5,909.98yuan as shown in Table 5 below. Through comparative analysis between Table 4 and Table 5 it can be observed that regardless of whether submerged resources (land requisition comprehensive area price) and non-submerged resources (land transfer and custody price) or submerged resources and non-submerged resources (land requisition comprehensive area price) are used as calculation methods, the per capita annual rental income of immigrants exceeds their per capita agricultural net income from the base year to the planning year.

5 Conclusion and Discussion

This article analyzes the basic elements of resettlement for reservoir immigrants as the starting point. It delves into three key elements and builds an analytical framework based on "production resettlement-living resettlement-institutional arrangements". Based on this, it focuses on the core issue of production resettlement, and endeavors to propose an enclave economic model in areas with more developed secondary and tertiary industries. This model refers to the purchase of industrial park standard factories in economically strong towns across township enclaves by calculating the total value of land resources in the immigrant reservoir area, and set up or rely on the relevant rural stock economic cooperatives to operate and manage them, through renting to achieve sustainable income increase of immigrant groups. Simultaneously, this model is applied to the "NA" reservoir in Zhejiang, calculating the value of land resources in the reservoir area, proposing specific purchase plans for the resettlement area by cross-township, and analyzing the effects of immigrant resettlement. The main conclusions are as follows: 1) The base year production resettlement population of Zhejiang "NA" reservoir is 4432 people, and the planning year production resettlement population is 5550 people. 2) By trading submerged resources (land requisition comprehensive area price) in the reservoir area plus non-submerged resources (land transfer and custody price) for industrial land, the preferred purchase solution for industrial land is 1298.83 acres of industrial land plus 0 acres of standard factories, while the preferred purchase solution for standard factory is 0 acres of industrial land plus 116.80 acres of standard factories. By trading submerged resources in the reservoir area plus non-submerged resources (land requisition comprehensive area price) for industrial land, the preferred purchase solution for industrial land is 3,827.03 acres of industrial land plus 0 acres of standard factories, while the preferred purchase solution for standard factory is 0 acres of industrial land plus 344.15 acres of standard factories. 3) The submerged resources (land requisition comprehensive area price) and non-submerged resources (land transfer and custody price) in the reservoir

Table 4. Comparison of rental income under different calculation purchase methods

Purchase plan	Standard factory area (mu)	Rental income situation in the base year (2023)				Rental income situation in the planning year (2028)			
		Rent per unit area (yuan/m^2 · month)	Production resettlement population (people)	Total rental income (10000 yuan/year)	Per capita annual rental income (yuan/year)	Rent per unit area (yuan/m^2 · month)	Poduction resettlement population (people)	Total rental income (10000 yuan/year)	Per capita annual rental income (yuan/year)
Submerged resource (land requisition comprehensive area price) and non-submerged resource (land transfer and custody price)	115.03~122.02	28.3	4432	2605.58~2763.91	5879.02~6236.27	36.1	5550	3323.46~3527.53	5991.81~6355.91
Submerged resource and non-submerged resource (land requisition comprehensive area price)	338.93~359.52			7677.21~8143.60	17322.23~18374.55			9798.28~10393.53	17654.57~18727.08

Table 5. Comparison of net income of immigrants under different purchase methods

Town	Village	Per capita net income in the base year (2023)	Among		Among		Per capita net incomein the planning year (2028)	Among		Among	
			Agricultural income (yuan)	Proportion (%)	Secondary and tertiary industries income (Yuan)	Proportion (%)		Agricultural income (yuan)	Proportion (%)	Secondary and tertiary industries income (Yuan)	Proportion (%)
Town 1	Village 1	12208.42	4346.80	35.60	7861.62	64.40	17914.03	5170.69	28.86	12743.34	71.14
	Village 2	12248.84	4451.22	36.34	7797.62	63.66	17905.27	5294.87	29.57	12610.4	70.43
	Village 3	11386.56	4314.80	37.89	7071.76	62.11	16645.84	5132.7	30.83	11513.14	69.17
	Village 4	12976.40	4786.36	36.89	8190.04	63.11	18968.92	5693.74	30.02	13275.18	69.98
	Village 5	12059.53	4249.78	35.24	7809.75	64.76	18358.74	5057.83	27.55	13300.91	72.45
Town 2	Village 6	13599.53	4980.04	36.62	8619.49	63.38	19817.79	5924.59	29.90	13893.2	70.10
	Village 7	11864.17	4382.62	36.94	7481.55	63.06	17343.76	5211.80	30.05	12131.96	69.95
	Village 8	11930.54	4198.59	35.19	7731.95	64.81	17440.65	4993.89	28.63	12446.76	71.37
	Average	12284.25	4463.78	36.34	7820.36	63.66	18049.38	5909.98	29.42	12739.15	70.58

area can be purchased with 115.03 to 122.02 mu of standard factories. The per capita annual rental income in the base year is 5879.02 to 6236.27 yuan, and the per capita annual rental income in the planning year is 5991.81 to 6355.91 yuan. The submerged resources and non-submerged resources (land requisition comprehensive area price) in the reservoir area can be purchased with 338.93 to 359.52 mu of standard factories. The per capita annual rental income in the base year is 173,222.3 to 183,744.5 yuan, and the per capita annual rental income in the planning year is 176,545.7 to 187,270.8 yuan.

Therefore, based on the above analysis, whether using the calculation method of submerged resources (land requisition comprehensive area price) and non-submerged resources (land transfer and custody price) or submerged resources and non-submerged resources (land requisition comprehensive area price), the per capita rental income of immigrants exceeds the per capita agricultural net income of immigrants from the base year to the planning year. In addition, before land acquisition, the main source of income for immigrants came from cultivated land, as well as some wages from working outside. After land acquisition, the per capita rental income obtained by immigrants can ensure a stable income without the need for productive expenses and labor output. Due to the relatively developed industrial economy in the resettlement area, immigrants can have more opportunities to engage in the secondary and tertiary industries after land acquisition. This not only expands the sources of income for immigrants in the secondary and tertiary industries, but also further increases their income, freeing them from the constraints of land. Therefore, the enclave economic model has a great promoting effect on the future production recovery and development of immigrants and can fully ensure the improvement of their production income level and sustainable development after resettlement.

However, the paper argues that regardless of the chosen resettlement type, immigrants' own perspectives play a decisive role, and ultimately, their opinions should be respected. Due to variations in immigrant families' resource endowment and individual circumstances, not all immigrants are suitable for urbanization. Thus, it is essential to establish an evaluation index system for assessing the adaptability of migrant groups following urbanization resettlement. Based on the evaluation results, immigrants with high, moderate, and low adaptability levels can be identified and provided with support in various aspects (Chen, Vanclay, & Yu, 2020; Jiang, Wang, & Zhang, 2021)." Furthermore, the development of reservoir resettlement areas should not solely depend on external inputs such as hydropower efficiency funds and government support funds, and should not adopt a "development affected" attitude, as this may further increase development dependence and regional positioning differentiation. The author suggests that a portion of the rental income can be derived from the enclave economic model to establish a connection between the development of resettlement and non-resettlement areas, thereby enhancing social and economic activities for both immigrants and non-immigrants within the same administrative unit. Additionally, considering the specific characteristics of each reservoir's location, it is essential to pre-plan the industrial development mode, pathway, and implementation strategies post-resettlement. This includes local policy preferences, industrial development promotion, logistics system integration, as well as gradual incorporation of asset ownership, utilization methods, and management plans into official documents. In addition, regular supervision and evaluation by higher-level governments and third-party agencies are necessary to ensure comprehensive sustainable development for migrant groups and resettlement areas (Jiang, 2022; Li & Sun, 2022; Fang & Zhao, 2023).

Author Contributions. Conceptualization, J.W. and S.C.; data curation, J.W.; formal analysis, J.W. and L.X.; investigation, J.W.; methodology, L.X. and J.S.; project administration, J.W.; resources, S.C.; software, L.X. and J.S.; supervision, S.C.; validation, J.W.; writing—original draft, J.W.; writing—review and editing, J.W. and S.C. All authors have read and agreed to the published version of the manuscript.

Funding. This research was funded by the Key Research Project of the National Foundation of Social Science of China (Fund No. 21&ZD 183) and the Jiangsu Province Graduate Research and Practice Innovation Program Project (Project Approval Number: KYCX23:0643).

Institutional Review Board Statement Not applicable.

Informed Consent Statement Informed consent was obtained from all subjects involved in the study.

Data Availability Statement The original contributions presented in the study are included in the article, and further inquiries can be directed to the corresponding author.

Conflicts of Interest. The authors declare that the research was conducted in the absence of any commercial or financial relationships that could be construed as a potential conflict of interest. Publisher's note: All claims expressed in this article are solely those of the authors and do not necessarily represent those of their affiliated organizations or those of the publisher, the editors, and the reviewers. Any product that may be evaluated in this article, or claim that may be made by its manufacturer, is not guaranteed or endorsed by the publisher.

References

1. Yao Yuqin: 70 Years of land acqusition and resettlement for water conservancy and hydropower projects in China. Water Power **46**(5), 8–12+55 (2020)
2. Wang Yahua, Hu Angang: The road of China water development: Retrospect and prospect (1949–2050). Journal of Tsinghua University (Philosophy and Social Sciences) **26**(5), 99–112+162 (2011). https://doi.org/10.13613/j.cnki.qhdz.002012
3. Wilmsen, B.: After the deluge: A longitudinal study of resettlement at the three Gorges Dam, China. World Dev. **84**, 41–54 (2016)
4. E. Jingping: Deeply implementing the general tone of water conservancy reform and development, writing a new chapter in water control at a new historical starting point-speech at the 2021 national water conservancy work conference. Water Resour. Dev. Res. **21**(1), 1–14 (2021). https://doi.org/10.13928/j.cnki.wrdr.2021.01.001
5. Fan Qixiang, Lu Youmei, Zhang Chaoran, Li Guo: Innovations in technology and management of dam construction and their application to Xiluodu hydropower station on Jinsha River. J. Hydroelectr. Eng., 39(7), 21–33 (2020)
6. Wu Liangyong, Zhao Wanming: Three Gorges project and human habitation environment science. Yangtze River **2**, 1–5 (1997). https://doi.org/10.16232/j.cnki.1001-4179.1997.02.002

7. Jia Yongfei, Shi Guoqing: Optimization of Population Allocation for Reservoir Resettlement. Social Science Literature Publishing House, Beijing (2012)
8. Zheng Ruiqiang, Zhang Chunmei, Shi Guoqing: Reflections on the innovation model of resettlement and city building resettlement. China Rural Water and Hydropower **6**, 160–172 (2011).
9. Rozelle, S., Guo, L., Shen, M.: Leaving China's farms: Survey Results of new paths and remaining hurdles to rural migration. The China Quarterly **6**(158), 367–393 (1999). https://doi.org/10.1017/S0305741000005816
10. Banister, J., Taylor, J.R.: China: Surplus labour and migration. Asia-Pac. Popul. J. **4**(4), 3–20 (1989)
11. Cai Fang: Two processes and institutional barriers in migration. Sociological Studies, 4, 44–51. https://doi.org/10.19934/j.cnki.shxyj.2001.04.006
12. Solinger, D.J.: Citizenship issues in China's internal migration: Comparison with Germany and Japan. Polit. Sci. Q. **114**(3), 455–478 (1999)
13. North, D.C.: Institutions, Institutional Change and Economic Performance. Cambridge University Press, Cambridge (1990)
14. Vendryes, T.: Migration constrains and development: Hukou and capital accumnlation in China. China Econ. Rev. **22**(4), 669–692 (2011)
15. Cai Fang: Causes, trends, and policies of population migration and mobility. Chin. J. Popul. Sci. **6**, 8–16 (1995)
16. Yang Wenjian: Research on the Resettlement Model of rural Immigrants in Chinese Reservoirs. Hohai University, Nanjing (2004)
17. Du Yunsu, Li Fei: Strategies for resettling rural reservoir migrants in the context of urbanization. Rural. Econ. **6**, 109–112 (2016)
18. Zheng Ruiqiang, Zhang Chunmei, Shi Guoqing. Study on the mechanism of diversified portfolio mode for reservoir resettlement. Water Power **37**(9), 1–4 (2011)
19. Sun Haibing: Research on Non-agriculturalization of reservoir migrants under the background of new urbanization. China Rural Water and Hydropower **3**, 123–125 (2014)
20. Jedwab, R., Vollrath, D.: Urbanization without growth in historical perspective. Explor. Econ. Hist. **58**(1), 1–21 (2015)
21. Friedmann, J.: Four theses in the study of China's urbanization. Int. J. Urban Reg. Nal Res. **30**(2), 440–451 (2006)
22. Wirth, L.: Urbanism as a way of life. Am. J. Sociol. **44**(1), 1–24 (1938)
23. Zhu Nong: Migration of the Three Gorges Project and Economic Development of the Reservoir Area, pp. 67–80. Wuhan University Press, Wuhan (1996)
24. Han Guanghui: Implementing non-agricultural transfer is the fundamental way out for the resettlement project in the reservoir area. Journal of Peking University (Philosophy and Social Sciences) **1**, 33–41 (1997)
25. Yang Dingguo: A discussion on the migrators from village of the three Gorges reservoir area to be resettled in dities and towns. Resour. Environ. Yangtze Basin **3**, 209–215 (1995)
26. Gu Shengzu, Zhang Yongsheng: Rural urbanization and resettlement of immigrants in the three Gorges reservoir area. Hubei Soc. Sci. **12**, 28–29 (1992)
27. Zhang Hongyan: China's new urbanization theory and strategic innovation. Sociol. Stud. **3**, 1–14, 241 (2013). https://doi.org/10.19934/j.cnki.shxyj.2013.03.001
28. Zhou Xiaojun: Research on Urbanization Resettlement of Rural Immigrants in Reservoir Projects. Hohai University, Nanjing (2017)
29. Shi Guoqing, Yan Dengcai, Sun Zhonggen: A study of impact and reconstruction of resettlement social system pertainging to water conservancy and hydropower project constrction. Journal of Hohai University (Philosophy and Social Sciences), **17**(1), 36–41+90 (2015)
30. Zhao Xu, Xiao Jiaqi, Duan Yuefang: Relocation, farmland transfer and livelihood transformation of reservoir resettlement. Resour. Sci., 40(10), 1954–1965 (2018)

31. Fang Zijie, Qiu Qunyi, Dai Shunguang: On systematic governance of small and medium-sized river basins and the restructuring of land space layout with large-scale reservoirs as a starting point--taking the Nanxi River Basin as an example. Water Resour. Dev. Res., **22**(4), 89–95 (2022)
32. Zhou Yi, Gu Mengsha: Planning and practice of reservoir resettlement in new-type urbanization mode: Case of Nan'an Reservoir in Yongjia County of Wenzhou City. Yangtze River **46**(22), 107–111 (2015). https://doi.org/10.16232/j.cnki.1001-4179.2015.22.025
33. Zhang Jiarong, Liu Jianlin, Zhu Jiwei: The calculation of rural migrants settlement population based on the improved method in the reservoir. Chinese J. Agric. Resour. Reg.Nal Plan. **38**(4), 21–27 (2017)
34. Chen, X., Vanclay, F., Yu, J.: Evaluating Chinese policy on post-resettlement support for dam-induced displacement an resettlement. Impact Assement Proj. Apprais. **3**, 1–9 (2020)
35. Jiang, T., Wang, M., Zhang, Y.: What about the "Stayers"? Examining China's resettlement induced by Large Reservior Projects. Land, 10 (2021)
36. Jiang Tianhe: Analysis and Example of Reservoir Placement Mechanism from the Perspective of "Three Lives" Integration. Hohai University, Nanjing (2022)
37. Li Guoping, Sun Yu: Research on the construction of new urbanization with people as the core. Reform **12**, 36–43 (2022)
38. Fang Chuanglin, Zhao Wenjie: Facilitating Chinese path to modernization through new-type urbanization and integrated urban-rural development. Econ. Geogr. **43**(1), 10–16 (2023)

Multiobjective Operation of Cascade Reservoirs Considering Different Ecological Flows

Kunhui Hong[1], Aixing Ma[2,3](✉), Yin Hu[2], Wei Zhang[1], and Mingxiong Cao[2,3]

[1] College of Harbour Coastal and Offshore Engineering, Hohai University, Nanjing 210098, China
hongkunhui@yeah.net

[2] Nanjing Hydraulic Research Institute, Nanjing 210029, China
{axma,huying,mxcao}@nhri.cn

[3] Key Laboratory of Port, Waterway & Sedimentation Engineering Ministry of Communications, Nanjing 210029, China

Abstract. Reservoir scheduling is an important and effective measure to optimize water resource allocation. It effectively mitigates issues such as regional water scarcity and ecological degradation and plays a crucial role in supporting sustainable water resource development. The Yellow River has experienced severe water scarcity, and the increasing human water demand in recent decades has significantly reduced the river's ecological flow, causing significant ecological damage. Balancing the benefits of ecological flow, hydropower generation, and water supply and selecting appropriate ecological flow levels for different hydrological years is a major challenge for reservoir operators. This paper aims to integrate ecological flow into reservoir management standards by investigating the competition and reasonable coexistence between reservoir benefits and ecology in water-scarce regions. A multiobjective reservoir ecology management model was proposed to study the effects of different ecological flow levels on hydropower generation, water supply, and storage. Using the LYX and LJX reservoirs as a case study, the results illustrate the applicability and effectiveness of the model in balancing conflicts among hydropower generation, water supply, storage, and ecology in multi-objective reservoir management. Under basic ecological flow management, hydropower generation and water supply benefits are fully met in typical years, with a significant increase in hydropower generation. Under suitable ecological flow management, hydropower generation and water supply benefits are met only in abundant water years. Considering the impact of storage on management, in typical normal and dry years, under suitable flow management standards, the hydropower generation and water supply guarantee rate targets are approximately 12.49 billion kWh, 96.71%, and 8.22 billion kWh, 96.71%, respectively, promoting efficient water resource utilization and enhancing future drought resilience.

Keywords: Yellow River basin · Multiobjective evolutionary algorithm · Cascade reservoirs · Ecological flow

S. Zheng et al. (Eds.): IHDC 2024, LNCE 487, pp. 446–461, 2025.
https://doi.org/10.1007/978-981-97-9184-2_37

1 Introduction

Rapid urbanization and industrialization have led to water scarcity, which is a major challenge for regional sustainable development [1]. It is crucial to rationally allocate water resources in transboundary river basin to improve water use efficiency and reduce potential conflicts between economic growth and environmental protection [2]. Traditional reservoir management tends to prioritize economic benefits and often overlooks ecological and environmental impacts [3]. Therefore, there is a need to further optimize reservoir operation management. It is essential to ensure the benefits of hydropower generation while meeting the environmental demands for water supply, ecological flows, and the social demands for water in residential, industrial, and agricultural uses [4–6]. Water use and allocation in the design, operation, or management of hydraulic projects can facilitate hydropower generation, flood control, agricultural irrigation, and ice-flood control [7–9].

Since the beginning of the 21st century, the construction and operation of numerous reservoir power stations in major river basins in China have led to a general pattern of joint development and utilization of reservoir groups. While cascade reservoir projects have brought significant economic benefits, they have also had a negative impact on river and lake ecosystems, severely altering natural flow patterns. Excessive water extraction in the Yellow River basin has severely affected the ecological environment. In recent decades, the water use environment in the Yellow River basin has undergone significant changes. Frequent irrigation, increased grain demand, and decreased water supply have created immense pressure. According to Tang et al. [10], climate change is the dominant factor influencing annual runoff changes in the upper and middle reaches, while human activities, such as irrigation water use, dominate runoff changes in the lower reaches. The North China Plain, located in the lower reaches of the Yellow River, is one of China's major grain-producing areas and is highly dependent on water resources. Over the past half century, excessive groundwater extraction and drying have led to a sharp decline in groundwater levels and increased salinization of the land [11–14].

Since the completion of large-scale irrigation projects in 1969, the Yellow River has been experiencing frequent flow interruptions due to intense competition between supply and demand [15, 16]. The irrigation water ratio, which is the ratio of total annual irrigation water use to annual natural runoff, has increased from 21% to 68% over the past 50 years. Various engineering and non-engineering measures have been implemented to address the negative impacts on the ecosystem, with reservoir ecological scheduling being a prominent non-engineering measure. The objective of reservoir ecological scheduling is to strike a balance between economic benefits such as hydropower generation, irrigation, flood control, and navigation, and ecological and environmental protection. By adjusting reservoir operation methods, it is possible to minimize negative ecological impacts while maximizing economic benefits. Scientific and rational optimization of reservoir group scheduling can improve water use efficiency and overall reservoir benefits. At present, China's hydropower projects are in a critical transition from construction to management and operation. Reservoir scheduling tasks have evolved from their original focus on benefits and harm mitigation to promoting the sustainable development of river basins and maintaining river health. However, the requirements for realizing reservoir benefits and protecting river ecology have different operational demands, requiring a balanced approach through optimized reservoir group scheduling. This approach will maximize the comprehensive benefits of water resources in the basin, and balance water resource development with socio-economic progress and river ecological protection.

The multiobjective operation of cascade reservoirs is a complex task that typically involves multiple conflicting objectives, numerous decision variables, and uncertainties [17, 18]. The multiobjective evolutionary algorithm (MOEA) has been considered an efficient way to address multiobjective problems. The recently popular MOEA based on group search has demonstrated excellent practical advantages in finding Pareto optimal solutions for high-dimensional decision variables and multiple nonlinear objective functions [19, 20]. MOEA is an approach that simulates intergenerational natural selection and biological evolution to achieve global optimization. According to the different selection mechanisms, MOEAs can be broadly classified into three groups: Pareto dominance-based MOEA [21, 22], indicator-based MOEA [23], and decomposition-based MOEA. The MOEA based on decomposition with a differential evolution operator (MOEA/D [24]) is considered one of the most efficient algorithms, especially for solving complex multiobjective problems. Therefore, we made efforts to successfully establish a multiobjective model for LYX and LJX reservoirs based on MOEA/D.

The field of reservoir scheduling models has a rich history, marked by significant advances since the mid-20th century. In 1953, the U.S. Army Corps of Engineers developed a joint scheduling model for multiple reservoirs along the Missouri River that effectively addressed seasonal irrigation, water supply, flood control, and hydropower generation. Subsequently, researchers have continued to improve reservoir scheduling optimization methods. Tang et al. [25] integrated cascade stochastic runoff and time delays into hydropower optimization models, emphasizing the importance of accounting for real-world engineering considerations. Rui Hui et al. [26] investigated an optimal allocation model for total flood storage in parallel reservoirs and successfully applied it to the Oroville and New Bullards Bar reservoirs, demonstrating its practicality and effectiveness. Moridi et al. [27] developed a mixed-integer linear programming model with the objective of minimizing flood and hydropower generation losses. When applied to the Karkheh reservoir system in Iran for 25-year and 50-year flood scenarios, this model resulted in reduced flood losses and reduced hydropower generation risk, highlighting the benefits of coordinated operation. Bai et al. [28] used a constraint method to address the multiobjective optimization scheduling problem of reservoir groups, taking into account factors such as water supply, hydropower generation, flood control, ice-flood control, and ecology. They transformed this complex problem into a single-objective problem for different time periods and tackled it using the POA-DPSA approach. Olofintoye et al. [29] integrated artificial neural networks with a multiobjective differential evolution model to improve inflow forecasting and real-time multiobjective optimization scheduling at the Vanderkloof Reservoir in South Africa. This integration resulted in a significant improvement in scheduling capabilities. Liu et al. [30] applied a sliding support vector machine to establish optimal spillway operation rules for the Three Gorges Reservoir. They simplified the multiobjective problem into a single-objective optimization using the weighting method and found that the sequence and number of spillways have a significant impact on the reservoir's multiobjective benefits.

Afshar [31] proposed a hybrid autonomous and coordinated search approach using genetic algorithms to improve the operational efficiency of cascade reservoir groups. Wang et al. [32] introduced the concept of subjective trade-off rate and proposed an optimal decision-making method that considers ecological risk for multiobjective optimization of the Three Gorges Reservoir for ecology and hydropower generation. Uen et al. [33] developed a multiobjective joint optimization model for Taiwan's Shimen Reservoir and irrigation pools to maximize hydropower generation and storage, and achieved satisfactory results using the NSGA-II algorithm. Zhang et al. [35] presented an improved multiobjective moth-flame optimization algorithm to solve the cascade reservoir group multiobjective scheduling model, taking into account hydropower generation, ecology, and navigation, and achieved a well-converged and evenly distributed Pareto front. In previous studies on joint scheduling for reservoir groups, researchers have extensively examined hydropower generation, ecology, and water supply. Some models consider only water-sediment issues in scheduling and overlook ecological impacts. Others discuss the trade-off between hydropower generation and ecological flow, or target overall reservoir benefits, potentially compromising aspects such as year-end reservoir storage levels. This compromise could affect future hydropower generation and water supply.

The Yellow River is currently facing significant ecological and water resource challenges. The rapid economic and social development in the basin has led to an increase in water demand, while there is a decreasing trend in runoff. Inefficient water use and low agricultural water use efficiency have exacerbated water scarcity and ecological problems, which have had a significant impact on the ecology of Yellow River. The Longyangxia (LYX) and Liujiaxia (LJX) reservoirs, which are the upstream cascade reservoirs, control more than half of the Yellow River's flow, providing clean energy and water supply to more than 420 million people. Therefore, the scheduling of the LYX and LJX reservoirs must not only meet the annual hydropower generation, water supply, and storage requirements but also comply with upstream ecological flow standards in order to minimize the ecological impact on the Yellow River.

This paper addresses the issues of hydropower generation, water supply, and future water scarcity risk for the LYX and LJX reservoirs in the Yellow River basin. By incorporating ecological flow into mandatory reservoir management standards, an ecological management scheduling model is developed for the LYX and LJX reservoirs. The paper examines the impact of ecological flow management on reservoir scheduling and overall benefits, and demonstrates the feasibility of the multiobjective scheduling model incorporating ecological flow standards. It discusses the impact of different ecological flow standards on the hydropower generation and water supply of the LYX and LJX reservoirs and examines the feasibility of these standards. In addition, the paper investigates the competitive and cooperative relationship between hydropower generation, water supply, and storage under suitable ecological flow conditions for the LYX and LJX reservoirs to provide decision makers with the best trade-offs between human needs and ecological flow maintenance.

2 Study Area and Data

2.1 Study Areas

The Yellow River originates in the Bayan Har Mountains of Qinghai Province in western China and flows through nine provinces before emptying into the Bohai Sea [35]. With a total length of 5463 km, it is revered as the "Mother River of China" due to its historical significance as the birthplace of northern Chinese civilization and its importance in early Chinese history [36]. In recent years, however, the river has been severely impacted by pollution, hydropower development, and intensive water extraction for human consumption, agriculture, and industry. These problems in the Yellow River basin are largely attributed to unsustainable human activities and the accelerated alteration of the river's ecological flow [37].

The upper reaches of the Yellow River mainly receive runoff from areas upstream of Lanzhou. The section from Lanzhou to Toudaoguai mainly carries transit water, with the LYX and LJX reservoirs controlling over 40% of the natural runoff of the upper Yellow River. The management of these reservoirs significantly influences flood control, water supply, and ecology in the upper reaches. The main characteristic water levels and power station parameters of the cascade reservoirs are presented in Table 1, and the overall layout of the Yellow River basin's cascade reservoir system is shown in Fig. 1.

The agricultural regions in the upper Yellow River, especially the Ningxia and Inner Mongolia irrigation areas downstream of the reservoirs, rely heavily on irrigation for crop growth. In the Ningxia and Inner Mongolia irrigation areas, the peak irrigation periods are from April to July and September to November due to low rainfall. The total irrigated areas in these districts are 6,573 km^2 and 21,300 km^2, respectively. Approximately 80% of off-channel water is used for agricultural irrigation, making it the main water consumer in the region. Downstream water uses include forestry, livestock, fisheries, and domestic consumption, which are collectively categorized as Yellow River water extraction. Therefore, the scheduling of the LYX and LJX reservoirs is critical to hydropower generation, water supply, and ecological sustainability in the upper Yellow River.

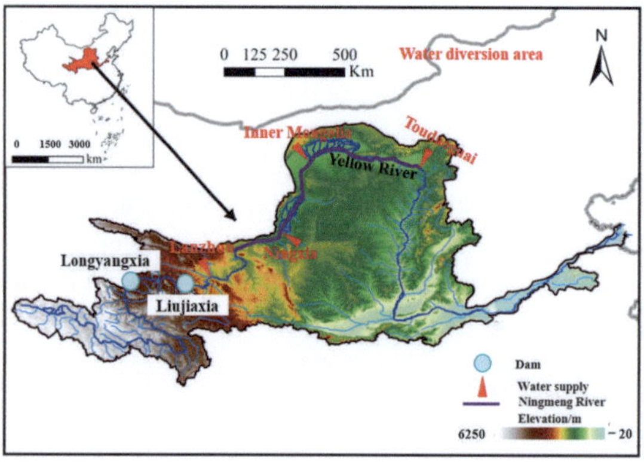

Fig. 1. Diagram of the study area

Table 1. The characteristics of Longyangxia and Liujiaxia reservoirs

Characteristics	Longyangxia	Liujiaxia
Normal water level (m)	2600	1735
Flood limit water level (m)	2594	1726
Dead water level (m)	2560	1694
Total storage (10^8 m^3)	247	57
Power generation capacity (MW)	128	139

2.2 Data

The paper uses data on reservoir water supply, hydropower generation, runoff, and reservoir inflow. The water supply and hydropower generation data for the upper Yellow River were obtained from the Hydrological Bureau of the Yellow River Conservancy Commission (YRCC). Daily observed inflow and outflow, water level, and storage data for the LYX and LJX reservoirs from 2007 to 2018 were obtained from the "Yellow River Basin Hydrological Yearbook" compiled and published by the Yellow River Conservancy Commission.

3 Method

3.1 Multiobjective Framework

With the rapid pace of societal development, reservoirs are required to perform flood control functions while meeting the demands for energy and water supply [38, 39]. LYX and LJX cascades, which serve as important regulatory hubs, assume the mission of comprehensive resource utilization. Therefore, this paper identifies hydropower generation, water supply, and ecology as primary objectives.

3.1.1 Hydropower Generation

Hydropower generation stands as a critical function of the reservoir and constitutes one of the primary research objectives of this study, shown as follows:

$$\max F = \sum_{i=1}^{N}\sum_{t=1}^{T} k_i q_{i,t} h_{i,t} \tau_t$$

where F is the total hydropower generation in one year (kWh), k_i is the output coefficient specific to the ith hydropower, $q_{i,t}$ is the average discharge of the ith reservoir in the tth month (m^3/s), $h_{i,t}$ is the average water level of the ith reservoir in the tth month (m), τ_t is the time interval (month), N is the number of reservoirs, and T is the number of operation periods ($T = 12$).

3.1.2 Water Supply

To evaluate the water supply case, minimizing the water supply shortage rate was set as the criterion for evaluating water supply.

$$\min S = \begin{cases} \sum_{t=1}^{T} \sum_{i=1}^{N} \frac{Rd_t - Rt_t}{d_t}, & \text{if } Rt_t < Rd_t \\ 0, & \text{if } R_t t_t \geq Rd_t \end{cases}$$

where S represents the water shortage rate in one year, and Rt_t and Rd_t denote the water supply and water demand for the ith reservoir in the tth month (m^3).

3.1.3 Ecological Flow

To alleviate the adverse effects on river ecology caused by the operation of cascade reservoirs, the minimum shortage rate of ecological flow was adopted as the criterion for ecological assessment. Efforts have been made to minimize instances of insufficient ecological flow by adjusting reservoir discharge.

$$\min E = \begin{cases} \sum_{t=1}^{T} \sum_{i=1}^{N} \frac{EF_t - E_t}{EF_t}, & \text{if } E_t < EF_t \\ 0, & \text{if } E_t \geq EF_t \end{cases}$$

where E is the ecological flow shortage rate in one year, while EF_t and E_t represent, respectively, the demand for ecological flow and the actual ecological flow in the tth month (m^3/s).

3.1.4 Constrain

The above objectives are subject to the following constraints.

(1) Water balance constraint

$$V_{i,t+1} = V_{i,t} + (Q_{i,\text{in},t} - Q_{i,\text{out},t})\tau_t$$

where $V_{i,t}$ and $V_{i,t+1}$ represent the ith reservoir storages at tth and $(t+1)$th, respectively. $Q_{i,\text{in},t}$ and $Q_{i,\text{out},t}$ represent the ith average inflow and average outflow of the reservoir in the tth month (m^3/s), respectively.

(2) Water release capacity constraints

$$Q_{i,t}^{\min} \leq Q_{i,t} \leq Q_{i,t}^{\max}$$

where $Q_{i,t}^{\min}$ and $Q_{i,t}^{\max}$ are the minimum and maximum discharges in the tth month (m^3/s).

(3) Power generation output constraints

$$q_{i,t}^{\min} \leq q_{i,t} \leq q_{i,t}^{\max}$$

where $q_{i,t}^{\min}$ and $q_{i,t}^{\max}$ are the minimum and maximum hydraulic turbine discharges in the tth month (m^3/s), respectively.

(4) Water-level constraints

$$Z_{i,t}^{\min} \leq Z_{i,t} \leq Z_{i,t}^{\max}$$

$$Z_{i,t}^{\min} \leq Z_{i,t} \leq Z_{i,t}^{\max}$$

where $Z_{i,t}^{\min}$ and $Z_{i,t}^{\max}$ are the minimum and maximum water levels in the tth month, respectively (Fig. 2).

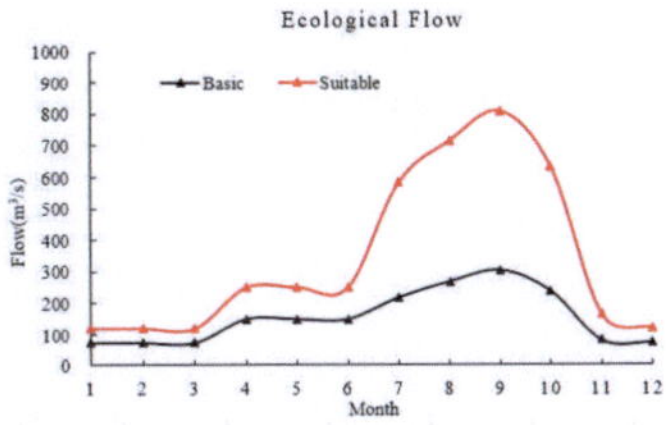

Fig. 2. Monthly flow rates required for basic and sustain

This paper examines the scheduling of two major reservoirs, LYX and LJX reservoirs, located on the mainstream of the Yellow River. The main objective is to meet the requirements for hydropower generation, water supply, and ecological water use. Two schedule plans have been proposed:

Plan I: Prioritize the maximization of hydropower generation, while ensuring the essential flow requirements of the river and meeting downstream water supply needs for both the LYX and LJX reservoirs.

Plan II: Prioritize the maximization of hydropower generation, while ensuring the appropriate flow requirements of the river and meeting downstream water supply needs for both the LYX and LJX reservoirs.

4 Results and Discussion

4.1 Typical Year Analysis

Using guarantee rates of 30%, 50%, and 80%, and analyzing the annual natural inflow data of the Yellow River from 2003 to 2021, we selected three hydrological years to trigger the model: a wet year (2018), a normal year (2013), and a dry year (2016). We then analyzed the impact of incorporating basic and suitable ecological flow guarantees on hydropower generation and water supply for the LYX and LJX reservoirs based on the ecological management scheduling model. We used the initial water levels for the typical wet, normal, and dry years as model conditions, and the results of the reservoir scheduling, hydropower generation, and water supply are shown in Tables 2, 3, 4 and 5. The inflow, outflow, and water levels during the scheduling process are shown in Figs. 3 and 4.

According to the data in Table 2, the hydropower generation of the LYX and LJX reservoirs showed a significant increase under the basic ecological flow in wet, normal, and dry years compared to the actual values. Specifically, in wet years, hydropower generation increased by 5.24% and 4.74% respectively, in normal years, the increase was 1.39% and 1.10%, and in dry years, LYX's hydropower generation decreased by 0.08% to meet basic ecological flow requirements, but the total hydropower generation of the LYX and LJX reservoirs increased by 0.08%. As depicted in Table 4, the water supply targets for all scheduling periods were achieved under the basic ecological flow conditions. The improved availability of water resources in wet and normal years facilitated optimized scheduling, resulting in significant increases in hydropower generation while meeting water supply and ecological needs. Table 3 shows that under the appropriate ecological flow, the hydropower generation of the LYX and LJX reservoirs increased significantly by 5.00% in wet years. Specifically, LYX and LJX reservoirs experienced hydropower generation increases of 4.14% and 4.39%, respectively. However, in normal and dry years, the hydropower generation of the cascade reservoirs decreased noticeably by 1.29% and 0.22%, respectively, to meet the suitable ecological flow requirements. Table 5 shows that the water supply targets in wet and normal years were achieved for all scheduling periods, but the water supply guarantee rate in dry years was 95.39%. This can be attributed to reduced upstream inflows, where water scarcity emerged as the primary factor affecting hydropower generation and water supply, which could not be solved by optimization.

In summary, under the basic ecological flow, the hydropower generation and water supply goals were met in wet, normal, and dry years, with a slight increase in hydropower generation. When transitioning from the basic ecological flow to the appropriate ecological flow, the total hydropower generation from the cascade reservoirs decreased by 0.70%, 2.50%, and 0.30% in wet, normal, and dry years, respectively. The water supply guarantee rate in dry years decreased to 95.39%. The increase in ecological flow requirements limited downstream water demand, which in turn prevented the reservoirs from achieving optimal hydropower generation efficiency.

Table 2. Hydropower generation of meeting basic ecological flow unit: 10^8 kWh

Reservoir	Optimization scheme			Actual scheme			Changing rate (%)		
	Wet	Normal	Dry	Wet	Normal	Dry	–	–	–
LYX	81.94	67.97	37.83	77.86	67.04	37.86	5.24	1.39	−0.08
LJX	77.98	68.83	44.91	74.45	68.08	44.81	4.74	1.10	0.22
Total	159.92	136.8	82.74	152.31	135.12	82.67	5.00	1.24	0.08

4.2 Relationship of Hydropower Generation, Water Supply and Water Storage

Due to its location in a semi-arid region, the upper reaches of the Yellow River suffer from water scarcity. The LYX and LJX reservoirs, which serve as the main regulating

Table 3. Hydropower generation of meeting suitable ecological flow unit: 10^8 kWh

Reservoir	Optimization scheme			Actual scheme			Changing rate (%)		
	Wet	Normal	Dry	Wet	Normal	Dry	–	–	–
LYX	81.08	67.09	37.92	77.86	67.04	37.86	4.14	0.07	0.16
LJX	77.72	66.29	44.58	74.45	68.08	44.81	4.39	−2.63	−0.51
Total	158.80	133.38	82.49	152.31	135.12	82.67	4.26	−1.29	−0.22

Table 4. Water supply guarantee rate of meeting the basic ecological flow

Reservoir	Guarantee rate of water supply (%)			Water shortage单位?		
	Wet	Normal	Dry	Wet	Normal	Dry
LYX	–	–	–	–	–	–
LJX	100	100	100	0	0	0

Table 5. Water supply guarantee rate of meeting the suitable ecological flow

Reservoir	Guarantee rate of water supply (%)			Water shortage单位?		
	Wet	Normal	Dry	Wet	Normal	Dry
LYX	–	–	–	–	–	–
LJX	100	100	95.39	0	0	5.57

reservoirs in this area, play an important role in meeting ecological and water supply requirements, as well as in replenishing water downstream. Therefore, it is essential to strike a balance between ecological flow, water supply, and reservoir storage in the upper reaches of the Yellow River, and to determine suitable scheduling intervals.

Figure 5 shows that during normal years, as power generation increases, the water supply guarantee rate also increases, while reservoir storage decreases. A significant inflection point is observed at a power generation of 12.49 billion kWh and a water supply guarantee rate of 96.71%. When power generation is below 12.49 billion kWh, both power generation and the water supply guarantee rate increase rapidly as reservoir storage decreases. When power generation exceeds 12.49 billion kWh, the year-end reservoir storage decreases at a similar rate, but the increases in power generation and the water supply guarantee rate slow down, indicating lower water resource utilization efficiency.

As shown in Fig. 5, during dry years, a decrease in reservoir water demand leads to a gradual increase in hydropower generation and water supply guarantee rate. A distinct inflection point appears at 8.22 billion kWh of power generation and a water supply guarantee rate of 97.74%. Below this threshold, both power generation and the water supply guarantee rate increase rapidly as reservoir storage decreases. When power

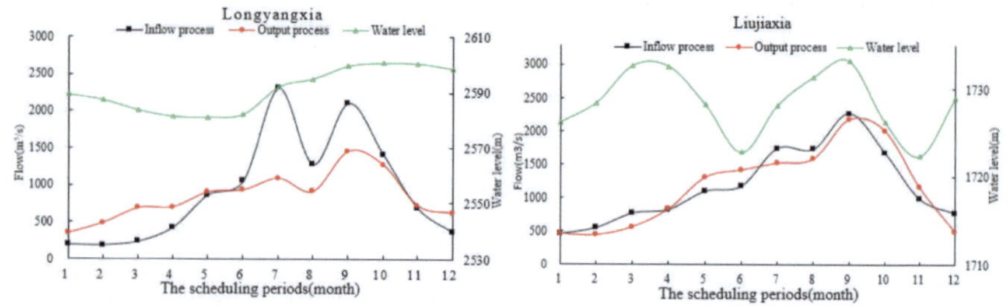

(1) Wet year

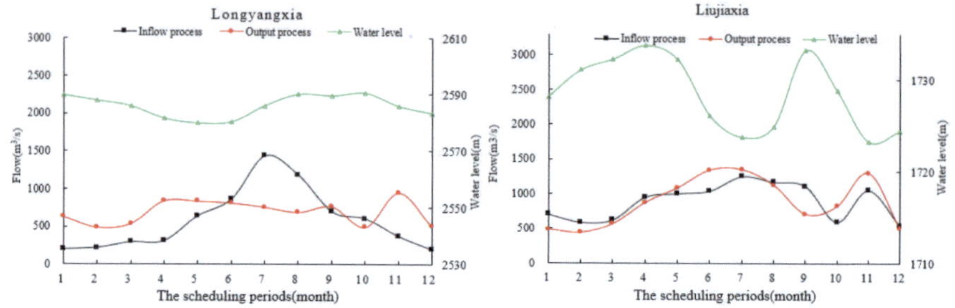

(2) Normal year

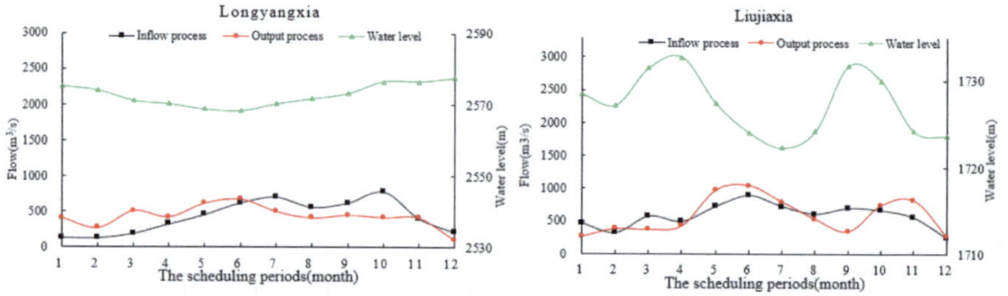

(3) Dry year

Fig. 3. The operation of LYX and LJX reservoirs to meet the basic ecological flow

generation exceeds 8.222 billion kWh, the year-end reservoir storage decreases at a similar rate, but the growth in power generation and the water supply guarantee rate slows, indicating a less efficient use of water resources.

When considering water resource management, it is important to incorporate suitable ecological flow for management standards. There are noticeable differences in the growth rates of hydropower generation and water supply guarantee rates between normal and dry years for different reservoir storage intervals. In both scenarios, the rates initially increase rapidly with decreasing reservoir storage, and then gradually decrease. This is because the initial constraint on hydropower generation and water supply is primarily the water volume. However, as the reservoir storage decreases, the constraint shifts to optimizing reservoir scheduling. LYX and LJX reservoirs maintain hydropower generation and water supply guarantee rates of about 124.90 10^8 and 96.71%, respectively, in normal years, respectively, considering the scarcity of Yellow River water resources and aiming for their rational and efficient utilization while meeting downstream ecological flow

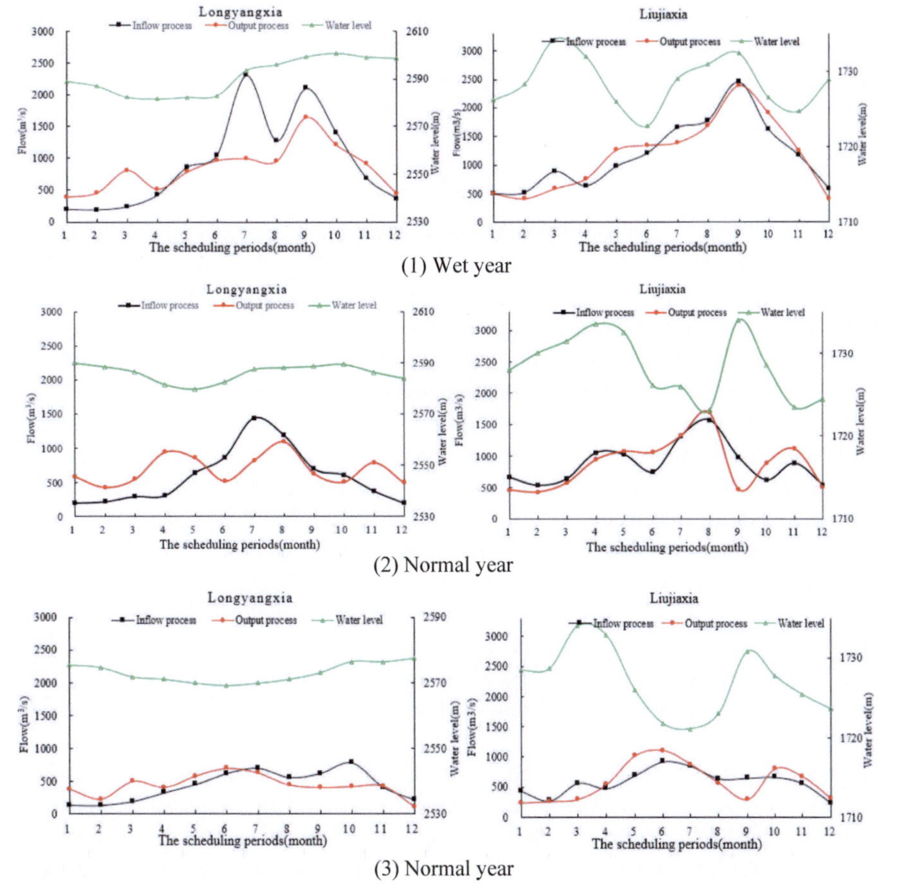

Fig. 4. The operation of LYX and LJX reservoirs to meet the suitable ecological flow

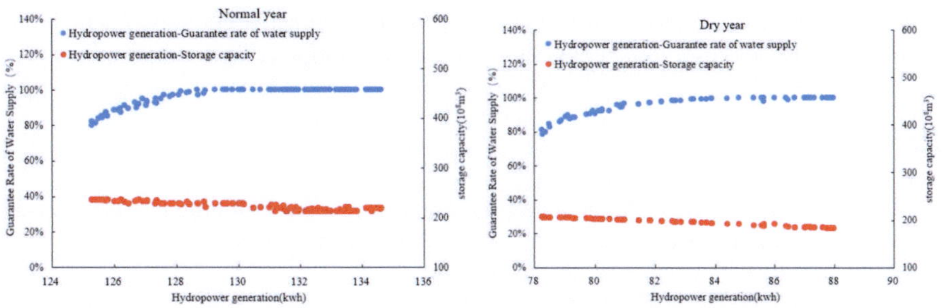

Fig. 5. Relationship of Hydropower generation, Water supply and water storage in wet and dry years

management standards. This indicates that the water resource utilization rate of LYX and LJX reservoirs is relatively high. In dry years, the water supply guarantee rate remains at 96.71%, which is more conducive to efficient water resource utilization, with a hydropower generation of 9.77 billion kWh, exceeding multi-year average hydropower generation of the LYX and LJX reservoirs.

5 Conclusion

This paper has established a multiobjective ecological management model for the LYX and LJX reservoirs to examine the competing interests of hydropower generation, water supply, reservoir storage, and ecological flow. It presents ecological flow targets for different water level scenarios in the upper reaches of the Yellow River, allowing water resource managers to operate within defined parameters and achieve a satisfactory balance between human water consumption and ecosystem needs.

The results suggest that by implementing reservoir ecological management scheduling models and meeting basic ecological flow standards, the LYX and LJX reservoirs experienced respective increases of 5.00%, 1.24%, and 0.08%, respectively, in typical annual hydropower generation. When suitable ecological flow standards were met, there was an increase of 4.26% in wet years, while decreases of 1.29% and 0.22% were observed in normal and dry years, respectively. These results indicate that with scheduling, the LYX and LJX reservoirs can meet the river's basic ecological flow requirements while still meeting objectives such as hydropower generation and water supply. However, due to the limited water resources of the Yellow River, it is only in wet years that the river can attain suitable ecological flow without compromising hydropower generation and water supply objectives.

Under the fulfillment of downstream river ecological flow management standards, the LYX and LJX reservoirs maintains hydropower generation and water supply guarantee rates at around 12.49 billion and 96.71%, respectively, in normal years, indicating relatively high water resource utilization efficiency. In dry years, the targets for hydropower generation and water supply guarantee rates are around 8.22 billion kWh and 96.71%, respectively, which is more conducive to efficient water resource utilization. It is evident that the reservoir ecological management scheduling model proposed in this paper has the ability to provide an effective approach to solve the multiobjective ecological scheduling problem of cascade reservoirs, considering different ecological flow requirements.

References

1. Li, J.W., Liu, Z.F., He, C.Y., et al.: Water shortages raised a legitimate concern over the sustainable development of the drylands of northern China: Evidence from the water stress index. Sci. Total. Environ. **590**, 739–750 (2017)
2. Afshar, M.H., Hajiabadi, R.: A novel parallel cellular automata algorithm for multi-objective reservoir operation optimization. Water Resour. Manage **32**, 785–803 (2018)
3. Jiang, Z.Q., Liu, P., Ji, C.M., et al.: Ecological flow considered multi-objective storage energy operation chart optimization of large-scale mixed reservoirs. J. Hydrol. **577**, 123949 (2019)
4. Liu, B., Zhang, F.L., Wan, W.Y., et al.: Multi-objective decision-making for the ecological operation of built reservoirs based on the improved comprehensive fuzzy evaluation method. Water Resour. Manage **33**, 3949–3964 (2019)

5. Chen, D., Leon, A.S., Chen, Q.W., Li, R.N.: Aderivative-free hybrid optimization model for short-term operation of a multi-objective reservoir system under uncertainty Water Resour. Manage **32**, 3707–3721 (2018)
6. Hatamkhani, A., Moridi, A.: Multi-objective optimization of hydropower and agricultural development at river basin scale Water Resour. Manage **33**, 4431–4450 (2019)
7. Wang, C., Yu, Y., Wang, P.F., Sun, Q.Y., Hou, J., Qian, J.: Assessment of the ecological reservoir operation in the Yangtze Estuary based on the salinity requirements of the indicator species River Res. Appl. **32**, 946–957 (2016)
8. Razurel, P., Perona, P., Gorla, L., Tron, S., Niayifar, A., Perona, P.: Improving the ecohydrological and economic efficiency of small hydropower plants with water diversion Adv. Water Resour. **113**, 249–259 (2018)
9. Tang, Q., Oki, T., Kanae, S., Hu, H.: Hydrological cycles change in the Yellow River basin during the last half of the twentieth century. J. Climate **21**, 1790–1806 (2008). https://doi.org/10.1175/2007JCLI1854.1
10. Brown, L.R., Halweil, B.: China's water shortage could shake world food security. World Watch **11**(4), 10–18 (July/August 1998)
11. Shimada, J.: Proposals for the groundwater preservation toward 21st century through the view point of hydrological cycle. J. Jpn. Assoc. Hydrol. Sci. **30**, 63–72(in Japanese) (2000)
12. Chen, J.Y., Tang, C.Y., Shen, Y.J., Sakura, Y., Kondoh, A., Shimada, J.: Use of waterbalance calculation and tritium to examine the dropdown of groundwater table in the piedmont of the North China Plain (NCP). Environ. Geol. **44**, 564–571 (2003)
13. Nakayama, T., Yang, Y., Watanabe, M., Zhang, X.: Simulation of groundwater dynamics in the North China Plain by coupled hydrology and agricultural models. Hydrol. Process. **20**(16), 3441–3466 (2006). https://doi.org/10.1002/hyp.6142
14. Yang, Z.S., Milliman, J.D., Galler, J., Liu, J.P., Sun, X.G.: Yellow River's water and sediment discharge decreasing steadily. Eos **79**(48), 589–592 (1998)
15. Fu, G., Chen, S., Liu, C., Shepard, D.: Hydro-climatic trends of the Yellow River basin for the last 50 years. Clim. Change **65**, 149–178 (2004)
16. Chang, L.C., Chang, F.J., Wang, K.W., Dai, S.Y.: Constrained Geneti-c algorithms for optimizing multi-use reservoir operation. J. Hydrol. **390**, 66–74 (2010). https://doi.org/10.1016/j.jhydrol.2010.06.031
17. Guo ShengLian, G.S., Zhang HongGang, Z.H., Chen Hua, C.H., Peng DingZhi, P.D., Liu Pan, L.P., Pang Bo, P.B.A.: Rese-rvoir flood forecasting and control system for China. Int. Assoc. Sci. Hydrol. Bull. **49**, 959–972 (2004). https://doi.org/10.1623/hysj.49.6.959.55728
18. Yang, G.; Guo, S.; Liu, P.; Li, L.; Xu, C.: Multi-objective reservoir operating rules based on cascade reservoir input variable selection method. Water Resour. Res. **53**, 3446–3463. https://doi.org/10.1002/2016WR020301 (2017)
19. Wang, K.W., Chang, L.C., Chang, F.J.: Multit-interactive genetic algorithms for the optimization of long-term reservoir operation. Adv. Water Resour. **34**, 1343–1351 (2011). https://doi.org/10.1016/j.advwatres.2011.07.004
20. Deb, K., Pratap, S.A., Meyarivan, T.: A fast and elitist multiobjective genetic algorithm: NSGA-II. IEEE Trans. Evol. Comput. **6**, 182–197 (2002). https://doi.org/10.1109/4235.996017
21. Zitzler, E., Künzli, S.: Indicator-based selection in multiobjective search. In Parallel Problem Solving from Nature PPSN VIII, Proceedings of the 8th International Conference, Birmingham, UK, 18–22 September (2004)
22. Yao, X., Burke, E.K., Lozano, J.A., Smith, J., Merelo-Guervós, J.J., Bullinaria, J.A., Rowe, J.E., Tiňo, P., Kabán, A., Schwefel, H.-P. (eds.): Lecture notes in computer science; Springer: Berlin/Heidelberg, Germany, 2004; Volume 3242. https://doi.org/10.1007/978-3-540-30217-9_84.

23. Beume, N., Naujoks, B., Emmerich, M.: SMS-EMOA: Multi-objective selection based on dominated hypervolume. Eur. J. Oper. Res. **181**, 1653–1669 (2007). https://doi.org/10.1016/j.ejor.2006.08.008
24. Zhang, Q.; Li, H. Moea/d: A multiobjective evolutionary algorithm based on decomposition. IEEE Trans. Evol. Comput. 2007, 11, 712–731. https://https://doi.org/10.1109/TEVC.2007.892759
25. Tang, J., Huang, w.V.: A decomposition method for generation scheduling of hydro-Systems with delays and unpredictable changes in natural inflows. Comput. Ind. Eng. **22**(2), 147–155 (1992)
26. Hui, R., Jay, R., Lund.: Flood storage allocation rules for parallel reservoirs. J. Water Resour. Plan. Manag. **5**(141), (2015)
27. Moridi, A., Yazdi, J.: Optimal allocation of flood control capacity for multi-reservoir systems using multi-objective optimization approach. Water Resour. Manag. **31**(14), 4521–4538 (2017)
28. Bai, T., Chang, J.X., Chang, F.J., et al.: Synergistic gains from the multi-objective optimal operation of cascade reservoirs in the Upper Yellow River basin. J. Hydrol. **523**, 758–767 (2015)
29. Olofintoye, O.,Otieno, F., Adeyemo, J.: Real-time optimal water allocation for daily hydropower generation from the Vanderkloof dam, South Africa. Appl. Soft Comput. **47**, 119–129 (2016)
30. Liu, X.Y., Chen, L., Zhou, Y.H., et al.: Multi-objective reservoir operation during flood season considering spillway optimization. J. Hydrol. **552**, 554–563 (2017)
31. Afshar, M.H., Azizipour, M., Oghbaeea, B., et al.: Exploring the efficiency of harmony search algorithm for hydropower operation of multi-reservoir systems: A hybrid cellular automat-harmony search approach. Adv. Intell. Syst. Comput. **514**, 252–260 (2017)
32. Wang, X.J., Dong, Z., Ai, X.S., et al.: Multi-objective model and decision-making method for coordinating the ecologic al benefits of the Three Gorger Reservoir. J. Clean. Prod. **270**(10), 122066 (2020)
33. Uen, T.S., Chang, F.J., Zhou, Y.L., et al.: Exp loring synergistic benefits of Water-Food-Energy Nexus through multi-objective reservoir optimization schemes. Sci. Total. Environ. **633**, 341–351 (2018)
34. Wang, Y., Tang, F., Jiang, E., Wang, X., Zhao, J.: Optimizing hydropower generation and sediment transport in Yellow River basin via cooperative game theory. J. Hydrol. **614**, 128581 (2022). https://doi.org/10.1016/j.jhydrol.2022.128581
35. Zhang, Z.D., Qin, H., Yao, L.Q., et al.: Improved multi-objective moth-flame optimization algorithm based on R-domination for cascade reservoirs operation. J. Hydrol. **581**, 124431 (2020)
36. Yang, T., Zhang, Q., Chen, Y.D., Tao, X., Xu, C.Y., Chen, X.: A spatial assessment of hydrologic alteration caused by dam construction in the middle and lower Yellow River. China. Hydrol Process **22**(18), 3829–3843 (2008)
37. Reddy, M.J., Kumar, D.N.: Optimal reservoir operation using multi-objective evolutionary algorithm. Water Resour. Manag. **20**(6), 861–878 (2006)
38. Hou, Y., Guo, S.: Incorporating ecological requirement into multipurpose reservoir operating rule curves for adaptation to climate change. J. Hydrol. 498, 153–164. https://doi.org/10.1016/j.jhydrol.2013.06.028 (2013)
39. Zhou, Y., Chang, L.C., Uen, T.S., Guo, S., Xu, C.Y., Chang, F.J.: Prospect for small hydropower installation settled upon optimal water allocation: An action to stimulate synergies of water-food-energy nexus. Appl. Energy **238**, 668–682 (2019). https://doi.org/10.1016/j.apenergy.2019.01.069

Comparison of Ecological Value Before and After the Construction of Hydraulic Engineering Projects: A Case Study of Lianhu Reservoir

Zelong Qu[1(✉)], Jianfeng Li[1], and Guofu Yang[2]

[1] Huadong Engineering Corporation Limited, Hangzhou 311122, China
qu_zl@hdec.com

[2] School of Art and Archaeology, Hangzhou City University, Hangzhou 310015, China

Abstract. Hydraulic engineering projects provide essential functions and services, including flood regulation, hydroelectric power generation, and agricultural irrigation. Additionally, they offer benefits such as navigation, water purification, recreational opportunities, and biodiversity maintenance for upstream and downstream regions. To comprehensively describe the overall functioning of hydraulic engineering as an ecosystem, evaluate its contributions to human well-being, assess its support for economic and social development, and understand its ecological linkages across regions, it is essential to scientifically account for the comprehensive benefits of these projects. This study uses the Lianhu Reservoir in Lishui, Zhejiang Province, as a case study to assess the Gross Ecosystem Product (GEP) before and after the construction of the hydraulic engineering project. The results show that before the construction of the Lianhu Reservoir (in 2021), the total GEP within the construction area was 206 million yuan. The value of regulating services was 197 million yuan, accounting for 95.6% of the total GEP, with climate regulation and flood control contributing 175 million yuan and 21 million yuan, respectively. The value of provisioning services was 9.06 million yuan, accounting for 4.4% of the total GEP. After the construction of the Lianhu Reservoir, the total GEP within the construction area increased to 1.42 billion yuan. The value of regulating services was 781 million yuan, accounting for 55.0% of the total GEP, with flood control and climate regulation contributing 480 million yuan and 297 million yuan, respectively. The value of cultural services was 498 million yuan, accounting for 35.1% of the total GEP, while the value of provisioning services was 141 million yuan, accounting for 9.9% of the total GEP. The results indicate that compared to the pre-construction period, the GEP after the construction of the Lianhu Reservoir increased by 1.21 billion yuan, representing a 588% increase. Except for the services of agricultural product supply, carbon sequestration and oxygen release, and air purification, all other service items were enhanced.

Keywords: Ecosystem services · Gross Ecosystem Product · Sustainable development

S. Zheng et al. (Eds.): IHDC 2024, LNCE 487, pp. 462–475, 2025.
https://doi.org/10.1007/978-981-97-9184-2_38

1 Introduction

The construction of hydraulic and hydroelectric engineering projects has played a significant role in social development. The impact of these projects on water ecosystem services has always been a focal point for various sectors of society. Hydraulic engineering provides essential functions and services to humans, such as flood regulation, hydroelectric power generation, and agricultural irrigation. Beyond these direct benefits, hydraulic engineering also supports navigation, water purification, recreational activities, and biodiversity maintenance in upstream and downstream regions. However, by altering the surrounding ecosystems, hydraulic projects may also bring about some negative impacts. Therefore, to comprehensively describe the overall functioning of hydraulic engineering as an ecosystem, evaluate its contributions to human well-being, assess its support for economic and social development, and understand its ecological linkages across regions, it is crucial to scientifically account for the comprehensive benefits of these projects.

Since the concept of Gross Ecosystem Product (GEP) was introduced, numerous studies have been conducted on GEP accounting at various administrative levels, including national, provincial, municipal, and county levels. In March 2021, the United Nations Statistical Commission officially included GEP in the latest System of Environmental-Economic Accounting, recognizing it as an indicator for ecosystem service and ecological asset valuation and as a measure for the UN Sustainable Development Goals (SDGs) for 2050. GEP accounting not only describes the overall functioning of ecosystems but also evaluates the effectiveness of ecological protection, assesses contributions to human well-being, supports economic and social development, and helps understand ecological linkages between regions.

For instance, Academician Lu Youmei regards the Three Gorges Project as an ecological project beneficial to global environmental protection and a development opportunity for the Three Gorges Reservoir area and Hubei Province. Conversely, Pan Liwu et al. argue that the reservoir's impoundment has altered the hydrological conditions of the Yangtze River, causing adverse effects. From a dialectical perspective, hydraulic and hydroelectric projects fulfill human needs and prevent flood disasters, but they also inevitably affect local ecosystem services by altering natural hydrological patterns. Traditionally, the construction of hydraulic and hydroelectric projects focused primarily on social and economic benefits. However, with increasing emphasis on ecological benefits in China, there is a pressing need to quantitatively assess the impact of such projects on water ecosystem services and promote their coordinated development.

Previous studies have calculated the changes in river ecosystem function value due to hydroelectric development and used these calculations to analyze wind-hydro comprehensive projects quantitatively. Xiao Jianhong et al. established an evaluation system to assess the impact of dams on river ecosystem services and evaluated the impact of dams nationwide in 2002. Wei Guoliang et al. developed an accounting system to evaluate the impact of hydroelectric energy development on river ecosystem services based on the characteristics of river ecosystem services. Chen et al. conducted a comprehensive analysis of the ecological benefits and losses of the Xiaolangdi Reservoir from 2000 to 2012, proposing a three-step framework for assessing the ecological gains and losses of hydroelectric projects and quantitatively evaluating the changes in service functions caused by the project. Jia Jianhui et al. used the functional value method to evaluate the impact of hydroelectric energy development on the main stream of the Wujiang River on river ecosystem services.

Although some exploratory studies on the impact of hydraulic and hydroelectric projects on various water ecosystem services exist, these studies have used different accounting systems according to their research directions. The lack of a unified evaluation standard makes it challenging to compare results across different projects, hindering the quantitative assessment of the contribution of hydraulic engineering to water ecosystem services. Moreover, the existing standardized GEP evaluation norms face several challenges when applied to hydraulic engineering:

First, current GEP accounting norms lack focus on hydraulic aspects. Most GEP assessments target terrestrial ecosystems, neglecting the ecological products related to hydraulic engineering. Important functions such as flood regulation, navigation enhancement, reduction of carbon emissions through hydroelectric power, and the contribution to the value of surrounding lands are not adequately reflected in current accounting standards. Second, existing accounting methods do not consider the practical application needs of hydraulic engineering. Traditional GEP accounting systems for terrestrial ecosystems rely mainly on land use type data and ecological service value coefficients. These methods need to be optimized and improved to account for the ecological products of hydraulic projects. For example, the significant function of flood regulation by hydraulic projects, which reduces downstream flood risks, is not accounted for in the current system. Third, the application scenarios for existing accounting results are limited. Currently, the results of GEP accounting are primarily used by management departments for planning and managing ecological development within their jurisdictions, lacking practical application in project design, decision-making, and operational evaluation.

Therefore, this study aims to construct an accounting framework for the total value of hydraulic engineering ecological products and, using the Lianhu Reservoir in Lishui, Zhejiang Province, as a case study, calculate the GEP before and after the construction of this hydraulic project to evaluate its comprehensive benefits.

2 GEP Accounting Method for Hydraulic Engineering Projects

2.1 GEP Accounting Framework for Hydraulic Engineering Projects

To comprehensively describe the overall functioning of hydraulic engineering projects as ecosystems, evaluate their contributions to human well-being, assess their support for economic and social development, and understand their ecological linkages across regions, this guide constructs a GEP accounting system for hydraulic engineering based on the method used for terrestrial ecosystem product accounting. This system clarifies the indicators and accounting methods for hydraulic engineering ecosystem services, explores the direct and indirect contributions of GEP to GDP, and provides foundational support for the decision-making, evaluation, and operational management of hydraulic projects.

Based on the characteristics of hydraulic engineering and the requirements for ecosystem product accounting, the framework for the GEP accounting of hydraulic engineering projects is proposed as shown in the Fig. 1.

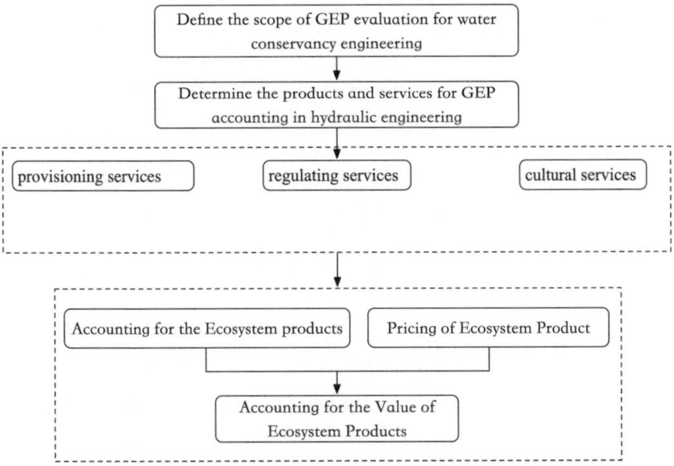

Fig. 1. Technical Route for GEP Accounting in Hydraulic Engineering Projects.

2.2 Boundary for GEP Accounting in Hydraulic Engineering

To objectively reflect the ecological, social, and economic contributions of hydraulic engineering, it is necessary to scientifically delineate the boundaries of the services provided by the project. Based on the watershed where the hydraulic engineering is located, and considering the type and accounting purposes of the project (such as reservoir construction, river enhancement, lake restoration, etc.), the direct or indirect impacts and service areas of the project should be reasonably determined. This includes upstream areas benefiting from improved navigation and backwater inundation, downstream areas affected by flood regulation and agricultural irrigation, as well as regions promoting related ecological tourism. Comprehensive accounting of the value of various ecosystem products provided by hydraulic engineering should be conducted within these boundaries.

2.3 Compilation of Hydraulic Engineering Ecosystem Product List

Determine the types, areas, and distribution of ecosystems within the assessment scope of the hydraulic engineering project and map the distribution of these ecosystems. Investigate and analyze the types of ecosystem products in the region, clarifying specific indicators in the three major categories of provisioning services, regulating services, and cultural services. Focus on ecosystem service items related to the functions and purposes of hydraulic engineering, such as flood control and disaster reduction, navigation, and carbon emission reduction. Combine the actual situation of the accounting region to refine the direct utilization and transformation utilization of provisioning services directories and compile a comprehensive list of ecosystem products (Table 1).

Table 1. GEP accounting indicators

Category	Indicator	Accounting content
Provisioning services	Hydraulic electrogenerating	Power generation capacity
	Water supply	Amount
	Agricultural product	Grain
		Fruits
Regulating services	Flood regulation and storage	Urban flood regulation
		Rural flood regulation
	Water purification	Purification of COD
		Purification of nitrogen
		Purification of phosphorus
	Air purification	Purification of SO2
		Purification of NO
	Carbon sequestration	Carbon storage
		Oxygen release
	Reduce carbon emissions	Carbon reduction
	Climate regulation	Cooling by water
		Cooling by forest
		Cooling by grassland
Cultural services	Ecotourism	Number of tourists
	Environmental premium	Land premium appreciation

2.4 Calculation of Ecological Product Functional Quantity

The functional quantity of ecological products and services refers to the physical quantity or functional quantity of the final products obtained directly or indirectly from ecosystems by humans, such as shipping volume, aquatic product supply, flood control and disaster reduction capacity, pollution purification volume, soil and water conservation capacity, and the number of tourists attracted by landscape. Among them, the measurement of physical quantity for provisioning services and cultural services uses statistical survey methods, while the measurement of physical quantity for regulating services uses methods such as water balance and pollution purification models. Functional quantity provides a clear and specific quantity of ecological products, but due to different measurement units, the functional quantities of different ecological products and services cannot be summed. Therefore, relying solely on functional quantity indicators makes it difficult to obtain the total output of ecosystem products and services over a period of time.

2.5 Determination of Ecological Product Prices

Only by using the prices of ecological products can the functional quantities of different ecological products and services be converted into monetary units of output. Market value method is used for calculating the value quantity of various indicators of provisioning services, while alternative cost method is used for soil conservation, water purification, air purification, and climate regulation indicators in regulating services. Shadow engineering method is used for water conservation, opportunity cost method is used for flood control and carbon sequestration, and protective cost method is used for habitat provision. Various methods such as travel cost method, hedonic pricing method, and research investment method are used for calculating the value quantity of cultural services.

2.6 Calculation of Ecological Product Value

Finally, the values of provisioning services, regulating services, and cultural services obtained separately are summed to obtain the total GEP of the hydraulic engineering project.

3 Case Overview—Lianhu Reservoir in Lishui City

3.1 Overview of Lianhu Reservoir

Lianhu Reservoir is located on Xuanpingxi river in Liandu District, Lishui City, Zhejiang Province. Primarily designed for flood control, it also serves purposes such as improving aquatic ecological environment and power generation (refer to Fig. 1). The dam site of the project is approximately 15 km away from the urban area and situated about 2 km upstream from the main stream of Xuanpingxi river in Gangkou Village. The watershed area above the dam site of Lianhu Reservoir is 828 km^2, with a long-term average flow rate of 25.2 m^3/s and a long-term average inflow volume of about 795 million m^3. The design flood level is 94.73 m, the design flood level is 91.42 m, the normal water level is 85.00 m, the dead water level is 68.00 m, the extreme dead water level is 63.00 m, and the total storage capacity is 130.55 million m^3. Below the normal water level, the capacity is 63.38 million m^3, with a utilization capacity of 62.28 million m^3 and a flood control capacity of 69.00 million m^3. The total population in the protected area is 451,500, with a farmland area of 7,800 mu. The installed capacity of the supporting power station is 21.6 MW, with an average annual power generation of 33.72 million kW·h (Fig. 2).

3.2 Pricing of Ecological Products

Only by utilizing the prices of ecological products can the functional quantities of different ecological products and services be converted into monetary units of output. Market value method is used for calculating the value quantity of various indicators of provisioning services, while alternative cost method is used for soil conservation, water purification, air purification, and climate regulation indicators in regulating services.

Fig. 2. Location of Lianhu reservoir.

Water conservation is calculated using the shadow engineering method, and flood control and carbon sequestration are calculated using the opportunity cost method. Various methods such as travel cost method and hedonic pricing method are used for calculating the value quantity of cultural services (Table 2).

4 Pre-construction GEP Composition of Lianhu Reservoir Project

Combining the calculations of ecological product functional quantities and the determination of ecological product prices as mentioned earlier, we further calculate the value of each category of ecological products and then sum up the values of all ecological products to obtain the total ecological product value before the construction of Lianhu Reservoir (Table 3).

4.1 Provisioning Services Value

In 2021, the Lianhu Reservoir area provided agricultural products. Calculated based on market prices, the value of grain is 497.8 million RMB (1386.6 tons × 3.59 RMB/ton), and the value of fruits is 408.0 million RMB (1025.1 tons × 3.98 RMB/ton). The total value of agricultural products is 905.78 million RMB.

4.2 Regulation Service Value

In 2021, the regulatory service value of the Lianhu Reservoir was 197 million RMB. The largest contribution is from climate regulation, valued at 175 million RMB (3.24

Table 2. Pricing methods and prices of ecological products in Lianhu reservoir

Ecosystem product	Pricing method	Price
Hydraulic electrogenerating	Market valuation	0.54 yuan/kWh
Water suppley	Market valuation	1 yuan /m^3
Grain	Market valuation	3.59 yuan /kg
Fruits	Market valuation	3.98 yuan /kg
Flood regulation and storage	Replacement cost	25.85 yuan /m^3
Purification of COD	Replacement cost	8000 yuan /t
Purification of nitrogen	Replacement cost	9572 yuan /t
Purification of phosphorus	Replacement cost	10000 yuan /t
Purification of SO2	Replacement cost	2000 yuan /t
Purification of NO	Replacement cost	2518 yuan /t
Carbon storage	Opportunity cost	23.72 yuan /t
Oxygen release	Opportunity cost	1200 yuan /t
Carbon reduction	Opportunity cost	0.020 yuan /kWh
Climate regulation	Replacement cost	0.54 yuan /kWh

billion kWh/a × 0.54 RMB/kWh) for cooling the surrounding environment. The next significant contribution is from the flood control function provided by the ecosystem, valued at 21 million RMB. The combined value of other regulating services such as carbon sequestration, oxygen release, water purification, and air purification are 1.866 million RMB, accounting for only 0.68% of the total regulatory service value.

4.3 Cultural Service Value

In 2021, within the construction boundary of the Lianhu Reservoir, it was difficult to attract local and foreign visitors for sightseeing, and it did not contribute to eco-tourism in the surrounding areas. Therefore, the cultural service value within the boundary for that year is 0. Additionally, there were no commercial real estate developments in the vicinity. Therefore, for comparison with the post-construction period, this value is also 0.

4.4 Total GEP Value Before Construction

In 2021, the total GEP value in the area before the construction of the Lianhu Reservoir was 206 million RMB. The value of regulating services was 197 million RMB, accounting for 95.60% of the total GEP. The value of provisioning services was 9.06 million RMB, accounting for 4.4% of the total GEP. Since there were no tourism facilities in the area, the cultural service value was 0.

Table 3. GEP calculation before the construction of Lianhu reservoir

Category	Indicator	Accounting content	Ecosystem product			Gross ecosystem product (10^4 yuan)		
			amount	units	prices	values	subtotal	total
Material goods	Hydraulic elcctrogenerating	Power generation capacity	0	10^4 kWh/a	0,54 yuan/kWh	0.00	0.00	
	Water suppley	Amount	0	10^4 m^3/a	1.63 yuan/m3	0.00	0.00	905.78
	Agricultural product	Grain	1386.6	t/a	3.59 yuan/kg	497.79	905.78	
		Fruits	1025.1	t/a	3.98 yuan/kg	407.99		
	Flood regulation and storage	Flood regulation	79.9	10^4 m^3/a	25.85 yuan/m^3	2065.42	2065.42	
		Purification of COD	49.86	t/a	8000 yuan/t	39.89		
	Water purification	Purification of nitrogen	59.13	t/a	9572 yuan/t	56.60	101.00	
		Purification of phosphorus	4 51	t/a	10000 yuan/t	4.51		
Regulating	Air purification	Purification of SO_2	0.35	t/a	2000 yuan/t	0.07	0.42	
		Purification of NO	1.4	t/a	2518 yuan/t	0.35		19739.10
Services	Carbon Sequestration	Carbon storage	221.45	t/a	24 yuan/t	5.31	76.27	
		Oxygen release	591.28	t/a	1200 yuan/t	70.95		
	Reduce carbon emissions	Carbon reduction	0	10^4 kWh/a	0.020 yuan/kWh	0.00	0.00	
		Cooling by water	1.84	10^8 kWh/a	0.54 yuan/kWh	9936.00		
	Climate regulation	Cooling by forest	1.22	10^8 kWh/a	0.54 yuan/kWh	6588.00	17496.00	
		Cooling by grassland	0.18	10^8 kWh/a	0.54 yuan/kWh	972.00		
Cultural services	Ecotourism	Number of tourist	0	10^4 people	–	0.00	0.00	0
								20644.88

5 Lianhu Reservoir GEP Composition After Construction

Based on the calculations of ecological product quantities and the determination of ecological product prices mentioned earlier, we can further calculate the value of each type of ecological product and then sum up these values to obtain the total value of ecological products after the completion and operation of the Lianhu Reservoir.

5.1 Provisioning Services Value

After the completion of Lianhu Reservoir, the expected annual provisioning services value is 141 million yuan. Hydroelectric power generation and water supply contribute 18 million and 123 million yuan annually, accounting for 13% and 87% of the provisioning services value, respectively. The agricultural product value provided by the cultivated land retained through land resource integration is 40,700 yuan annually.

5.2 Regulation Service Value

As a flood control project, the main purpose of the Lianhu Reservoir construction is to regulate floodwaters downstream. After calculation, it is estimated that the total regulation service value per year after the reservoir's operation is 781 million yuan. The largest contribution is flood control and disaster reduction value, reaching 480 million yuan, accounting for 61.5% of the total regulation service. This result aligns with the project's functional orientation. The next significant contribution is climate regulation function, valued at 297 million yuan. With the enhancement of water quality purification due to the reservoir's construction, it amounts to 3.02 million yuan annually. However, the original air purification and plant carbon sequestration and oxygen release functions are weakened, totaling only 117,900 yuan annually.

5.3 Cultural Service Value

This report refers to the initial tourist reception volume at the beginning of the operation of Geyan Painting Township in 2009, which was 335,000 person-times per year, as the initial tourist reception volume for the Lianhu Reservoir after its completion and operation. Considering the average tourism expenditure per person visiting Liandu District in the first half of 2022 was 1,153 yuan/person-time, and with an annual consumption increase rate of 3%, it is estimated that the average tourist expenditure will reach 1,297 yuan/person-time by 2026. Therefore, it is estimated that the annual ecological recreational value in the initial period of operation of the Lianhu Reservoir will be 434 million yuan. With the development of the "Health Care Lianhu National Tourism Resort" and based on the preliminary development plan, an average of 64 million yuan of land premium will be generated annually. Consequently, the total annual cultural value of the Lianhu Reservoir is estimated to be 498 million yuan. As the surrounding tourist facilities are further improved and the social reputation increases, the future ecological tourism value and environmental premium function of the Lianhu Reservoir will be further enhanced.

5.4 Total GEP Value After Completion

After calculation, the total annual GEP within the construction scope of the Lianhu Reservoir after its completion and operation is 1.42 billion yuan. Among them, the value of regulation services is 781 million yuan, accounting for 55.0% of the total GEP; the value of cultural services is 498 million yuan, accounting for 35.1% of the total GEP; and the value of provisioning services is 141 million yuan, accounting for 9.9% of the total GEP. The values of provisioning services and cultural services can be directly included in the GDP accounting system, meaning that after the completion of the Lianhu Reservoir operation, there will be 639 million yuan of GEP converted directly into local GDP annually, accounting for 45.0% of the total GEP of the Lianhu Reservoir (Table 4).

6 Conclusion

Comparing the GEP calculations before and after the construction of the Lianhu Reservoir in two periods, the results show that the GEP of the Lianhu Reservoir will increase from 206 million yuan in 2021 to 1.42 billion yuan per year after the reservoir is operational, an increase of 588%. This clearly demonstrates the positive contribution of the construction and operation of the Lianhu Reservoir to the local ecology, economy, and society.

6.1 Maximum Increment in Regulation Service Value

An important construction goal of the Lianhu Reservoir is to provide flood control and disaster reduction functions and maintain local ecological balance. Compared to before construction, after completion, there will be an additional 584 million yuan in regulation services each year. Among them, flood control and disaster reduction and climate regulation will increase by 460 million and 122 million yuan per year, respectively. Although some forest and grassland areas were submerged by reservoir backwater, resulting in a decrease of 645,000 yuan in carbon sequestration and oxygen release services, this proportion of value is very low compared to the total regulation services. This indicates that the Lianhu Reservoir is expected to achieve its goals well after construction and operation, enhancing urban flood control capacity in the context of global climate change and frequent extreme weather events, and improving the stability and resilience of the Ou River Basin ecosystem.

6.2 Great Improvement in Cultural Services

Before the construction of the Lianhu Reservoir, there were almost no cultural services within the construction scope. According to tourism planning, it is estimated that in the initial period of operation of the Lianhu Reservoir, the annual ecological recreational value will be 434 million yuan, with an annual land appreciation value of 64 million yuan. This will significantly enhance the comprehensive utilization value of the reservoir area and effectively realize the transformation of water ecological product value in water conservancy engineering construction. With the promotion of the Lianhu Reservoir to a

Table 4. GEP calculation after the construction of Lianhu reservoir

Category	Indicator	Accounting content	Ecosystem product			Gross ecosystem product (10^4 yuan)		
			Amount	Units	Prices	Values	Subtotal	Total
Material goods	Hydraulic electrogenerating	power generation capacity	3372	10^4 kWh/a	0.54 yuan/kWh	1820.88	1820.88	
	Water supply	Amount	7298	10^4 m^3/a	1.68yuan/m^3	12260.64	12260.64	14085.59
	Agricultural product	Grain	11.35	t/a	3.59yuati/kg	4.07	0.00	
		Fruits	0	t/a	3.98yuan/kg	0.00		
	Flood regulation and storage	Urban flood regulation	79.9	10^4 yuan	–	47843.90	48038.10	
		Rural flood regulation	194.2	10^4	–	194.20		
		Purification of COD	150.75	t/a	8000yuan/t	120.60		
	Water purification	Purification of nitrogen	178.76	t/a	9572yuan/t	171.11	302.33	
		Purification of phosphorus	10.62	t/a	10000yuan/t	10.62		
Regulating	Air purification	Purification of SO_2	0	t/a	2000yuan/t	0.00	0.42	
		Purification of NO	0	t/a	2518yuan/t	0.00		19739.10
Services	Carbon sequestration	Carbon storage	34.23	t/a	24yuan/t	0.82	11.79	
		Oxygen release	91.40	t/a	1200yuan/t	10.97		
	Reduce carbon emissions	Carbon reduction	33372	10^4 kWh/a	0.02yuan/kWh	67.44	67.44	
		Cooling by water	5.5	10^8 kWh/a	0.54yuan/kWh	29700.00		
	Climate regulation	Cooling by forest	0	10^8 kWh/a	0.54yuan/kWh	0.00	29700.00	
		Cooling by grassland	0	10^8 kWh/a	0.54yuan/kWh			
Cultural services	ecotourism	Number of tourist	33.5	10^4 people	129/yuan/	43449.50	49792.10	49792.10

(continued)

Table 4. *(continued)*

Category	Indicator	Accounting content	Ecosystem product			Gross ecosystem product (10^4 yuan)		
			Amount	Units	Prices	Values	Subtotal	Total
	Environmental premium	Land premium appreciation	1503	ha	–	6342.60		
								141997.35

national tourist resort, it will further drive the development of real estate, tourism, agritainment, and other industries, promoting the sustainable development of the economy and society of Lishui City.

6.3 Steady Improvement in Provisioning Services

After the construction and operation of the Lianhu Reservoir, due to changes in land use types, it will no longer provide provisioning services. Although the agricultural products worth 9.06 million yuan within the construction scope in 2021 will be mostly lost after the reservoir is built, the Lianhu Reservoir will provide two services, hydroelectric power generation and water supply, totaling 141 million yuan, greatly enhancing its role in promoting the local economy. When determining the water supply price of the Lianhu Reservoir, this report adopts the recommended price of the GEP calculation technical specification of ecological product value. If regional water rights trading, or even cross-regional trading, can be promoted in the future, the price of high-quality water resources of the Lianhu Reservoir will be further increased.

Evolution Laws and Spatial Differentiation Characteristics of Climate and Extreme Climate Before and After the Impoundment of the Three Gorges Reservoir

Ruirui Liu(✉), Xiaomei Kou, Wei Song, and Chuang Dong

Power China Northwest Engineering Corporation Limited, Xi'an 710065, Shaanxi Province, China
574579863@qq.com

Abstract. Human-induced climate change has affected weather and extreme climate events, the Three Gorges Hydropower Project is the largest hydropower project in the world, which must inevitably have some impacts on the regional climate and extreme climate. Based on the data of precipitation, temperature, sunshine hours, relative humidity, minimum temperature and maximum temperature of 14 meteorological stations in the study area for 59 years from 1961 to 2019, this paper adopts the climate tendency rate, Mann-Kendall test, ordered clustering method, Kriging difference method to analyze the climate change trend and spatial differentiation characteristics before and after the impoundment of the Three Gorges Reservoir. The results indicated that the impact on precipitation is weak, there is no significant trends; sunshine hours and relative humidity all showed a significant decreasing trend at 11 stations. However, Except Gaoping, Badong, Enshi and Laifeng, the temperature of the other 10 stations has changed significantly rise trend from a cooling trend to a warming trend. The Three Gorges Reservoir has a slowing effect on the rise of minimum temperature at Wanyuan, Badong, Wufeng, Yichang, Jingzhou, Wanzhou, Shapingba, Laifeng and Yibing, the impact on the ecosystem is beneficial. The increase of maximum temperature at 13 stations except Yichang will inevitably change the regional ecosystem. The abrupt changes of temperature, relative humidity and maximum temperature all occurred after impoundment of the Three Gorges Reservoir. After the impoundment of the Three Gorges Dam, the precipitation variability increased in the west and decreased in the east; North-central temperatures rise more; The decrease range in the east is greater than that in the west; Relative humidity in the west decreased more than that in the east. The temporal and spatial changes of lacal climate will inevitably have a certain impact on the local ecosystem.

Keyword: Three Gorges reservoir · Climate and extreme climate · Evolution law · Spatial differentiation characteristics

S. Zheng et al. (Eds.): IHDC 2024, LNCE 487, pp. 476–488, 2025.
https://doi.org/10.1007/978-981-97-9184-2_39

1 Introduction

The IPCC's Sixth Assessment Report pointed out that it is an indisputable fact that global warming is caused by human activities, and that the global surface temperature in 2011–2020 was 1.1 °C higher than that of 1850–1900. Human-induced climate change has affected weather and extreme climate events[1] in every region of the globe. Changes in the water cycle caused by global warming have led to frequent and intensified extreme weather events[2] such as high temperatures, droughts, and heavy rainfall. As one of typical human activities, the construction of hydropower projects directly changes the underlying surface conditions and increases the area of the water body, while water body has such properties as low reflectivity, high heat capacity and low surface roughness, shows different laws of water-vapor exchange and energy balance compared with other underlying surface conditions, and thus affects the climate and extreme climate events in the surrounding areas. Scholars both at home and abroad have carried out a great number of studies on the impacts of hydropower project construction on regional climate. Overseas studies on the climate effect of the Aswan Dam in Egypt have shown that the temperature near the reservoir decreased and the relative humidity and evaporation increased[3–4] after the impoundment; The air humidity of the completed Cabora Bassa Dam increased to some extent[5] after the completion of the Cabora Bassa Dam in Mozambique; the temperature of Itaipu Hydropower Project reduces daytime temperatures in the lake area by 0 to 1.5 °C from July to January of the following year, and by 2 to 3 °C in other months[6]; The average annual temperature rise in the neighboring Mengjin County slows down after the impoundment of Xiaolangdi Reservoir by the linear propensity test[7]; WU et al. [8]adopted the independent satellite data sets and numerical simulation clearly indicate that the land use change associated with the TGD construction has increased the precipitation in the region between the dam and Qinling mountains and reduced the precipitation in the vicinity of the TGD after the TGD water level abruptly rose from 66 to 135 m in June 2003. WU Huiling, et al. [9]analyze the local climate change trend and abrupt change before and after the impoundment of the Three Gorges Reservoir by the combination of the Mann-Kendall nonparametric test and the cumulative anomaly method.The results show that the impoundment of the Three Gorges Reservoir has an influence on the temperature and precipitation,and the influence degree has an obvious geographical distribution law and a stabilizing effect on the temperature trend. Zhang Jianmin, et al. [10]analyzed the change in monthly precipitation patterns under the climate scenario derived from a coupled general circulation model(CGCM), the Results indicated that the flood risk in the running will increase in the early and mid summer. Although the drought risk in mid and late winter due to climate average change will be generally lessened, the frequency of extreme drought occurrence will increase as the result of changes in precipitation variability, which will lead to an increase in risk.The AR6 report and related studies[11] indicate that the ecological impacts of climate warming are significant, including advancement of phenology, extension of plant growing season; species range shift towards high latitude or elevation, tree line moving towards the top of the hill; local extinction of species or habitat loss; increase in frequency, severity and range of disease outbreaks.

The Three Gorges Hydropower Project is the largest hydropower project in the world, which will inevitably have some impacts on the regional climate and extreme climate while bringing huge economic benefits. Therefore, it is of great significance to study the change law and spatial differentiation of regional climate and extreme climate elements after the impoundment of the Three Gorges Reservoir.

2 Overview of the Study Area and Study Methodology

2.1 Overview of the Study Area

The Three Gorges Reservoir is a typical river channel reservoir[12]. The reservoir area is located in 28.5°N-31.7°N, 105.8°E-111.7°E. The administrative region starts from Yichang, Hubei Province in the east and ends at Jiangjin of Chongqing in the west. The reservoir area has a total length of over 600 km. The construction of the Three Gorges Hydropower Project was officially started in 1994. The river closure for first phase of the dam works was achieved in 1997 and the water level was raised by 10-75m; In June 2003, the second phase of the dam works began to impound water for power generation, and the water level in front of the dam rose significantly to EL135m; The acceptance of normal impoundment (at water level of EL175m) in August 2009 marked that the Three Gorges Reservoir entered the normal operation stage. The scope of this study is the both banks of the reservoir area and the urban administrative region within 100km in the upstream and downstream (Fig. 1).

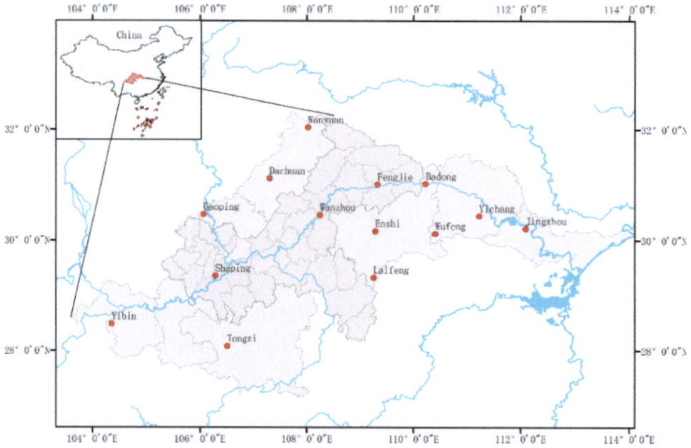

Fig. 1. Studied stations in the Three Gorges Reservoir area

2.2 Data Source

The data source used in this study is China Meteorological Data Service Center, and 14 meteorological stations in the vicinity of the Three Gorges Reservoir area and within 100 km upstream and downstream were taken as the research objects. In order to eliminate the impacts of inconsistent data series, the data ranging from January 1, 1960 to December 31, 2019 are selected in this study, including precipitation, temperature, sunshine hours, relative humidity, minimum temperature and maximum temperature.

2.3 Research Methodology

This paper adopts the climate tendency rate, Mann-Kendall test to analyze the climate change trend before and after the impoundment of the Three Gorges Reservoir. Mann-Kendall test and ordered clustering method are used to calculate the years when the factors have a sudden change, and to judge whether the factors have a sudden change during the impoundment years. The kriging difference method is an optimal, linear and unbiased spatial interpolation method, which gives a certain weight coefficient to each observation data after fully considering the interrelationship between the observation data, and the weighted average is used to obtain the estimated value. Kriging difference method is used to analyze the spatial differentiation characteristics of the factors before and after the impoundment.

3 Evolution Law of Climatic and Extreme Climatic Elements

3.1 Evolution Law

3.1.1 Evolution Law of Climatic Elements

The average precipitation of the Three Gorges Dam is 944.4–1,366.1mm before impoundment and 1,001.1–1,354.4mm after impoundment. Except for Enshi, Fengjie and Yibin, the precipitation at other stations has somewhat increased after impoundment. According to Mann-Kendall trend test (see the Table 3.1), except for Yibin($Z = -2.05$) station upstream of the reservoir area, where the precipitation shows a significant upward trend through the significance test of confidence level $\alpha = 0.05$(If $|Z| > 1.96$, the change trend is considered to be significant, and when Z is positive, it is considered to be a significantly upward, when Z is negative, it is considered to be a significant decrease.) and the other stations show no significant change. Among them, 7 stations show an insignificant downward trend and 6 stations show an insignificant upward trend. According to the analysis of Variation range of climate tendency rate(the difference between the climate tendency rates after the impoundment (1998–2019) and before the impoundment (1961–1997)),there are 7 stations has a decreasing trend and 7 stations has an upward trend. In general, after the impoundment of the Three Gorges Reservoir has effects on precipitation in the surrounding areas, but the effects are weak. Therefore, precipitation has little impact on the total amount of runoff and ecosystems after the impoundment.

The average temperature of the Three Gorges Dam is 13.1mm-18.2mm before impoundment and 15.1 mm-18.8 mm after the impoundment. The temperature of 14 meteorological stations has somewhat increased after the impoundment. According to Mann-Kendall trend test (see the Table 3.1), the temperature of 14 meteorological stations shows a temperature rise trend. Except Gaoping, Badong, Enshi and Laifeng, the temperature of the other 10 stations shows a significant temperature rise trend. The analysis results of the climate tendency rate also show an increasing trend, with a growth rate of 0.06–0.51°C/10a. Before the impoundment of the Three Gorges Dam, there are 13 meteorological stations with the climate tendency rate showing a temperature drop trend, with the climate tendency rate ranging from -0.02 to -0.18°C/10a. After the impoundment of the Three Gorges Dam, there are 10 meteorological stations with the climate tendency rate showing a temperature rise trend, with the climate tendency rate ranging from 0.00 to 1.37 °C/10a. This indicates that after the impoundment of the Three Gorges Reservoir, the temperature change trend around the reservoir area has changed significantly, from a temperature drop trend to a temperature rise trend. The increase in temperature will have a certain degree of impact on the survival of local species and the ecological environment.

The average sunshine hours of the Three Gorges Dam are 1,060–1,805h before impoundment and 902–1,493h after the impoundment. According to the Mann-Kendall trend test results(see the Table 1), the sunshine hours of all stations around the reservoir area show a decreasing trend, and the decreasing trend is significant. According to the analysis of climate tendency rate, the decrease range of the Three Gorges Reservoir is -44.84 to -157.71h/10a before impoundment and -37.30 to -262.18h/10a after impoundment. Among them, the decrease range of sunshine hours at 11 stations(Dachuan, Badong, Enshi, Wufeng, Yichang, Jingzhou, Wanzhou, Shaping- ba, Tongzi, Fengjie, Laifeng) increases, indicating the intensified decrease of sunshine hours after impoundment.

The relative humidity of the Three Gorges Dam is 70–82% before impoundment and is 71–80% after impoundment. The relative humidity of Badong, Wufeng and Fengjie stations has somewhat increased after impoundment and that of the other 11 stations has somewhat decreased. Through Mann-Kendall trend test(see the Table 1), the relative humidity of Badong, Wufeng and Fengjie stations has insignificant increasing trend, and that of the other 11 stations has a decreasing trend. Among them, the relative humidity of 10 stations has a significant decreasing trend and that of one station has no significant decreasing trend. According to the analysis of climate tendency rate, the climate tendency rate ranges from -1.46 to 1.14%/10a before impoundment and from -3.73 to 1.38% after impoundment. The relative humidity variability of 11 stations shows a decreasing trend, indicating that after the impoundment of the Three Gorges Dam has a certain impact on the decrease of relative humidity.

3.1.2 Evolution Law of Extreme Climatic Elements

The minimum temperature around the Three Gorges Reservoir Area ranges from -7.2 °C to 0.7 °C before the impoundment and from - 4.8 °C to 1.8 °C after the impoundment. The minimum temperature after the impoundment shows a general upward trend. According to Mann-Kendall trend test(see the Table 3.2), the minimum temperature in the Three Gorges Reservoir Area shows an upward trend. Among them, 9 stations(Dachuang, Badong, Enshi, Wufeng, Wanzhou, Shapingba, Tongzi, Fengjie, Laifeng) have a significant upward trend through the significance test of confidence level $\alpha = 0.05$ and other 5 stationshave no significant upward trend. The minimum temperature and climate tendency rate range from -0.07 °C to 1.25 °C/10a before the impoundment of the Three Gorges Dam, and from -0.74 °C to 1.97 °C/10a after the impoundment of the Three Gorges Dam. There are 9 stations(Wanyuan, Badong, Wufeng, Yichang, Jingzhou, Wanzhou, Shapingba, Laifeng, Yibing) with a downward trend in the minimum temperature variability, and other 5 stations with an upward trend in the minimum temperature variability. It can be seen from this that the reservoir impoundment has a certain slowing-down effect on the overall rise of the lowest climate.

Table 1 Statistics of PRCP and TEMP and SD and RH by Mann-Kendall test and variation range of climate tendency rates

S.N	Description of meteorological station	Mann-Kendall test				Variation range of climate tendency rate /°C/10a			
		Z value							
		PRCP	TEMP	SD	RH	PRCP	TEMP	SD	RH
1	Wanyuan	0.57	2.94	−3.72	−2.99	50.93	1.41	70.57	−0.80
2	Gaoping	0.10	0.32	−4.77	−1.87	43.02	0.25	17.23	−2.96
3	Dachuan	0.62	2.01	−5.32	−4.35	−29.54	0.52	−97.57	−2.88
4	Badong	-0.07	0.29	−4.65	0.81	23.64	0.10	−168.12	−1.50
5	Enshi	−1.27	1.88	−3.01	−6.54	−16.45	0.45	−140.38	−0.98
6	Wufeng	−0.34	4.88	−5.61	1.68	−44.12	0.22	−26.43	0.02
7	Yichang	0.57	2.94	−3.72	−2.99	−74.94	−0.43	−150.87	0.75
8	Jingzhou	1.00	5.37	−6.86	−6.82	−79.93	−0.21	−143.57	2.84
9	Wanzhou	−1.11	3.35	−5.94	−5.51	8.46	0.41	−66.92	−2.70
10	Shapingba	0.97	2.57	−5.79	−3.32	123.42	0.27	−13.90	−4.18
11	Tongzi	−1.51	2.78	−3.96	−3.70	−45.55	0.39	−72.12	−0.82
12	Fengjie	−1.60	4.00	−4.17	0.14	−114.35	1.26	−195.17	−2.52
13	Laifeng	−0.30	1.63	−4.64	−2.21	14.30	0.14	−117.93	−1.30
14	Yibin	−2.05	2.60	−4.92	−5.67	170.59	−0.05	52.68	−2.81

The maximum temperature around the reservoir area ranges from 34.3 °C to 39.5 °C before the impoundment, and from 34.5 °C to 40.4 °C after the impoundment. After the impoundment of the Three Gorges Reservoir, the maximum temperature shows a general temperature rise trend. According to Mann-Kendall trend test(see the Table 2), the maximum temperature except for Badong station shows a temperature rise trend. There are 6 stations(Dachuang, Wufeng, Wanzhou, Shapingba, Tongzi, Fengjie) show a significant temperature rise trend and other 7 stations show no significant temperature rise trend. The maximum temperature and climate tendency rate of the Three Gorges Reservoir range from -0.19 to 0.43 °C/10a before impoundment, and from -0.18 to 1.44°C/10a after the impoundment. The temperature rise trend of the maximum temperature after the impoundment of the Three Gorges Reservoir is more obvious. It can be seen from this that reservoir impoundment has the intensified rise of maximum temperature to a certain extent.

The increase of extreme temperature around the Three Gorges Reservoir area will inevitably change the regional ecosystem. The Three Gorges Reservoir has a slowing effect on the rise of minimum temperature, and the impact on the ecosystem is beneficial. The specific impact needs to be further studied.

Table 2 Statistics of minimum temperature and maximum temperature by Mann-Kendall test and Variation range of climate tendency rates

S.N	Description of meteorological station	Mann-Kendall test		Variation range of climate tendency rates /°C/10a	
		Z value			
		MIN TEMP	Max TEMP	MIN TEMP	Max TEMP
1	Wanyuan	1.70	1.58	−0.12	0.25
2	Gaoping	0.56	1.67	0.15	0.84
3	Dachuan	2.10	2.58	0.32	0.66
4	Badong	2.09	-0.42	−0.34	0.22
5	Enshi	3.68	1.25	0.26	0.58
6	Wufeng	4.30	5.08	−0.78	0.60
7	Yichang	1.70	1.58	−0.64	−0.15
8	Jingzhou	1.79	1.03	−1.99	0.45
9	Wanzhou	3.58	2.14	−0.28	0.63
10	Shapingba	3.08	1.99	−0.17	1.29
11	Tongzi	3.86	0.81	0.16	0.32
12	Fengjie	4.45	4.06	1.75	2.52
13	Laifeng	2.52	1.95	−0.34	1.29
14	Yibin	0.87	3.00	−0.45	0.49

3.2 Analysis of Abrupt Change

The Mann-Kendall method and ordered clustering algorithm are used to analyze the abrupt change points of climate elements and extreme climate elements at the Three Gorges Reservoir. See Fig. 2 and Fig. 3 for details. According to the Mann-Kendall test pattern, the precipitation time series of most stations in the study area have no abrupt change, but temperature, sunshine hours, relative humidity, minimum temperature and maximum temperature have abrupt change, which the abrupt change time of temperature and relative humidity and maximum temperature time series most occurred after 2000. According to the ordered clustering analysis, except for the minimum temperature, the change-point time of time series for precipitation, temperature, sunshine hours, relative humidity and maximum temperature at most stations is concentrated after the impoundment of the Three Gorges Reservoir. It is known by both methods that the abrupt changes in temperature, relative humidity and maximum temperature all occurred after impoundment of the Three Gorges Reservoir.

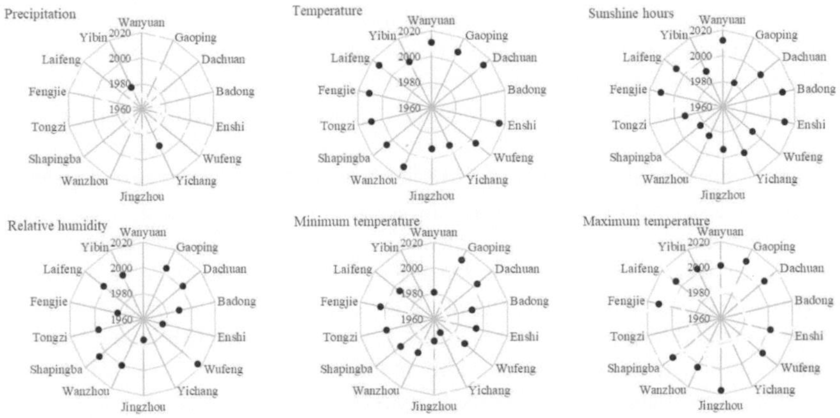

Fig. 2. Mann-kendall method for abrupt change of climatic factors

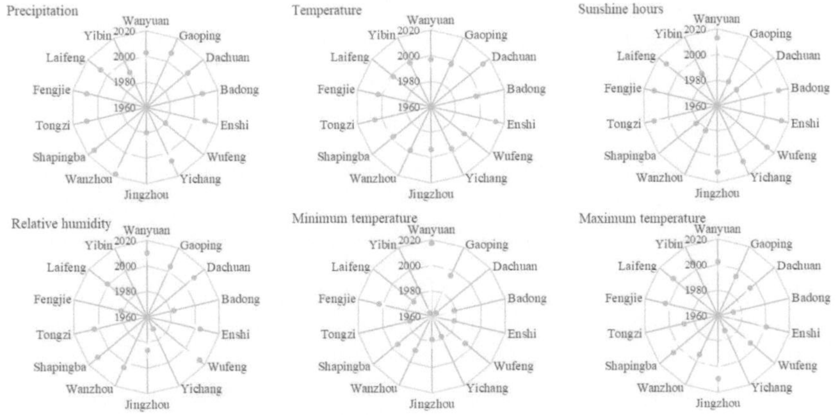

Fig. 3. Ordered clustering method for abrupt change of climatic factors

4 Spatial Differentiation Characteristics of Climate and Extreme Climate Elements

4.1 Spatial Differentiation Characteristics of Climate Elements

The spatial distribution of climate elements in the reservoir area and the spatial differentiation characteristics of the variation range of climate tendency rate before and after the impoundment of the Three Gorges Reservoir (the difference between the climate tendency rates after the impoundment (1998–2019) and before the impoundment (1961–1997)) are analyzed. See Fig. 4 ~ Fig. 7 for details. The precipitation in the reservoir area is generally small in the west and large in the east. The precipitation in Enshi and Laifeng is large, which is related to the altitude; After the impoundment of the Three Gorges Project, the precipitation variability increased in the west and decreased in the east. Contrary to the spatial distribution of precipitation. The average annual temperature in the reservoir area is higher in the west and lower in the east. After the impoundment of the Three Gorges Reservoir, the temperature variability is generally positive. Fengjie and Wanyuan in the north-central part have the largest temperature rise variability, with a variation range of 1.3–1.4°C/10a, indicating that north-central temperatures rise more after the impoundment of the Three Gorges Project. The sunshine hours show the spatial distribution characteristics of being less in the west and more in the east. The average sunshine hours in Yibin in the west are the least, 1,003h/a, and the average sunshine hours in Jingzhou in the east are the most, 1,692h/a; After the impoundment of the Three Gorges Reservoir, the decrease range in the east is greater than that in the west. The spatial distribution of relative humidity is bigger in west and smaller in the east. After the impoundment of the Three Gorges Reservoir, the relative humidity shows a general decreasing trend. Relative humidity in the west decreased more than that in the east after the impoundment of the Three Gorges Reservoir.

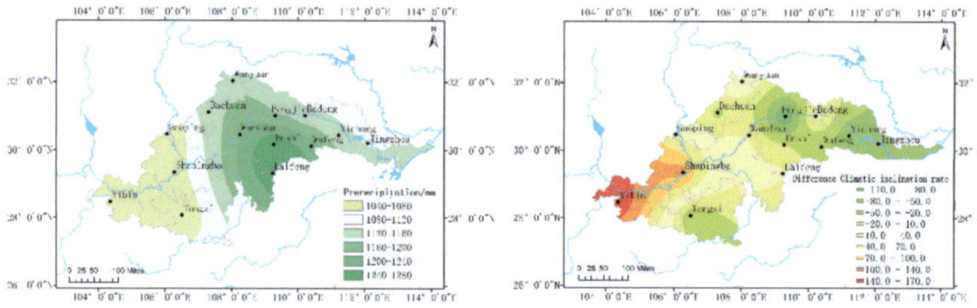

Fig. 4. Average annual precipitation and differences climate inclination rate of precipitation

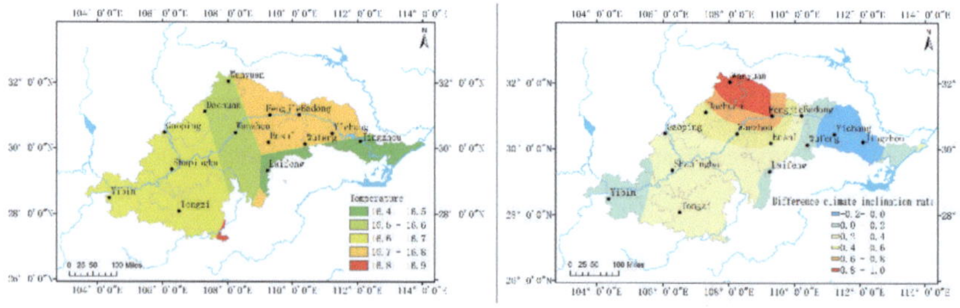

Fig. 5. Average annual temperature and differences climate inclination rate of temperature

4.2 Spatial Differentiation Characteristics of Extreme Climate Elements

The spatial distribution of extreme climate elements (minimum and maximum temperatures) at 14 meteorological stations in the reservoir area and the spatial differentiation characteristics of the variation range of climate tendency rate before and after the impoundment of the Three Gorges Reservoir are analyzed. See Fig. 8 and Fig. 9 for details. The minimum temperature is high in the west and low in the east. The minimum temperature in Yibin and Shapingba in the west was higher than 0.0 °C, and the

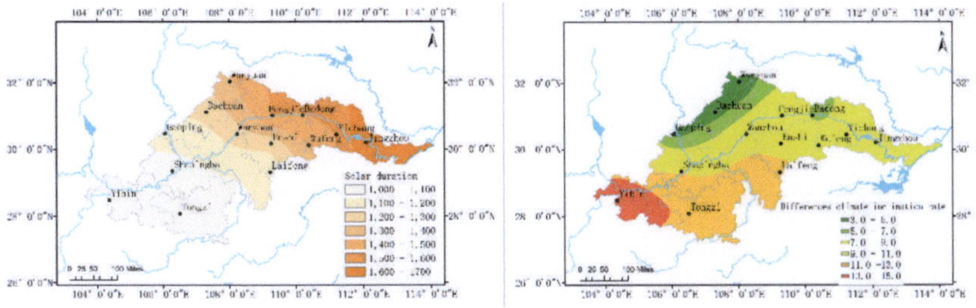

Fig. 6. Average annual sunshine hours and differences climate inclination rate of solar duration

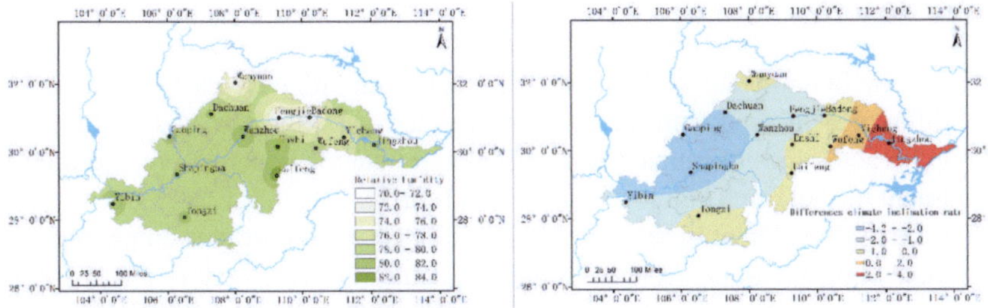

Fig. 7. Average annual relative humidity and differences climate inclination rate of relative humidity

minimum temperature in Jingzhou and Wufeng in the east was lower than -3.3 °C. The lowest temperature caused the slow warming in the east, and the warming in the west intensified after the impoundment. The spatial differentiation characteristics between maximum temperature and the minimum temperature is consistent, which is high in the west and low in the east. The variability of the maximum temperature shows an overall increasing trend after the impoundment, Wufeng and Fengjie have the largest warming amplitude, which is 2.5 ~ 3.1 °C. Overall, the west heats up more than the east.

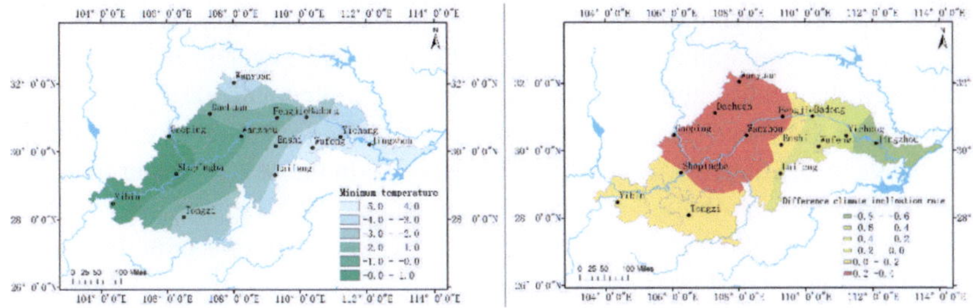

Fig. 8. Minimum temperature and differences climate inclination rate of minimum temperature

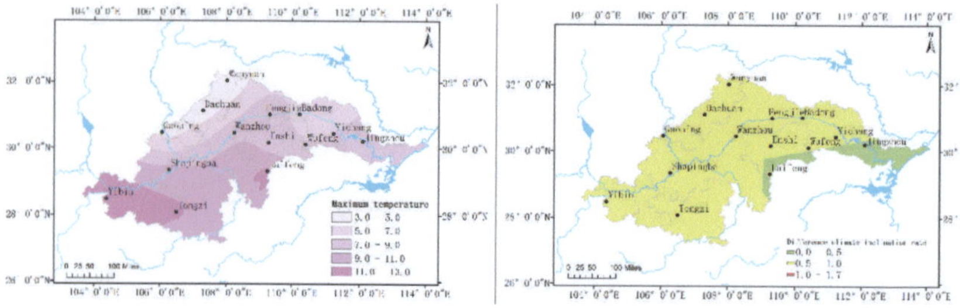

Fig. 9. Maximum temperature and differences climate inclination rate of maximum temperature

5 Conclusions

The climatic inclination rate method, Mann-Kendall test, ordered clustering method and kriging difference method were used to analyze the variation law and spatial differentiation characteristics of climate and extreme climate elements before and after the impoundment of the Three Gorges Reservoir as follows:

After the impoundment of the Three Gorges Reservoir has effects on precipitation, temperature, Sunshine hours, relative humidity, minimum temperature and maximum temperature in the surrounding areas. The impact on precipitation is weak, there is no significant trends; sunshine hours and relative humidity all showed a significant decreasing trend at 11 stations. However, the temperatures(including minimum and maximum temperatures) around the reservoir area all have a significant warming trend, from a cooling trend to a warming trend. The Three Gorges Reservoir has a slowing effect on the rise of minimum temperature, and the impact on the ecosystem is beneficial. The increase of extreme temperature around the Three Gorges Reservoir area will inevitably change the regional ecosystem. The specific impact needs to be further studied.

Through the mutation analysis, there was no obvious abrupt change in precipitation after the impoundment of the Three Gorges Reservoir. However, the sudden changes in temperature increase, relative humidity decrease, and maximum temperature rise all occurred after the impoundment of the Three Gorges Reservoir.

The precipitation variability in the reservoir area increased in the west and decreased in the east after the impoundment, which was opposite to the spatial distribution of precipitation. The temperature was higher in the west and lower in the east, and the temperature in the central and northern parts of the country increased more after the impoundment of the Three Gorges Reservoir. The spatial distribution of sunshine hours is less in the west and more in the east, and the decrease in the east is greater than that in the west. The spatial distribution of relative humidity showed a spatial pattern of large in the west and small in the east, and the decrease of relative humidity in the west was greater than that in the east after impoundment.

This paper presents a brief results of the construction of hydropower projects on lacal climate factors. The lacal climate will change the ecosystem, we need adapt strategies to mitigate the impact. Such as studying the criticality of regional ecosystems under climate change tipping point, taking a targeted guarantee protection measures, and improving early warning and governance capacity.

Notes

1. IPCC.Climate Change 2022: Mitigation of Climate Change[R].2022.
2. Zhang Lili, Zhou Junju, Zhang Hengwei, Wang Bei, Cao Jianjun. Study on the spatial-temporal pattern characteristics of climate dry and wet changes and drought events in Shiyang River basin based on SPI. Acta Ecologica Sinica, 2017, 37 (3): 996–1007.
3. HAFEZ M,SHENOUDA W K.The environmental impacts of the Aswan High Dam[C]//Water Management and Development, Proceedings of the United Naters Water Conference, New York: Pergamon Press,1977.
4. MOUSSA A, SOLIMAN M,AZIZ M. Environmental evaluation for High Aswan Dam since its construction until present[C]//Sixth International Water Technology Conference,Egypt, 2001:301–311.
5. HENRIQUES A G,SILVA H S. Cahora bassa dam[C]//7th Congress on Large Dams. Austria: Vienna,1991: 427–442.
6. STIVARI S M S, DE OLIVEIRA A P,SOARES J,et al. On the Climate Impact of the Local Circulation in the Itaipu Lake Area[J]. Climatic Change, 2005,72(1/2): 103–121.
7. Wang Lingling. Study on the Impact of Xiaolangdi Reservoir on the Yellow River on the Climate of Mengjin [D]. Zhengzhou: Henan Agricultural University, 2012.
8. WU L, ZHANG Q, JIANG Z. Three Gorges Dam affects regional precipitation[J]. Geophysical research letters, 2006, 331(13): 338–345.
9. WU Huiling, Zhou Jianzhong, Tian Mengqi,et el. Analysis of Climate Change before and after the Impoundment of the Three Gorges Reservoir [J]. Water Power, 2021, 47(5): 30–35.
10. Zhang Jianmin, Huang Chaoying, Wu Jindong. Impacts of Climate Chane on Risk in Running of the Three Gorges Reservoir[J]. Acta Geographica Sinica, 2000, 55(Supplement): 26–33.
11. Wan X R, Cheng C Y, Bai D F, et al. Ecological impacts of climate change and adaption strategies. Bulletin of Chinese Academy of Sciences, 2023, 38(3): 518–527, https://doi.org/10.16418/j.issn.1000-3045.20220815002. (in Chinese).
12. Wu Jia, Gao Xuejie, Zhang Dongfeng, et al. Climate effect of the Three Gorges Reservoir and regional climate simulation of high temperature and drought events in Sichuan and Chongqing in summer 2006 [J]. Journal of Tropical Meteorology, 2011, 27 (1): 44–52.

Exploration of the Application of Fine Reconstruction of Unmanned Aerial Vehicles in Landslide Disaster Investigation and Management

Daiyao Zhao(✉), Jingnan Han, Xianggang Liu, Zhouchang Zuo, and Kexun Zheng

Guiyang Engineering Corporation Limited, Power China, Guiyang 550081, China
1057016856@qq.com

Abstract. The mountainous terrain in the southwest region is characterized by significant topographic relief and abundant rainfall. When constructing new energy infrastructure such as booster stations in mountainous areas, it is extremely easy to induce landslide instability due to the excavation of mountain slopes. This article takes the Wangmo landslide in the southwestern region of Guizhou Province as the research object, uses multi-rotor drones as platforms, and utilizes techniques such as ground-based photogrammetry and close-range photogrammetry to detect and obtain relevant parameters such as the shape, deformation and damage characteristics, and zoning of the Wangmo landslide. The specific application methods and application effects of fine reconstruction of drones in landslide hazard detection and management are elaborated and discussed. The conclusions are as follows: Compared with traditional orthophotography, the high-definition image data obtained through fine reconstruction of ground-based photogrammetry and close-range photogrammetry generates more comprehensive orthophotographs and 3D model textures with higher resolution. Using the orthophotographs and 3D models produced by fine reconstruction, the structural surface combination, surface crack development, and landslide deformation zone are interpreted, and the formation mechanism of the landslide is determined in conjunction with ground surveys, providing a data foundation for prevention and control design. UAV aerial survey has the characteristics of short cycle, flexibility, and strong emergency response, and has high promotion value in landslide hazard investigation of booster stations.

Keywords: UAV · Landslide · Simulated ground flight · Close-up photogrammetry · Refined reconstruction

1 Introduction

The southwest mountainous region has a high altitude, developed valley terrain, and large topographic relief. It is rich in regional wind resources and is currently a hot area for wind power construction. However, due to active geological activities and complex and changing natural environments, landslide disasters often occur during the construction of wind power projects due to various reasons (Hong Yebing 2016). Traditional landslide

S. Zheng et al. (Eds.): IHDC 2024, LNCE 487, pp. 489–502, 2025.
https://doi.org/10.1007/978-981-97-9184-2_40

investigations are mainly based on manual field surveys, which are often inefficient, laborious, and time-consuming, and sometimes pose a threat to the personal safety of investigators (Xie Muwen et al. 2014). In this case, how to quickly, accurately, and efficiently obtain data from disaster areas and provide accurate and detailed information for subsequent disaster mitigation plans is the focus.

In recent years, satellite remote sensing technology and drone oblique photogrammetry technology have been widely used in geological hazard investigation and monitoring (Xu Qiang et al. 2024). Satellite remote sensing technology has the disadvantages of long data cycle and being affected by cloudy and foggy weather (Li Qiang et al. 2019). The drone oblique photogrammetry technology has the advantages of high accuracy and flexibility due to its non-contact measurement method, providing an efficient on-site image acquisition and remote sensing results processing solution for geological hazard investigation (Li Yin 2012; Guo Chen 2020). However, due to the often rugged terrain in areas where wind power is constructed, there are still some shortcomings in the resolution and accuracy of oblique photogrammetry technology, which cannot fully capture information on geological hazards. In order to meet the needs of refined investigation, relevant scholars have proposed imitation flight technology and close-up photogrammetry technology. Imitation flight technology has the advantage of obtaining high-precision 3D real-world models with consistent resolution, which can achieve high-precision rendering and reconstruction of 3D models in areas with large elevation differences (Huang Lizhang 2022); close-up photogrammetry technology takes the surface as the photographic object and can take multiple-angle shots close to the surface of the object, obtaining millimeter-level high-resolution images and highly restoring the refined structure of the object (Yan Si et al. 2019)

This article quickly obtained high-precision geological impact data of the landslide disaster area in the booster station through the simulation of ground flight technology and close-up photogrammetry technology, and used refined reconstruction technology to produce high-precision orthophoto and real-world 3D model products. The structural plane combination and surface crack development in the landslide disaster area were obtained, and the deformation characteristics of the landslide were analyzed, providing a data foundation for prevention and control design.

2 UAV Technology and Data Acquisition and Processing Methods

2.1 Imitation of Ground Flight Technology

Ground-imitating flight refers to the UAV flying at a constant relative height according to the ground relief of the flight area (Fig. 1). In flight operations, the existing three-dimensional surface data (DSM) is used to keep the UAV at a constant height with the ground target, overcoming the problems of large terrain elevation differences and occlusion by protruding surfaces (Zhang Yilin 2023). As shown in Fig. 1(a), the three-dimensional real-world model established by traditional contour flight photography technology has inconsistent resolution at the top and bottom due to the same flight path line in the same plane for areas with large terrain elevation differences. However, as shown in Fig. 1(b), the UAV ground-imitating flight technology can ensure consistent data resolution and reflect more detailed surface relief and microtopographic features under the relatively large terrain elevation differences in the reservoir area (Pang Xin et al. 2023).

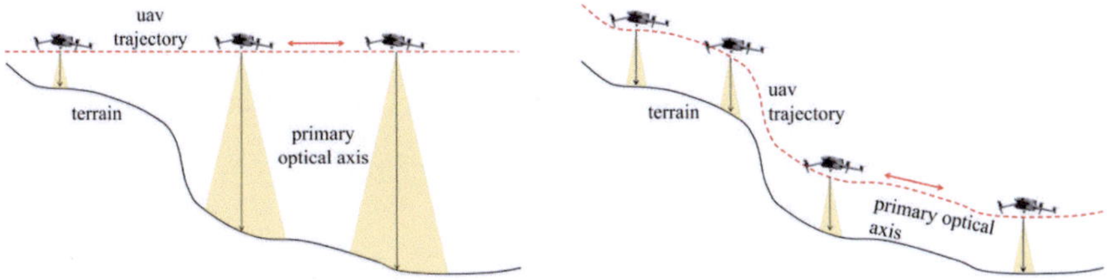

a.Traditional equal height flight b.ground imitation flight

Fig. 1. Ground-like flight principle diagram

2.2 Close to Photogrammetry Technology

Close-up photogrammetry technology is a technique that uses a rotary-wing drone to closely approach the surface of the subject being photographed to obtain ultra-high-definition images, precise coordinates, and detailed shapes of the object. The camera direction is perpendicular to the surface of the object, allowing for the acquisition of image information of steep terrain, as shown in Fig. 2. Compared to traditional aerial photogrammetry methods, close-up photogrammetry has the advantages of fine and multi-angle photography (Yao Futan 2023). This technology has the following characteristics: it can obtain high-resolution images at the millimeter level; the camera angle can be dynamically adjusted according to the shape of the object; before performing intelligent close-up photography, it is necessary to reconstruct the image using conventional photography and other means to obtain the initial terrain. Close-up photogrammetry mainly relies on high-precision positioning technology of drones and the ability of drone gimbal attitude control to achieve, and it takes a long time. Its shooting objects are mostly steep rock masses (He Kuangyu 2022).

2.3 Data Acquisition and Processing Method

The research data was collected using the DJI Mavic3E drone. When the RTK module is activated, the horizontal accuracy reaches ($\pm$10) cm and the vertical accuracy reaches ($\pm$10) cm, which can effectively meet the positioning accuracy requirements of both flight technologies. The drone-related data is shown in Tables 1 and 2.

First, a conventional flight with an average resolution of 8 cm was conducted over the study area with an area of about 0.64 km^2 to obtain coarse topographic data of the survey area. Secondly, based on the coarse topographic data, the landslide area was delineated for ground-based flight, and the exposed bedrock on the road side was photographed for close-up photography. Then, Context Capture was used to complete the detailed 3D

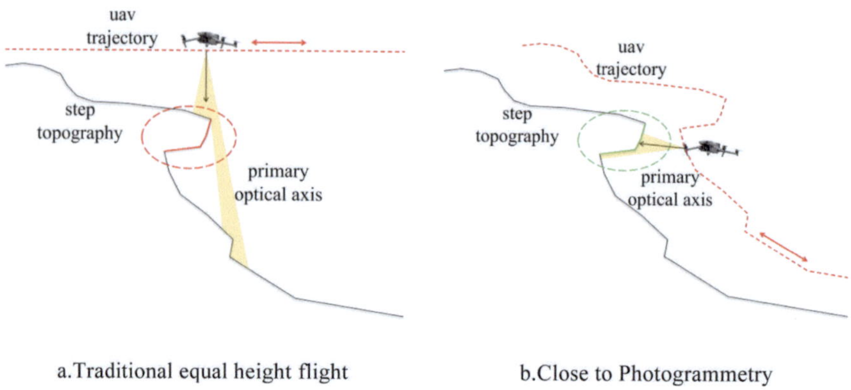

Fig. 2. Closer to the principles of photogrammetric flight

real-time modeling. Finally, DOM and DEM data obtained from ground-based analysis were used to analyze the surface characteristics of the landslide, and the high-precision 3D point cloud model obtained from close-up photography was used to analyze the structural plane combination in the landslide area. The process is shown in Fig. 3.

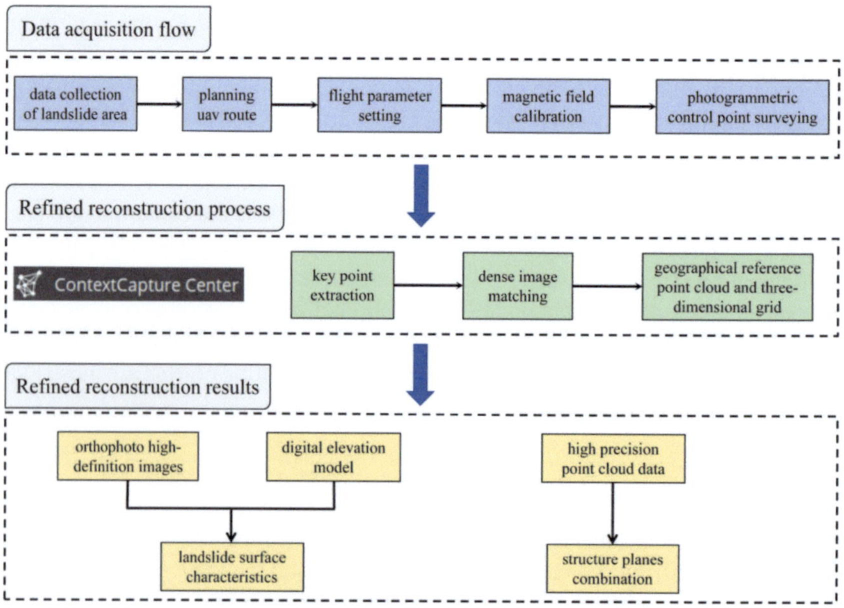

Fig. 3. Technology roadmap

Table 1. Related parameters of DJI Mavic3E

	Parameter
Wheel base	380.1 mm
Weight	915 g
Flight load	1050 g
Capability	45 min
Acceptable wind speed	12 m/s
RTK accuracy	±0.1m (Perpendicular direction) ±0.1m (Horizontal direction)

Table 2. Main technical parameters of digital camera

	Parameter
Pixel	20 million
Sensor size	4/3 CMOS
Image size	5280 × 3956
Aperture range	f/2.8 ~ f/11
Focal-length	24 mm
ISO range	100 ~ 6400

Table 3. Statistical table of structural plane

Structural surface code	dip angle(°)	Tendency(°)	Number of structural planes
J1	88	168	29
J2	83	218	50
J3	80	350	9
J4	42	69	18
J5	55	230	32
J6	87	36	102

3 Engineering Examples

3.1 Overview of the Study Area

As shown in Fig. 4, the study area is located in the Wangmo County of the Southwest Buyei and Miao Autonomous Prefecture of Guizhou Province. The area is characterized by eroded medium and low mountainous landforms, with dendritic gullies that are all seasonal gullies with large flow during the rainy season. The geological structure of the study area is clearly developed, mainly manifested as faults, folds, and joints. The exposed bedrock is mainly the Triassic Middle Biyang Formation (T_2b) stratum, with a top of siltstone mixed with mudstone and a bottom of sandy mudstone mixed with sandstone. This landslide was caused by continuous heavy rain, and the front edge of the landslide has already penetrated into the northwest side of the booster station, seriously threatening the safety of the site.

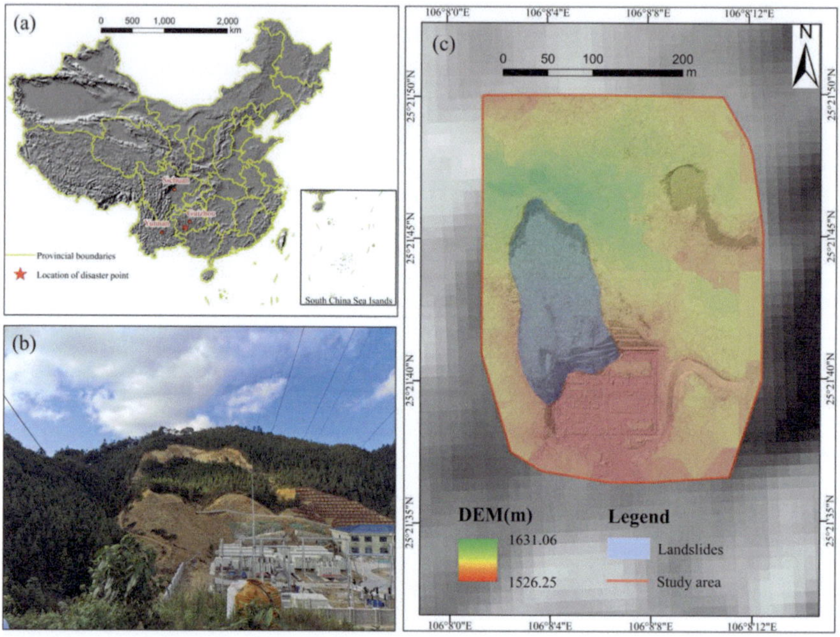

Fig. 4. Overview of the study area (a)Geographical location of landslide;(b)Photos of the landslide site(angle of view 6°);(c)Terrain of study area(8 cm resolution)

3.2 Investigation Results of Ground-Imitating Flight

3.2.1 Basic Information of Landslide

The DJI Mavic3E multi-rotor drone was used on site. The resolution of this model of drone is GSD = H/36. Three flights were set up for simulated ground flight, with a relative ground height of 60 m. High-precision images with a resolution of 1.6 cm within a range of 9×10^4 m^2 in the landslide area were obtained. Through Context Capture processing and calculation, high-resolution 3D images and 3D real-life models of the disaster were obtained, providing a comprehensive understanding of the basic situation of the landslide. The disaster is located on the slope to the north of the booster station. The front and rear edges of the landslide have elevations of approximately 1550 m and 1582 m respectively, with an overall height difference of 0 m between the front and rear edges. The landslide movement direction is 290 m long and approximately 150 m wide in the sliding direction. The average thickness of the landslide source area is about 8 m, and the average thickness of the accumulation body is 18 m. The landslide volume is about 41.3 × 104 m3, covering the northwest area of the booster station, as shown in Fig. 5.

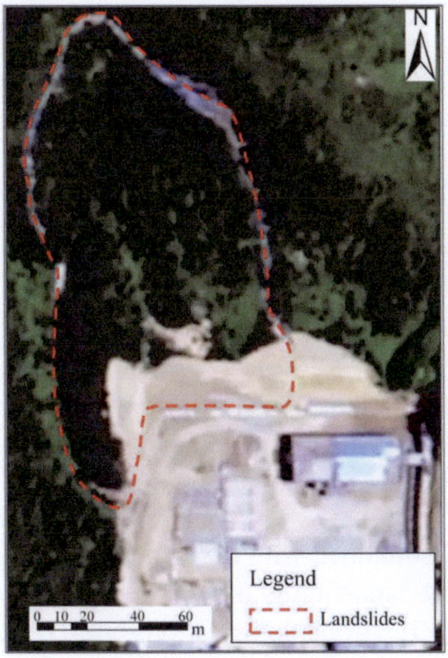

(a)Pre-slip satellite images

(b) Slide-back UAV orthographic image

Fig. 5. Images before and after sliding before Wangmo landslide

3.2.2 Investigation of Landslide Cracks

Before the overall instability and failure of a landslide, it generally undergoes a long deformation development and evolution process (Xu Qiang 2008). A large number of examples show that cracks corresponding to its mechanical properties will be generated at different locations due to stress concentration during different deformation stages of the landslide. The development and evolution of these cracks follow certain rules, namely the staged and coordinated characteristics of crack development. When the landslide enters the accelerated deformation stage, these cracks will gradually connect with each other and eventually tend to trap. When the landslide sliding surface is completely connected and the surface cracks are completely trapped, the landslide may occur. Therefore, the investigation of landslide cracks is of great significance for evaluating the stability of the landslide. The Wangmo landslide has experienced a large-scale collapse, and tensile cracks have occurred in multiple parts of the slope, with a rapid deformation rate. If the cracks are connected, there is a threat of larger-scale collapse. This article obtained a high-precision 3D model and DEM through drone data processing collected by simulated flight, and combined with ground surveys to identify surface cracks. A total of 49 cracks were identified, as shown in Fig. 6.

Fig. 6. Landslide fissure identification results

3.2.3 Rapid Zoning of Landslide

Based on high-precision drone orthophoto data, combined with topographic conditions and deformation and damage characteristics, the landslide is divided into four zones (Fig. 7), namely Zone I (main sliding zone), Zone II (strong deformation zone), Zone III (accumulation zone), and Zone IV (disturbance zone), A brief description is as follows:

① Zone I (main sliding zone)

The main sliding area of the landslide is located in the middle and upper part of the landslide's trailing edge, with an elevation range of 1587 m–1637 m. It is located in the core of a syncline structure. Overall, due to the influence of fold structures, the rear part of the slope is relatively steep, and the slope body is inclined (stratum dip angle of 25°–40°). The slope body tends to be steep upward and forms a broad and well-developed gentle slope terrain downward. Due to the sliding of the main sliding area, a relatively flat trailing edge platform is formed, which is prone to form catchment areas under rainfall conditions.

② Zone II (strong deformation zone)

The strong deformation area on the west side of the slope has a slope gradient of 16°–42°. The right boundary of the area is bounded by the landslide downslope scarp, and the left boundary is bounded by the vertical tension fractures generated in the middle of the landslide. The overall soil mass in the deformation area is pulled and shifted downward, with longitudinal fractures developing on the slope surface, resulting in severe surface deformation.

③ Zone III (accumulation area)

After moving for a certain distance, the landslide material of Wangmo landslide gradually entered the accumulation stage due to the gradual relaxation of topography and the continuous dissipation of energy. The accumulation area at the slope toe was pushed into the site area, which is the main area affecting the safety of engineering structures in the site area.

④ Area IV (disturbed area)

The disturbed area is located on the east side of the slope, near the left boundary of the ancient landslide. When a new landslide occurs, the area is disturbed and locally deformed. After this equilibrium evolution, it has been in a relatively stable state with a clear overall outline, but it still has potential dangers.

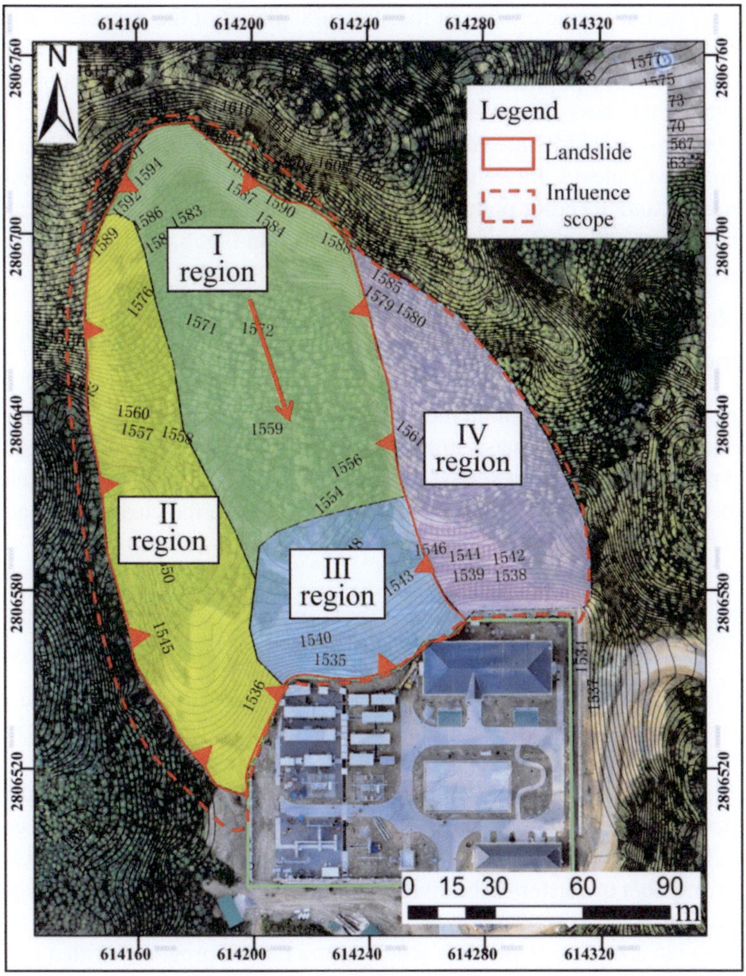

Fig. 7. Landslide zoning map

3.3 Data Acquisition and Processing for Close-Range Photogrammetry

3.3.1 Data Acquisition and Processing

In order to analyze the structural characteristics of the landslide, a section of exposed rock mass on the west side of the Wangmo landslide was selected for close-up photogrammetry. To ensure the standardization of the acquired image data, the flight route was planned and designed using the professional version of UAV Manager software. Based on the development characteristics of the structural planes of the dangerous rock mass, multiple-angle intelligent routes were planned and the intelligent routes were imported into the flight control equipment to automatically complete the close-up photography of the dangerous rock mass. The average design GSD of this intelligent route is 3 mm, the heading overlap rate is 85%, the lateral overlap rate is about 85%, and the flight height is about 15m. The Context Capture software was used to reconstruct the collected images into 3D models and 3D color point cloud models. The reconstructed models accurately restore the texture information of the slope, as shown in Fig. 8.

3.3.2 Information Extraction of Rock Mass Structural Plane

(1) Spherical k-means clustering analysis

After obtaining the point cloud model, k-means clustering analysis is performed on the information of the rock mass structural planes in the area to achieve grouping of structural planes with different spatial distributions. The similarity standard calculation formula between the maximum standard vector involved in the clustering analysis and its related data vector is (Dhillon et al. 2001):

$$\sum_{n}\sum_{b\in C_n}\cos(x_b, p_n) \tag{1}$$

In the formula, n is the number of groups in the clustering division; p_n is the standard vector in n groups; x_b is the bth vector in n groups; C_n is the group of the nth group; and $\cos(x, p)$ is the cosine value of the similarity between two vectors.

Based on the above principle, this article uses the relevant program written in the open source software R to construct a matrix of normal information for the generated point cloud structure surface. By calling the spherical k-means clustering package and entering the number of clusters to be divided, the normal matrix is grouped and analyzed, as shown in Fig. 9.

Based on the above principle, this article uses a related program written in the open source software Rr to construct a matrix of normal information for the generated point cloud structure surface. By calling the spherical k-means clustering package and entering the number of clusters to be divided, the normal matrix is grouped and analyzed, as shown in Fig. 9.

(2) Extraction of structural plane occurrence

After the completion of the clustering and grouping calculation, combined with the three-dimensional color point cloud model obtained, the effect of each group of structural

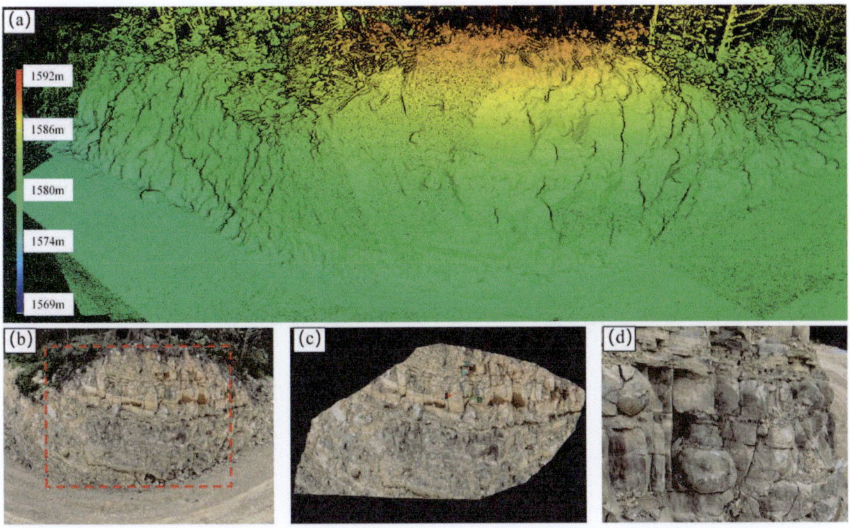

Fig. 8. Three-dimensional model of rock mass (2 mm resolution) (a) Digital elevation of rock mass; (b) 3d real scene model; (c) 3d color point cloud model; (d) partial model

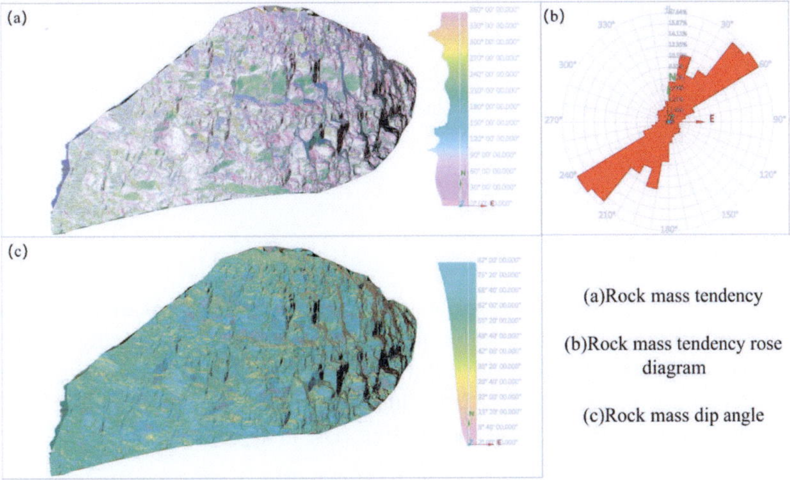

Fig. 9. The overall dip angle HSV diagram of rock mass

planes can be directly tested by extracting the color of each group of structural planes based on the grouping. Each group of structural planes should be pure color without other colors after clustering and grouping. After clustering and calculating the point cloud in the collapse area, six groups of pure-colored structural planes were separately separated, so the inclination and trend of the structural planes can be directly extracted based on the assigned colors. As shown in Fig. 10, there are six groups of different colored structural planes, where J1 is purple (168°∠88°), J3 is yellow (350°∠80°), and the trend difference is greater than 180°. This is due to the effects of weathering and erosion on the slope surface, which makes it impossible for the slope structural planes to be a smooth and flat plane. Therefore, such deviations in trend are also in line with the actual development of slope structural planes. J2 is purple (218°∠83°), which is also a steep-inclined structural plane and is conjugate with J1 and J2. J6 is light pink (36°∠87°), which is the rock layer of the rock mass and is consistent with the stratigraphic occurrence of the landslide area.

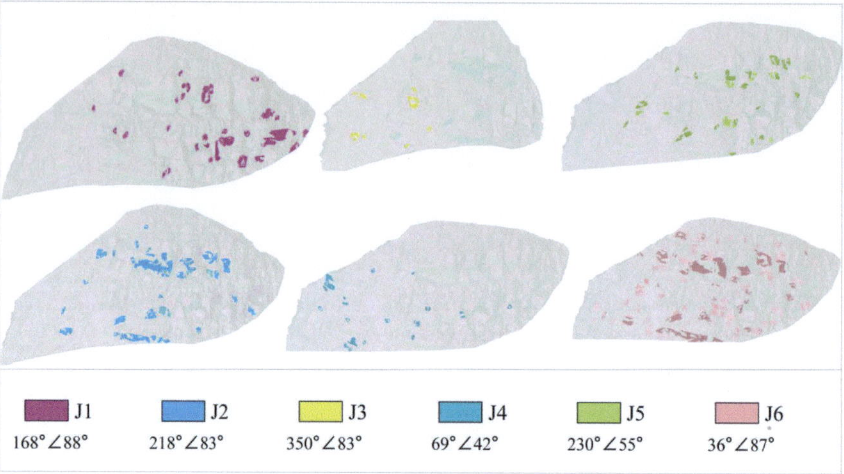

Fig. 10. Statistical diagram of structural plane

4 Conclusions

This article uses multi-rotor drones to conduct disaster detection work on the landslide at the Wangmo booster station using techniques such as ground-simulating flight and close-range photogrammetry. It obtains low-altitude high-precision remote sensing data, and provides detailed investigation of the landslide at the booster station, providing basic data for disaster management and prevention design. The main conclusions are as follows:

(1) Compared with traditional oblique photogrammetry, ground-simulating flight and close-up photogrammetry can obtain higher-resolution remote sensing data. Ground-simulating flight is suitable for large-scale landslides with a resolution of cm level, while close-up photogrammetry is suitable for small-scale steep rock masses with a resolution of mm level.
(2) Using the imitation of ground flight technology to reconstruct the DOM and 3D real-time model (1.6 cm resolution) for production, 49 cracks were identified, and five deformation zones were divided based on deformation characteristics, greatly reducing ground work. Using close-range photogrammetry technology, a total of six structural planes were identified,

(3) The safety of the booster station is the key to the normal operation of the wind power project. When a disaster with a large impact area such as a landslide occurs in the booster station area, drone flight simulation and close-up photogrammetry can assist in on-site investigation to identify the development characteristics of the disaster.

References

1. Dhillon, I.S., Modha, D. S.: Concept decompositions for large sparse text data using clustering. Mach. Learn **42**(1/2), 143–175 (2001)
2. Guo Chen, et al.: Application of UAV photogrammetry technology in the emergency rescue of Baige landslide in Jinsha River. J. Disaster Stud. **35**(1), 203–210 (2020)
3. He Kuangyu: Identification method based on drone close-up photography for high and steep slope rock surface information. Hunan University (2022)
4. Hong Yebing: Analysis and treatment measures for slope stability of 220kV Qingshui Langshan step-up substation. Shanxi Archit. **42**(23), 68–70 (2016)
5. Huang Lizhang: Application of uav imitation flight in complex terrain reservoir inclined photogrammetry. J. Geograp. Geograp. **37**(4), 44–46 (2022)
6. Li Qiang, Zhang Jingfa: General survey of jiuzhaigou valley scenic and historic interest area earthquake landslides using high-resolution third-generation satellite polarization SAR images. J. Remote Sens. **23**(5), 883–891 (2019)
7. Li Yin: Research on emergency processing of UAV images and disaster information extraction technology in the Wenchuan earthquake disaster area. Chengdu University of Technology, Chengdu (2012)
8. Pang Xin, et al.: A rapid identification method for high and steep slope dangerous rock masses based on LiDAR imitation flight technology of unmanned aerial vehicles. Bull. Geol. Sci. Technol. (2023)
9. Xie Mowen, Hu Man, Du Yan, et al.: Progress of TLS technology and its application in landslide monitoring. Remote Sens. Land Resour. **26**(3), 8–15 (2014)
10. Xu Qiang, et al.: Research on the Spatiotemporal evolution law and early warning and prediction of landslides. Chin. J. Rock Mech. Eng. **27**(6), 1104–1112 (2008)
11. Xu Qiang, et al.: Characteristics and formation mechanism of liquefaction-type landslide-mudflow triggered by the M6.2 Jishishan Earthquake in Gansu Province and Its Impact on Zhongchuan Township in Qinghai Province. J. Wuhan Univ. (Inf. Sci. Edn.), 1–18 (2024)
12. Yao Futan, et al.: A method for investigating the structural planes of high, steep and dangerous rock masses based on close-range photogrammetry technology. J. Chengdu Univ. Technol. (Nat. Sci. Edn.) **50**(02), 218–228 (2023)
13. Yan Si: Pioneering and continuously innovating to approach photogrammetry: the birth of the third photogrammetric method—an exclusive interview with Zhang Zuxun, academician of the School of Remote Sensing Information Engineering at Wuhan University. China Surv. Mapp. **28**(10), 31–37 (2019)
14. Zhang Yilin: Application of UAV imitation flight in large-scale mapping in complex mountainous areas. Heilongjiang Sci. 14(10), 114–116 (2023)

Spotlight on Groundbreaking Sustainable Energy Technologies

Research on the Deflection Deformation of Photovoltaic Modules Caused by Low-Temperature Environment

Lian Chunxing, Wang Shusheng, and Sun Zhenyu(✉)

China Water Northeastern Investigation, Design & Research Co.,Ltd., Changchun, China
zhenyusun@126.com

Abstract. The purpose of this study is to conduct a preliminary study on the flexural deformation of photovoltaic modules in low-temperature environments. By analyzing the characteristics and influencing mechanisms of flexural deformation, theoretical basis and technical guidance are provided for the design, manufacturing, and application of photovoltaic modules, and the performance and reliability of photovoltaic modules in low-temperature environments are improved. We have developed a warping deformation testing plan for photovoltaic modules under different temperature environments using a true type test method, and measured and analyzed the warping deformation of photovoltaic modules under different temperature environments. The results indicate that low-temperature environment is the main cause of deflection deformation of photovoltaic modules, and the strength of the frame structure and materials also have a certain impact on the degree of deformation. This study can provide assurance for the long-term operation of photovoltaic modules, reduce maintenance costs and failure rates of photovoltaic systems.

Keywords: Photovoltaics · Modules · Low temperature · Deformation

1 Introduction

As one of the most core components in solar power generation systems, photovoltaic modules directly affect the power generation efficiency and reliability of the entire system [1, 2]. The frame of photovoltaic modules is an important component of photovoltaic systems, which not only plays a supporting and protective role, but also plays a crucial role in the performance and efficiency of the modules.

The selection of photovoltaic module frame structure type has a direct impact on its performance. Typical frame structures mainly include aluminum alloy frames and frameless structures. Aluminum alloy frames are widely used due to their excellent oxidation resistance and mechanical strength [3], while frameless structures are receiving increasing attention due to their advantages such as reducing weight and improving light utilization.

The strength requirement of the border is directly related to the stability and durability of photovoltaic modules. In the design phase, it is necessary to consider the mechanical

S. Zheng et al. (Eds.): IHDC 2024, LNCE 487, pp. 505–514, 2025.
https://doi.org/10.1007/978-981-97-9184-2_41

loads of components under different environmental conditions (such as wind loads, snow loads, etc.) [4]. The strength requirement depends on various factors, including climate conditions, installation position of pressure blocks [5], fixing form of components [6], and design and material of frames.

The deformation of the border can also affect the performance of the component [7]. Research has shown that frame deformation may lead to stress concentration in glass, thereby increasing the risk of component rupture. At the same time, the deformation of the frame may also affect the sealing of the photovoltaic module, leading to moisture and impurities infiltration, which affects the performance of the battery cells.

However, research on the strength of component frames by domestic and foreign scholars has mainly focused on resisting static loads [8] and wind and snow loads [9], with little research on the mechanical performance and deformation characteristics of components under temperature loads.

The objective of this study is to conduct a preliminary study on the flexural deformation of photovoltaic modules in low-temperature environments, and to explore the reasons and influencing factors that cause module deformation. By analyzing the characteristics and influencing mechanisms of flexural deformation, reliable theoretical basis and technical guidance can be provided for the design, manufacturing, and application of photovoltaic modules, thereby improving the performance and reliability of photovoltaic modules in low-temperature environments.

2 Engineering Design Parameters and Photovoltaic Module Dimensions

The photovoltaic plant site of this project is located in Xishechang, Wanghua District, Fushun City, Liaoning Province. The terrain is a plain area, with an altitude of 120–180 m. The center point coordinates are 123°48′34.18″ E and 41°49′09.20″ N. The site area is approximately 312.4 hectares (2345 acres), and the terrain is flat.

The total installed capacity of this project is about 213.58441 MWp, and the installation method is all fixed brackets with a component inclination angle of 30 degrees. Through a comprehensive comparison of technology and economy, it is proposed to use single crystal silicon P-type 545 Wp double-sided photovoltaic modules, with a total quantity of 391898 pieces. The photovoltaic module grid is 2278 mm × 1134 mm × 35 mm, using a fixed inclination angle bracket with a module inclination angle of 30°.

According to the module specifications provided by the photovoltaic module manufacturer, the main technical parameters of the module are as follows:

The component is 2278 mm long, 1134 mm wide, and 35 mm thick. The component frame is made of aluminum alloy material, and the frame section adopts a hollow frame structure. The interface dimensions of the short side aluminum alloy frame are shown in Fig. 1, and the interface dimensions of the long side aluminum alloy frame are shown in Fig. 2.

The weight of the components, number of battery cells, operating temperature, maximum bearing capacity of the front frame, maximum bearing capacity of the back frame, and wall thickness of the aluminum alloy frame used in this project are shown in Table 1.

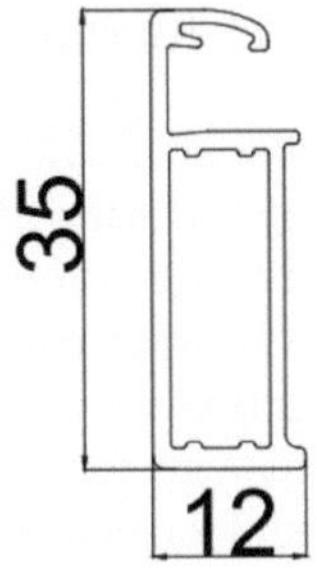

Fig. 1. Cross section of short aluminum alloy frame

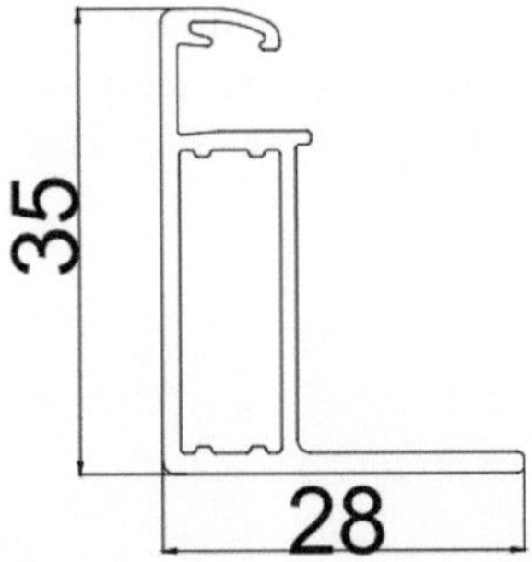

Fig. 2. Cross section of long edged aluminum alloy frame

Table 1. Specification parameters of photovoltaic modules

Project	Parameter
Battery type	Single crystal
Component weight	31.6 kg ± 3%
Front glass/back glass	2.0 mm/2.0 mm
Number of battery cells	144 (6 × 24)
Maximum system voltage	1500 VDC
Working temperature	–40 °C ~ + 85 °C
Maximum rated current of fuse	30 A
Rated battery operating temperature	45 ± 2 °C
Fire resistance performance	UL Type29
Maximum load-bearing capacity of frame (front)	5400 Pa
Maximum load-bearing capacity of frame (back)	2400 Pa
Aluminum alloy frame wall thickness	1 mm

3 Discovery of Warping and Deformation of Three Components

To ensure grid connection by December 30, 2022, this project will organize the installation of photovoltaic modules in winter. During the quality acceptance process of the general contracting unit, it was found that the flatness of the components on the same group of supports did not meet the specification requirements. After investigation, it was found that the severe warping and deformation of the components were the cause.

According to the technical requirements for purchasing photovoltaic modules, the degree of warping of photovoltaic modules cannot exceed 6 mm. To measure the degree of warping of photovoltaic modules, on-site personnel used a simple method of pulling wires for measurement. Measurement personnel work in groups of three, with each person holding the two ends of a thin nylon wire and placing the two ends of the wire at the ends of the long side of the photovoltaic module. The third person held a steel tape measure and searched for the area with the largest gap between the fine nylon thread and the component frame, then measured the curvature. The measurement result is shown in Fig. 3.

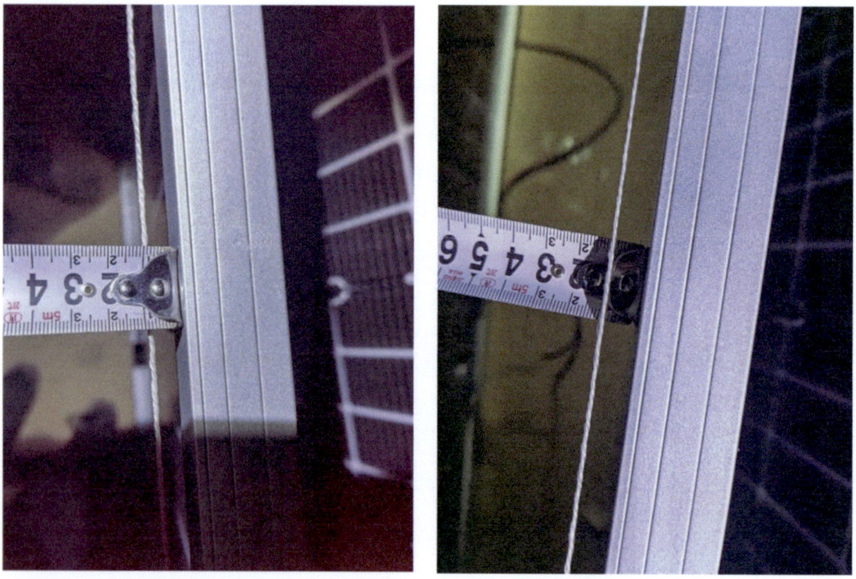

Fig. 3. On site measurement of photovoltaic module warpage

The photovoltaic modules of this project are supplied by two manufacturers. After discovering the warping phenomenon of the modules, the quality inspection personnel of the general contractor conducted sampling inspections on the modules of both manufacturers. The inspection results are shown in Table 2.

After spot checks, it was found that photovoltaic modules from both manufacturers generally exhibit warping, with a minimum warpage of 5 mm and a maximum warpage of 12 mm. The majority of modules have a warpage of 7 mm–8 mm, and the overall distribution is normal.

Table 2. Warpage tables of photovoltaic modules provided by two manufacturers

Component number	Manufacturer 1 Warpage (mm)	Manufacturer 2 Warpage (mm)
1	10	7
2	8	6
3	9	5
4	12	8
5	7	10
6	8	12
7	6	7
8	8	8
9	7	8
10	6	7

4 Analysis of the Causes of Warping Deformations

4.1 Component Size and Border Strength Impact

Larger photovoltaic modules are more prone to warping. This is because when the surface of the component is exposed to light, the temperature at the center of the component is usually higher than the edge, causing the center of the component to expand and causing the edge of the component to tilt upwards. When photovoltaic modules are exposed to sunlight, the surface of the modules absorbs solar radiation, generates heat, and thus raises the temperature of the module surface. Due to the uneven distribution of solar radiation on the surface of the module, the temperature at the center of the module surface is usually higher than at the edges. In this case, the material at the center of the component surface will expand, while the material at the edge of the component will remain relatively stable, which will cause the component edge to tilt upwards. In addition, larger photovoltaic modules are exposed to more light and heat due to their larger surface area, making them more prone to warping. Therefore, in order to reduce the degree of warping of photovoltaic modules, the size of the modules can be appropriately reduced. This can make the solar radiation on the surface of the component more uniform, thereby reducing the temperature rise at the center of the component surface and reducing the possibility of the component warping upwards. Of course, reducing the size of modules may also affect the power generation efficiency of photovoltaic modules. Because a smaller surface area of the component means that it can receive less solar energy, the power generation capacity of the component may decrease. When designing photovoltaic modules, it is necessary to comprehensively consider factors such as module warping and power generation efficiency, and find a balance point.

The strength of the border has a significant impact on the degree of warping of photovoltaic modules. The stronger the border, the smaller the degree of warping of the photovoltaic module. This is because the frame provides support and protection for the photovoltaic module. When the surface of a component is exposed to light, the temperature at the center of the component is usually higher than the edge, causing the center of the component to expand and causing the edge of the component to tilt upwards. If the border is not strong enough, the edges of the component will bend, causing deformation of the component shape and affecting the photoelectric conversion efficiency.

Therefore, using strong borders is an effective method to reduce the degree of warping of photovoltaic modules. Here are some specific measures for strong border design:

(1) Increase the thickness of the border material: The thicker the border material, the stronger the support it provides, which can effectively reduce the degree of component warping. However, it should be noted that increasing the thickness of the frame will increase the material cost and weight, which may affect the installation and transportation of photovoltaic modules.
(2) Adopting stronger border design: Some modern border designs use complex structures and material combinations to provide stronger support. For example, some frame designs use a composite structure of aluminum alloy and steel, or use a beam structure to increase the strength of the frame.
(3) Consider the bonding strength between the frame and glass: The bonding strength between the frame and glass can also affect the degree of warping of photovoltaic modules. If the adhesion between the frame and the glass is not strong enough, the frame cannot effectively support the glass, resulting in component deformation. Therefore, stronger adhesives or improved bonding processes can be considered to enhance the bonding strength between the frame and glass.

4.2 Component Transportation Impact

Photovoltaic modules may be affected by various mechanical and environmental factors during transportation, leading to module warping. The following are some possible causes of component warping:

(1) Pressure: During transportation, photovoltaic modules may be subjected to pressure from surrounding objects or other components. If the pressure is uneven or exceeds the limit that the component can withstand, the component may warp.
(2) Vibration: Vibration during transportation may cause loose connections between internal components or components, resulting in component warping.
(3) Improper stacking: During transportation, photovoltaic modules may be improperly stacked or stored, causing irregular pressure between modules and resulting in module warping.

Therefore, during transportation, a series of measures need to be taken to reduce component warping, such as using appropriate packaging materials and methods, controlling transportation temperature and humidity, avoiding excessive stacking and transportation vibration, etc. At the same time, sufficient inspections should be conducted on the components before transportation to ensure their quality and stability.

4.3 Environmental Temperature Impact

The thermal expansion coefficient of photovoltaic power generation modules varies at different temperatures, and they may undergo warping deformation when heated or cooled. Specifically, when photovoltaic power generation modules are heated by light, the surface temperature of the module will be higher than the internal temperature of the module, causing the surface of the module to expand while the internal temperature of the module relatively shrinks. In this way, the center of the component will rise upwards, causing the component to warp and deform. On the contrary, when photovoltaic power generation modules are cooled, the surface temperature of the modules will be lower than the internal temperature, causing the surface of the modules to shrink while the internal temperature of the modules will relatively expand. In this way, the center of the component will compress downwards, which may also cause the component to warp and deform.

To reduce the warping deformation of photovoltaic power generation modules, the following methods can be adopted:

(1) Choose materials with a smaller coefficient of thermal expansion to make components, such as tempered glass or composite materials, which can reduce the thermal expansion of the components, thereby reducing their warping deformation.
(2) Adopting more stable component support structures, such as thicker aluminum frames or steel brackets, to provide better support and reduce component warping deformation.
(3) Consider temperature changes in component design, such as using larger gaps or more flexible support structures to adapt to component warping deformation, thereby reducing the risk of component damage.

5 True Type Test of Warpage of 5 Components Under Different Temperature Environments

The photovoltaic modules delivered in this project showed significant warping deformation after being unpacked. Therefore, it is speculated that the high temperature at the production site of the modules and the low environmental temperature after the modules arrived may have caused severe warping deformation of the modules due to the large temperature difference.

To verify the above conjecture, three photovoltaic modules with different degrees of warping were selected and moved to a greenhouse with an ambient temperature of 24 °C. After the deformation stabilized, their warping degree was measured. After the measurement is completed, move the test components to an outdoor environment for freezing, with an outdoor temperature of about minus 20 °C. After the deformation of the components in the low-temperature environment stabilizes, measure their warpage. Repeat this process three times and record the degree of warping. The test data is shown in Table 3.

The following conclusions can be drawn from the measurement data of the above cold and hot cycles:

(1) Temperature changes are the main cause of photovoltaic module warping, with modules contracting in low temperature environments and modules stretching in high temperature environments.
(2) For components with a warpage of 12 mm, the degree of warping of the components is greatly alleviated in indoor environments; For components with smaller

Table 3. Warpage of photovoltaic modules during cold and hot cycle testing

Component number	First cold and hot cycle		Second hot and cold cycle		Third hot and cold cycle	
	Indoor	Outdoor	Indoor	Outdoor	Indoor	Outdoor
1	6	12	5	11	5	12
2	5	8	6	9	4	7
3	3	5	4	4	4	5

warpage, temperature changes have little effect on the recovery deformation of the components.

(3) After the component undergoes warping deformation, even at high temperatures, it cannot return to a straight state. The warping caused by low temperature environments will result in residual deformation.

6 Conclusion and Outlook

Through on-site cold and hot cycle tests on the modules, it was verified that temperature changes are the main cause of warping deformation in the photovoltaic modules of this project. The warping deformation of the components discovered in this project in the low-temperature environment of Northeast China is still the first discovery in China, and no scholars have studied this phenomenon in China. The warping and deformation of components caused by temperature is only the surface cause of the problem, and its essence still lies in the size and strength of the photovoltaic module frame.

In order to pursue larger single chip capacity and economic benefits, photovoltaic module manufacturers are increasing the size of photovoltaic modules and thinning the thickness of aluminum alloy frames. At present, component manufacturers are jointly launching large-sized components with a frame thickness of 30mm in order to reduce costs. The larger the component size, the thinner the border thickness. However, the wall thickness of the border section remains unchanged, resulting in insufficient stiffness of the component border to resist stress and deformation caused by temperature changes.

Based on the practical experience of this project, the following measures can be taken to avoid or reduce the problem of component warping and deformation caused by temperature:

(1) Choose appropriate materials: The materials of photovoltaic modules should have a lower coefficient of thermal expansion to mitigate the effects of warping deformation.

(2) Optimization design: The design of photovoltaic modules should consider thermal and mechanical factors to minimize warping deformation caused by temperature. For example, designs with thinness, flexibility, and bending curvature can be used to adapt to the thermal expansion needs of materials under temperature changes. When purchasing photovoltaic modules, try to choose aluminum alloy frame components with larger thickness and higher strength.

(3) Strengthening heat dissipation: Improving the heat dissipation performance of photovoltaic modules is an important way to reduce temperature induced warping deformation. The following methods can be used to increase heat dissipation: selecting substrate materials and packaging materials with high thermal conductivity to enhance heat conduction. Add heat sinks or fans to improve heat dissipation capacity and reduce component temperature.
(4) Temperature compensation: Based on the actual working temperature of photovoltaic modules, corresponding temperature compensation methods are adopted during module manufacturing. For example, adding materials that match the thermal expansion coefficient or setting a temperature compensation layer during packaging can reduce the internal stress of photovoltaic modules caused by temperature changes.
(5) Installation method improvement: When installing photovoltaic modules, attention should be paid to maintaining appropriate gaps between the modules to ensure that there is enough space for thermal expansion and contraction of the modules during temperature changes. In addition, adjustable mounting brackets or suspension devices can be considered to reduce the stress on photovoltaic modules caused by temperature.
(6) Adopting a distributed power generation system: dispersing the installation of photovoltaic modules to reduce the temperature load borne by individual modules, thereby reducing the risk of warping and deformation.
(7) Prevent local hot spot effect: Local hot spot effect can cause the local temperature of photovoltaic modules to be too high, increasing warping. Local thermal spot effects can be reduced by strengthening the cleaning and maintenance of photovoltaic modules and avoiding obstruction of light sources.
(8) Monitoring the temperature of photovoltaic modules: Using temperature sensors to monitor the real-time temperature of photovoltaic modules, taking necessary measures to adjust the structure or external conditions to a certain extent, thereby reducing the temperature of photovoltaic modules and alleviating warping deformation.

The next stage plans to use finite element analysis software to analyze the internal forces and deformations of the photovoltaic modules in this project under different temperature fields, in order to verify the correctness of the above conjectures.

The discovery of this project has accumulated valuable experience for future photovoltaic projects in cold regions. In the subsequent research process, the technology of the frame is continuously improved to avoid such situations.

References

1. Boxian, D., Poplar W.: Optimization of pressure block position for thin film photovoltaic modules. Solar Energy 4:101–104 (2017)
2. Zhongjiang, F., Shilin, Y., Gang, C.: Mechanical load testing and numerical simulation research on photovoltaic modules. J. Solid Mech. 35(S1):107–112 (2014)
3. Shengjuan, H., Rong, T., Lijun, T.: Research on power attenuation analysis of photovoltaic modules. Solar Energy 06:1–5 (2015)
4. Hao, H., Zhixue W.: Research on the influence of photovoltaic module deformation on module performance. Mech. Eng. Autom. 4:102–104 (2011)

5. Dawei, L., Hongxiang, T., Hu, G., Wenbo, P., Zhiguo, Z.: Simulation calculation study on the mechanical static load resistance of photovoltaic modules. Mech. Eng. Autom. 4:112–115 (2011)
6. Qingfa, M., Yanping, H., Jun, D.: Analysis of the causes of "mold spots" on aluminum alloy frames used in photovoltaic modules. China's New Technol. Prod. 17:334 (2017)
7. Qingfa, M., Yanping, H., Jun, D.: Design code for photovoltaic power stations GB50797-2012. Beijing: China Planning Press (2012)
8. Xiaofeng, W., Fan, Y., Ying, Y.: Testing and finite element modeling of photovoltaic modules with different fixed spacing. Mechanical Strength 40(01):113–116 (2018)
9. Xingang, W.: Research progress on factors affecting power attenuation of photovoltaic modules. Appl. Energy Technol. 12:190–195 (2018)

Wide-Area Long Sequence Photovoltaic Power Simulation Based on ERA5 Reanalysis Data

Siwei Tang[1], Xu Wang[2(✉)], Jie Gao[2], Fangliang Zhu[2], and Jianzan Yang[1]

[1] PowerChina Guiyang Engineering Corporation Limited, Guiyang, China
[2] China Renewable Energy Engineering Institute, Beijing, China
2818312759@qq.com

Abstract. The long-sequence hourly photovoltaic power simulation sequence is an important reference information in the stage of power station planning, designing and dispatching operation. Based on the ERA5 reanalysis data, this paper adopts the tilted plane radiation model and the photoelectric conversion model to construct a photovoltaic power physical simulation model, and conducts hourly power simulation study for a total of 23 years from 2000 to 2022 for 30 existing photovoltaic power stations in China, and analyzes the simulation effect and the interannual volatility of power generation capacity in different temporal and spatial ranges. The results show that: on different time scales, the power simulation results are different, and the long time scale is better than the short time scale. The root mean square errors of the monthly average, daily average and hourly scales are 0.89–20.8, 0.43–9.61, and 0.3–3.42 respectively; in different spatial ranges, the power simulation results are also different, and the power stations with relatively low simulation effect are mainly distributed in the western region with smaller longitude; there are great differences in the interannual power generation capacity volatility of each power station, and the volatility between years of the power stations at low latitudes is relatively large. In general, the power simulation result of the ERA5 reanalysis data can better reflect the actual operation law of the power station, and carrying out long-sequence power simulation through the ERA5 reanalysis data can provide relatively reliable data support for the planning and operation of the power station.

Keywords: ERA5 reanalysis data · wide-area · long sequence · power simulation

1 Introduction

With the intensification of the climate change trend and the frequent occurrence of extreme weather, as a clean and renewable energy source, photovoltaic power stations have continuously accelerated the construction pace and continuously increased in scale in China in recent years, and have made important contributions to the reduction of carbon dioxide emissions. In the actual operation process of photovoltaic power stations, they are easily affected by the weather, and there is great uncertainty in the output. The output sequence on a short time coverage is difficult to effectively reflect the true power generation capacity of the power station. By simulating the output sequence of the power

S. Zheng et al. (Eds.): IHDC 2024, LNCE 487, pp. 515–525, 2025.
https://doi.org/10.1007/978-981-97-9184-2_42

station through the long-sequence historical meteorological data, it can reliably reflect its true power generation capacity and has good guiding significance for the planning, designing and dispatching operation of the power station [1].

At present, the photovoltaic power sequence is mainly simulated and calculated based on two types of meteorological data sources, one is the meteorological stations built on the ground, and the other is the meteorological reanalysis data released by meteorological institutions. Among the two types of data, the data accuracy of meteorological stations is relatively high, but the coverage area of meteorological stations is limited and there are problems of missing data and a short monitoring sequence, which is difficult to meet the needs of large-area long-sequence research; the monitoring range of reanalysis data covers the whole world, and the sequence length is also more than several decades, so this type of data is widely used in large-area long-sequence research. In terms of power simulation methods, there are mainly two types [2]: statistical and physical models. Statistical models mainly include methods such as regression analysis, machine learning, and deep learning. This type of method realizes power simulation calculation by establishing the potential mapping relationship between meteorological elements, historical operation information of the power station and the power of the power station. The training data of the model is an important basis of this method, so this type of model has higher requirements for the sequence length and quality of the training data; the physical model is based on physical concepts and methods, and is essentially the reflection of physical laws. No training data is required in the construction process. Reliable and accurate meteorological sequence and basic information such as the coordinates of the power station and equipment parameters are the keys to carrying out power simulation using this model. Reanalysis data and physical models have the advantages of long data sequence and no need for training respectively. Conducting power simulation through reanalysis data and physical models is the main way to obtain the long-sequence power sequence of the power station in the stage of power station planning and dispatching.

There are many sources of reanalysis data. Among them, the ERA5 data set released by the European Centre for Medium Range Weather Forecasts (ECMWF) has good accuracy in the Chinese region [3, 4]. This data set is widely used in solar energy resource assessment, but there are relatively few photovoltaic power simulation studies related to it. In this paper, taking the ERA5 reanalysis data as the data source and through the physical model of photovoltaic power simulation, the power simulation study of 30 photovoltaic power stations in China will be carried out, and the accuracy and applicability of ERA5 reanalysis data in power simulation will be analyzed, in order to provide a reference for the long-sequence hourly photovoltaic power simulation study of ERA5 reanalysis data in China.

2 Data and Methods

2.1 Data

In this paper, ERA5 reanalysis data and photovoltaic site data are used to carry out relevant research. The former is used to generate power simulation sequences, and the latter is used to test the simulation effect.

2.1.1 ERA5 Reanalysis Data

ERA5 is the fifth-generation meteorological reanalysis data released by ECMWF. This data is generated based on ground observation data and satellite data through advanced data assimilation technology, and has high accuracy among similar global data sets. ERA5 provides rich hourly gridded meteorological elements with a grid resolution of $0.25° \times 0.25°$ and a time coverage length from 1950 to the present. The meteorological elements used in this paper are the total horizontal irradiance, the direct horizontal irradiance and the temperature, and the selected period is from January 1, 2000 to December 31, 2022.

2.1.2 Photovoltaic Power Stations Data

This paper selects 30 photovoltaic power stations in China to verify the accuracy and reliability of the power simulation sequence. The above power stations are distributed in various regions in China (Fig. 1), and the selected operating data time range is from January 1, 2022 to December 31, 2022.

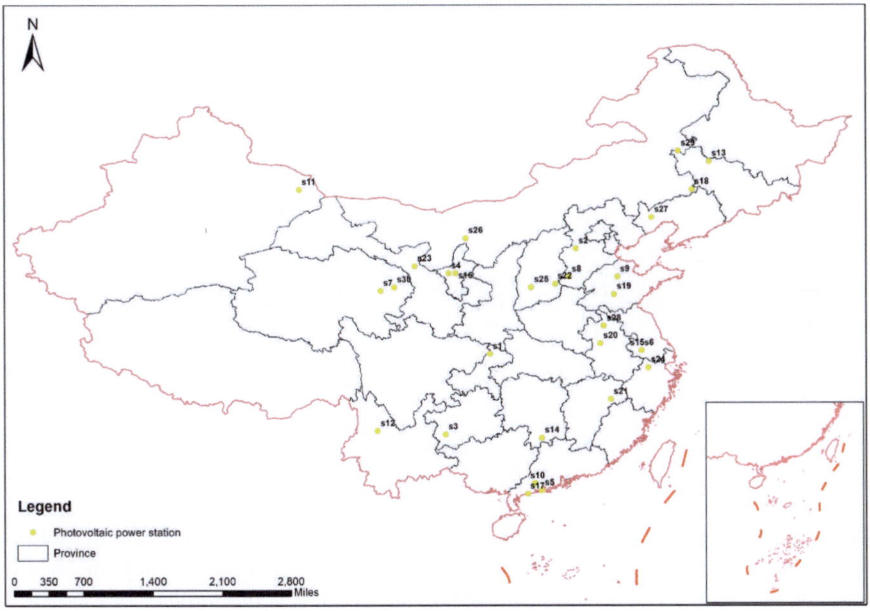

Fig. 1. Spatial distribution of photovoltaic power station

2.2 The Photovoltaic Power Simulation Model

The photovoltaic power simulation model mainly consists of two parts: the tilted plane radiation model and the photoelectric conversion model. The tilted plane radiation model is used to calculate the irradiance received by the photovoltaic module, and the photoelectric conversion model is used to calculate the power generation capacity of the photovoltaic module.

2.2.1 Tilted Plane Radiation Model

The irradiance of the photovoltaic module mainly consists of three parts: direct irradiance, diffuse irradiance, and reflection irradiance [5–7]. The calculation formula is as follows:

$$I_{st} = I_{sdir} + I_{sdif} + I_{sref} \tag{2.1}$$

where $I_{st}, I_{sdir}, I_{sdif}, I_{sref}$ are the global irradiance, the direct irradiance, the diffuse irradiance, and the reflection irradiance on the tilted plane, respectively, in units of W/m^2.

The direct irradiance on the tilted plane I_{sdir} can be calculated from the direct radiation on the horizontal plane I_{hdir} and the conversion coefficient R_b:

$$I_{sdir} = I_{hdir} \times R_b \tag{2.2}$$

$$R_b = \frac{\begin{array}{c}\sin\delta\sin\varphi\cos\beta - \sin\delta\cos\varphi\sin\beta\cos\gamma_t + \cos\delta\cos\varphi\cos\beta\cos\omega \\ \cos\delta\sin\varphi\sin\beta\cos\gamma_t\cos\omega + \cos\delta\sin\beta\sin\gamma_t\sin\omega\end{array}}{\sin\varphi\sin\delta + \cos\varphi\cos\delta\cos\omega} \tag{2.3}$$

$$\delta = 23.45\sin 360{*}(284 + n)/365 \tag{2.4}$$

$$\omega = \left[\frac{\pi}{12}\left(t + eot + \frac{\lambda - 120}{15}\right) - \pi\right]{*}\frac{180}{\pi} \tag{2.5}$$

$$eot = 9.87\sin(2\theta) - 7.53\cos\theta - 1.5\sin\theta \tag{2.6}$$

$$\theta = \frac{2\pi(n - 81)}{364} \tag{2.7}$$

where I_{hdir} is the direct irradiance on the horizontal plane (W/m^2); φ is the latitude (°); β is the inclination angle of the photovoltaic module (°); γ_t is the azimuth angle of the photovoltaic module (°); δ is the declination angle (°), n is the number of days in the year; ω is the solar hour angle (°), t is the Beijing time (h), λ is the longitude of the photovoltaic power station (°); eot is the time difference (h) [8].

The diffusion radiation on the tilted plane adopts the Hay-Davies model, and the calculation formula is as follows:

$$I_{sdif} = AR_b + (1 - A)\left(\frac{1 + \cos\beta}{2}\right) \tag{2.8}$$

$$A = I_{hdir}/I_0 \tag{2.9}$$

$$I_0 = \gamma E_{sc}(\sin\varphi \sin\delta + \cos\delta \cos\varphi \cos\omega) \tag{2.10}$$

$$\gamma = 1 + 0.033\cos(360 \times n/365) \tag{2.11}$$

where A is the anisotropy index of the sky; I_0 is extraterrestrial horizontal solar irradiance; E_{sc} is solar constant, $E_{sc} = 1367W/m^2$; n is the number of the day in the year.

The reflected radiation on the slope can be calculated from the global radiation on the horizontal plane I_{ht}:

$$I_{sref} = I_{ht}\rho\left(\frac{1-\cos\beta}{2}\right) \tag{2.12}$$

where ρ is the albedo of the surface, generally 0.2.

2.2.2 Photoelectric Conversion Model

The amount of irradiance received by a photovoltaic power station is the main factor determining its output. The relationship between them can be approximately expressed as follows:

$$P = KI_{st}S \tag{2.13}$$

where P is the power generation (W); K is the photovoltaic conversion coefficient; S is the area of the photovoltaic panel (m^2).

The photovoltaic conversion coefficient reflects the comprehensive efficiency of converting the irradiance received by the power station into output. Since the photoelectric conversion efficiency, array aging, dust occlusion, temperature, etc. are all important factors affecting its power generation during the power generation process, the photoelectric conversion coefficient can be summarized as follows:

$$K = \eta_s[1 - \alpha(T_c - 25)]K_1K_2K_3K_4 \tag{2.14}$$

where η_s is the photoelectric conversion efficiency under standard test conditions; α is the temperature coefficient (°C^{-1}), and the value of crystalline silicon material is 0.003–0.005(°C^{-1}); T_c is the photovoltaic array panel temperature (°C); I_T is the global irradiance on the slope (W/m^2); S is the effective area of the photovoltaic module of the power station (m^2); K_1 is the loss coefficient of photovoltaic array aging; K_2 is the loss coefficient of photovoltaic array mismatch; K_3 is the loss coefficient of dust occlusion; K_4 is the loss coefficient of the DC circuit line.

2.3 Evaluation Metrics

In this paper, the actual operating power of the selected photovoltaic power station in 2022 is used as the benchmark, and the correlation coefficient (CORR) and the root mean square error (RMSE) are selected to test and analyze the power simulation effect of different time scales for the simulated power sequence in the corresponding period [9].

$$CORR = \frac{\sum_{i=1}^{n}(y_{ri} - \overline{y_r})(y_{si} - \overline{y_s})}{\sqrt{\sum_{i=1}^{n}(y_{ri} - \overline{y_r})^2 \sum_{i=1}^{n}(y_{si} - \overline{y_s})^2}} \tag{2.15}$$

$$RMSE = \sqrt{\frac{1}{n}\sum_{i=1}^{n}(y_{si} - y_{ri})^2} \tag{2.16}$$

where $\overline{y_r}$, $\overline{y_s}$ are the average values of the actual operating power and the simulated power, respectively; y_{ri}, y_{si} are the actual operating power and the simulated power, respectively; n is the sequence length. Both CORR and RMSE reflect the closeness of the simulated value to the actual value. The closer the CORR value is to 1, the better the simulation effect; the closer the RMSE value is to 0, the better the simulation effect.

3 Results and Discussions

3.1 Simulation Accuracy and Applicability Analysis

The photovoltaic power simulation results of 30 photovoltaic power stations at different time scales of hours, days, and months are evaluated respectively. As shown in Fig. 1, the simulation accuracy of photovoltaic power varies at different time scales. The RMSE at the hourly scale is 0.89–20.8, the RMSE at the daily scale is 0.43–9.61, and the RMSE at the monthly scale is 0.3–3.42. The accuracy of photovoltaic power simulation is overall good; as the time scale of power simulation increases, the accuracy of power simulation increases. The reason is that the output of photovoltaic is affected by meteorological factors such as irradiance. The smaller the time scale, the greater the randomness of irradiance, ERA5 is more difficult to capture the real irradiance information, and then transmits this part of the error to the power simulation value, resulting in an increase in simulation error.

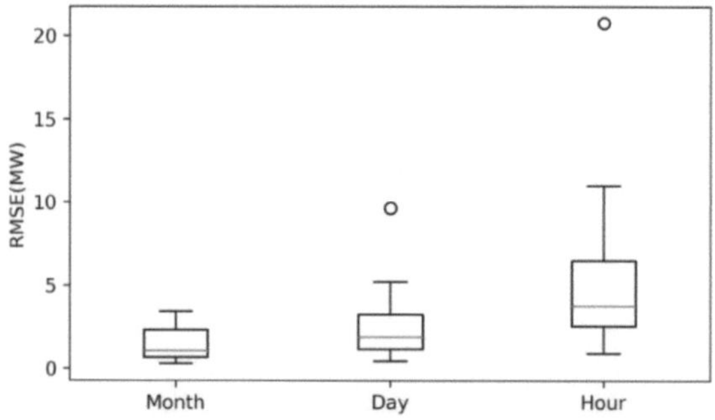

Fig. 2. Multi-time scale power simulation results

The correlation coefficient between the power simulation value and the real value at the hourly scale of 30 photovoltaic power stations was calculated, and the power simulation effect in different spaces was analyzed in combination with its spatial distribution. As shown in Fig. 2, the hourly power simulation correlation coefficient of 30 power stations is 0.81–0.95, and the correlation coefficient of most power stations is higher than 0.9, only 10 power stations have a correlation coefficient lower than 0.9, indicating that the overall accuracy of power simulation results is high. From the spatial distribution of power stations in Fig. 2, it can be known that the power stations with low simulation effect are mainly distributed in the western region with a smaller longitude. For example, S1 and S3 power station are located in the southwest region of China, S4 and S7 power station are located in the northwest region of China. The climate and terrain conditions in the above two regions are complex, and the randomness of irradiance and other meteorological factors is large, resulting in relatively large simulation difficulties.

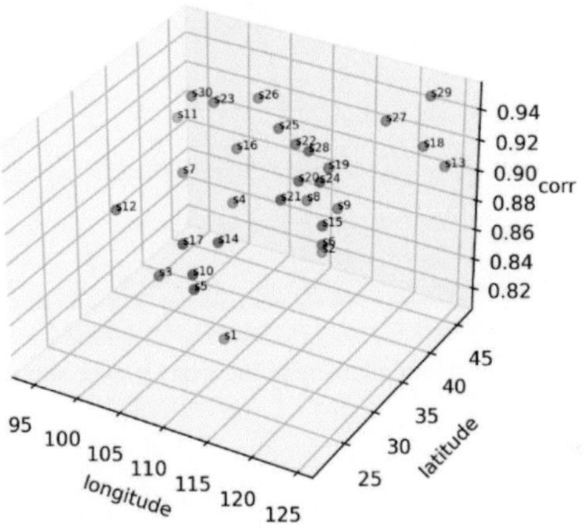

Fig. 3. Spatial distribution of power simulation results

The output of photovoltaic shows a relatively regular change characteristic within a day, so the annual average intra-day process of 30 photovoltaic power stations is analyzed, and the correlation coefficient between the simulation value and the real value is calculated. As shown in Fig. 3, the range of its correlation coefficient is 0.815–0948. Combined with the results of Figs. 2 and 3, it shows that the overall accuracy of power simulation is high. Further analysis of its intra-day output characteristics shows that the change trend of the power simulation sequence and the occurrence time of the peak are basically the same as the real situation. The simulation error in the morning and afternoon is relatively small, and the error is mainly in the peak stage at noon. The reason is that there are errors in various theoretical parameters adopted in the power station simulation model compared with the actual situation. And at this stage, the irradiation is also at the peak stage and the irradiation and the output are basically in a linear relationship, so the simulation error at this stage is also relatively large.

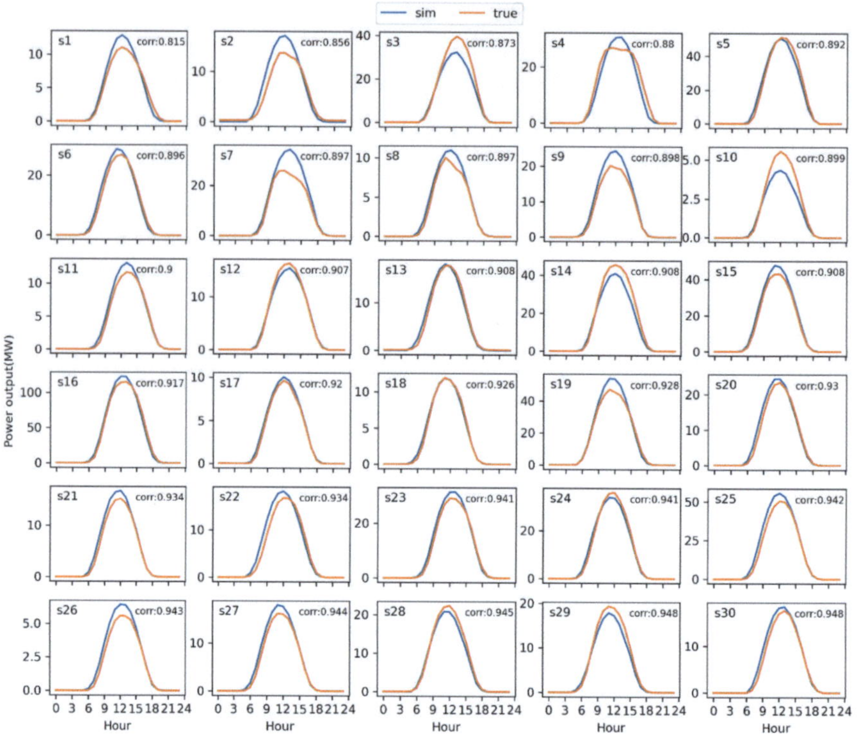

Fig. 4. Comparison of annual average intra-day power simulation results

3.2 Interannual Variation

In order to clarify the changing rule of the power generation capacity of the photovoltaic power station under a long time scale, a total of 23-year power simulation sequence was calculated, and the interannual fluctuation of the power generation capacity of each power station was calculated by taking the standard deviation as the evaluation index. As shown in Fig. 4, the inter-annual volatility of 30 photovoltaic power stations varies greatly, among which S14 has the largest volatility of 0.77, and S26 has the smallest volatility of 0.02. In addition, the volatility of each power station shows a certain distribution pattern in space. As shown in Fig. 5, the higher the latitude at which the power station is located, the relatively smaller its interannual volatility. The reason for this phenomenon is that the terrain and climate conditions where the low-latitude power station is located are more complex, resulting in more significant interannual changes in irradiation [10], and then resulting in relatively large changes in the interannual output of photovoltaic (Fig. 6).

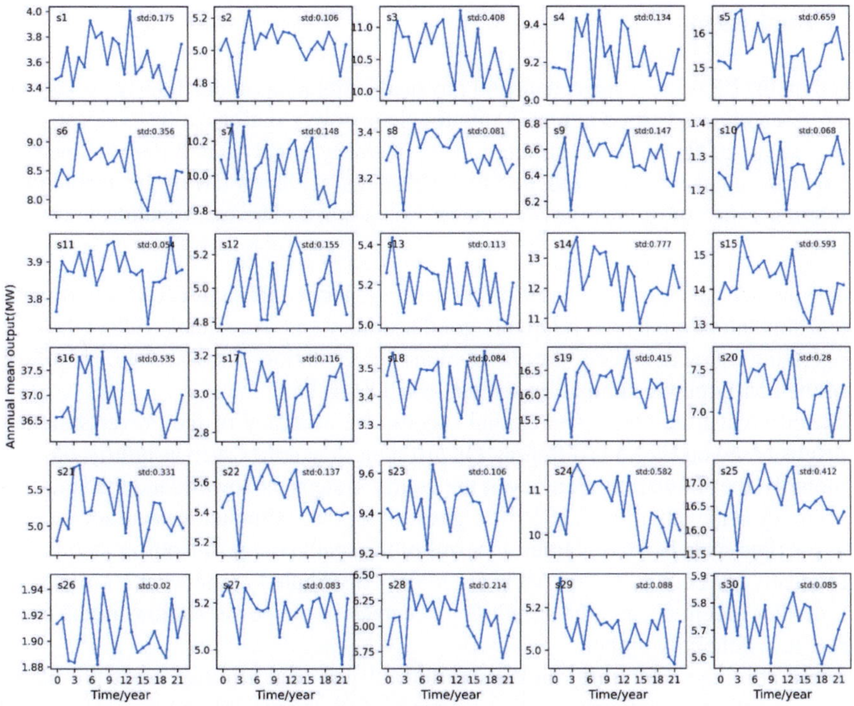

Fig. 5. Interannual fluctuation of photovoltaic power

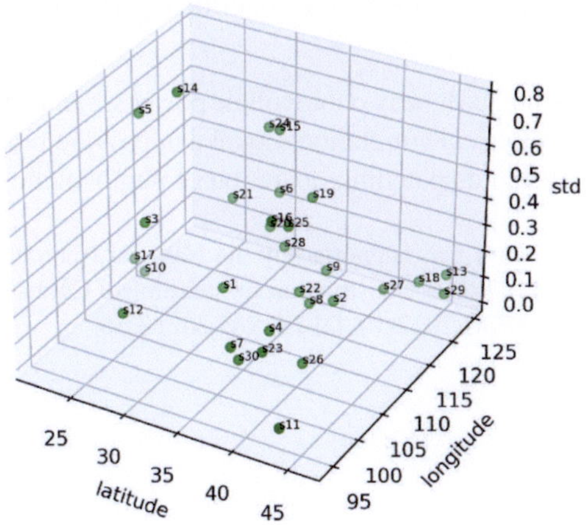

Fig. 6. Spatial distribution of photovoltaic power station fluctuation

4 Conclusions

Based on the ERA5 reanalysis data and physical simulation model, hourly power simulation research was carried out on 30 photovoltaic power stations in the Chinese region from 2000 to 2022, the power simulation applicability of ERA5 in different time scales and spatial ranges was analyzed, and the interannual variation law of the annual power generation capacity of each power station was studied. The main conclusions are as follows:

1. The photovoltaic power simulation sequence generated through ERA5 and the physical simulation model has relatively good accuracy and can reliably reflect the actual operating characteristics of the photovoltaic power station. The correlation coefficient of the hourly power simulation sequence is 0.81–0.95.
2. On different time scales and spatial ranges, the accuracy of the power simulation results based on ERA5 is different. On different time scales such as hourly, daily and monthly, the simulation accuracy at the hourly scale is the lowest, and the simulation accuracy at the monthly scale is the highest, The power simulation accuracy increases as the time scale becomes larger; in terms of the spatial range, the power stations with relatively lower simulation effects are mainly distributed in the western region with a smaller longitude, and the simulation accuracy of the power stations in the eastern region is generally higher.
3. There are large differences in the interannual power generation capacity volatility among different photovoltaic power stations, with the minimum being 0.02 and the maximum being 0.77; there is a certain distribution pattern in the spatial distribution of the interannual power generation capacity volatility of the power stations. Compared with the low-latitude power stations, the volatility of the high-latitude power stations is relatively small.

Overall, the long-sequence hourly photovoltaic power simulation sequence generated through ERA5 and the physical simulation model can provide reliable data support for the planning, designing and dispatching operation of photovoltaic power stations.

Acknowledgement. This work was supported by the National Natural Science Foundation of China (Grant No. U2243232), Power Construction Corporation of China, Ltd Technology Project (DJ-HXGG-2022-01, DJ-ZDXM-2022-10, DJ-ZDXM-2021-26).

References

1. Peng, B., Sun, Z., Liu, M.: Medium and long term scenario generation method based on autoencoder and generation adversarial network. In: 2023 3rd International Conference on Neural Networks, Information and Communication Engineering (NNICE). IEEE, pp. 639–645 (2023)
2. Li, Y., Li, Z., Wang, X., et al.: Prediction methods of short-term photovoltaic power based on inclined plane solar radiation algorithm. J. Arid Meteorol. **38**(5), 869–877 (2020)
3. Zhang, S., Li, X.: Application of ERA5 data to solar energy resource assessment in China. Acta Energiae Solaris Sinica **44**(5), 280–285 (2023)

4. Li, Z., Yang, X., Tang, H.: Evaluation of the hourly ERA5 radiation product and its relationship with aerosols over China. Atmos. Res. **294**, 106941 (2023)
5. Jiang, W., Zhao, Y., Wang, B., et al.: Photovoltaic power prediction method based on NWP irradiance inclination conversion. J. Shandong Univ. (Engineering Science) **51**(5), 114–121 (2021)
6. Jinhuan, Y., Jiajun, M., Zhonghua, C.: Calculation of solar radiation on variously oriented tilted surface and optimum tilt angle. J. Shanghai Jiao Tong University **07**, 1032–1036 (2002)
7. Enyu, W., Qiang, G., Xueyou, Z., et al.: Research on the calculation method of solar radiation on inclined surfaces. J. Therm. Sci. Technol. **18**(1), 35–41 (2019)
8. Li, W., Zhao, Y.C.: The improvement in solar position calculations in the ellipsoid model of the earth. J. Univ. Chinese Acad. Sci. **36**(3), 363–375 (2019)
9. Jiao, D., Xu, N., Yang, F., et al.: Evaluation of spatial-temporal variation performance of ERA5 precipitation data in China. Sci. Rep. **11**(1), 17956 (2021)
10. Chuanhui, W., Yanbo, S., et al.: Applicability of three types of reanalysis data in solar energy resource assessment. Acta Energiae Solaris Sinica **43**(8), 164–173 (2022)

Effective Models for the Integration of Green Energy Systems Within Existing Infrastructures

Application of Flow Prediction Models to the HPP Castro Alves for Planning Preventive Actions Against Extreme Events

Beatriz Sepulveda Pires(✉), Xinjian Chen, and Huiyi Zhang

CPFL Energia, Campinas, Brazil
beatriz.pires@cpfl.com.br

Abstract. Today, the effects of climate change are increasingly evident. In the southern region of Brazil, significant impacts have been observed, such as the peak flows in November 2023, when a state of emergency was declared in the areas of the municipalities of Rio das Antas, Rio Grande do Sul, due to the impacts of heavy rainfall in the region. Such events may become more frequent in the coming years, so that the forecasting of hydroelectric inflows is necessary to plan the operation of the plants and to carry out preventive actions to deal with extreme events. The Ceran complex is located on the Antas River and consists of three hydroelectric plants, the Castro Alves, Monte Claro and Quartorze de Julho plants, and with a view to the sustainable operation of the projects in order to make conscious use of water resources, optimize operations and minimize the impact on the well-being of local populations, it has been studied and applied that rainfall-runoff models to forecast the inflows to the projects on the Antas River cascade. Rainfall-runoff models are essential for predicting extreme flows and taking preventive action, providing the operations center and the team involved with information for prior decision-making and, if necessary, evacuation of operators and potentially affected communities. In view of the above, this paper will apply the MEL model to convert rainfall into flow and compare the predicted results with the observed ones.

Keywords: mathematical model · hydrological forecasting · rainfall-runoff model · operation planning · heavy rainfall

1 Introduction

As the impacts of climate change become increasingly evident, discussions around these issues are reaching a crucial juncture. We are facing a scenario in which climate resilience and adaptability are becoming imperative not only for communities and ecosystems, but also for hydroelectric projects, the fundamental pillars of our energy infrastructure. In this context, the operational planning of hydroelectric plants becomes a vital tool in the search for the "survival" and efficiency of these projects in an ever-changing climate scenario. Anticipating and responding to extreme weather events has become not only a precautionary measure, but also an essential strategy to guide the actions and decisions of entrepreneurs.

S. Zheng et al. (Eds.): IHDC 2024, LNCE 487, pp. 529–536, 2025.
https://doi.org/10.1007/978-981-97-9184-2_43

In view of the above, the objective of this paper was to implement the linear stochastic model, MEL, at the Castro Alves Power Station, which is part of the CERAN complex of power stations, estimate the average daily inflows and then estimate the daily peak flow, comparing the observed data with the predicted data. HPP Castro Alves is a hydroelectric power plant located on the Antas River, sub–basin of the Taquari River, South Atlantic watershed, southeastern section, has an installed capacity of 130 MW and has been in operation since 2018.

The model showed satisfactory results in the calibration period and acceptable and satisfactory results in the validation period. In relation to the calibration period, the value of the CNS objective function was 0.84 and an R^2 of 0.84, which means that it is properly calibrated. The validation period showed a CNS coefficient and R^2 of 0.63 for the period 2017 to 2019, which according to the reference indicates that the correlation is acceptable and 0.97 for the period 01/2024 to 05/2024. From the calculated and observed maximum flow data, it was possible to see that the calculated maximum flow showed a good correlation, with a value of 0.85.

This paper explores the importance of planning the operation of hydroelectric plants as a fundamental approach to meeting the challenges posed by climate change, with a view to ensuring the safety of structures and local communities.

2 Conceptualization of the MEL Model

The MEL model, a tool for simulating hydrological processes, is a linear stochastic model designed to convert rainfall data into predictions of river flow. It operates based on the principles of transfer function models as outlined by the ONS in 2007. The MEL model operates within a stochastic framework and integrates time series analysis techniques into its functioning. Using observed discharge and observed and predicted rainfall data from the upstream area, the model equation is structured to capture these hydrological dynamics (ONS, 2007).

In the calibration phase, the MEL model estimates parameters through multiple linear regression analysis using the least squares method. This calibration process aims to capture lateral contributions and damping effects within the river basin by adjusting equation parameters accordingly. Equation (1) provides insight into these parameter specifications.

$$Q^t_{\text{Cal}} = a_0 + a_1 Q^{t-1}_{\text{Obs}} + a_2 Q^{t-2}_{\text{Obs}} + a_3 Q^t_{\text{Prev,P1}} + a_4 P^{t-1}_{\text{Pr ev,P1}} + a_5 P^{t-2}_{\text{Prev,P1}} + a_6 P^t_{\text{Prev,P2}} + a_7 P^{t-1}_{\text{Prev,P2}} + a_8 P^{t-2}_{\text{Prev,P2}} \quad (1)$$

Analysis of Eq. (1) reveals that current flow predictions are dependent on past observed and calculated flows, as well as observed and calculated rainfall. Consequently, longer-term forecasts are inherently more uncertain because they rely on estimates derived from previous forecasts (Fadiga Júnior et al. 2008).

3 Methodology

3.1 Case Study

The present work studied the HPP Castro Alves, a hydroelectric power plant located on the Antas River, sub–basin of the Taquari River, South Atlantic watershed, southeastern section. The Castro Alves HPP is located between the municipalities of Nova Roma do Sul and Nova Pádua, with latitude coordinate –29,0058 and longitude –51,3853. It has an installed capacity of 130 MW and has been in operation since 2018. The power plant operates on a run-of-river basis, and the entire affluent flow is discharged, with no water storage in the reservoir. The Antas River basin, in the Castro Alves HPP basin region, does not have a well-defined seasonality in terms of precipitation.

3.2 Model Input Data and Calibration

Rainfall data from two stations and flow data were used to calibrate the model. The period analyzed was from 2016 to 2024. Since the data for the most recent period is of better quality due to the lower number of failures and gross errors, this data was used to calibrate and validate the model. The observed rainfall and flow data were acquired from the National Water Agency (ANA) website, and the stations used were UHE Castro Alves Barramento and UHE Castro Alves Rs-122 and Stattio caxias do Sul from the SISDAGRO system. SISDAGRO is an agricultural decision support system developed by the National Institute of Meteorology – INMET (Table 1).

Table 1. Model input data

Data/Model	MEL
Calibration parameters	9 parameters
Rain gauge stations used	UHE Castro Alves Barramento Caxias do Sul
Flow Station used	UHE Castro Alves RS-22

The objective functions chosen for analysis and model calibration were the Nash-Sutcliffe efficiency coefficient (CNS). Table 2 shows the objective function with the reference parameters.

Data from 11/2020 to 12/2023 was chosen for the calibration period. When calibrating the model, it was found that for calculated flows greater than or equal to 1500 m^3/s the values were underestimated, so the relationship between the calculated values and the observed values was evaluated in order to define an increase factor to make these values more representative. For higher values, above 4000, it was found that the increment factor should not be the same, so after analysis and testing, a different increment factor was adopted for these values. In addition to the above, in order to implement this increase factor, we also assessed whether the rainfall on the day of the forecast was above 100 mm, i.e. whether it would be a day with moderate to heavy rainfall.

Table 2. Ojective function and reference parameters

Objective function	Equation	Reference value	Reference
CNS	$Cns = 1 - \frac{\sum_{t=1}^{N}(Q_{Obs}(t) - Q_{Cal}(t))^2}{\sum_{t=1}^{N}(Q_{Obs}(t) - \overline{Q}_{Obs})^2}$	Properly calibrated: Cns > 0.75 Acceptable: 0.36 < Cns < 0.75 Unsatisfactory: Cns < 0.36	Gottschalk and Motovilov (2000)

To estimate the maximum flow for the day, a correlation was made between the average daily flows and the maximum daily flows. Using the linear correlation, it was possible to estimate the maximum daily flow from the daily flow data.

3.3 Peak Flow Estimate

To calculate the peak flow, was selected the hourly observed flow data for the entire data period, from 2020 to 2024. The average daily flow and maximum daily flow data were then calculated. These data were then correlated to determine the periods in which the data was best correlated.

According to the correlation of the data, increment values were used to multiply the average flow and obtain the maximum daily flow. The Excel solver tool was used to estimate the increment factor values.

4 Results

After executing the presented methodology, the calibration and validation parameters of the model were obtained, as shown in Table 3. The coefficients of the equation are shown in Table 4.

Table 3. Calculated parameters and reference parameters

Parameters	Period	Value	
CNS calibration	11/2020 to 12/2023	0.84	Properly calibrated
R^2 calibration	11/2020 to 12/2023	0.84	–
CNS validation	01/2017 to 12/2019	0.63	Acceptable
R^2 validation	01/2017 to 12/2019	0.64	–
CNS validation	01/2024 to 05/2024	0.96	Properly calibrated
R^2 validation	01/2024 to 05/2024	0.97	–

Table 4. Values of the equation coefficients

Coefficient	Value
a_0	2.50
a_1	0.68
a_2	0.00
a_3	2.57
a_4	5.37
a_5	0.00
a_6	0.00
a_7	0.00
a_8	0.97
Increase factor $1500 \leq Q < 4000$	2.60
Increase factor $4000 \leq Q$	1.50

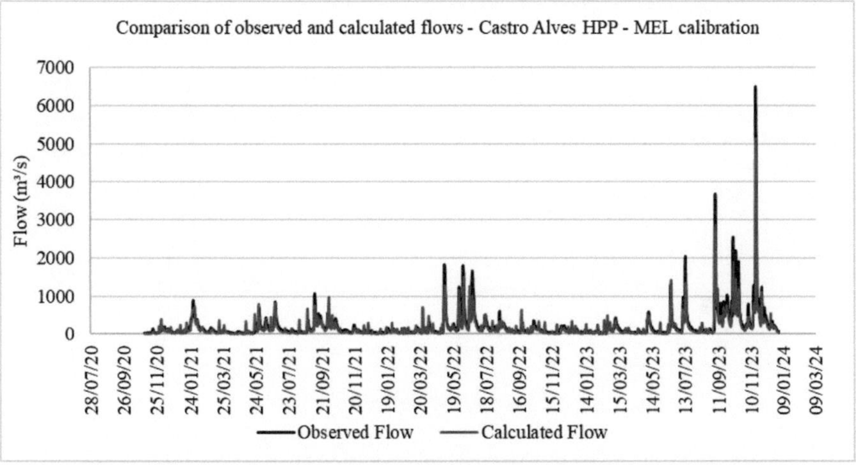

Fig. 1. Comparison of observed and calculated flows – calibration 2020 to 2023

From the analysis of the graphs and parameters presented, it can be seen that the calibration period had a CNS of 0.84 and an R^2 of 0.84, which means that it is properly calibrated. The validation period showed a CNS coefficient of xx and R^2 of 0.6, for the period 2017 to 2019, which according to the reference indicates that the correlation is acceptable. For the period from 01/2024 to 05/2024, the CNS and the R value were 0.97, showing an excellent result. Analysis of the data showed that the series for previous years presented some data gaps, which may be a relevant factor in reducing the accuracy of the model for the validation period (Figs. 1, 2 and 3).

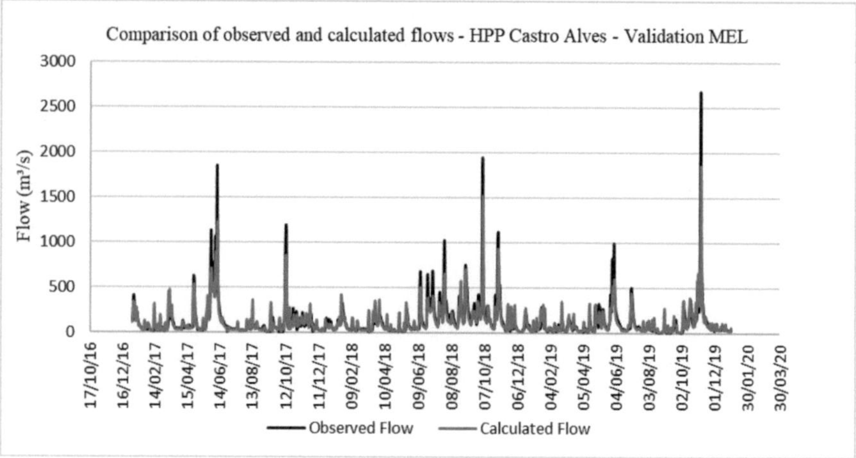

Fig. 2. Comparison of observed and calculated flows – Validation 2017 to 2019

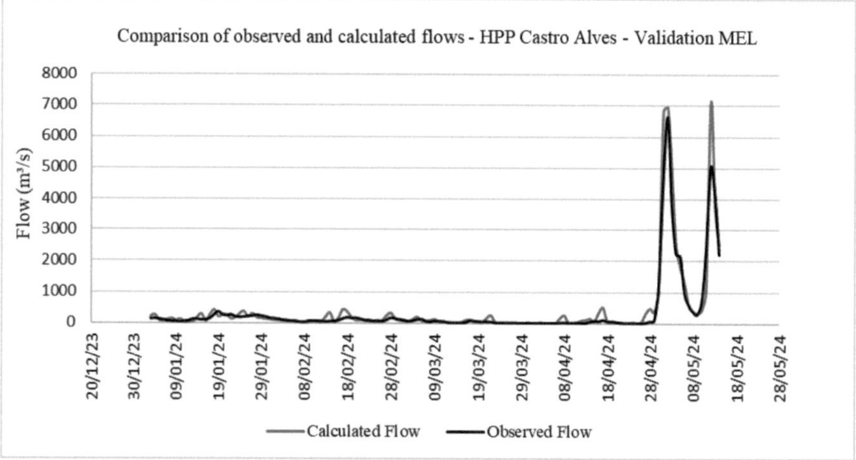

Fig. 3. Comparison of observed and calculated flows – validation 2024

When calculating the peaks, the series from 2020 to 2024 was divided into greater than 2000 m^3/s, between 400 and 2000 m^3/s and less than 400 m^3/s because the individual stretches had different correlations with each other and provided a higher quality estimate when assessed separately. Table 5 shows the increment values for each stretch analyzed and Fig. 4 shows the result of the estimated maximum daily flow when the factor is applied to the average daily flow calculated by the MEL model and the maximum observed flow. The series with data from 2017 to 2019 was not used to analyze the peaks, as the lack of hourly data could lead to possible inconsistencies.

From the calculated and observed maximum flow data, it could be seen that the calculated maximum flow showed a good correlation, with a value of 0.85.

Table 5. Analysis range and increment factor value

Analysis range	Increment factor value
Greater than 2000	1.38
Greater than 400	1.18
Less equal than 400	1.14

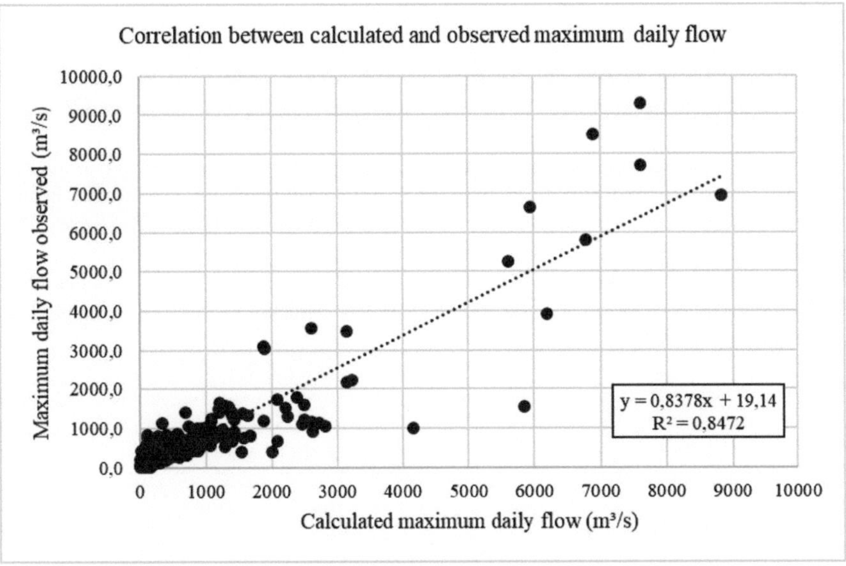

Fig. 4. Comparison maximum daily flow calculated and observed

5 Conclusion

Adapting to climate change is a crucial component in ensuring the smooth development and operation of hydroelectric plants. In a scenario where climate patterns are constantly evolving, the ability to adapt and respond to these changes becomes fundamental to ensuring the efficiency and safety of these important structures.

In addition, planning for the operation of hydroelectric plants plays an essential role in the management of water resources and in the short-term forecasting of hydrological conditions. By using models such as MEL, it is possible to accurately estimate the average flow and daily peak flows, providing valuable information to guide decision-making and ensure the safety of the structures and adjacent communities.

This work demonstrated the successful application of the MEL model at the Castro Alves HPP, achieving promising results with satisfactory coefficients of efficiency (CNS) and coefficients of determination (R^2) in calibration and acceptable coefficients in validation phases. The results obtained indicate that the model is capable of providing reliable estimates for the average daily flow and peak flows.

Although there is room for improvement in the performance of the model, especially with regard to improving the input data, it is encouraging to note that the most recent data tends to present a higher quality of information. Therefore, even with these considerations, it is feasible and recommendable to apply the MEL model to obtain accurate and useful results for estimating the flow of hydroelectric plants, thus contributing to more efficient and sustainable management of water resources in the face of ongoing climate change.

References

1. Fadiga Júnior, F.M., Lopes, J.G., Martins, J.B.D.F., et al.: Modelos de Previsão de Vazões para a Bacia Incremental à UHE Itaipu. Revista Brasileira de Recursos Hídricos **13**(2), 65–75 (2008)
2. Gottschalk, L., Motovilov, Y.: Macroscale hydrological modelling: a Scandinavian experience. In: International Symposium on "Can science and society save the water crisis in the 21 century"—Reports from the World 'Japan Society of Hydrology and Water Resources. Tokyo, pp. 38–45 (2000)
3. OPERADOR NACIONAL DO SISTEMA ELÉTRICO: NT 173/2007: Novo modelo de previsão de vazões com informação de precipitação para o trecho incremental de itaipu. Rio de Janeiro, 63 pp. (2007)

Research on the Design Method of Ultimate Pressure for Compressed Air Energy Storage in Hydroelectric Tunnels

Yue Han(✉), Yaohui Gao, Hao Cui, Ning Liu, and Pingzhi Chen

PowerChina Huadong Engineering Corporation Limited, Hangzhou 310014, Zhejiang, China
han_y2@hdec.com

Abstract. The design of storage pressure is an indispensable step for reusing the existing hydropower tunnels into compressed air storage cavern, which directly affects the safety and economy of the plant. To design appropriate pressure, firstly the three anti lifting theoretical models are adopted to establish the analytical ultimate pressure methods in this paper and the ultimate pressure obtained under different parameters are compraed and anyasided. Based on multi-layer thick-walled cylinder theory, an improved calculation method for the ultimate pressure is proposed. Then parameter sensitivity analysis are conducted to reveal the key coefficient affecting the ultimate pressure corresponding to the model. Finallly, the proposed method is applied in the traffic tunnel of Baihetan Hydropower Station as an example.The results show that the straight failure plane (SFP) model is relatively conservative, resulting in lower ultimate pressure. The shear resistance of the conical failure surface surface will increase the ultimate pressure. The newly proposed model takes into account the influence of structures, and obtains a more reasonable ultimate pressure by limiting the tension stress of the steel lining. The surrounding rock's elastic modulus and burial depth exert the most significant influence on the ultimate pressure. In contrast, the effects of tunnel radius, steel grade, and concrete lining thickness are of moderate importance. The impact of steel lining thickness, concrete grade, Poisson's ratio of the surrounding rock, and surrounding rock density on the ultimate pressure is comparatively minor. The ultimate pressure of the Baihetan traffic tunnel by this method is 23.6 MPa. This study is of great significance for the renovation of compressed air storage in hydropower tunnles.

Keyword: Ultimate pressure · Compressed air energy storage · Hydroelectric tunnels

1 Introduction

Compressed air energy storage(CAES) is a long-term and large-scale physical energy storage technology with short construction period, pollution-free, and low cost [1]. Due to the large volume and high internal pressure of gas storage facilities, underground storage caverns for CAES can be advantageous from a safety and an economical perspective [2].

S. Zheng et al. (Eds.): IHDC 2024, LNCE 487, pp. 537–551, 2025.
https://doi.org/10.1007/978-981-97-9184-2_44

The underground caverns of hydropower stations with huge space and good geological conditions, can provide high-quality storage resources for compressed air. Reusing the hydroelectric tunnels is one of the innovative ways to achieve China's 'dual carbon' goals, which can save excavation and support costs for tunnels compared to newly-built tunnels. One of the important design aspects of underground pressurized caverns is the ultimate pressure against ground uplift [3]. If overburden rock masses are not sufficiently strong enough to resist the upward lifting pressure due to the internal high pressurized compressed air, crack may generate at the cavern periphery and result in overall instability of the cavern.Therefore, anti uplift safety is a method to constrain internal pressure.

The anti uplift criterion is mainly established in the field of hydroelectric tunnels, such as the common Norwegian criterion [4], which requires the weight of the overlying rock mass to be no less than the vertical uplift pressure acting on the surrounding rock area of the tunnel. However, this criterion ignores the influence of the strength and geostress of the rock mass. The rigid-cone model have been introduced in determining the depth of pressurized storage cavern to prevent from overburden rock mass failure [5], the cone angle (α) is taken as 30° or 45°. Except for the cone angle, the model does not account for any rock mass strength. The log-spiral model have proposed based on the response of soils in resisting a pull-out of soil anchor at shallow depth [6]. In addition to overburden rock mass weight, the model also includes the resistance from friction along the log-spiral failure surface. Kim [7] assume that failure plane is straight upward to ground surface, and shear resistant force on the failure plane and buoyant force were considered. Tunsakul [8] studied the failure plane of storage tunnel through model experiments, and the results showed that the failure was approximately conical in shape. On this basis, Xu et al. [9] derived the function of the uplift failure surface, refined the shape of the fracture surface, and considered factors such as rock friction and cohesion. Sun et al. [10] controlled the ultimate equilibrium state of the cone model through the initiation angle and the obliquity of failure plane. The above research results basically supported the assumption that the failure plane was conical. Collapse shape of shallow circular tunnel is derived by Yang and Huang [11] using a new curved failure mechanism based on Nonlinear Hoek–Brown failure criterion. But the failure mode for traffic tunnel is different from that of high internal pressure tunnels. Carranza-Torres [12] assumed the angle offailure plane is $45° - \varphi/2$, which conforms to the Mohr Coulomb criterion, and established a limit pressure model considering the normal stress and shear force.

Except that the imit equilibrium analysis against ground uplift, The stress state of the tunnel has also attracted attention, Wang et al. [13] established a theoretical analytical solution for a circular tunnel, and obtained the upper and lower limit pressure relationship equations based on the Mohr Coulomb strength criterion. But the above method did not consider the actual high-pressure gas acting on the lining structure. In the design of pressure steel pipes for hydropower stations, the steel lining is designed considering the bearing capacity of the steel lining structure under water pressure [14]. However, the model neglected the influence of geostress, which is not related to burial depth.

In this paper, three typical anti lifting theoretical models are listed, and three ultimate pressure methods are established for comparative analysis. A theoretical model is established based on the multi-layer thick-walled cylinder theory. Based on the analytical solutions for its stress of the steel lining, a new ultimate pressure solution is provided by limiting the hoop tension of steel lining.Parametric sensitivity analysis is conduted to reveal the key factor for ultimate pressure. Finallly, the ultimate pressure of Baihetan hydropower station is calculated by the new method.

2 Anti Lifting Theoretical Model

2.1 The Rigid-Cone Model

As presented in Fig. 1, the tunnel is assumed to have a circular cross-section. This model does not consider the influence of the strength and geostress of the rock mass, only the weight of the overlying circular abutment. The cone angle is taken as 30°. According to the ultimate equilibrium between the overlying gravity of the rock mass and the internal pressure of the tunnel, the expression for the ultimate pressure is obtained as:

$$P_{\max} = \frac{H\gamma\left(2r + H/\sqrt{3}\right)}{2rF_{safe}} \tag{1}$$

where P_{max} is the ultimate pressure, MPa; H is the Burial depth of the tunnel, m; γ is the volume-weight of rock mass, kN/m3; r is the tunnel radius, m; Fsafe is the safety factor.

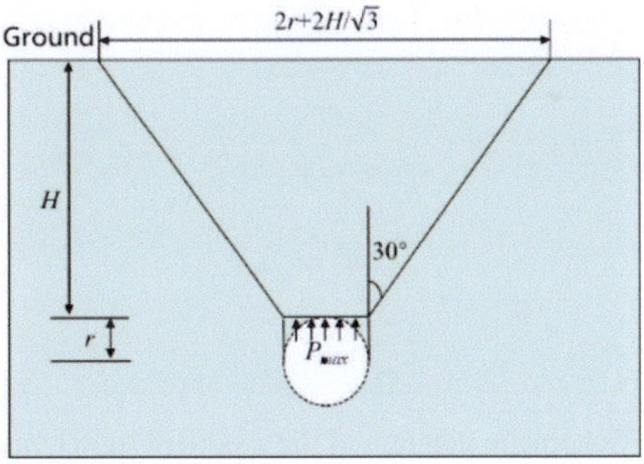

Fig. 1. Schematic diagram of the rigid-cone model

2.2 The Straight Failure Plane Model

Similar to Kim's model, the failure plane is straight upward to ground surface, and only vertical gravity, frictional force and internal pressure are considered. The expression for the ultimate pressure is:

$$P_{\max} = \frac{rH\gamma + cH}{rF_{safe}} \tag{2}$$

where c is the cohesive force of overlying rock mass, MPa.

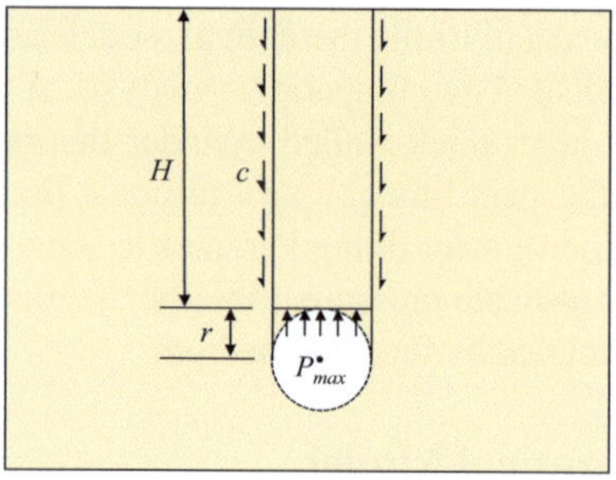

Fig. 2. Schematic diagram of the straight failure plane model

2.3 The Modified Rigid Cone Model

This model is an advanced version of the rigid-cone model, which takes into account the Mohr Coulomb criterion in failure and incorporates the fracture angle and cohesion.

The model diagram is shown in Fig. 2, the overlying rock mass is subjected to its own weight.The BC and B'C' surfaces subjected to pressure F_{out} and shear resistance T are the failure plane of the overlying rock, and the failure angle is 45° − φ/2. The CC surface is on the ground and is not affected by external forces, The BB 'surface is subjected to gas pressure, and the shear force at the fracture surface satisfies the Mohr-Coulomb strength criterion. The pressure can be expressed as (Fig. 3).

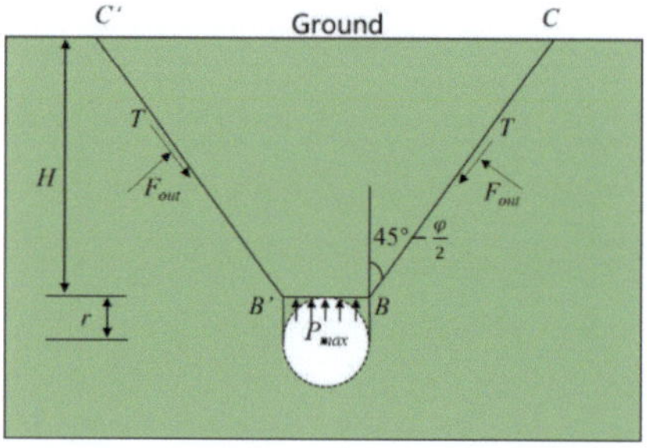

Fig. 3. Schematic diagram of the modified rigid cone model

$$P_{\max} = \frac{\alpha H^2 + \beta H}{F_{safe}} \tag{3}$$

The coefficient α and β are:

$$\alpha = \frac{\gamma \tan\left(45^\circ - \frac{\varphi}{2}\right)}{2r} + \frac{\lambda y\left[\tan\varphi - \tan\left(45^\circ - \frac{\varphi}{2}\right)\right]}{2r\left[1 + \tan\varphi \tan\left(45^\circ - \frac{\varphi}{2}\right)\right]};$$

$$\beta = \gamma + \frac{c \sec^2\left(45^\circ - \frac{\varphi}{2}\right)}{r\left[1 + \tan\varphi \tan\left(45^\circ - \frac{\varphi}{2}\right)\right]} \tag{4}$$

where λ is the coefficient of lateral pressure; φ is the internal friction angle of the surrounding rock, °.

2.4 Calculation Scheme and Result Analysis

Basic value used in the calculations for ground uplift models. The radius r of the CAES tunnel involved in the equation is 5m, and the burial depth H is 100 m. The volume-weight γ of the rock mass is 27 kN/m^3, the internal friction angle of the rock mass φ is 40°, and the cohesive force c is 1.5 MPa.

A factor of safety of 2.0 is chosen in the present analysis, and this is a common factor of safety both for slope stability design using limit equilibrium [7] and in tunnel design to ensure safety against potentially falling ground [10].

Compare the above three methods by substituting parameter values into the above equations. The ultimate pressures for the RC model, the SFP model and the MRC model are 9.14 MPa, 1.36 MPa and 11.99 MPa. The obtained ultimate pressures for the SFP model is the smallest of the three, which means the vertical failure surface is too conservative. The conical failure surface is more suitable for CAES shallow buried artificial tunnels. Comparing the RC model and the MRC model, MRC model has slightly higher values than the RC model, and the pressure and shear resistance of BC and B'C 'surfaces will increase the ultimate pressure. Essentially, both geostress and rock strength can promote the ultimate pressure.

Parametric sensitivity analysis are carried out to investigate the key influencing parameters. Figure 4(**a**) shows the ultimate pressures for three ground uplift models at different depths. The results of RC model and the MRC model are similar. As the burial depth increases, the pressure increases significantly. When the tunnel at a depth of 280m, the ultimate pressure is greater than 60 MPa, and even up to 90 MPa. It is too overstated.

Only RC model and the MRC model are selected to conduct sensitivity analysis, and the results are shown in the Fig. 4(b)-(f). The ultimate pressures of the two models decreases with the increase of radius, and the slope of decrease fast firstly and then slow. The ultimate pressures increase with the volume-weight of rock mass increases moderately. In the Fig. 4(d)-(f), Only the MRC model considers the lateral coefficient and rock strength parameters. As the lateral pressure coefficient or internal friction angle of the rock mass increases, the ultimate pressure increases. Among the two, the influence of friction angle is more significant. However, the cohesive force is exceptional,and the ultimate pressure is almost unaffected by the cohesive force.

In design, strong influence parameters should be given priority consideration. Therefore, in the anti lifting methods, the ultimate pressure design should prioritize the burial depth and radius.

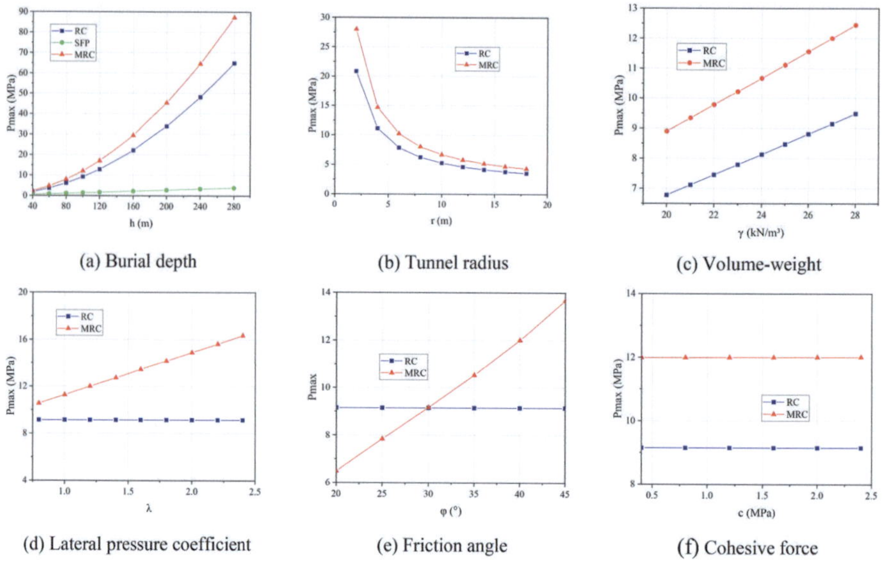

Fig. 4. The influence of different parameters on the ultimate pressure of anti lifting theoretical models

3 The Newly Proposed Model

Although the above anti lift models are simple and easy to generalize for solving ultimate pressure, the safety factor of ultimate equilibrium has uncertainty and the fine load-bearing structures inside the tunnel have not been considered. A theoretical model for the ultimate pressure is newly proposed based on the multi-layer stuctures.

3.1 Model Definition

Steel lining and concrete lining together with surrounding rock form a load-bearing structure to resist high internal pressure for artificial lining CAES cavern. Hence, a theoretical model is established using the multi-layer thick-walled cylinder theory.

The force model of lined tunnel is shown in the Fig. 5, simplify the CAES tunnel into a plane strain problem of a three-layer structure consisting of steel lining, concrete lining, and surrounding rock. The structures contact with each other, and the parameters (r_0, r_1, r_2, r_3) represent the inner and outer radius of each structure, and a sufficient distance of r_3 represents is not affected by the storage pressure.

To simplify theoretical calculations, the lateral pressure coefficient is set to be 1. The model is transformed into a symmetric solution problem for multi-layer thick walled cylinders under internal and external pressure.

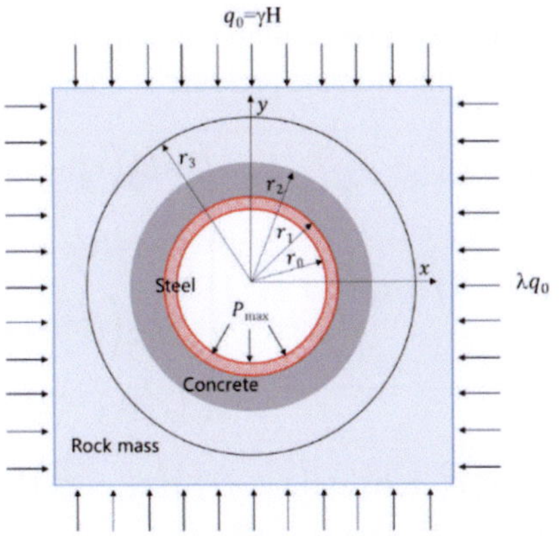

Fig. 5. Force diagram of tunnel type lined gas storage caverns

The analytical solutions for its stress and deformation of a three-layer thick walled cylinder subjected to axisymmetric loads are below:

$$\begin{cases} \sigma_{r1} = \frac{A_1}{r^2} + 2C_1; \sigma_{\theta 1} = -\frac{A_1}{r^2} + 2C_1;\ u_{r1} = \frac{1}{E_1'}\left[-\left(1 + v_1'\right)\right]\frac{A_1}{E_1'} + 2\left(1 - v_1'\right)C_1 r \\ \sigma_{r2} = \frac{A_2}{r^2} + 2C_2; \sigma_{\theta 2} = -\frac{A_2}{r^2} + 2C_2;\ u_{r2} = \frac{1}{E_2'}\left[-\left(1 + v_2'\right)\right]\frac{A_2}{E_2'} + 2\left(1 - v_2'\right)C_2 r \\ \sigma_{r3} = \frac{A_3}{r^2} + 2C_3; \sigma_{\theta 3} = -\frac{A_3}{r^2} + 2C_3;\ u_{r1} = \frac{1}{E_3'}\left[-\left(1 + v_3'\right)\right]\frac{A_3}{E_3'} + 2\left(1 - v_3'\right)C_3 r \end{cases} \tag{5}$$

where subscripts 1, 2, and 3 represent steel lining, concrete lining, and surrounding rock, respectively. σ_{r} is the radial stress; σ_{θ} is the hoop stress; u_{r} is the radial displacement. $E' = E/\left(1 - v^2\right)$ and $v' = v/(1 - v)$ are the elastic modulus and Poisson's ratio required for analyzing plane strain problems, respectively. And E,v are the elastic modulus and Poisson's ratio,repectively. A and C are the undetermined coefficients determined by the boundary conditions.

The stress boundary conditions and displacement boundary conditions of this model are following.

$$\begin{cases} (\sigma_{r1})_{r=r_0} = -P_{\max}; \quad (\sigma_{r3})_{r=r_3} = -q_0 \\ (\sigma_{r1})_{r=r_1} = (\sigma_{r2})_{r=r_1}; (\sigma_{r2})_{r=r_2} = (\sigma_{r2})_{r=r_2} \\ (u_{r1})_{r=r_1} = (u_{r2})_{r=r_1}; (u_{r2})_{r=r_2} = (u_{r2})_{r=r_2} \end{cases} \tag{6}$$

Substitute the stress function and displacement function into the boundary conditions, and solve for six coefficients based on six equations. Then, substitute the coefficients into the stress function and to obtain the stress analytical solution.

By constraining the hoop tensile stress of the steel lining, when $(\sigma_{\theta 1})_{r=r_0} \le [\sigma]$, the steel lining remains intact and the ultimate pressure is obtained as:

$$P_{\max} = \frac{\left(\begin{array}{l} E_2'[\sigma]\left(-r_1^2(v_1'-1)+r_0^2(1+v_2')\right)\left(\begin{array}{l} -E_3'\left(r_2^2-r_3^2\right) \\ \left(-r_2^2(v_2'-1)+r_1^2(v_2'+1)\right) \\ +E_2'\left(r_1^2-r_2^2\right) \\ \left(r_2^2(v_3'-1)-r_3^2(v_3'+1)\right) \end{array}\right) \\ +E_1'\left(\begin{array}{l} 8E_2'q_0r_1^2r_2^2r_3^2+E_3'[\sigma]\left(r_0^2-r_1^2\right)\left(r_1^2-r_2^2\right)\left(r_2^2-r_3^2\right)\left(v_2'^2-1\right) \\ -E_2'[\sigma]\left(r_0^2-r_1^2\right) \\ \left(r_1^2(v_2'-1)-r_2^2(v_2'+1)\right)\left(r_2^2(v_3'-1)-r_3^2(v_3'+1)\right) \end{array}\right) \end{array}\right)}{\left(\begin{array}{l} -E_2'\left(r_1^2(v_1'-1)+r_0^2(1+v_1')\right)\left(\begin{array}{l} -E_3'\left(r_2^2-r_3^2\right) \\ \left(-r_2^2(v_2'-1)+r_1^2(v_1'+1)\right) \\ +E_2'\left(r_1^2-r_2^2\right) \\ \left(r_2^2(v_3'-1)-r_3^2(v_3'+1)\right) \end{array}\right) \\ +E_1'\left(r_0^2-r_1^2\right)\left(\begin{array}{l} -E_3'\left(r_1^2-r_2^2\right)\left(r_2^2-r_3^2\right)\left(v_2'^2-1\right) \\ +E_2'\left(r_1^2(v_2'-1)\right)-\left(r_2^2(v_2'+1)\right) \\ \left(r_2^2(v_3'-1)-r_3^2(v_3'+1)\right) \end{array}\right) \end{array}\right)} \quad (7)$$

where $[\sigma]$ represents the allowable tensile stress of steel lining. For different grades of steel, this value varies.

3.2 Model Analysis

The values of basic parameters are listed in Table 1,the number of parameters is significantly greater than that of the anti lifting model. Substitute the parameter values into the Eqs. (6), and the calculated ultimate pressure is 11.81MPa, which is close to that of MRC model. Both of these are in line with the pressure design of actual engineering at a depth of 100m.

To invetigate the key the influence factors on the ultimate pressure, parametric sensitivity analysis are carried out such as burial depth, inner radius, steel thickness, allowable tensile stress, elastic modulus of rock(According to the actual situation, the deformation modulus data is used), poisson's ratio of rock, volume weight of rock, concrete type and lining thickness. Among them, allowable tensile stress, steel thickness, lining thickness and concrete type belong the design phase, and can be reconstructed according to the

Table 1. Basic parameters of joint bearing theory based on the multi-layer stuctures

Parameters	Units	Value	Parameters	Units	Value
Burial depth, H	m	100	Elastic modulus of steel,E_1	GPa	206
Inner radius, r_0	m	5	Elastic modulus of lining, E_2	GPa	30
Volume weight, γ	kN/m^3	27	Elastic modulus of rock,E_3	GPa	5
Steel thickness, r_1-r_0	mm	18	Poisson's ratio of steel, v_1	/	0.3
Lining thickness,r_2-r_1	m	0.5	Poisson's ratio of lining, v_2	/	0.2
Allowable tensile stress, $[\sigma]$	MPa	235	Poisson's ratio of rock,v_3	/	0.26

existing hydropower tunnels. The other parameters are related to site selection, that is, pick suitable tunnels from existing ones. The variation schemes are shown in the Table 2, and the results under various parameters are shown in Fig. 6.

Table 2. Calculation schemes for parametric sensitivity analysis

Parameters	Units	Calculation schemes	Phase
Burial depth, H	m	40: 20: 300	Site selection
Inner radius, r_0	m	3: 1: 20	Site selection
Steel thickness, r_1-r_0	mm	1: 2: 29	Design
Allowable tensile stress, $[\sigma]$	MPa	235, 305, 385	Design
Elastic modulus of rock,E_3	GPa	3: 1: 15	Site selection
Poisson's ratio of rock,v_3	/	0.15:0.05:0.4	Site selection
Volume weight of rock, γ	g/cm^3	2300: 100: 2800	Site selection
Concrete type	/	C20, C25, C30, C35, C40	Design
lining thickness,r_2-r_1	m	0.4: 0.1: 1.2	Design

In Fig. 6(a), the ultimate pressure increases linearly with increasing burial depth. The ultimate pressure corresponding to 50m is 9.66 MPa, while the ultimate pressure corresponding to 200 m increases to 16.07 MPa, with a growth rate of 4.27 MPa/100m. The change pattern is different from that of the MRC model which increases nonlinearly. Especially when the burial depth exceeds the intersection of two lines, the ultimate pressure of the MRC model rapidly increases. In contrast, the ultimate pressure of the new proposed model under large burial depths is more reliable.

As shown in the Fig. 6(b), the ultimate pressure gradually decreases as the radius increases. The change trend is similar to that of the MRC model, but the change magnitude is significantly smaller. The the ultimate pressure for the proposed model of a 10 m radius is 9.82 MPa, while the ultimate pressure of a 20m radius is 8.44 MPa. The smaller the radius, the more advantageous it is in terms of steel stress.

Three different steel types are selected, with corresponding tensile strengths of 500 MPa, 600 MPa, and 800 MPa, respectively. Considering the yield strength and structural coefficient, the allowable stress values for the three types are set to 235 MPa, 305 MPa, and 385 MPa, respectively. As shown in the Fig. 6 **(c)**, the ultimate pressure increases significantly with the increase of allowable stress. Compared with 500 MPa grade steel, the ultimate pressure of 800 MPa grade steel lining can be increased by about 4.8 MPa. The increase in operating pressure can result in more energy being stored and a larger installed capacity of the power plant under the same volume. Consequently, the installed capacity can be increased by upgrading the steel material under a limited tunnel volume.

In Fig. 6(d), the ultimate pressure is not highly sensitive to the steel thickness. When the thickness increases from 1mm to 29mm, the ultimate pressure only increases by about 1.3 MPa. When the thickness of the steel is infinitely close to 0, the ultimate pressure can also reach 11 MPa. It follows that, while ensuring welding sealing, the thickness of the steel lining can be minimized as much as possible.The ultimate pressure is positively correlated with the the lining thickness, seen in Fig. 6(e), and thickening the concrete lining will increase the ultimate pressure. When the thickness increases from 0.4m to 1.2m, the ultimate pressure can increase by 4 MPa. Figure 6(f) shows the effect of elastic modulus of surrounding rock. The ultimate pressure is very sensitive to the elastic modulus of the surrounding rock and shows a linear growth relationship. When the elastic modulus of the rock mass is higher than 10GPa, the ultimate pressure can exceed 15 MPa. However, when the elastic modulus of the rock mass approaches zero infinitely, and the ultimate pressure tends to 8.77 MPa. This means that in order to ensure high-pressure operation, it is essential to select good rock mass. Fortunately, the hydropower station has screened the site, and the tunnel has natural advantages when used for CAES.

Shown in the Table 3, different concrete grades correspond to different elastic modulus. The ultimate pressure increases with the increase of concrete elastic modulus, but the slope is small (seen Fig. 6(g)). When changing from C20 to C70, the ultimate pressure only increased by 1.2 MPa, but as the grade increases, the price increases significantly. Hence, it is not recommended to increase the maximum pressure by increasing the concrete grade.

Seen from Fig. 6(h), the ultimate pressure is also affected by the Poisson's ratio of the surrounding rock. The larger the Poisson's ratio, the more obvious the hoop deformation of the surrounding rock, resulting in a more significant tensile stress and a smaller ultimate pressure. When other mechanical parameters are consistent, the ultimate pressure increases with increasing volume weight of rock(in Fig. 6(i)). The mechanism is similar to that of burial depth, both of which increase the rock confining pressure, but the effect of rock volume weight is weak.

3.3 Comparison of Parameter Sensitivity

Comparing the effects of nine parameters on the ultimate pressure comprehensively, seen from the Fig. 7, that the elastic modulus of surrounding rock (E_3) and burial depth(H) are the most critical parameters. In existing hydroelectric tunnels, selecting good surrounding rocks and tunnels with larger burial depths have significant advantages for high-pressure storage of compressed air. Similarly, areas with strong geostress is also

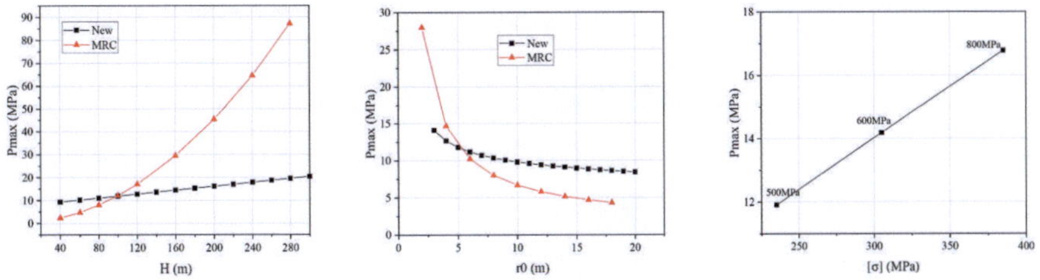

(a) Burial depth (b) Inner radius (c) Allowable tensile stress of steel

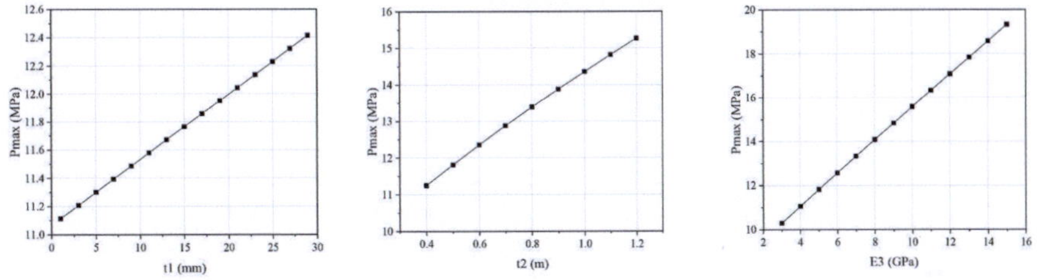

(d) Steel thickness (e) Thickness of the concrete lining (f) Elastic modulus of rock

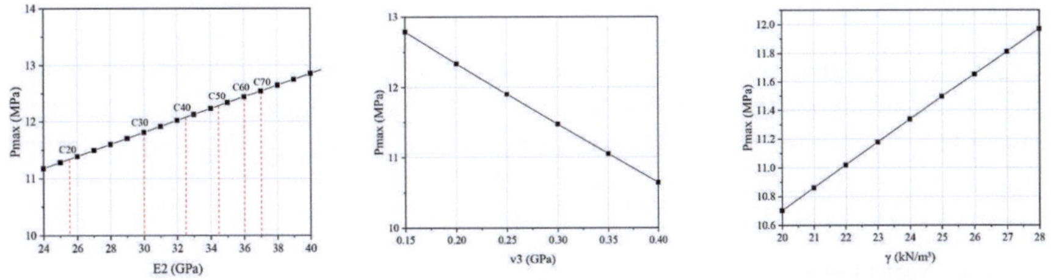

(g) Concrete type (h) Poisson's ratio of rock (i) volume weight of rock

Fig. 6. The influence of different parameters on the ultimate pressure of the proposed model

Table 3. Elastic modulus and ultimate pressure of different concrete

Rank	C20	C30	C40	C50	C60	C70
E2	25.5	30	32.5	34.5	36	37
$P_{\max}$	11.34	11.81	12.05	12.29	12.43	12.54

conducive to high-pressure storage.Secondly, the radius, allowable tensile stress, and the thickness of the concrete lining are sensitive factors that affect the ultimate pressure. Among the these, the allowable tensile stress of the steel (i.e. steel grades) and the thickness of the concrete lining have more designable space. Finally, the thickness of the steel lining, concrete grade, the Poisson's ratio of the surrounding rock, and the density of the surrounding rock have a weak effect on the ultimate pressure.

Based on the effects of above parameters, it can be found that the properties of the existing hydropower tunnels basically determine the ultimate pressure they can withstand, such as burial depth, tunnel radius, and surrounding rock quality. The renovation

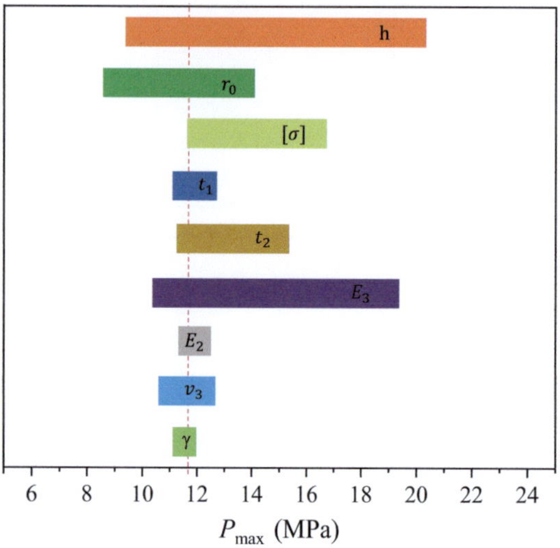

Fig. 7. Tornado chart of parameter sensitivity analysis

mainly focuses on the steel grade and concrete lining thickness. To select the optimal improvement and enhancement measures, it is necessary to compare the economic benefits brought by increasing the storage pressure and the cost of increasing lining thickness or upgrading the steel grade.

4 Case Analysis

The Baihetan Hydropower Station is the second largest hydropower station in the world, and also the largest underground power station in the world. The power station adopts an all underground factory layout, and the types of tunnels not only include water diversion and power generation buildings, but also some auxiliary tunnels such as construction support tunnels, traffic tunnels, irrigation and drainage corridors. After the power station is put into operation, a large number of tunnels are idle, especially construction and traffic support tunnels.

A tunnel with a length of 640m and a section size of 8.2 m × 8.3 m was constructed at a certain location of the power station. The section was initially sprayed with 5cm steel fiber concrete and then sprayed with 10cm thick C25 concrete, finally, a 50cm thick C25 concrete lining was formed. The inner wall city gate is 6.9 m wide and 7.65 m high.

Considering the circular shape of the steel lining and the uniform thickness of the concrete lining, the inner radius r_0 is designed to be 3.35 m (shown in Fig. 8). The thinnest part of C25 concrete lining is 75cm. The buried depth of the construction adit is about 240 m, and the third principal stress is nearly vertical, with a value equivalent to the weight of the overlying rock mass. The surrounding rock is basalt, with a density of 2.70 g/cm^3, a deformation modulus of 10-12GPa, and a Poisson's ratio of 0.25.

Based on the information of rock mass and cross-sectional dimensions, 500 MPa grade steel lining is used here, and the thickness of the steel lining can meet the welding sealing thickness (10mm). The ultimate pressure obtained by the MRC model is

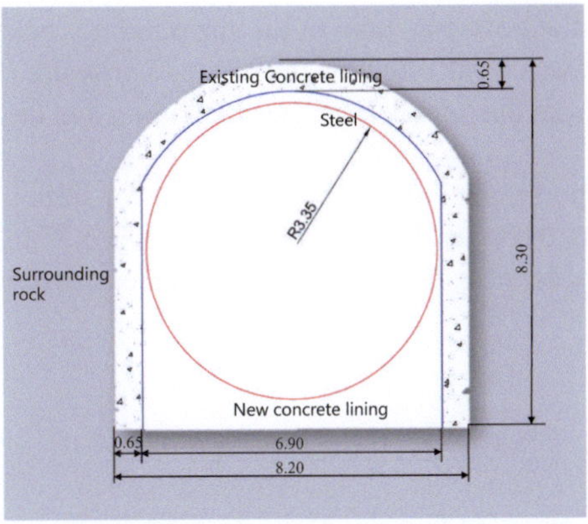

Fig. 8. Schematic diagram of tunnel renovation for CAES

94.72 MPa, which is unscientific. The ultimate pressure calculated by the proposed model that can be stored in the construction support tunnel is 23.6 MPa. Compared to the anti lifting theory, this result obtained by the proposed model is more reasonable.

5 Conclusions

In this study, three models based on anti lifting theoretical and a new proposed model based on the multi-layer thick-walled cylinder theory are summarized and compared to investigate the ultimate pressure for compressed air storage in hydroelectric tunnels. The influences of sensitive parameters such as the radius, rock quality and steel thickness on the ultimate pressure of the CAES are analyzed. The key findings are summarized as follows:

1. Among anti lifting theoretical models, the SFP model is relatively conservative, resulting in lower ultimate pressure. The conical failure surface is more reasonable, and the shear resistance of the failure surface will increase the ability of the rock mass to resist internal pressure and increase the ultimate pressure resistance to uplift. The ultimate pressure of anti lifting theoretical models is most affected by burial depth, while the radius, bulk density, friction angle/lateral pressure coefficient decrease in sequence. The influence of cohesion on ultimate pressure is almost negligible.
2. Comparing the anti lifting theoretical models, the newly proposed model takes into account the combined bearing capacity of steel lining, concrete lining and surrounding rock, as well as geostress. The allowable stress of steel lining is adopted for the critical condition, which is more in line with the actual failure situation. The ultimate pressure obtained by the new model is more reasonable.
3. The quality of surrounding rock and burial depth have the greatest impact on the ultimate pressure of the new proposed model, which are key factors in the selection of hydropower tunnels. Secondly, the ultimate pressure is sensitive to the tunnel radius, steel grade, and concrete lining thickness and these parameters should be carefully planned during the later renovation design. The thickness of the steel lining,

the concrete grade, the Poisson's ratio of the surrounding rock, and the density of the surrounding rock have a weak effect on the ultimate pressure. Therefore, cost control can be carried out from the thickness of the steel lining and the concrete grade during the renovation.

4. A construction support tunnel of Baihetan hydropower station is selected for CAES. After the renovation, the inner diameter of the tunnel is 3.35 m, using 500 MPa grade steel lining with a thickness of 10mm. The ultimate pressure obtained by the new model is 23.6 MPa.

References

1. Zhang, G., Li, Y., Daemen, J.J.K., et al.: Geotechnical feasibility analysis of compressed air energy storage (CAES) in bedded salt formations: a case study in Huai'an City, China. Rock Mech. Rock Eng. **48**(5), 2111–2127 (2015)
2. Li, H., Ma, H., Zhao, K., et al.: Parameter design of the compressed air energy storage salt cavern in highly impure rock salt formations. Energy 286 (2024)
3. Han, Y., Ma, H., Yang, C., et al.: A modified creep model for cyclic characterization of rock salt considering the effects of the mean stress, half-amplitude and cycle period. Rock Mech. Rock Eng. **53**(7), 3223–3236 (2020)
4. Basnet, C.B., Panthi, K.K.: Analysis of unlined pressure shafts and tunnels of selected Norwegian hydropower projects. J. Rock Mechanics Geotech. Eng. **10**(3), 27 (2018)
5. Brandshaug, T., Christianson, M., Damjanac, B.: Technical review of the lined rock cavern (LRC) concept and design methodology: mechanical response of rock mass. Itasca consulting group, Inc., Minnesota (2001)
6. Ghaly, A., Hanna, A.: Ultimate pullout resistance of single vertical anchors: reply. Can. Geotech. J. **32**(6), 1093–1094 (1995)
7. Kim, H.M., Park, D., Ryu, D.W., et al.: Parametric sensitivity analysis of ground uplift above pressurized underground rock caverns. Eng. Geol 135–136(none):60–65 (2012)
8. Tunsakul, J., Jongpradist, P., Soparat, P., et al.: Analysis of fracture propagation in a rock mass surrounding a tunnel under high internal pressure by the element-free Galerkin method. Comput. Geotech. **55**, 78–90 (2014)
9. Yingjun, X., Caichu, X., Shuwei, Z., et al.: Anti-uplift failure criterion of caverns for compressed air energy storage based on the upper bound theorem of limit analysis. Chin. J. Rock Mech. Eng. **41**(10), 1971–1980 (2022). (in Chinese)
10. Sun, G., Wang, Z., Wang, J., et al.: Limit equilibrium method for calculating the safe burial depth of underground caverns in compressed air energy storage [J]. Chin. Civil Eng. J. **56**, 67–77 (2023)
11. Yang, X.L., Huang, F.: Collapse mechanism of shallow tunnel based on nonlinear Hoek-Brown failure criterion[J]. Tunn. Undergr. Space Technol. **26**(6), 686–691 (2011)
12. Carranza-Torres, C., Fosnacht, D., Hudak, G.: Geomechanical analysis of the stability conditions of shallow cavities for Compressed Air Energy Storage (CAES) applications[J]. Geomechanics and Geophysics for Geo-Energy and Geo-Resources **3**(2), 131–174 (2017)
13. Wang, Z., Jia, W., Feng, X., et al.: Analytical solution of limit storage pressures for tunnel type lined gas storage caverns[J]. Chinese Journal of Theoretical and Applied Mechanics **55**(3), 710–718 (2023)
14. Released by the National Development and Reform Commission of the People's Republic of China, Design Specification for Pressure Steel Pipes for Hydroelectric Stations DL/T 5141–2001 (2008)

Water Level Calculation and Influencing Factors of Single-Step Locks with Water-Saving Basins

Duo Xu, Zhonghua Li(✉), and Jianfeng An

Nanjing Hydraulic Research Institute, Nanjing, China
zhli@nhri.cn

Abstract. During the water level design process of lock with water saving basins (WSBs), it's an important part that to determine the water level and head of the WSB, and there are many influencing factors in the process. This paper derives and establishes calculation formulas for evaluating the water level and the water head of the single-step lock with WSBs, which integrates the total head of ship lock, the upper and lower stream water level variation, the number of the WSB levels and area of the WSB, and the remaining head. On the basis of formulas, the influence of factors such as the upper and lower stream water level variation, the number of the WSB levels and the area of the WSB on the water level and the water head of the WSB are analyzed, and the control conditions of the integrated and decentralized arrangement of WSBs and water-saving rates of replenishment and overflow operations are calculated. The results show that when total head of the lock is constant, head of the WSB decreases with the increase of the area of the WSB and the number of the WSB levels, and the increase of the number of the WSB levels is greater than the increase of the area of the WSB; when the number of levels and area of the WSB increase to a certain level, the reduction of the water head of the WSB becomes slow; the highest and lowest water levels of the WSB are controlled by the maximum variation of the downstream and upstream water levels respectively. Finally, the water levels and the water head of difference WSB schemes are calculated and compared in a 60m single-level ship lock, and a reasonable water level scheme of WSB is proposed.

Keywords: navigation structure · lock with water-saving basins · water level scheme · water-saving basins (WSBs) · water-saving rate

Water-saving ship locks are usually equipped with more than one stage of water-saving basins on one or both sides of the lock. When the lock is in the discharge process, the lock chamber first discharges to the high water-saving basin (basin A), then discharges to low water-saving basins (basins B and C), and the remaining water discharges to the downstream channel; The order of water filling is opposite to the order of water discharge. First, the low water-saving basin (basin C) water is injected into the lock chamber, and then high water-saving basins discharge to the lock chamber in turn. The insufficient water is finally replenished by the upstream channel, as shown in Fig. 1. The water-saving ship lock can reduce the water consumption of lock operations to improve the rate of water resource utilization and reduce the working water head. For high-head locks, it has the advantage of reducing the technical difficulty of solving the working

S. Zheng et al. (Eds.): IHDC 2024, LNCE 487, pp. 552–564, 2025.
https://doi.org/10.1007/978-981-97-9184-2_45

conditions of the valve and simplifying the layout of the lock water delivery system. The water-saving ship lock has a good application prospect in navigating artificial canals and high dams.

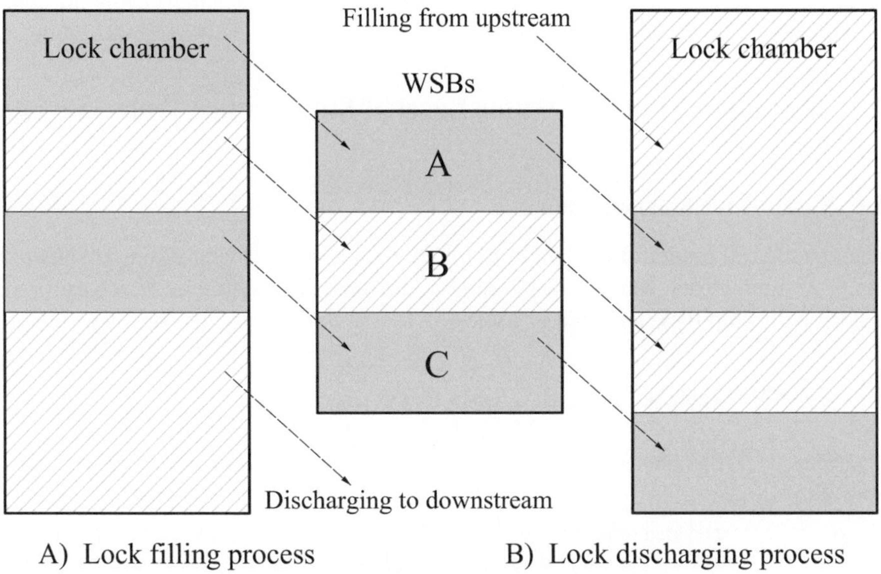

Fig. 1. Working principle of water-saving ship lock.

Water-saving ship locks have been used in Europe for hundreds of years. In the 19th century, the French Callelli first carried out the research for the water-saving ship lock on the Loire River [1]. Germany is the country with the largest number of water-saving ship locks in the world. On the Rhine ~ Danube canal, a total length of 171km from Bamberg on the Rhine River to Kelheim on the Danube River, with a water level difference of 243m, a total of 16 locks have been built, of which 14 are water-saving ship locks [2] There are fewer examples of water-saving ship lock project in China. There are almost no water-saving ship locks in the domestic graded waterway. In recent years, water resources have become increasingly scarce in our country, so the need to save water in ship locks is very urgent. In recent years, Jinjiayan Water-saving Ship Lock and Shuiniuhan Water-saving Ship Lock on Xiaoqing River in Shandong Province have been built, a large number of water-saving ship locks, such as Bajiangkou Second-line Ship Lock on Gui River in Guangxi Province, and the Baishi Lock on Qingshui River in Guizhou Province [3–7], have entered the design stage successively. At present, the influence of different basin areas and different stages of saving basin on water level, working head, and water-saving rate of each stage of water-saving basin is mainly considered in the design of water-saving ship lock [8, 9]. Comparing the characteristics of water-saving ship locks at home and abroad, we can see that the water-saving ship locks abroad are mainly built on canals where there are little variations in upstream and downstream water levels. Water-saving ship locks in China are mainly built in natural rivers, and variations of upstream and downstream water levels are large, the influencing factors for the classification of water-saving basins are more complicated, especially for the water-saving ship lock with integrated water-saving basins. The integrated design has more stringent requirements for upstream and downstream water levels variations of the lock, and the calculation of the water level of the water-saving ship lock and its influencing factors are more complicated.

1 Calculation Model of Water Level of Single-Stage Water-Saving Ship Lock

1.1 Basic Water Balance Equations

The generalized model of the single-step lock with n water-saving basins is shown in Fig. 2 Assuming each water-saving basin has the same area, symbolic variables for the water-saving ship lock are listed in Table 1.

Table. 1. Symbolic variables for the water-saving ship lock.

Symbolic Variables	Definitions
S_{lock}	The area of the lock (m^2)
S_{wsb}	The area of a water-saving basin (m^2)
k	The ratio of the area of the basin to the area of the lock chamber
i	The stage number of the water-saving basin
$Z_{wu}(i)$	The high water level of stage i water saving basin (m)
$Z_{wd}(i)$	The low water level of stage i water-saving basin (m)
H_{wsb}	The change of water level in the water-saving basin during each stage of water-saving operation (m)
Z_{up}	The upstream water level (m)
Z_{down}	The downstream water level (m)
H_c	The change of water level in the lock chamber during each stage of the water-saving operation (m)
ΔH	The residual head (m)
H_{CL}	The water head after the filling and discharging processes (m)

According to the conservation of water mass in the filling and discharging processes of the water-saving ship lock, as indicated in Fig. 2, the following relationships among the variables above are satisfied:

$$H_c = kH_{wsb} \tag{1.1}$$

$$\sum_{i=1}^{n} H_{c,i} + H_{cL} = nH_c + H_{cL} = Z_{up} - Z_{down} \tag{1.2}$$

$$H_{cL} = H_c + H_{wsb} + 2\Delta \tag{1.3}$$

From the above equations, it can be obtained that the water level variation for each stage of water-saving basins is:

$$H_{wsb} = \frac{Z_{up} - Z_{down} - 2\Delta H}{k(n+1)+1} \tag{1.4}$$

The corresponding water level variation of the lock chamber is:

$$H_c = \frac{k\left(Z_{up} - Z_{down} - 2\Delta H\right)}{k(n+1)+1} \tag{1.5}$$

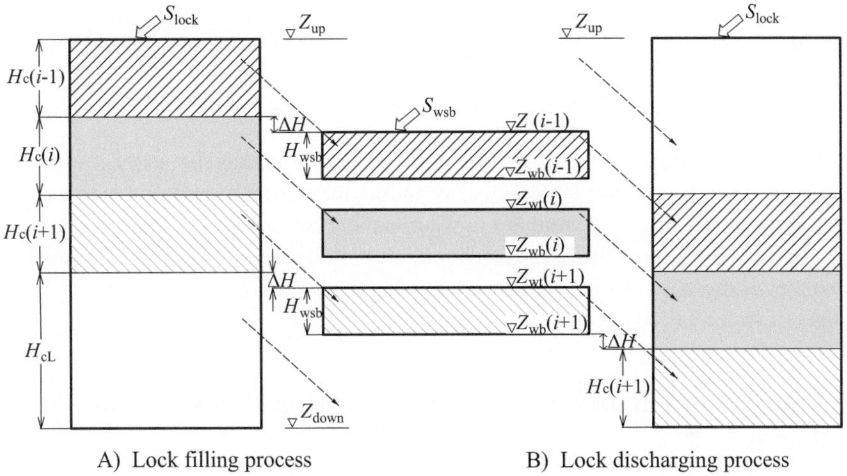

Fig. 2. Generalization of water mass balance in water-saving ship lock operation

1.2 Equations for Calculating Water Levels in Water-Saving Basins

From Fig. 2, the high water level of stage i water saving basin $Z_{wt}(i) = Z_{up} - iH_c - \Delta H$. By substituting Eq. (1.5), it can be concluded that under any downstream or upstream water levels, the high water level of stage i water-saving basin is:

$$Z_{wt}(i) = Z_{up} - \frac{ik\left(Z_{up} - Z_{down}\right) - [k(2i+n+1)+1]\Delta H}{k(n+1)+1} \tag{1.6}$$

In the same way, the low water level of stage i water-saving basin is:

$$Z_{wb}(i) = Z_{up} - \frac{(ik+1)\left(Z_{up} - Z_{down}\right) - [k(2i+n+1)+3]\Delta H}{k(n+1)+1} \tag{1.7}$$

1.3 Equations for Calculating Working Head of Filling and Emptying Operations

We define the working water head between the lock chamber and the water-saving basin as H_{c-wsb} (m), define the correlation coefficient of the water-saving ship lock as $\delta = \frac{1}{\frac{kn}{k+1}+1}$. It can be seen from Fig. 2 that $H_{c-wsb} = H_c + H_{wsb} + \Delta H$, by substituting Eqs. (1.4) and (1.5), the calculation equation of the working water head of the lock chamber and the water-saving basin can be obtained as follows:

$$H_{c-wsb} = \delta\left(H_{lock} - \Delta Z_{up} - \Delta Z_{down}\right) + (1 - 2\delta)\Delta H \tag{1.8}$$

In Eq. (1.8), H_{lock} refers to the maximum working head of the ship lock (m), $H_{lock} = Z_{up-max} - Z_{down-min}$, where Z_{up-max} refers to the upstream highest navigable stage (m), $Z_{down-min}$ refers to the downstream lowest navigable stage (m), ΔZ_{up} and ΔZ_{down} indicate variations of upstream and downstream water levels (m), respectively.

The equation for the water head H_{cL} of between the lock and the upstream or downstream navigation channels can be obtained similarly:

$$H_{cL} = \delta\left(H_{lock} - \Delta Z_{up} - \Delta Z_{down}\right) + (2 - 2\delta)\Delta H \tag{1.9}$$

According to Eqs. (1.6) ~ (1.9), the main factors affecting the water level of lock include maximum working head H_{lock}, variations of upstream and downstream water levels ΔZ_{up} and ΔZ_{down}, the number of water-saving basin stages n, the ratio of the area of the basin to the area of the lock chamber k, the residual head ΔH. The comprehensive influences of these factors are mainly reflected in the two aspects: the working head and heights of water-saving basins.

2 Analyses of Factors Affecting the Working Head of Water-Saving Basins

2.1 The Residual Head ΔH

From Eqs. (1.8) and (1.9), it can be seen that the working head of a single-stage water-saving ship lock is related to the residual head ΔH. From Eq. (1.8), it can be seen that when $\delta = 0.5$, that is, the number of stages n of the water-saving basin and the area ratio k satisfy the relation $k = \frac{1}{n-1}$, the working head H_{c-wsb} between the water-saving basin and the lock chamber is not affected by the residual head ΔH; When $\delta > 0.5$, that is, when $k < \frac{1}{n-1}$, H_{c-wsb} decreases as ΔH increases; When $\delta < 0.5$, that is, when $k > \frac{1}{n-1}$, H_{c-wsb} increases as ΔH increases. In actual engineering, the ratio k of the area between the water-saving basin and the lock chamber is generally greater than 1, and the number of stages n is greater than 2, under this condition ΔH will increase the working head H_{c-wsb} between the water-saving basin and the lock, and may reduce the working head H_{c-wsb} only in the case of the lock with one stage water-saving basin.

To improve the water-saving rate, the residual head value of the general water-saving ship lock is relatively small. When the residual head is 0.5m, k = 1 ~ 2, and n = 1 ~ 5 in a reasonable range of values. Therefore, the residual head has little influence on the operating water head of the water-saving ship lock.

2.2 Effect of Water-Saving Basin Stages N and Area Ratio k

The compositive coefficient δ comprehensively represents the influence of the number of water-saving basin stages n and the area ratio k on the working head. Ignoring the influence of the residual head presence, it can be seen from Eqs. (1.8) and (1.9) that the working head of the lock changes in direct proportion to δ.

As can be seen from Fig. 3, when the area ratio k is fixed, the compositive coefficient δ is inversely proportional to the number of the water-saving basin stages n. The more the number of the water-saving basin stages n is, the smaller the value of the comprehensive coefficient δ is. When the number of reservoir stages is less than 5, the change in the value of the comprehensive coefficient δ is more obvious. When the number of reservoir stages is greater than 5, the reduction of the water-saving working head slows down, and the effect of increasing the number of reservoir stages on reducing the graded working head is not obvious.

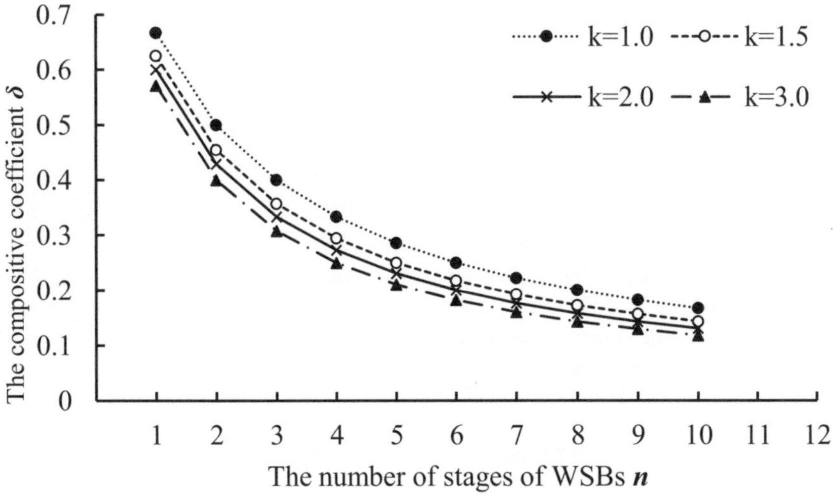

Fig. 3. The relationship between the water-saving basin stages n and the compositive coefficient δ

As can be seen from Fig. 4, when the number of the water-saving basin stages is constant, the compositive coefficient δ is inversely proportional to the area ratio k of the water-saving ship lock. When the k value is greater than 2.0, the effect of increasing the area ratio k on further reducing the working head is not obvious, and the effect of the water-saving basin stages n on the compositive coefficient δ is significantly greater than the k value.

According to the analysis of the compositive coefficient δ, the change in the water-saving basin stages multiplied $\frac{kn}{k+1}$, and $\frac{k}{k+1}$ only varies from 0 to 1. Therefore, the influence of water-saving basin stages n on the working water head of the lock is significantly greater than the area ratio k. Overall, when the water-saving basin stages n > 5, or the area ratio k > 2, the working head has decreased less.

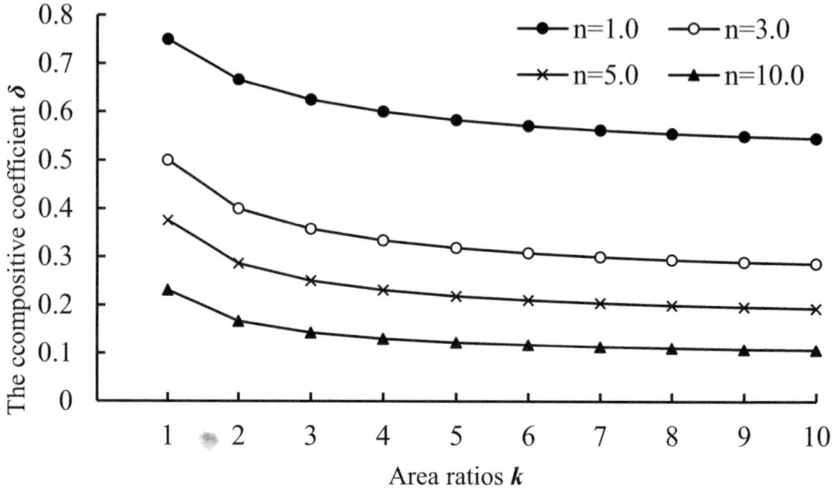

Fig. 4. The relationship between the area ratio k and the compositive coefficient δ

2.3 Effect of Upstream and Downstream Water Levels Variations

Equations (1.8) and (1.9) comprehensively reflect the impact of upstream and downstream water levels and their variations on the effective head of each stage of water-saving basins. The influence of upstream and downstream water levels on the working head of water-saving ship locks in water-saving basins is primarily determined by the maximum working head of the lock and is unrelated to the upstream and downstream water levels. The upstream and downstream water levels variations only affect the water-saving basin's minimum working head.

3 The Factors Affecting the Elevations of Water-Saving Basins

According to Eqs. (1.7) and (1.8), when both upstream and downstream water levels reach their maximum, substituting into Eq. (1.6) yields the maximum water level of stage i water-saving basin.

$$Z_{\text{wt-max}}(i) = Z_{\text{up-max}} - \frac{\text{`}ikH_{lock} - ik\Delta Z_{\text{down-max}} - [k(2i + n + 1) + 1]\Delta H}{k(n + 1) + 1} \tag{3.1}$$

When the upstream and downstream are the lowest water levels, the lowest water level of the stage i water-saving basin can be calculated according to the following equation:

$$Z_{\text{wb-min}}(i) = Z_{\text{up-max}} - \Delta Z_{\text{up-max}} - \frac{(ik + 1)\left(H_{\text{lock}} - \Delta Z_{\text{up-max}}\right) - [k(2i + n + 1) + 3]\Delta H}{k(n + 1) + 1} \tag{3.2}$$

From Eqs. (3.1) and (3.2), it can be seen that when the number of stages n, the area ratio, and the residual head ΔH are constant, the maximum water level of stage i water-saving basin is controlled by the variation of the downstream water level $\Delta Z_{\text{down-max}}$. The greater the variation of downstream water level, the higher the maximum water level of the water-saving basin, and the closer to the downstream (the value of i is smaller), the more obvious. The minimum water level of stage i water-saving basin is controlled by the upstream water level variation $\Delta Z_{\text{up-max}}$. The larger the upstream water level variation, the lower the minimum water level of the water-saving basin, and the closer to the upstream water-saving basin (the value of i is greater), the more obvious.

For a water-saving ship lock of integrated structure, the lowest level of stage i water-saving basin should be higher than the highest level of stage $i + 1$ and should meet Eq. (3.3), where H_r is the structure and safety distance (m) to be met between the upper and lower water-saving basins.

$$\mathrm{Z_{wb-min}(i) - Z_{wt-max}(i+1) \geq H_r} \tag{3.3}$$

By substituting Eqs. (3.1) and (3.2) into Eq. (3.3), the control requirement of the area ratio of the water-saving basin of the integrated structure of the water-saving ship lock is as follows:

$$k \geq \frac{(H_{\text{lock}} - 2\Delta H) + \Delta Z_{\text{up-max}} + \Delta Z_{\text{down-max}} + H_r}{(H_{\text{lock}} - 2\Delta H) - (i+1)\Delta Z_{\text{up-max}} - i\Delta Z_{\text{down-max}} - (n+1)H_r} \tag{3.4}$$

As can be seen from Eq. (3.4), the area ratio k increases with the increase of the variations of the upstream and downstream water levels, and the area ratio k of the last stage ($i = n$) is the largest, that is, the area ratio of the last stage water-saving basin as terms of controlling.

4 Water-Saving Rates of Water Replenishment and Overflow Operations

As previously discussed, the determination of the top and bottom heights of each WSB relies on the highest and lowest water levels upstream and downstream, respectively. However, for locks situated along natural rivers, significant variations in water levels upstream and downstream can pose challenges. Simply adhering to the highest and lowest navigable water levels upstream and downstream for setting the heights of WSBs necessitates extensive earth and stone excavation, thereby escalating construction costs. Moreover, maintaining consistent height across all levels of the WSB becomes unattainable, leading to unfavorable conditions in both design and maintenance.

To address this issue, an alternative method is proposed. Starting from the minimum navigable water level upstream, the bottom height of each level of the WSB is determined based on both upstream and downstream minimum navigable water levels. The top height of the WSB is selected according to the highest navigable water level downstream and the lowest navigable water level upstream plus the upstream water level whose height is equal to the magnitude of water level change downstream. This lowers the top height of the WSB at all levels.

However, when the upstream water level surpasses the minimum upstream navigable water level plus ΔZ_{down}, the inflow into the first level of WSBs exceeds the capacity of subsequent levels, resulting in excess head ΔH, which spills downstream. The water-saving rate of overflow operation of water-saving ship locks is shown in Eq. (4.1):

$$E_W = \frac{kn(H_c - \Delta H)}{[k(n+1)+1]H_c} \times 100\% \tag{4.1}$$

The second way to solve the problem is to start from the highest navigable water level upstream; determine the bottom height of the WSBs at all levels by the lowest navigable water level downstream and the highest navigable water level upstream minus the upstream level of the downstream water level variation ΔZ_{down}. The top heights of the WSBs at each level are determined by the highest navigable water levels upstream and downstream. In this way, the amount of construction work can be reduced. But in this case, when the upstream water level is lower than the above water level, when the water flowing into the upper chamber, will be less than the following chamber of the water required. Therefore, it is necessary to replenish the insufficient amount of water ΔH to the first stage of the WSB during the emptying operation.

Water saving rate of emptying operation:

$$E_W = \frac{kn(H_c + \Delta H)}{[k(n+1)+1](\frac{k}{k+1}\Delta H + H_c)} \times 100\% \tag{4.2}$$

Water saving rate of filling operation:

$$E_W = \frac{kn(H_c + \Delta H)}{[k(n+1)+1]H_c} \times 100\% \tag{4.3}$$

This analysis reveals that the water-saving rate of the water replenishment scheme is higher than that of the water overflow scheme. Compared to conventional WSB operation, this scheme enhances water-saving rate while reducing construction costs. It is worth noting that the make-up water operation scheme is also commonly used in engineering multi-stage ship locks [10]. Therefore, it is more cost-effective to use the make-up water arrangement scheme. This scheme is fully applicable in the design of water-saving ship locks.

5 Application

A lock, the highest upstream navigable water level is 60.0m, the upper reaches of the water level variation is 10.0m; The lowest downstream navigable water level is 0m, the downstream water level variation is 3.0m, and the residual water head is less than 0.2m. We consider setting 1 ~ 8 stages water-saving basins, and the area ratio k is 1.0 ~ 3.0 respectively. According to Eq. (1.9), the maximum working water head of the water-saving basin has nothing to do with upstream and downstream water levels variations. The maximum working water head of the water-saving basin corresponds to the different water levels of water-saving basin water classification schemes (Different water-saving basin stages n and different area ratios k) can be calculated. As shown in Fig. 5, when the number of the water-saving basin stages is the same, and $k = 1.0 \sim 3.0$, the maximum working head of the water-saving basin has little change. When the area ratio of the water-saving basin is increased from 1.0 to 3.0, the maximum working head of the water-saving basin only decreases from 24.04m to 18.53m. With the same area ratio of the water-saving basin, the stages of the water-saving basin are increased from 1 to 3, the working head of the water-saving basin is rapidly reduced from 39.93m to 24.04m, and the head is reduced by 15.89m. From 3 stages to 5 stages, the degree of reduction of the working water head of the water-saving basin becomes significantly slower, and the water head only decreases by 6.81m. The scheme of 3 stages of water-saving basins is adopted to synthesize the water-saving ratio and construction investment of the ship lock.

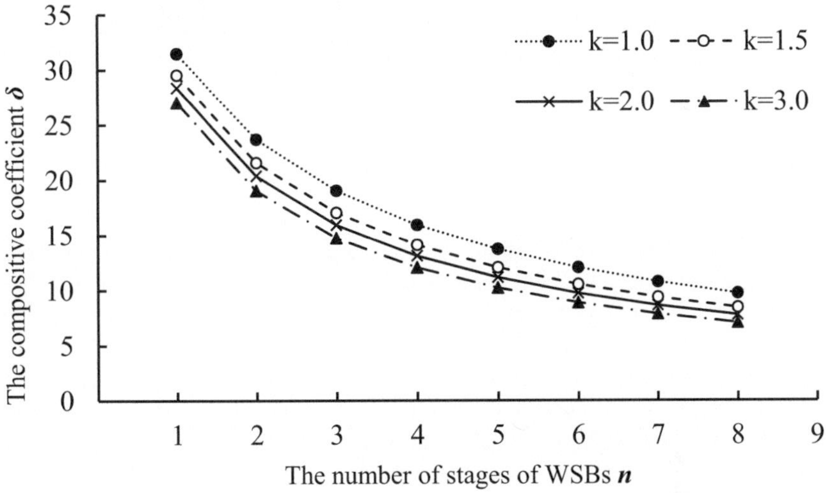

Fig. 5. The relationship between the maximum working head of the WSB and the water level classification scheme.

When there is no residual head ($\Delta H = 0$), according to Eqs. (3.1) and (3.2), we can calculate the minimum and maximum water levels of each water-saving basin under the scheme of $n = 3$, $k = 1.0$. As can be seen from Fig. 6, the area ratio of the water-saving basin is 1.0, the highest water level of the 2nd and 3rd water-saving basin is 37.2m and 25.8m, and the lowest water level of the 1st and 2nd water-saving basin is 30.0m and 20.0m, respectively. Therefore, when the area ratio of the water-saving basin $k = 1.0$, the scheme of integrated water-saving basin arrangement can not be used, the open arrangement scheme is appropriate.

To analyze the feasibility of the integrated arrangement scheme, we calculated the water level of each level of the water-saving basin under different area ratios and upstream water levels variations. As can be seen from 错误!未找到引用源。, under the condition that the upstream water level variation remains unchanged by 10m, the structural safety distance H_r is considered as 1.1m, when the area ratio k should not be less than 3.1, the integrated arrangement scheme can be adopted. In practical application, due to the cost

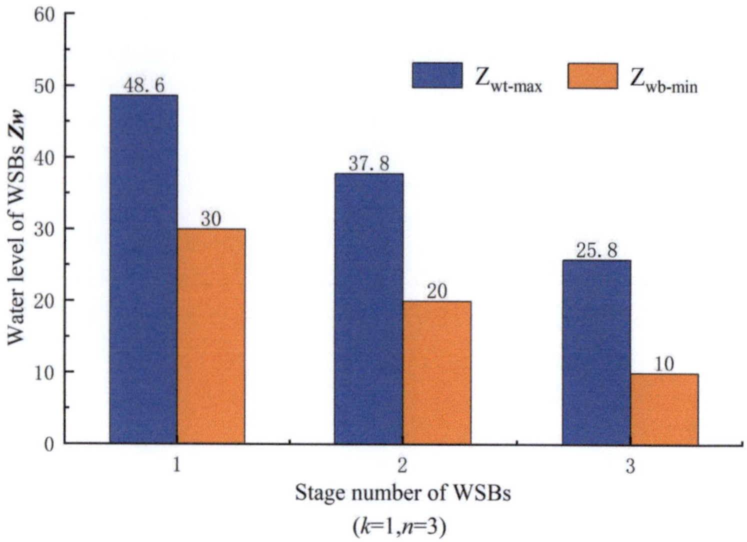

Fig. 6. Water levels of water-saving basins.

and other factors, the area ratio k should be appropriately reduced. In order to reduce the area ratio k, water replenishment operation mode can be used to reduce the influence of upstream water level change. Overflow operation mode is adopted to reduce the influence of downstream water level variation. In this case, the upstream water level variation is 10.0m, and the downstream water level variation is 3.0 m. In this operation mode, the variation of the upstream water level is reduced to 3.0m (that is, the upstream water level is lower than the highest navigable water level of 3.0m, and water refill operation is adopted). Water replenishment operation mode can be used to reduce the impact of upstream water level changes. The variation of the upstream water level can be reduced to 3.0m (that is, the upstream water level is 3.0m lower than the highest navigable water level, and the water replenishment operation is adopted). As seen in Table 2, the area ratio k of the water-saving basin is greater than 1.5, and the integrated arrangement scheme can be adopted.

6 Conclusions

1. Under the condition that the gross head of the lock is constant, the number of stages n, and the area ratio k are the main influence factors of the working head of the lock. The working head of the lock decreases with the increase of n, k, and the influence of n is greater than that of k. When the stage number of water-saving basins is greater than 5, the water head reduction is getting smaller; When the area ratio k is greater than 2, the rate of working head decline is slower.
2. The highest water level upstream of the lock determines the highest water level of each water-saving basin, and the lower water level determines the lowest water level of each water-saving basin. The higher the upstream water level variation, the lower

Table. 2. Influences of area ratios and upstream water level variations on water levels of water-saving basins.

Upstream water level variation /m		10.0			3.0		
WSBs		Stage 1	Stage 2	Stage 3	Stage 1	Stage 2	Stage 3
$k = 1.0$	Z_{wt}/m	48.6	37.2	25.8	48.6	37.2	25.8
	Z_{wb}/m	30	20	10	34.2	22.8	11.4
	safe distance /m	-7.2	-5.8	/	-3	-3	/
$k = 1.5$	Z_{wt}/m	47.8	35.6	23.4	47.8	35.6	23.4
	Z_{wb}/m	32.1	21.4	10.7	36.6	24.4	12.2
	safe distance /m	-3.4	-1.9	/	**1.1**	**1.1**	/
$k = 3.1$	Z_{wt}/m	46.8	33.6	20.4	46.8	33.6	20.4
	Z_{wb}/m	34.7	23.1	11.6	39.6	26.4	13.2
	safe distance /m	**1.1**	2.7	/	5.9	5.9	/

the lowest water level of the stage 1 water-saving basin; The larger the downstream water level variation, the higher the highest water level of the stage *n* water-saving basin.

3. To adopt the integrated arrangement of the lock water-saving basin, the water-saving basin area ratio *k* should meet the requirements of Eq. (3.4). When the upstream or downstream water level varies greatly, water refill and overflow can be used respectively to reduce the influence of water level variation, to reduce the water-saving basin area ratio *k*. When upstream and downstream water levels variations are large, the open arrangement of water-saving basins should be adopted.
4. The water-saving rate of the water replenishment scheme is higher than that of the overflow scheme and the conventional scheme. This scheme can be used in the design of water-saving ship locks.

References

1. Water Transport Planning and Design Institute, Ministry of Communictions, Science and Technology Information Institute, Ministry of Communictions: Special Information Materals, Foreign ship passing Buildings (Third Series 3). Water Transport Planning and Design Institute of Ministry of Communications, Institute of Science and Technology Information of Ministry of Communications, Beijing (1973)
2. PIANC: Innovations in navigation lock design NO. 106. PIANC, Brussels (2009)
3. Guo, C., Xuan, G.: Study on Key Hydraulics of Xiaoqinghe Provincial Water Lock Layout, Hydraulic Calculation and Physical Model Test of Jinjiayan Provincial Water Lock. Nanjing Institute of Hydraulic Science, Nanjing (2019)
4. Chen, Y., Li, Z.: Study on Key Hydraulics of Xiaoqinghe Provincial Water Lock Layout and Hydraulic Calculation Analysis of Buffuohan Ship Lock. Nanjing: Nanjing Institute of Hydraulic Scienc (2019)

5. Chen, Y., Li, Z., Xu, D.: Water Conveyance System Layout and Hydraulic Calclation of Xiaoqing River Buffalohan Water Lock. Water Transp. Eng. 63–69+102 (2020)
6. Xuan, G., Liu, B.: Experimental Study on Hydraulic Model of Water Delivery System of Economizing Ship Lock in Bajiangkou Ship Lock Energy Expansion Project of Guijiang in Guilin. Nanjing Institute of Hydraulic Science, Nanjing (2019)
7. Zhu, L., Xuan, G.: Experimental Research on Key Technologies of Shunew Ship Lock in Jiangbai City, Qingshui. Nanjing Institute of Hydraulic Science, Nanjing (2019)
8. PIANC: Final report of the international commission for the study of locks. PIANC, Brussels (1986)
9. Liu, B., Li, Y., Hu, Y., Xuan, G., et al.: Water saving arrangement and hydraulic calculation of large-scale ship lock with high head. Water Transp. Eng. **2016**(12), 42–46 (2016)
10. Chen, Z.: Hydraulic model tests of supplemental overflow water using water supply valves on multistage ship locks. People's Yangtze River **05**, 38–43 (1960). https://doi.org/10.16232/j.cnki.1001-4179.1960.05.007

Predictive Insights into the Future Landscape of Energy Technologies

A Review of Oscillating Buoy Devices in Wave Energy Power Generation

Jianchao Zhang, Le Wang(✉), and Haitao Ren

College of Water Conserwancy and Hydropower Engineering, North China Electric Power University, Beijing 102206, China
13893524527@163.com, lewang@ncepu.edu.cn, 13426066189@139.com

Abstract. Escalating energy demands and dramatic changes in global climate present a formidable challenge. As a clean, renewable, and unexploited large-scale energy source, ocean wave energy has attracted research attentions all over the world. Although the existing theoretical technologies of wave energy power generation are emerging rapidly, few of them can be applied to massive practical deployment. The current achievements of wave energy resources are reviewed in this paper; various forms and energy transmission modes of oscillating buoy wave energy converters are summarized, and their characteristics and scope of application are detailed; the different array arrangement and benefits of the oscillating buoy wave energy converter are summarized; the directions for future oscillating buoy device development are outlined. Particularly, enormous amount of wave energy from far offshores will be exploited through using advanced and sophisticated equipment and technology, the oscillating buoy energy capture method and more direct and efficient direct-drive power generation devices will be the main choice for wave energy development in the future; to optimize the selection, size, and array layout of oscillating buoy wave energy converter can effectively improve the energy conversion efficiency and lower the levelized cost.

Keywords: Wave Energy · Power Generation · Oscillating buoy

1 Introduction

In recent years, with the continuous increase of energy consumption, traditional fossil energy is difficult to fulfil energy demand for social development. As the largest electricity consumer, China is currently facing huge pressure and challenges in the energy field. According to the IEA-Electricity Market Report (2023), global electricity demand growth is projected to rise from 2.6% in 2023 to an average 3.2% in 2024–2025, by 2025 demand will increase by 2500TWh from 2022 levels and more than half of the increase will come from China [13]. Alicia et al. (2021) estimated the growth in renewable energy consumption in 2050, more than double that in 2018, from 28,000 TWh to 74,000 TWh, and is expected to outproduce natural gas and coal by 2030 and surpass oil by 2050, from 15% to 28% [9]. In this context, a series of renewable and clean energy sources, such as ocean wave energy, solar energy and wind energy, have attracted the attention of

S. Zheng et al. (Eds.): IHDC 2024, LNCE 487, pp. 567–578, 2025.
https://doi.org/10.1007/978-981-97-9184-2_46

scientific researchers, ccompared with the remarkable achievements of the development technology of land for solar energy and wind energy resources, people began to gradually deepen the research content into the ocean, among which wave energy is more and more favoured by researchers with its unique advantages.

Wave energy is mainly the periodic movement of sea water caused by the upper wind blowing on the sea surface and the change of atmospheric pressure. It has certain kinetic energy and potential energy. The kinetic energy refers to the energy caused by the movement of the fluctuating water quality point at a certain speed, and the potential energy refers to the energy caused by the movement of the water quality point and the displacement of the sea level. The continental coastline of mainland China is 1.8×10^4 km, and the renewable wave energy is infinite. It is converted by reasonable layout and use of wave energy conversion device, which can become a powerful boost for the global energy transformation. Compared with other energy sources, wave energy has the following advantages: (i) wave energy has a higher energy density than solar energy and wind energy, in which wave energy is 2.0 – 3.0 kW/m^2, wind energy is 0.4 - 0.6 kW/m^2, and solar energy is 0.1 - 0.2 kW/m^2, it is highly reliable; (ii) wave propagation distance is long and the energy loss is small, the effective working time of wave energy conversion device is relatively higher, which also provides a guarantee for efficient and large-scale development; (iii) wave energy reserves are immense and can be developed in coastal areas and even extended to offshore areas [8]; and (iv) waves change periodically, thereby serving as a law for standardized utilization [7]. Gradually, the wave energy was developed from the initial nearshore to the far offshore areas. In comparison, offshore devices can capture stronger wave energy due to their deployment in deeper waters, theoretically, there will be higher power generation efficiency, and the offshore arrangement will improve the impacts of shore-based devices on the reshaping of the coastline; however, the transportation and installation costs of the devices will increase due to the long distance, and the uncertainty of the impacts of wind and waves on the devices and the energy transmission costs will also increase accordingly. However, due to the long distance, the transportation and installation cost of the device, the difficulty for maintenance, the uncertainty of the effect of wind and waves on the device, and the cost of energy transmission will also increase.

2 Oscillating Buoy Devices in Wave Energy Power Generation

2.1 The Principle of Wave Power Generation

With the gradual deepening of wave energy research, various forms of wave energy converter (WEC) have appeared, although their energy conversion methods are different, the conversion principle is basically similar. Wave energy power generation is the main form of wave energy utilization, and the whole power generation process is generally completed through three-stage energy conversion: the first stage is the contact between the wave and the energy capture device (float, pendulum, etc.), which converts the wave kinetic energy into the mechanical energy, water potential energy or aerodynamic energy required by the power generation system; The second stage drives the mechanical devices in the power generation system, such as hydraulic pumps or turbines, which convert the mechanical energy into a rotational motion. The third stage is a rotating mechanical device driving the generator to convert the mechanical energy into electrical energy. The three levels of energy conversion are interconnected and interdependent on each other to jointly complete the conversion process from wave energy to electrical energy. The most important of these is the energy conversion of the first stage, and the amount of wave energy captured directly affects the conversion efficiency and power generation of the second and third stages. The second level is mainly to play a role in stabilizing energy, the energy smooth excessive, to meet the requirements for the next level of energy conversion, as shown in Fig. 1.

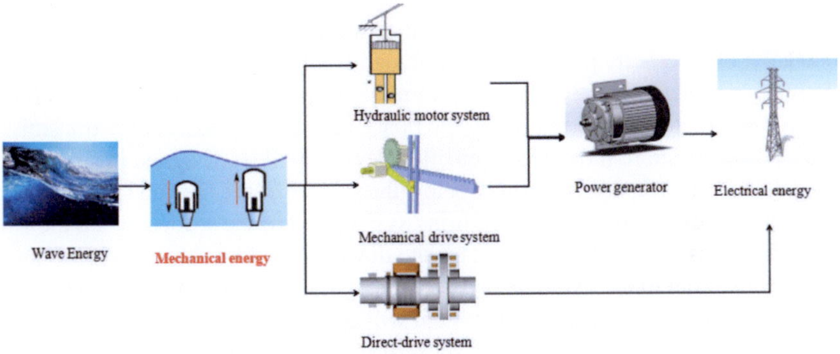

Fig. 1. Wave energy conversion process

2.2 Main Wave Energy Conversion Devices

Great progress has been made regarding wave energy power generation research, a variety of wave energy conversion devices continue to appear, such as oscillating water column WEC, oscillating buoy WEC, overtopping WEC, etc. Currently, the oscillating buoy WEC has become the main component of wave energy development system with the evident advantage of relatively low cost, the variety of forms, flexible arrangement, and reliability.

2.2.1 Classification of Oscillating Buoy Wave Energy Converters

Oscillating buoy WEC originated from oscillating water column WEC, and it is also the better developed converter at present, which has a wide range of applicable positions, and it can work from the shore position to the far sea position. According to the contact situation between float and water, it can be divided into floating surface type, partially submerged type, and submerged type.

Floating surface type structure was the earliest use of oscillating buoy devices, which were mounted on the coast or fastened to breakwaters and rocks above sea level. It is usually composed of a float, connecting rod, energy transmission device, power generation device and protection device, etc., the float is in contact with seawater when working, and the float moving with the wave drives the hydraulic cylinder to work and convert energy.

Its advantage is that it is close to the land, the float design is simple and easy to install and maintain and does not require anchoring devices and long power transmission cables and can capture wave energy of different frequencies and sizes. However, the wave energy resources in this location are relatively limited and will have an impact on the onshore environment. With the design and research of the offshore power generation platform, the floating surface type is also gradually developed from the original single shore fixed power generation to the far offshores in the form of a combination of floats and power generation platforms, such as the first megawatt floating wave energy generation device "Nan Kun" independently developed by China, which is a trilateral semi-submersible platform. Each side is equipped with five floating plates for energy conversion with a total area of more than 3,500 m^2 and weight of more than 6,000 t, and it can generate 24,000 kWh of electricity per day under full load conditions, as shown in Fig. 2(a).

Partially submerged type is different from the former, the assistance of the offshore platform can directly float in the ocean, the floats, connecting rods, energy transmission devices, power generation devices, protection devices will be designed in a centralized floating system. The device is partially below the surface of the flow and partially above the surface of the flow, through the mooring system tied to the seabed, when the waves work on the floats, the floats will move up and down in the flow, this movement through the mooring system to the hydraulic system, the hydraulic system for the conversion of energy to drive the power generation system will be converted from mechanical energy to electrical energy, due to the fact that most of the device is located in the submerged, by the wind and waves have less impact, improve the stability of the device and the efficiency of the operation. Such as the United States OPT company developed the "PB3 Power Buoy" conversion equipment, through the collection of energy from the waves, and constantly self-charging, can be more than 20 m to 3000 m in any depth of the ocean operation [15], as shown in Fig. 2(b).

Submerged type device structure and partially submerged type similar to the device connected to the seabed through the mooring device, the entire device is below the surface of the water, the working principle is to use the wave up and down fluctuations generated by the pressure difference to drive the float movement, when the wave upward movement, the float down, when the wave downward movement, the float upward, and so on and so forth to drive the power generation. As the float is located underwater, it is less affected by the wind and waves on the surface of the sea, so the stability of the power generation device is better. Such as the "CETO 6" designed by Carnegie Wave Energy Company of Australia is installed in the depth of 25 - 50 m underwater. The actuator starts the pump set and generator with the rise or fall of the wave, and the generated power is transmitted to the shore through the submarine cable and incorporated into the local power grid. The design not only protects the installation from storms at sea, but also does not affect the visual effect on the sea surface [2], as shown in Fig. 2(c).

With the increase of human energy demand and the research on wave energy development technology in the far offshores, the main area for wave energy development in the future will also be in the far offshores, where there is more wave energy and greater energy density than near shore. Compared with the characteristics of other conversion devices, the oscillating buoy WEC has the advantage of flexible position, can adapt

(a) (b) (c)

Fig. 2. (a) China's first megawatt floating wave power generation device "Nan Kun". (b) "PB3 Power Buoy" conversion device developed by OPT company in the United States. (c) "CETO 6" by Carnegie Wave Energy in Australia.

to different wave directions and sizes, and can work effectively in various sea conditions; The float is usually composed of a simple cylindrical or spherical structure, which reduces the complexity of mechanical parts and reduces the cost of manufacturing and maintenance; Although the energy output of a single float may not be high, this form can not only operate on a fixed platform scale, but also float the offshore array layout, and can improve the overall power generation efficiency by optimizing the design and increasing the number of floats; It can be combined with different forms of energy transmission to form a WEC with different conversion efficiency, which can provide a variety of options for the future wave energy development. Under the above outstanding advantages, oscillating float WEC will become the main way of wave energy development in the future.

2.2.2 Oscillating Buoy WEC in Varying Energy Transmission Modes

The energy is captured by the float and converted into mechanical energy for the second stage of energy conversion. The oscillating buoy WEC is combined with different energy transmission modes to form conversion devices with different efficiency, which are commonly hydraulic, mechanical, and direct drive.

The combination of oscillating float and hydraulic system is a form that is used more at present, and its working principle is to use the energy captured by the float to drive the reciprocating motion of the hydraulic cylinder, so that the piston squeezes the hydraulic oil into the energy storage device, and the energy storage device mainly plays a stable energy, and when the energy storage device reaches the set value, it is opened, so as to drive the hydraulic motor to rotate and drive the power generation system to generate electricity. The advantages of the hydraulic power generation system are flexible transmission, energy storage and pressure regulation, and large torque, which can store wave energy in the case of small wave speed and achieve continuous and stable conversion of wave energy to electric energy [20]; Hydraulic components have also been standardized and generalized [26], which is easy to use and has strong commercial applicability; The incompressible fluids used in hydraulic systems can produce higher efficiency; However, there is a risk of pollution, and it is often placed at sea, which is inconvenient for maintenance. Sun et al. (2021) designed a hydraulic transmission and accumulator system to replace the original mechanical transmission and flywheel system, the results show that the accumulator can effectively suppress the output power fluctuation, wave power into scheduling power supply, with the increase of the number of buoys and buoy phase difference, accumulator system energy conversion and output performance is better [18]. Hansen et al. (2013) showed that high pressure accumulators can store energy for a short period of time and partially filter out changes in pressure and flow in the system [11]. According to the ultimate pressure state of the hydraulic motor system, it is mainly divided into three categories, namely, variable-pressure HPTOs (VPHPTOs), constant-pressure HPTOs (CPHPTOs) and constant-variable pressure hydraulic systems [23], as shown in Fig. 3. The experimental results show that the power take off (PTO) efficiency of CPHTOs is higher than that of VPHTOs (30% higher), and the power generation is 2% lower than that of VPHTOs [16]. According to the circulation mode of oil, it can be divided into an open-circuit hydraulic system and closed-circuit hydraulic system. In an open hydraulic system, the hydraulic cylinder draws oil from the tank, delivers the oil to the actuator, and then returns the oil directly to the tank. In a closed hydraulic system, the oil output from the hydraulic cylinder directly enters the actuator and then returns to the suction port of the cylinder, forming closed circulation. In 2010, the Guangzhou Institute of Energy Conversion, Chinese Academy of Sciences, developed the "Eagle #1" floating wave energy power generation device, which was put into testing in the waters of Wanshan Island in Zhuhai City in 2012. The device adopts two sets of energy conversion systems, including hydraulic energy conversion system and direct-drive motor system, in which the hydraulic energy conversion system is the main, the direct drive motor system is supplemented, and the open hydraulic system is selected to facilitate the cooling of hydraulic oil. When the wave is small, the accumulator stores energy, and when the set value is reached, the energy is released to drive the power generation system to generate electric energy, and the power generation is intermittent in turn; when waves are large, the power generation system works continuously.

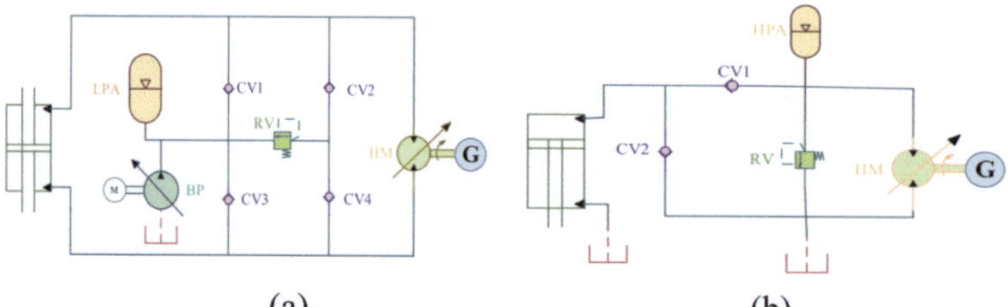

(a) (b)

Fig. 3. Hydraulic motor system configuration [23]: (a) typical CPHPTOs configuration and (b) typical VPHPTOs configuration

Mechanical drive system is connected to the oscillating buoy to connect the rack, rope wheel or connecting rod to absorb wave energy, and then use the equipment such as the transcendence clutch, ratchet, rack or sprocket to convert the up and down motion of the float into the unidirectional rotation motion of the rotating shaft, increase the speed by using the speed increase gearbox, use the flywheel to store energy, and finally drive the generator to generate electricity. The mechanical transmission device has the advantages of simple structure and low cost, but at the same time, the mechanical structure reduces the energy conversion efficiency and reliability, is easy to be corroded by seawater, increases the maintenance cost, and is not easy to realize the shortcomings of control and regulation. For example, Shandong University proposed in 2013 a floating rope wheel type wave energy power generation device, one end of the rope is tied to the gravity anchor of the seabed, and the other end is wound on the drum of the generator set through the rope guide, when the wave pushes the floating body to rise, the rope drags the drum to rotate, and the drum directly drives the low-speed synchronous permanent magnet alternator to generate electricity [27]; A mechanical transmission device designed and improved by Sun et al. (2021) is shown in the figure, which mainly includes a horizontal rotation shaft, a bidirectional transmission gear set, an overspeed gearbox, a flywheel, a permanent magnet generator (PMG), etc., and the relative motion between the buoy and the central platform is converted into the mechanical energy of the horizontal rotation shaft through the bidirectional transmission gear, and finally converted into the electrical energy output by PMG [18], as shown in Fig. 4.

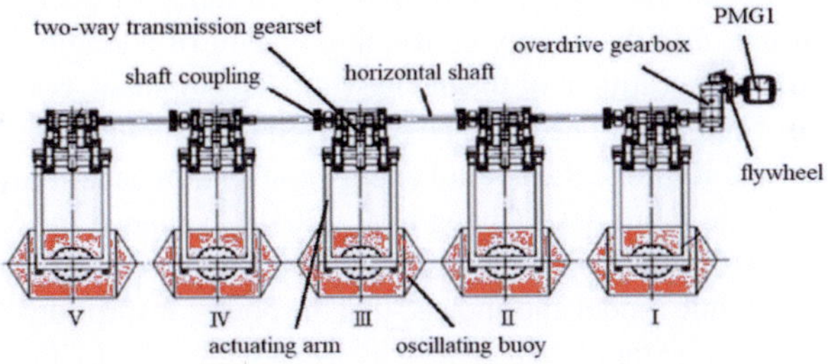

Fig. 4. Mechanical drive system structure diagram [18].

The direct-drive power generation device is to directly couple the captured mechanical energy to the moving part of the linear generator for power generation, and its working principle is that the float drives the linear generator to do reciprocating motion under the action of waves, so that the relative motion between the mover and the stator is generated, and the magnetic inductance line is cut to generate electric energy. The device reduces the intermediate mechanical drive system, so it has higher efficiency and reliability, reduces the probability of failure, and reduces the need for maintenance; However, because the whole device directly generates electricity with the wave, the process of outputting electricity is slow, and the output voltage is unstable. According to the principle of power generation, there are three main types of linear generators currently used in wave energy power generation: permanent magnet synchronous generators, induction generators and switched reluctance generators [12]. Permanent magnet synchronous generators are widely used in the field of wave energy power generation, and the power generation characteristics are closely related to their mechanical structure, permanent magnet material, effective wave height and wave period. The energy conversion efficiency of induction generators is lower than that of permanent magnet synchronous generators. The material cost of switched reluctance generators is low, and the output is relatively stable [6]. For example, the Archimedes Wave Swing (AWS) in the United Kingdom consists of an upper float for swinging motion and a lower structure fixed to the sea floor, with permanent magnets placed on the lower part and the upper float connected to the generator coil. The two parts are submerged under water when working, and the floating body moves downward when the wave peaks, and upward when the wave troughs, thus driving the linear motor movement.

Different from the three-stage energy conversion of mechanical system and hydraulic system, the direct-drive power generation device is designed with a more concise two-stage energy conversion structure for energy transmission, which saves the loss of energy in the intermediate transmission process, simplifies the device structure, and increases the reliability of use, especially in China's coastal waters with low wave density and short wave energy movement cycle, which is of great significance for the future country to solve the problem of marine energy exploration and large-scale development [19].

2.2.3 Efficiency-Oriented Arrangement of Oscillating Buoy WEC

Due to the randomness and uncertainty of waves, the transmission of long-distance energy makes the device have higher requirements in terms of energy storage, transmission and final use, and the energy conversion benefit of a single device cannot be balanced with the cost investment of the device from development, commissioning, and later maintenance. Therefore, it needs to be optimized in the collective arrangement and individual devices and improve the overall energy conversion benefit of the layout optimization to balance the input cost. Chang et al. (2018) evaluated the levelized cost of energy (LCOE) of 50 different configurations of wave energy conversion devices based on the technical economic model and showed that the power absorption of different array shapes can be increased by more than 20%, which can reduce the LCOE of commercial-scale WEC projects [4]. Chandrasekaran and Sricharan (2021) studied the effect of the distribution of different WECs on LCOE and found that LCOE gradually decreased as the number of WECs increased [3].

According to the number of oscillating floats that absorb wave energy, they can be divided into single float WEC, multi-float WEC and float array WEC. Single-float WEC uses a float to capture the energy of the wave, which is relatively simple in design, but does not make full use of the energy of the wave. To improve the efficiency of energy harvesting, some WEC designs use multiple floats, which can respond more effectively to waves in different directions and frequencies, thereby improving the overall power generation efficiency. A float array WEC is a special multi-float system in which floats are arranged in a specific geometry, such as a circle, square, or other complex array. This layout is designed to optimize wave energy absorption and may include interactions between floats to enhance energy conversion efficiency [24]. The advantages of multi-float and array wave energy devices are that they can use multiple oscillating floats to collect wave energy at the same time, generate electricity on a large scale, and the wave energy collection is more continuous and uniform [25], and the volume and weight of the oscillating floats can also be reduced, which is convenient for the manufacture and transportation of floats.

Under the same conditions, different arrays can be arranged to capture different amounts of energy. Malin et al. (2015) compared the total power of four different arrangements of about 250 conversion units and found that the geometry of the circular arrangement was the most beneficial [10]. Sinha et al. (2016) analyzed the energy conversion effects of five forms, including linear arrangement, grid, circle, concentric circle, and random arrangement, and showed that the conversion of concentric circles and circular arrays was higher than that of grid lines and linear arrays [17]. Abdulkadir and Abdelkhalik (2023) optimized the performance of multipoint absorbers for non-uniform arrays, and their results showed that optimal heterogeneous arrays could provide up to 40% performance improvement compared to homogeneous arrays [1]. Xu et al. (2021) compared the power output and interaction utility of the rectangular array float arrangement with the circular array arrangement under the condition of ignoring the central platform and verified the superiority of the circular array arrangement [22]. In summary, optimizing the overall array layout has an important impact on improving the energy conversion effect, in addition to the number and size of the devices, the distance between the devices, and the mixed arrangement of various wave energy conversion devices, etc., to fully capture the wave energy per unit area, to achieve the optimization of economic cost. For example, the Danish "Wave Star" array wave power generation device, when the total length of the power generation device is 70 m, with 20 oscillating floats with a diameter of 5.0 m, installed in the offshores of 10 ~ 15 m water depth, when the wave height is 2.5 m, the output electric power is 600 kW, and when its total length is 140 m, with 20 oscillating floats with a diameter of 5 m, installed in the offshores of 20 ~ 30 m water depth, when the wave height is 5 m, the device can output 6 MW electric power. That is, a 1-fold increase in the size of the device will result in a 10-fold increase in the output of electrical power [21], as shown in Fig. 5.

Fig. 5. The "Wave Star" array-type wave power generation device [21].

Based on the analysis of the power generation principle and characteristics of some existing wave energy power generation devices, the main problems faced by the current wave energy development are as follows. The first is the huge cost investment, which is the main reason why wave energy devices have not been developed on a large scale. At this stage, more attention is paid to the development of new energy, although there are constantly new technical and theoretical outputs, only a few parts are really used for practical development. The second is that the conversion efficiency is low, and the process of converting wave energy into electrical energy usually involves the conversion of mechanical energy to electrical energy, and the energy loss in this process is huge [5]. The reliability of the power generation device is low, and the wave energy power generation equipment operated in the marine environment, which is affected by harsh conditions such as waves, salt spray, corrosion, etc., which is likely to cause equipment damage and maintenance difficulties; meanwhile, wave energy resources are affected by weather and other factors, and have certain instability, which may affect the stability and reliability of wave energy power generation system. Through the array arrangement of power generation devices, the energy conversion efficiency can be effectively improved, and the ratio of input cost to output benefit can be reduced. Optimize the scale, layout and number of the devices, materials, energy conversion methods to enhance the conversion efficiency and reliability.

3 Directions of Oscillating Buoy Wave Energy Converters Development

It has been shown that the combination form of oscillating buoy device and simple and efficient direct drive energy transmission mode has certain advantages compared with other forms. The development of wave energy in the far offshores will eventually become the main battlefield of ocean energy development. In the development process of oscillating buoy WEC, in addition to the cost, efficiency and reliability of the device itself, the impact on the Marine environment is also comprehensively considered. Therefore, the content of future research on oscillating buoy WEC is summarized as follows: (i) Improving technology and optimizing efficiency: the float is the part of the device in direct contact with the wave, and it is also the most important link of energy conversion. By optimizing the design of the float, it can convert energy to the maximum extent under different wave conditions; For the overall design of the device, lighter and stronger materials can be used, such as carbon fibre and high-strength steel alloys, which can improve the weather resistance and corrosion resistance of the float and extend the service time of the device; Develop more efficient energy converters to improve the efficiency of converting wave energy into electricity, (ii) Economy-friendly: Huge cost input has always been the major obstacle to the large-scale development of wave energy, therefore, reducing manufacturing and maintenance costs, improving economic benefits through large-scale production, and making this device more competitive in the market is the development direction of the oscillating buoy WEC in the future, (iii) Intelligent development: the introduction of artificial intelligence and Internet of Things technology to realize the intelligent development of the device, respond to some risks in advance, improve the adaptive ability and fault diagnosis ability of the device, and enhance the reliability of the device; For power generation installations in the far offshores, real-time monitoring can be carried out by equipping remote monitoring systems and maintenance systems to reduce the need for manual maintenance and reduce operating costs, and (iv) Multi-energy complementarity: In the development process of marine energy, due to the essential differences of different energy sources and the different conversion methods, the conversion and utilization of energy are carried out in a single way, so that not only the energy utilization rate is low, and the cost is high, but also the different energy development technologies are difficult to achieve mutual benefit and complementary advantages. Therefore, strengthening multi-energy interconnection and promoting the energy coupling and synergy of various offshore energy sources such as wind, solar, and wave energy are the main directions of new energy development and research in the future.

4 Conclusions

For the large-scale development of wave energy, cost, conversion efficiency and reliability are the focus of research, but also the difficulty. Starting from the current energy demand, this paper summarizes the advantages of wave energy development. The different forms and energy transfer modes of the oscillating buoy WEC are introduced, and the development trend of the device is analysed. The summary is as follows:

To fulfil the increasing energy demand for sustainable development, and wave energy development will be an important means of energy output in the future. The far offshores will be the main site for wave energy development, the oscillating buoy energy capture method and more direct and efficient direct-drive power generation devices will be the main ways of wave energy development in the future. In nearshores and offshores with abundant and dense wave energy, optimizing the selection, size, and array layout of oscillating buoy WEC can effectively improve the energy conversion efficiency and reduce the levelized cost of energy.

Wave intermittency and the uncertainty of the device away from the coast could affect the large-scale development of wave energy and cost. In the future research, the impact on the Marine environment should be comprehensively considered, advanced intelligent technology and multi-energy complementarity can be used to improve the efficiency of wave energy conversion, strengthen the reliability of the device, and reduce the cost of wave energy conversion.

References

1. Abdulkadir, H., Abdelkhalik, O.: Optimization of heterogeneous arrays of wave energy converters. Ocean Eng. **272**, 113818 (2023)
2. CETO 6.: Available online. https://www.carnegiece.com/. Accessed May 2024
3. Chandrasekaran, S., Sricharan, V.V.S.: Numerical study of bean-float wave energy converter with float number parametrization using WEC-Sim in regular waves with the Levelized Cost of Electricity assessment for Indian sea states. Ocean Eng. **237**, 109591 (2021)
4. Chang, G., Jones, C.A., Roberts, J.D., et al.: A comprehensive evaluation of factors affecting the levelized cost of wave energy conversion projects. Renew. Energy **127**, 344–354 (2018)
5. Chen, J.J., Wang, L.W.: On the engineering feasibility of wave-generating systems. Ocean Eng. (1), 75–84 1995.https://doi.org/10.16483/j.issn.1005-9865.1995.01.011
6. Chen, J., et al.: Review on wave energy power generation control technology. Electric Power Automation Equipment/Dianli Zidonghua Shebei **43**(6) (2023)
7. Chen, W., et al.: Research status and development trend of wave power generation technology. Energy Environ. **3**, 83–84 (2014)
8. Clément, A., McCullen, P., Falcão, A., et al.: Wave energy in Europe: current status and perspectives. Renew. Sustain. Energy Rev. **6**(5), 405–431 (2002)
9. Gonzalez, A.T., Dunning, P., Howard, I., et al.: Is wave energy untapped potential? Int. J. Mech. Sci. **205**, 106544 (2021)
10. Göteman, M., Engström, J., Eriksson, M., et al.: Optimizing wave energy parks with over 1000 interacting point-absorbers using an approximate analytical method. Int. J. Marine Energy **10**, 113–126 (2015)
11. Hansen, R.H., Kramer, M.M., Vidal, E.: Discrete displacement hydraulic power take-off system for the wavestar wave energy converter. Energies **6**(8), 4001–4044 (2013)

12. Hong, Y., et al.: A review on linear generator based wave energy conversion systems. Proc. CSEE **39**(7), 1886–1899 (2019)
13. IEA-electricity market report, 2023. Available online. https://www.sgpjbg.com/baogao/114626.html. Accessed on May 2024
14. López, I., Andreu, J., Ceballos, S., et al.: Review of wave energy technologies and the necessary power-equipment. Renew. Sustain. Energy Rev. **27**, 413–434 (2013)
15. PB3 PowerBuoy.: Available online. https://oceanpowertechnologies.com/platform/opt-pb3-powerbuoy. Accessed on May 2024
16. Penalba, M., Ringwood, J.V.: A high-fidelity wave-to-wire model for wave energy converters. Renew. Energy **134**, 367–378 (2019)
17. Sinha, A., Karmakar, D., Soares, C.G.: Performance of optimally tuned arrays of heaving point absorbers. Renew. Energy **92**, 517–531 (2016)
18. Sun, P.Y., Li, Q., He, H.Z., et al.: Design and optimization investigation on hydraulic transmission and energy storage system for a floating-array-buoys wave energy converter. Energy Convers. Manage. **235**, 113998 (2021)
19. Wang, B.H.: Wave Energy Conversion Technology of Direct Drive Wave Energy Conversion System. Dalian Maritime University (2022). https://doi.org/10.26989/d.cnki.gdlhu.2022.001315
20. Wang, F.Y., et al.: The research situation and development prospect of wave energy power generation device and its hydraulic system. Equip. Manufact. Technol. **3**, 29–31 (2015)
21. Wave Star.: Available online. https://wavestarenergy.com/about/. Accessed on May 2024
22. Xu, Y.K.: Simulation and Optimization Design of an Array of Cantilever Oscillating Float Wave Energy Conversion Device. North China Electric Power University, Beijing (2021)
23. Yang, B., Duan, J., Chen, Y., et al.: A critical survey of power take-off systems based wave energy converters: summaries, advances, and perspectives[J]. Ocean Eng. **298**, 117149 (2024)
24. Yang, S.H., et al.: Research status and prospect of point absorber wave power generation technology. J. Ocean Technol. **35**(3), 8–16 (2016)
25. Yang, S.H., He, H.Z.: Simulation analysis and experiment research of multipoint direct-drive wave power generation system. Acta energiae solaris sinica **35**(8), 1375–1380 (2014)
26. Zhao, L.J., et al.: A novel wave energy conversion device of hydraulic—the designation of bi-directional output of high-pressure oil circuit. Renew. Energy Resour. **1**, 120–122 (2012)
27. Zhu, L.S., et al.: Wave power efficiency of buoy-rope-drum. Acta energiae solaris sinica **35**(8), 1381–1386 (2014)

Application of Smart Terminals in the Power Industrial Internet

Huang Fuqiang[1(✉)], Chen Zhengpu[2], Xiao Jiangtao[1], Jin Zhaoan[3], Chen Jun[4], and Yang Hai[1]

[1] Xiluodu Hydropower Station, China Yangtze Power Co., Ltd., Yunnan Province, Yongshan County 657300, China
346008715@qq.com
[2] State Grid International Development Co., Ltd., Hong Kong, China
[3] Shanghai Liqian Power Technology Co., Ltd., Shenzhen, China
[4] College of Electrical Engineering and Control Science, Nanjing Tech University, Nanjing, China

Abstract. The Power Industrial Internet serves as an essential platform for the digital transformation of the electric power industry, during its implementation, the integration of Information Technology (IT) and Operational Technology (OT) in power production and maintenance has posed significant challenges, necessitating the development of intelligent equipment based on Cyber-Physical Systems (CPS) to bridge this gap. Microcomputer platforms are well-established and reliable systems capable of running various complex algorithms. However, traditional microcomputer-based protection devices face limitations such as firmware programming, limited types of signal sampling, and insufficient data transmission capabilities. In the current era of digital transformation, there is an urgent need for intelligent devices and smart terminals that can overcome these shortcomings through cloud-edge collaboration to assist the digital transformation of the power industry, continuously creating effective applications that enhance quality and efficiency, and fostering new productive forces. This article starts from the strategic orientation of industrial development, the problem orientation of digital transformation, and the demand orientation for quality improvement and efficiency enhancement. It focuses on the key issues in the advancement of the industrial internet, proposes solutions based on microcomputer platforms, and explores new directions in the development of intelligent and digital transformation in the field of relay protection specialization.

Keywords: Industrial Internet · Cyber-Physical Systems · Cloud-Edge Collaboration · Power Industry · Smart Terminal · Microcomputer Platform

1 Introduction

Digital technology, synonymous with information technology, is one of the fastest growing and influential technological fields now. It is rapidly permeating and integrating into various industries, transforming the ways humans produce and live. Digital transformation is a key area for creating new productive forces today. Using digital thinking to

S. Zheng et al. (Eds.): IHDC 2024, LNCE 487, pp. 579–590, 2025.
https://doi.org/10.1007/978-981-97-9184-2_47

integrate traditional industries is the most effective way to build new productive forces. Through cross-industry integration and collaborative innovation, digital transformation promotes the upgrade of traditional industries and accelerates the formation of new productive forces. A new growth path is forged with less input of production factors, high efficiency, low resource and environmental costs, and good economic benefits. This paper delves into the application of intelligent terminals based on microcomputer platforms in the industrial internet.

2 Development of the Industrial Internet

Before delving into the industrial internet in the electric power sector, it's necessary to understand the development history of the industrial internet. Industrial Internet is the application of IT (Information Technology), DT (Digital Technology), CT (Communication Technology), and OT (Operational Technology) in the industrial sector. Essentially, it represents a revolution in tools and decision-making, enabling both physical and cognitive substitution. By leveraging network platforms, the Industrial Internet connects various elements such as people, machines, materials, and environments, facilitating efficient sharing of resources and significantly enhancing overall productivity.

Industrially developed countries such as the United States and Germany have established reference systems or standard systems related to the Industrial Internet, leading the development of infrastructure for the Industrial Internet. The Industrial Internet Reference Architecture (IIRA) of the United States, released by the Industrial Internet Consortium (IIC) on June 19, 2019. The latest version v1.9 consists of four viewpoints: Business, Usage, Functional and Implementation. The IIRA emphasizes nine system characteristics: system security, information security, resilience, interoperability, connectivity, data management, advanced data analytics, to drive optimization of end, edge, and cloud systems in the Industrial Internet.

The Reference Architectural Model Industrie 4.0 (RAMI 4.0) of Germany was released in April 2015. RAMI 4.0 is a three-dimensional model structure that focuses on the lifecycle of manufacturing processes and value chains. This model specifically addresses functional analysis, interoperability, and standardization of various units in the industrial environment. One of the most innovative aspects of RAMI 4.0 is the introduction of the Cyber-Physical Systems (CPS) model, which provides a unified digital interface for industrial physical assets. Through CPS, physical assets are transformed into information-rich digital assets, enabling the traversal from the physical world to the digital world.

China proposed its own industrial internet framework and jointly established the Alliance of Industrial Internet (AII) on February 1, 2016. The AII Alliance established a technical standards working group at its inception and released the "Industrial Internet Standard System (Version 3.0)" [2] in December 2021. This standard system framework consists of three major systems: network, platform, and security, where the network is the foundation, the platform is the core, and security is the safeguard. The model emphasizes that edge computing is a crucial support and key hub for the collaboration of the industrial internet's network and platform.

3 CPS Model and Edge Computing

By reviewing the development of the industrial internet in the United States, Germany, and China, a notable characteristic emerges—Germany's Industry 4.0 defines the Cyber-Physical System (CPS) model as its core, however, China's emphasis on edge computing in its industrial internet. Both share close connections yet exhibit certain differences, which are discussed in depth below:

3.1 Analysis of the CPS Model

Countries around the world have varying concepts of Cyber-Physical Systems (CPS), but in general, the essence of CPS is to establish a communication channel between the digital world and the physical world. This channel extracts data and establishes a closed-loop system for state awareness, diagnostic analysis, intelligent decision-making, automatic control, and precise execution. The goal is to address the complexity and uncertainty in production and life-related scenarios, optimizing system resources and improving efficiency [1].

State awareness relies on sensors to perceive the operating state of the physical world. Diagnostic analysis involves collecting data, extracting information, and generating knowledge through software. Intelligent decision-making is achieved by enabling data flow, information sharing, and knowledge reuse through a platform. Precise execution involves implementing feedback through actuators to carry out decisions. This process depends on a real-time, reliable, and secure network, and can be summarized as "one hardware," "one software," "one network," and "one platform," collectively referred to as the "new four fundamentals" [1].

Perception and execution are the starting and ending points of the Industrial Internet. The essence of perception is the digital extraction of information from the physical world, utilizing various sensors to read and identify hidden information throughout the production process. Execution is precise control based on data collection, transmission, storage and analysis, manifested in a series of actions or behaviors applied to people, machines, materials, methods, and environments [1].

Software is the coded programming of algorithms, which are abstract solutions to real-world problems described in computer language. Software can be seen as the data processing pipeline, fundamentally creating a closed-loop of "state awareness - diagnostic analysis - intelligent decision-making - automatic control - precise execution" to eliminate system uncertainty, reduce entropy, and achieve efficient resource allocation [1].

The network is the information network connecting industrial production systems, increasingly characterized by flatness, wireless, and flexibility. Technologically, it connects various devices, controllers, and information systems through fieldbuses, Ethernet, and 5G networks, serving as the infrastructure for system interconnection and communication [1].

The backend includes industrial clouds and intelligent service platforms, characterized by highly integrated, open, and shared cross-system, cross-platform, and cross-domain features. It plays a role in data collection, storage, analysis, and sharing. On the backend, various algorithm integration models are constructed, forming a data processing production line of digital production elements [1].

3.2 Essence of Edge Computing

Edge computing as the convergence point of Internet technology (IT), communication technology (CT), digital technology (DT), and operational technology (OT). Edge computing is crucial to achieving coordinated development among cloud, network, edge, and terminal.

The related technical standards for edge computing primarily include edge data collection and processing, edge devices, edge platforms, edge intelligence, edge-cloud collaboration, and computing power networks. The standards of them are shown below:

a) Edge Data Collection and Processing Standards: These mainly standardize the technical requirements for data collection from various devices/products, including protocol parsing, data conversion, edge data processing, data storage, data and application interfaces, and related application guidelines.
b) Edge Device Standards: These primarily define the functional, performance, and interface requirements of edge computing devices, including edge servers/integrated machines, edge gateways, edge controllers, and edge computing instruments.
c) Edge Platform Standards: These primarily standardize the technical requirements for edge clouds and edge computing platforms, including computation, storage, network resource management, device management, application management, and operational management.
d) Edge Intelligence Standards: These mainly standardize the technologies for intelligent edge computing capabilities, including virtualization and resource abstraction technologies, real-time operating systems, distributed computing task scheduling strategies and technologies, and open edge intelligence services.
e) Edge-Cloud Collaboration Standards: These primarily standardize the architectural requirements for edge-cloud collaboration, including interfaces and protocols for resource collaboration, application collaboration, service collaboration, and data collaboration.
f) Computing Power Network Standards: These mainly standardize the architecture of computing power networks, including computing power tracing, computing power measurement, and trust in computing power.

3.3 Alignment of CPS Model with Edge Computing

The Cyber-Physical Systems (CPS) model can be summarized as "one hardware," "one software," "one network," and "one platform." China's Industrial Internet standard system integrates edge computing into both the network and platform. It can be understood as different expressions of the same concept.

Understanding CPS and the cloud-edge collaborative model requires a deep grasp of their developmental evolution [1]. Specifically, CPS has distinct hierarchical characteristics. It can range from a single intelligent component or product to an entire intelligent production system. The construction of CPS progresses from localized small systems to complex systems. By analogy, edge devices can be positioned as unit-level small intelligent components or terminals, aligning perfectly with the role and function of CPS unit-level components. Cloud-edge collaboration involves networking edge terminals to integrate into a platform, forming small, large, and super systems, aligning seamlessly with CPS-related concepts.

The unit level represents the smallest indivisible unit of CPS and can be seen as the endpoint of a cloud-edge collaborative system. A terminal component or product, equipped with "one hardware" and "one software", can form a closed-loop system with capabilities such as awareness, control, programmability, computation, expansion, connectivity, and policy enforcement. Examples include intelligent circuit breakers, smart switch cabinets, intelligent robots, edge all-in-one machines, edge controllers, or smart terminals. Every unit is an information carrier, that can form a digital representation be mapped in the virtual space [1].

The system level is an organic combination of "one hardware, one software, one network," which can be viewed as a small or large cloud-edge collaborative system. Multiple unit-level CPS interconnected through a network can form intelligent production lines and processing plants [1]

The system-of-systems level (SoS level) represents an organic combination of multiple systems and can be compared to a super system in cloud-edge collaboration, encompassing "one hardware, one software, one network, and one platform." SoS-level CPS or cloud-edge collaborative super systems achieve cross-system and cross-platform integration through big data platforms, facilitating the integrated, exchanged, and shared closed-loop flow of multi-source heterogeneous data. This enables comprehensive information awareness, in-depth analysis, scientific decision-making, and precise execution on a global scale. Software systems and big data platforms like Siemens' Mindsphere, GE's Predix, PTC's ThingWorx, and Haier's COSMO achieve horizontal, vertical, and end-to-end integration, fostering an open, collaborative industrial ecosystem. This reflects the development direction of CPS and cloud-edge collaboration models at the SoS level [1].

4 Trends and Challenges in the Digital Transformation of the Power Industry

The Power Industrial Internet, as a key infrastructure for the digital transformation of the power industry, provides support with its network architecture and technological system. It has a profound impact on the power sector. By leveraging Industrial Internet technology, the power industry can achieve automation and intelligence in its production processes, promoting a shift from mere power production to a combination of production and services.

Compared to the thriving development of the consumer Internet, the Industrial Internet differs primarily in two aspects: first, the consumer Internet deals with mass life consumption scenarios where the density of IT talent is higher; second, the widespread adoption of smartphones has greatly advanced the standardized data collection for the consumer Internet. Currently, the power industry is actively pursuing digital transformation through Industrial Internet applications, but it faces a series of challenges. These include a lack of standardized terminals, incomplete cloud-edge collaboration, difficulties in data processing and high complexity in algorithm. The following are the reasons behind these challenges:

The Power Industrial Internet terminals encompass a variety of microcomputer devices, PLC devices, edge all-in-one machines, and edge controllers with different models and dispersed data. Every segment involves security risks related to devices, platforms, transmission, and data. The lack of unified technical standards, transmission protocols, and data interfaces for various terminals leads to differences in data formats, sampling frequencies, synchronization characteristics, and communication methods, making data sharing difficult. Each system is built independently with low integration, significantly affecting system reliability. Additionally, the various customized terminals severely lack openness for in-depth user development.

Despite overcoming many difficulties to upload second-level sampled data to the big data platform, the current low data quality allows only for the development of simple trend observation and analysis applications. In terms of talent, it is challenging and risky for electrical engineering employees in the power industry to develop advanced applications or algorithms on the big data backend using Internet information technology (IT). On the other hand, information software professional algorithm developers lack deep understanding and experience in power production and operations technology (OT), creating a talent mismatch and resulting in a talent predicament.

The collaboration between cloud and edge applications, data processing, and AI capabilities all require significant cost and manpower for development and exploration, greatly restricting the development of the Power Industrial Internet. Therefore, the market needs intelligent industrial Internet terminals that integrate the advantages of the Cyber-Physical Systems (CPS) model, support the improvement of industrial production quality and efficiency, and empower new industrialization.

5 Smart Terminals Based on Microcomputer Platforms and Their Application

5.1 Comparison of IPCs, MCUs, PLCs, and MPUs

The development of intelligent applications in the Industrial Internet urgently requires Industrial Internet terminals similar to smartphones. These terminals should feature open programming, unified electrical and process control quantities, standardized data formats, easily expandable signal types, and a modular structure. They should be capable of matching various industrial scenarios with low cost and high reliability.

The China Industrial Internet Industry Alliance launched the country's first edge computing industry project—the Edge Computing Standard Part Plan—which recommends four types of edge computing terminals: edge computing platforms/servers/integrated machines, and edge controllers. Among these, edge computing platforms/servers/integrated machines can collectively be referred to as industrial computers, abbreviated as IPCs. Edge controllers are divided into PLCs and MCUs, with a comparative analysis of their characteristics as follows (Table 1).

Industrial control computers have important computer attributes and features. They are powerful but have the drawback of being highly specialized, requiring in-depth

Table 1. Comparison of IPCs, MCUs, PLCs, and MPUs

Classification	Industrial PC (IPC)	Microcontroller Unit (MCU)	Programmable Logic Controller (PLC)	Microcomputer Protection Unit (MPU)
1. Processing Unit	CPU	MPU	MPU	MPU
2. Function	Realize various complex functions, the most powerful	Realize various complex functions, mediocre	Realize various complex functions, the weakest	Realize various complex functions, powerful
3. Reliability	Low failure rate	Low failure rate	Low failure rate	Low failure rate
4. Environmental Adaptation	High environmental requirements	High environmental requirements,	General industrial environment	General industrial environment
5. Anti-Interference Ability	Weak anti-interference ability	Relatively Strong anti-interference ability	Strong anti-interference ability	Strong anti-interference ability
6. Operation and Maintenance	Difficult	Difficult	Easy	Easy
7. System Development	The system design is complex and debugging calculations are difficult	The system design is complex and debugging calculations are difficult	Easy design, simple installation, short debugging cycle	Easy design, simple installation, short debugging cycle, The programming function is not open
8. Working Mode	Interrupt processing, fast response	Interrupt processing, fast response	Sequential scanning, slow response	Interrupt processing, fast response
9. Data Acquisition Interface	Special interface needs to be designed	Special interface needs to be designed	Many interfaces	few interfaces
10. Data Processing	Requires external acquisition device	Good with digital and analog processing	Good with digital,bad with analog processing	Good with digital and analog processing,Built-in sampling interface, high sampling rate (4kHZ)

(continued)

hardware and software development skills. Microcontrollers (MCUs) are specialized devices consisting of various chips, components, and PCB boards on the hardware side, and typically programmed in assembly language or C language on the software side. They have poor generality and long development and testing cycles.

Table 1. (*continued*)

Classification	Industrial PC (IPC)	Microcontroller Unit (MCU)	Programmable Logic Controller (PLC)	Microcomputer Protection Unit (MPU)
11. Communication Capability	Powerful data processing and communication capabilities	Strong data processing capabilities and weak communication capabilities	Weak data processing and communication capabilities	Strong data processing capabilities and weak communication capabilities
12. Response	Quick response	Quick response	Slow response	Quick response
13. Cost	High	Low	Relatively High	Relatively High

Based on microcontrollers, PLCs with better reliability and generality were developed. PLCs are products of microcontroller control systems, initially developed from sequential control, with powerful functions and stable performance. However, they have notable shortcomings in analog processing, electrical computation, high-speed synchronous sampling, and rapid response.

Microcomputer protection devices overcome these shortcomings of PLCs. They can stably, efficiently, reliably, and in real-time run various protection functions and model algorithms. Table 1 compares their performance across 13 dimensions. Microcomputer protection devices excel in 10 dimensions, with some weaknesses in system development, data acquisition interfaces, and communication capabilities. Overall, they have a significant competitive performance advantage and are an excellent carrier for core CPS units of the Industrial Internet and can be termed intelligent edge computing terminals.

5.2 Key Technologies Required for a Universal Smart Terminal

The key technologies required for developing a general-purpose intelligent terminal based on a microcomputer platform for Industrial Internet Cyber-Physical Systems (CPS) can be summarized as follows:

a) Develop graphical programming software to reduce the difficulty for non-professionals in developing algorithm.
b) Enhance communication and data processing capabilities to support various cascading and networking modes, addressing data cleaning challenges such as high-frequency sampling, time synchronization, high-speed computation, protocol compatibility, format standardization, and structural uniformity.
c) Expand the types of signal acquisition and configure various sensors as needed to flexibly adapt to different working conditions and scenarios.

By implementing these three measures, the practical application of the Industrial Internet can be actively supported. The main challenge lies in the extensive engineering effort required for the development of key technology "a". The development of key technologies "b" and "c" is technically mature, and upgrading these on the traditional microcomputer platform poses minimal implementation difficulty while maintaining the reliability of the microcomputer devices. This approach is highly feasible.

Drawing a parallel with the consumer Internet industry: traditional cell phones could only support a limited number of specific applications, whereas smartphones expanded on the capabilities by adding sensor signals such as cameras, infrared, and GPS, allowing users to install various apps. This cross-industry comparison provides insight: by adding graphical programming software, enhancing communication and data processing capabilities, and expanding the types of signal acquisition on the microcomputer platform, it can have broad application scenarios. This approach can effectively address the challenges of IT and OT integration and create multi-scenario solutions.

5.3 Conceptual Features of a Universal Smart Terminal

China Yangtze Power Co., Ltd.'s Xiluodu Hydropower Plant has proposed an industrial Internet modular intelligent terminal invention patent [3]. This invention achieves various purposes and functions by configuring control modules, communication power modules, and various I/O modules, which are then connected and combined through a base. This modular combination intelligent IoT terminal encapsulates the entire system's hardware and software independently into modules. Interaction between modules is achieved through standardized interfaces and protocols, shielding the underlying complexity of various software and hardware interfaces and operations, allowing the different modules to combine automatically and intelligently. Furthermore, peripheral components can be matched to various scenarios according to on-site requirements and unified into standard data types through transmitters. As a key node platform for the Industrial Internet and IoT, the intelligent terminal has distributed computing power and flexible configuration. It can achieve master-slave collaborative work modes through the selection of different control modules. It can also connect to the Internet via communication technology to achieve a joint working mode for different intelligent terminals and can enhance computing power through 5G communication and cloud computing.

The intelligent terminal described in this invention possesses basic programming capabilities and provides graphical algorithm editing software, reducing the difficulty for personnel of power sector in developing various algorithms. This platform allows for the customization, development, and use of various applications, displaying and controlling them via the network. It offers a deployable solution for building a flexible, modular, highly distributed IoT network. The modular, and standardized design can effectively reduce the production and inventory costs of intelligent terminals. Personalized solutions for various types of peripheral components can meet on-site needs. This significantly lowers the operation and maintenance costs of equipment and supporting facilities for state sensing or monitoring systems, reduces the difficulty of personnel training.

5.4 Application of Universal Smart Terminals

Intelligent terminals based on microcomputer platforms have a wide range of application scenarios. However, the challenge lies in the need to strengthen network security protection capabilities. Additionally, intelligent terminals developed on traditional microcomputer protection platform architectures require more testing and practical application. It is essential to control power safety production risks, and in the initial stage, these terminals should be considered as a supplement to rather than a replacement for traditional microcomputer protection devices. General-purpose intelligent terminals based on microcomputer platforms address real production site issues, have broad application scenarios, and possess high market value.

5.4.1 Application Scenario 1: Real-Time Expert Diagnostic Intelligent Terminal for GCB

To address various issues found in GCBs such as aging analog components, loosened linkage mechanisms, oil leakage and pressure drops, insufficient SF6 pressure, unclear node status and poor contact, a customized intelligent terminal based on the microcomputer platform is developed with various state monitoring functions. These functions include the integration and replacement of analog electrical components, synchronous monitoring of electrical and mechanical quantities (including contact linkage position, node position, contact temperature, current, etc.), and the addition of sensors to monitor GCB oil pressure, SF6 gas pressure, and other related characteristic indicators to provide early warnings on GCB status. By accumulating extensive operational data of GCBs and enhancing GCB operation and maintenance management through big data and algorithms, the reliability of system operation is effectively improved.

5.4.2 Application Scenario 2: Expert Diagnostic Intelligent Terminal for 400V Systems

Currently, 400V systems generally lack protection functions, with overcurrent protection integrated within the circuit breaker. The measurement and Automatic Switch Function of 400V systems are handled by a combination of different analog measurement, transmitters, PLCs, and relays, resulting in high failure rates and low reliability. The development of an intelligent terminal based on the microcomputer platform for 400V systems aims to solve these issues by replacing high-failure-rate components, reducing control circuits, and enhancing system reliability through big data and predictive models.

5.4.3 Application Scenario 3: Real-Time Expert Diagnostic Intelligent Terminal for 10kV Systems

This real-time expert diagnostic intelligent terminal integrates various functions on the same platform, including Automatic Switch Function, harmonic elimination, interlocking control, status monitoring, and oscillographic analysis. It aims to streamline the circuits and components of the 10kV system. By replacing high-failure-rate components, simplifying control circuits, integrating multiple functions, and exploring big data diagnostic, the system's reliability and scalability are enhanced.

5.4.4 Application Scenario 4: Intelligent Meters and Controllers

Power stations widely use analog meters or transmitters for measuring voltage, current, power, and frequency. However, there are reliability issues with the long-term operation of these measurement devices. Intelligent meters based on the microcomputer platform can effectively address these issues. A set of intelligent meters can integrate multiple measurements such as voltage, current, power, and frequency, simplifying the circuit and diagnostic models such as can be loaded, for example, after installing this intelligent meter on the PT cabinet, various intelligent algorithms can be loaded, including synchronization monitoring, harmonic elimination, and operational sequence control.

6 The Wave for the Implementation of the Power Industry Internet

The core of the industrial internet's implementation lies in Cyber-Physical Systems (CPS), and intelligent terminals that bridge the IT with OT are key. Edge computing is essentially a tool for intelligent upgrading of application scenarios, reducing costs, and increasing efficiency. However, many industrial sites in China have not yet completed their digital upgrades. Therefore, customized modifications are needed, providing comprehensive solutions. Intelligent terminals based on microcomputer platforms are such high-value, high-potential products. They compensate for the shortcomings of traditional microcomputer protection fields, tailor CPS systems for each equipment, create intelligent monitoring, early warning, and diagnostic solutions.

The industrial internet is currently in its early stages of explosive growth, and the industry's maturity is relatively low. International giants are accelerating the construction of industrial clouds and intelligent service platforms, applying AI technology to seize the rights to rule-making leadership. Various advantageous enterprises should be encouraged to focus on the industrial internet and digital transformation, to expand the scenarios of the industrial internet in the power sector.

References

1. An Xiaopeng, White Paper on Cyber-Physical Systems (2017). Preface
2. Industrial Internet Industry Alliance, Industrial Internet Standard System (Version 3.0)
3. China Yangtze Power Co., Ltd. Modular Combination IoT Terminal: China, 219918970U[P]. 2023–10–27

Influence of Water Molecules on the Interfacial Structures and Energy Storage Behavior of Ionic Liquid Electrolytes

Chenxuan Xu(✉), Xu Qian, Xingxing Gu, and Junjie Yang

Power China Huadong Engineering Corporation Limited, Zhejiang Province, Hangzhou 310000, China
xu_cx@hdec.com

Abstract. Ionic liquids have been considered as promising electrolytes for supercapacitors due to the wide electrochemical stability window. However, water molecules inevitably damage the electrochemical properties of ionic liquids due to the hygroscopic property. This paper reveals the effect of water molecules on the interfacial structure and energy storage performance of ionic liquids using the atomistic simulations. Unlike neat ionic liquids, the Helmholtz region for humid ionic liquids is mainly composed of BMI cations and water molecules. Importantly, water molecules primarily accumulate in the buffer region between BMI cation and graphene electrode, especially at the high negative charges, which is the crucial factor to induce the hydrogen evolution reactions for the decreased electrochemical stability window. More interestingly, the dielectric properties of water molecules in the buffer layer are beneficial for lowering the electric potentials for higher capacitive performance. The differential capacitance of $[BMI^+][BF4^-]/H_2O$ electrolyte exhibits a bell-shaped curve with a maximum value of ~5.0 F/cm^2 at 0.75 V. The revealed insights are important for understanding the water effect in ionic liquid-based supercapacitor energy storage.

Keywords: moist ionic liquid · charge storage mechanism · energy storage behavior · molecular dynamics simulation · supercapacitor

1 Introduction

Ionic liquids, composed of neat cations and anions without solvents, have attracted great research interests due to the obvious advantages of good thermal/chemical stability, wide electrochemical stability window (3.0–6.0 V) and non-flammable properties [1–3]. As such, ionic liquids have been widely used in the energy storage devices including supercapacitors and batteries [4, 5]. For example, Naguib et al. increased the interlayer spacing of MXene electrode and demonstrated a high capacitance value of 257 F/g and an ultrahigh energy density of 370 Wh/kg in neat ionic liquids of $[BMI^+][BF_4^-]$ electrolyte [6], which is mainly due to the good electrochemical stability of ionic liquids. Serrapede et al. developed a new combination of MoS_2/rGO and ionic liquid, which exhibits a superior capacitance of 210 F/g even at a high working temperature of 200 °C [7], which correlates with the excellent thermal stability of ionic liquids.

S. Zheng et al. (Eds.): IHDC 2024, LNCE 487, pp. 591–597, 2025.
https://doi.org/10.1007/978-981-97-9184-2_48

The outstanding physical, chemical, and electrochemical properties of ionic liquids primarily come from their cation-anion composition, strong intermolecular interactions, and unique microscopic structures. Thus, many researchers have investigated the effect of cation/anion species, their combinations, alkyl chain effects, and their intermolecular interactions on the dynamic properties (e.g., diffusion coefficient, ionic conductivity, wetting, and viscosity), electrowetting behaviors, low-temperature property (e.g., freezing point), electrochemical stability window, and thermal stability (e.g., volatility and flammability) [8–12].

In addition to the above progress, more attention has recently been paid to the inherent hygroscopic property of ionic liquids in the practical energy storage applications [13–16]. It is found that ionic liquids easily adsorb water molecules and become humid during the preparation process due to the hygroscopic property, leading to the obvious degradation of electrochemical stability window, which significantly sacrifices their inherent advantages [17]. Molecular simulation studies revealed that the population of $[BMIM^+][BF_4^-]$ ionic liquids near the Pt surface was decreased with the increase of water content, which further suggested that the thermal conductivity of $[BMIM^+][BF_4^-]$ was also decreased [18]. Using the coarse-grained model, Fan et al. showed that the distribution of water in the $[BMIM^+][BF_4^-]$ ionic liquids was asymmetric, which was shifted toward the charged surface at higher charge densities [19]. Therefore, an in-depth study of the effect of water on the microscopic structures of ionic liquids is highly needed.

This paper investigates the influence of water molecules on the microscopic structure of typical $[BMI^+][BF_4^-]$ ionic liquids using molecular dynamics simulation. This paper focuses mainly on the effect of water on the interfacial structure, electric potential, dielectric properties, and capacitive behavior. The results show that water molecules primarily act as a buffer layer between the $[BMI^+][BF_4^-]$ ionic liquid and the electrode surface, especially at a higher negative charge density of $>-8\ \mu C/cm^2$, which further influences the capacitive behaviors. These unveiled molecular insights are important for understanding, designing and optimizing the electrochemical properties of ionic liquids.

2 Results and Discussions

The atomic model composed of graphene, $[BMI^+][BF_4^-]$ ionic liquid and water molecules is shown in Fig. 1a. Near the neutral electrode, water molecules prefer to accompany the ions, and BMI^+ cation and BF_4^- anion have the similar density distributions due to the strong intermolecular electrostatic interactions. Upon charging the graphene electrodes (from 0 to $-$ 16 $\mu C/cm^2$), the population of BMI^+ cation in the Helmholtz region (~6 Å from the electrode surface) is significantly enhanced, while the corresponding density of BF_4^- anion is largely suppressed. Specifically, the peak of BMI^+ cation is increased by ~4.73 times (from 0.0038 #/Å^3 at 4.2 Å to 0.018 #/Å^3 at 3.7 Å). Meanwhile, the water molecules are also attracted to the closer position with higher density towards the electrode surface (from 0.03 #/Å^3 at 3.2 Å to 0.06 #/Å^3 at 3.0 Å for O atoms; from 0.041 #/Å^3 at 3.8 Å to 0.055 #/Å^3 at 2.3 Å for H atoms). In contrast to the neat ionic liquids [20, 21], we can conclude that water molecules primarily serve as a buffer layer between the $[BMI^+][BF_4^-]$ ionic liquid and the electrode surface, especially at a higher negative charge density of $>-8\ \mu C/cm^2$. As a result, the interfacial water molecules as the buffer layer is the key factor to induce the hydrogen evolution reaction of anode and thus reduce the electrochemical stability window of ionic liquids.

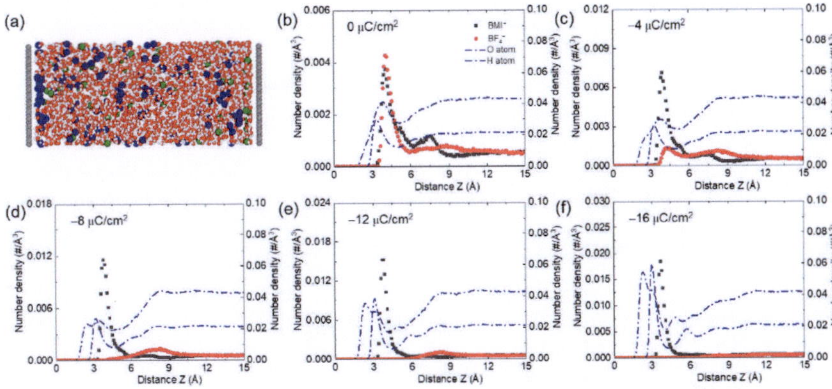

Fig. 1. (a) Simulation model and (b–f) density profiles of $[BMI^+][BF_4^-]/H_2O$ electrolyte. Red, white, blue, green and grey colors represent the O atom, H atom, BMI^+ cation, BF_4^- anion, and carbon atoms, respectively.

To more clearly show the microscopic structure of $[BMI^+][BF_4^-]/H_2O$ electrolyte, the corresponding representative side-view snapshots from molecular simulations are shown in Fig. 2. We can clearly see the layered and ordered structure of BMI^+ cations. More importantly, the Helmholtz region is mainly composed of BMI^+ cations and water molecules, exposing reactive H atoms towards the negative electrode, especially at the higher negative charges.

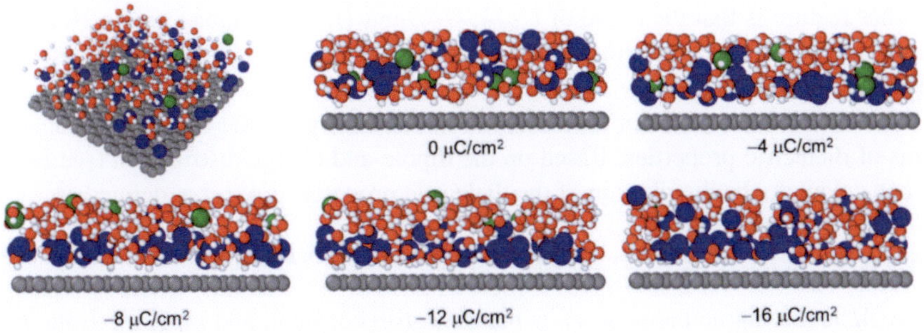

Fig. 2. Representative side-view snapshots of $[BMI^+][BF_4^-]/H_2O$ electrolyte at different negative charge densities

Using the Poisson equation, this work calculates the electric potential distributions based on the electrolyte number/charge density from Fig. 1. To quantitatively describe the contribution of ionic liquids and water molecules to the electric potentials, the total electric potential U_{total} is decomposed into two parts, the ionic-liquid-induced electric

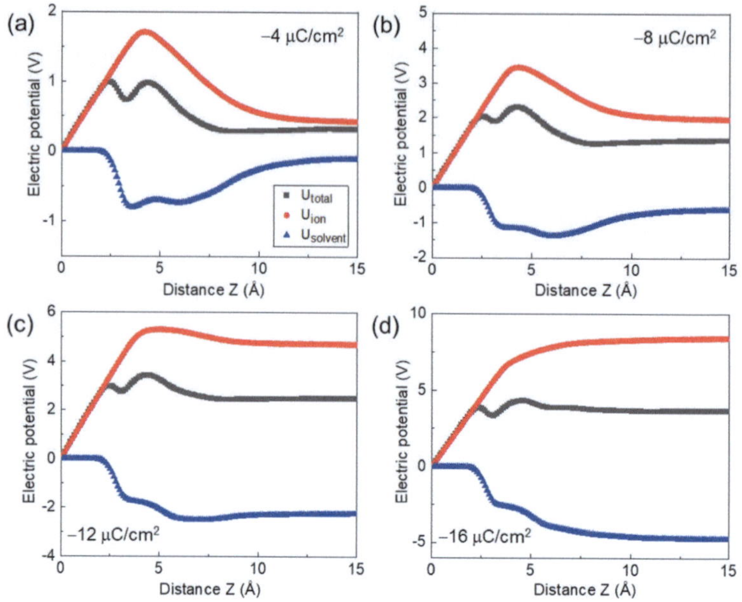

Fig. 3. Electric potential of $[BMI^+][BF_4^-]/H_2O$ electrolyte at different charge densities of (a) −4 μC/cm^2, (b) −8 μC/cm^2, (c) −12 μC/cm^2 and (d) −16 μC/cm^2

potential U_{ion} and the water molecule-induced electric potential $U_{solvent}$ (Fig. 3). An important feature is that the electric potential value of U_{total} and U_{ion} is positive while the corresponding value of $U_{solvent}$ is negative. Thus, the dielectric property of water molecules helps reduce the electric potentials for higher capacitance. For example, the electric potential induced by ion U_{ion} is ~2.0 V, and the electric potential induced by the water $U_{solvent}$ is ~−0.6 V, resulting in a total electric potential value U_{total} of ~1.4 V. Another feature is that the curve of U_{total}, U_{ion} and $U_{solvent}$ flattens within 10 Å from the electrode surface, indicating that the screening of electrode field is completed in a very thin interfacial region.

The contribution of water molecules to the electric potential can be understood in terms of dielectric properties. Based on the dipole and charge distribution (see Eq. 1), we can calculate the distribution of the dielectric property of water ε, defined as:

$$\varepsilon = 1 + (< M^2 > - < M >^2)/3Vk_B T\varepsilon_0 \quad (1)$$

where M is the dipole moment, ε_0 is the dielectric constant, and k_B is the Boltzmann constant. The water dielectric property ε shows a layered structure, which gradually increases to the bulk property with its position far away from the electrode surface (Fig. 4a). In particular, the water dielectric property ε exhibits a Helmholtz peak of ε = 42.5 at ~3.0 Å, as do the number density distributions in Fig. 1. Thus, the Helmholtz layer of water molecules helps to screen the electric field from electrodes for better capacitive energy storage, while it also easily induces the hydrogen evolution reaction and thus the limited electrochemical stability window of $[BMI^+][BF_4^-]$ ionic liquids.

In addition, the capacitive performance of graphene in $[BMI^+][BF_4^-]/H_2O$ electrolyte is quantitatively described. The differential capacitance curve of $[BMI^+][BF_4^-]/H_2O$ electrolyte exhibits a bell-shaped curve with a maximum value of 5.0 μF/cm^2 at 0.75 V (Fig. 4b). Moreover, the capacitance values of anode and cathode mainly fall in the range of 4~5 μF/cm^2 (Fig. 4c), which is slightly higher than that of neat ionic liquids [20], mainly due to the effect of water molecules.

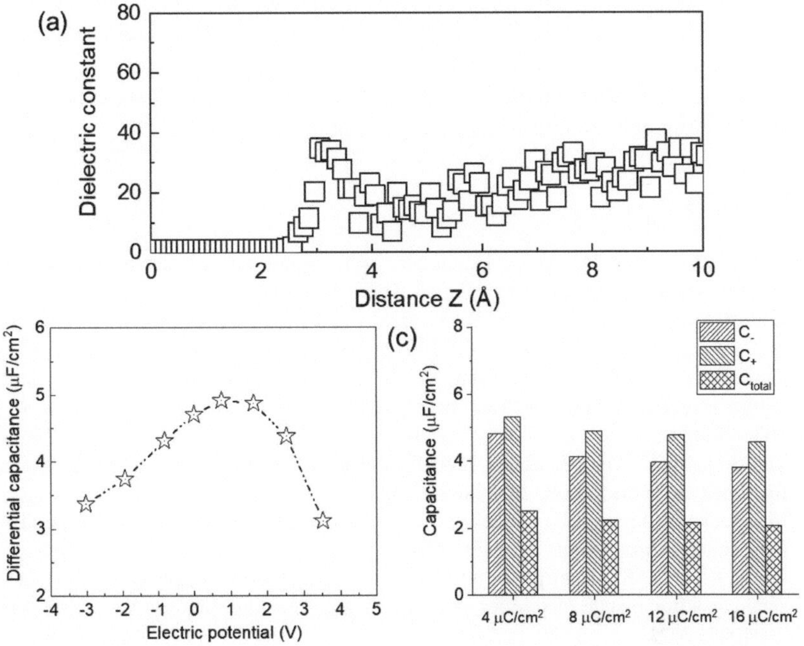

Fig. 4. (a) Dielectric property of water molecule in the $[BMI^+][BF_4^-]/H_2O$ electrolyte. (b) Differential capacitance and (c) electrode capacitance of $[BMI^+][BF_4^-]/H_2O$ electrolyte

3 Conclusion

This paper reveals the influence of water molecules on the interfacial structure and capacitive behavior of ionic liquids using the atomistic simulations. It is shown that water molecules primarily serve as the buffer layer between $[BMI^+][BF_4^-]$ ionic liquid and electrode surface, especially at higher negative charge density of >-8 μC/cm^2. We demonstrate that the dielectric property of water molecules helps to decrease the electric potentials for higher capacitive performance, meanwhile it also easily induces the hydrogen evolution reaction for decreased electrochemical stability window of ionic liquids. Moreover, the differential capacitance curve of $[BMI^+][BF_4^-]/H_2O$ electrolyte exhibits a bell-shaped curve with a maximum value of 5.0 μF/cm^2 at 0.75 V. The capacitance value of graphene electrode in $[BMI^+][BF_4^-]/H_2O$ electrolyte mainly falls in the range of 4~5 μF/cm^2.

4 Simulation Method

The atomic simulations are performed using LAMMPS software. The simulation model is built with two layers of graphene, water molecules and $[BMI^+]$ $[BF_4^-]$ ionic liquids. The dimensions of simulation box along *X*, *Y* and *Z* directions are 33.50 Å, 33.85 Å and 80.0 Å. The periodic boundary conditions are set on the *X* and *Y* directions. An additional 10 Å vacuum region is set along the *Z* direction to avoid periodic interaction. 50 pairs of $[BMI^+]$ $[BF_4^-]$ and 1795 pairs of water molecules are used as the representative electrolyte. The temperature of the electrolyte is kept at 300 K and the time step is set to 1 fs. The force fields for $[BMI^+]$ $[BF_4^-]$ with coarse-grained models, water molecules and graphene are taken from previous studies [22]. The simulation is run for at least 5 ns for equilibration and then rerun for at least 10 ns for data analysis.

References

1. Ruan, Q.Q., Yao, M., Yuan, D., et al.: Ionic liquid crystal electrolytes: fundamental, applications and prospects. Nano Energy **106**, 108087 (2023)
2. Zhang, W.X., Gao, Y.R., Xue, R., et al.: Liquid formulations based on ionic liquids in biomedicine. Mater. Today Phys. **30**, 100925 (2023)
3. Choudhary, G., Dhariwal, J., Saha, M., et al.: Ionic liquids: environmentally sustainable materials for energy conversion and storage applications. Environ. Sci. Pollut. Res. **31**, 10296–10316 (2024)
4. Lan, S.Q., Yu, C., Yu, J.H., Zhang, X.B:. Recent advances in low-temperature liquid electrolyte for supercapacitors. Small, 2309286 (2024)
5. Matuszek, K., Piper, S.L., Brzęczek-Szafran, A., et al.: Unexpected energy applications of ionic liquids. Adv. Mater. 2313023 (2024)
6. Liang, K., et al.: Engineering the interlayer spacing by pre-intercalation for high performance supercapacitor mxene electrodes in room temperature ionic liquid. Adv. Func. Mater. **31**, 2104007 (2021)
7. Serrapede, M., et al.: The combination of MoS_2/reduced graphene oxide composite electrode and ionic liquid for high-temperature supercapacitor. J. Energy Storage **73**, 109180 (2023)
8. Chen, W., et al.: Insight of alkyl imidazolium tetrafluoroborate ionic gels as supercapacitors and motion sensors: effects of alkyl chain length and intermolecular interactions. J. Power Sources **599**, 234224 (2024)
9. Song, F., et al.: Wetting and electro-wetting behaviors of $[Bmim][BF_4]$ ionic liquid droplet on lyophobic and lyophilic solid substrates. J. Mol. Liq. **347**, 118405 (2022)
10. McDaniel, J.G.: Capacitance of carbon nanotube/graphene composite electrodes with $[BMIM^+][BF_4^-]$/Acetonitrile: fixed voltage molecular dynamics simulations. J. Phys. Chem. C **126**, 5822–5837 (2022)
11. McDaniel, J.G., Son, C.Y.: Ion correlation and collective dynamics in $BMIM/BF_4$-based organic electrolytes: from dilute solutions to the ionic liquid limit. J. Phys. Chem. B **122**, 7154–7169 (2018)
12. Amiri, M., Bélanger, D.: Intermolecular interactions and electrochemical studies on highly concentrated acetate-based water-in-salt and ionic liquid electrolytes. J. Phys. Chem. B **127**, 2979–2990 (2023)
13. Dick, L., et al.: Hygroscopic protic ionic liquids as electrolytes for electric double layer capacitors. Energy Storage Mater. **53**, 744–753 (2022)
14. Meng, J., et al.: Recovering the electrochemical window by forming a localized solvation nanostructure in ionic liquids with trace water. Sci. China Chem. **65**, 96–105 (2022)

15. Zhang, C., et al.: Suppressing water clusters by using "hydrotropic" ionic liquids for highly stable aqueous lithium-ion batteries. J. Mater. Chem. A **10**, 20545–20551 (2022)
16. Zheng, Q., et al.: Water in the electrical double layer of ionic liquids on graphene. ACS Nano **17**, 9347–9360 (2023)
17. Chen, M., et al.: Adding salt to expand voltage window of humid ionic liquids. Nat. Commun. **11**, 5809 (2020)
18. Cai, S., Li, Q., Li, M., Liu, C.: Molecular characteristics of [BMIM][BF_4] ionic liquids and water mixtures on the Pt surface. J. Mol. Liq. **304**, 112782 (2020)
19. Song, F., Wang, F., Ma, J., Xue, J., Fan, J.: Microstructure of ionic liquids mixed with water on the charged graphene surface: a coarse-grained molecular dynamics simulation study. J. Mol. Liq. **391**, 123253 (2023)
20. Chaban, V.V., Andreeva, N.A., Fileti, E.E.: Graphene/ionic liquid ultracapacitors: does ionic size correlate with energy storage performance? New J. Chem. **42**, 18409–18417 (2018). https://doi.org/10.1039/C8NJ04399J
21. Xu, C., et al.: Molecular Dynamics Simulation of the Interfacial Structure and Differential Capacitance of [BMI^+][PF_6^-] Ionic Liquids on MoS_2 Electrode. Processes **11**, 380 (2023)
22. Merlet, C., Salanne, M., Rotenberg, B., Madden, P.A.: Influence of solvation on the structural and capacitive properties of electrical double layer capacitors. Electrochim. Acta **101**, 262–271 (2013)

Author Index

S. Zheng et al. (Eds.): IHDC 2024, LNCE 487, pp. 599–601, 2025.
https://doi.org/10.1007/978-981-97-9184-2

MIX
Papier aus verantwortungsvollen Quellen
Paper from responsible sources
FSC® C105338

If you have any concerns about our products,
you can contact us on
ProductSafety@springernature.com

In case Publisher is established outside the EU,
the EU authorized representative is:
Springer Nature Customer Service Center GmbH
Europaplatz 3, 69115 Heidelberg, Germany

Printed by Libri Plureos GmbH
in Hamburg, Germany